ANALOG AND DIGITAL
CONTROL SYSTEMS

ANALOG AND DIGITAL CONTROL SYSTEMS

Ramakant Gayakwad

Leonard Sokoloff

PRENTICE HALL, Englewood Cliffs, NJ 07632

Library of Congress Cataloging-in-Publication Data

Gayakwad, Ramakant
  Analog and digital control systems.

  Includes index.
  1. Automatic control.   2. Analog-to-digital
converters.   3. Digital-to-analog converters.
I. Sokoloff, Leonard.   II. Title.
TJ213.G377   1988   629.8   87–19350
ISBN 0–13–033028–0

Editorial/production supervision and
interior design: *Ellen Denning*
Manufacturing buyer: *Peter Havens*

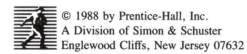 © 1988 by Prentice-Hall, Inc.
A Division of Simon & Schuster
Englewood Cliffs, New Jersey 07632

Printed in the United States of America

10 9 8 7 6 5 4 3 2 1

ISBN   0-13-033028-0   025

PRENTICE-HALL INTERNATIONAL (UK) LIMITED, *London*
PRENTICE-HALL OF AUSTRALIA PTY. LIMITED, *Sydney*
PRENTICE-HALL CANADA INC., *Toronto*
PRENTICE-HALL HISPANOAMERICANA, S.A., *Mexico*
PRENTICE-HALL OF INDIA PRIVATE LIMITED, *New Delhi*
PRENTICE-HALL OF JAPAN, INC., *Tokyo*
SIMON & SCHUSTER ASIA PTE. LTD., *Singapore*
EDITORA PRENTICE-HALL DO BRASIL, LTDA., *Rio de Janeiro*

This book is dedicated to our wives,
Helen *and* Pratibha.

CONTENTS

3 MOTORS   112

4 REPRESENTING CONTROL SYSTEMS   190

PREFACE

Analog and Digital Control Systems is a textbook intended to support courses in control system theory and practical applications. Although it focuses primarily on the needs of electronic technology programs, its scope is sufficiently broad to accommodate the EET and EE curriculum control system course requirements in many colleges and universities. To a practicing engineer, this book may serve as a useful reference.

Over the years, as many of us are aware, the rapidly growing computer technology has made a profound impact in practically all segments of the industrial community. This influence has been felt particularly by control system technologists. Computer- or microprocessor-controlled robot arms which provide a variety of motions in an industrial process control environment are as common today as they were uncommon a decade or two ago. The availability and cost-effectiveness of the computer or a microprocessor-operated control systems has created an attractive climate that supports the growing trend toward the use of digital control systems. This does not mean that analog control systems are becoming obsolete. On the contrary, many analog control systems are presently in use, and many more are being installed (for example, digital-to-analog and analog-to-digital converters are far from being obsolete). In many industrial applications, digital and analog control systems coexist in harmony to produce optimum performance. In recognizing these realities and trends, the authors felt it important to write a book that offers broad coverage of the analog and the digital control system technologies with practical applications.

Emphasis on practical applications is one of the key features of this book. Laboratory-tested analog and digital control systems based on practical applications together with detailed interface circuits and associated software are included in the body of the text. Real-time applications demonstrate the use of a microprocessor as a tool in solving practical problems. The emphasis, again, is not on microprocessor operation but rather on the use of the microprocessor as a tool in achieving automatic control. This type of approach and philosophy are uncommon in most control system textbooks.

Although inclusion of practical systems is an important feature of this book, no less important are the laboratory experiments based on practical systems. The ultimate goal of any lab experiment is to demonstrate in a practical manner the theoretical concepts presented in class. Students must therefore appreciate and recognize the importance of the correlation of theory and measurements, and must be able to reconcile any differences between them through additional investigation or experimentation. This philosophy (which is inherent in each experiment) requires students to extract meaningful results from the experiment objectives and offers them an opportunity to work on practical systems, thus providing them with a type of experience that cannot be duplicated in class.

Comprehensive coverage (in Chapter 3) of stepper motor theory and operation is a very special feature of this book. In the past, applications for stepper motors were limited. However, as computer technology grew and expanded, so did use of the stepper motor and the scope of its practical applications. Today, the stepper motor is widely used in computer peripheral equipment, such as printers and disk drives; in industrial process applications the stepper motor provides very accurate motion control; and in consumer products the stepper motor is used frequently, for example, for accurate positioning of the daisywheel print head in typewriters. Numerical control and many other applications depend on the stepper motor as an essential control system component. Most control system textbooks omit stepper motors.

The stepper motor controller is a chip or in some cases a board containing circuits that provide the necessary logic and drives to operate a stepper motor. In this respect the controller simplifies for the user the task of operating the stepper motor. In order to reveal to students the details of a typical controller structure and its operation, a step-by-step design of a sequencer using the sequential logic design is included in Chapter 3. The sequencer is then incorporated in a practical stepper motor control system, which is finally converted into an experiment allowing students to use a sequencer or microprocessor as an input control source in investigating the motion performance of a stepper motor.

The future of control systems will, no doubt, be affected a great deal by the developments in microprocessor technology. Most control system applications do not require the very powerful and versatile 16- and 32-bit microprocessors. The 4- and 8-bit microprocessors meet most of the general requirements. The growing trend in the use of microprocessors in control systems is responsible for focusing special attention on this very important topic in Chapter 10. Use of the microprocessor also appears in other sections of the book and in some experiments. Particular emphasis is directed toward the interfacing circuits and the nature of software used in control system applications.

The content of the book is organized as follows:

Chapter 1 focuses attention on introductory concepts and on the general characteristics of control systems.

Chapter 2 describes various types of transducers and control system components. This chapter serves as a foundation and reference for other chapters.

Chapter 3 presents the theory and operation of the dc PM generator, dc PM motor, and stepper motors.

Chapter 4 is concerned with the representation of a control system. This includes Laplace transform review, transfer functions, block diagrams, and flow graphs.

Chapter 5 is on feedback system characteristics. System characteristics such as sensitivity, accuracy, the effect of disturbances, and the effect of negative feedback on transient response are considered in this chapter. Stability criteria according to Bode, Nyquist, and Routh–Hutwitz are also discussed in a general manner.

Chapter 6 includes the time-domain analysis of first- and second-order systems. Also discussed is the relationship between the time domain, the frequency domain, and the s plane.

Chapter 7 pertains to frequency-domain analysis. Included in this chapter are polar and Bode plot construction and the use of such plots in stability analysis.

Chapter 8 directs its attention to analog control system description and analysis. Four systems are presented in this chapter: speed control, position control, temperature control, and liquid-level control. Practical systems are also included.

Chapter 9 is based on digital control systems. Digital techniques (not the microprocessor) are used as the basis of control. The systems discussed in this chapter include temperature control, dc motor speed control, stepper motor position control, appliance timer, and digital light intensity controller.

Chapter 10 concentrates on use of the microprocessor in control system applications. The systems discussed in this chapter include home heating and cooling system, dc motor speed control, control of an appliance timer, and light intensity controller.

ACKNOWLEDGMENTS

We are grateful to George Dean, Vice President of Research and Curriculum Development, DeVry Inc., for his generous attention in writing the foreword. We are also grateful to the following reviewers for their helpful suggestions and technical advice:

At DeVry Los Angeles:

Many thanks to President Paul McGuirk for his continued support and encouragement. Thanks to the General Education Dean, Ed Galyen, and Ismael Lopez, EET student at DeVry, for their help in getting the manuscript ready. We also wish to thank Professors Zahzah and Bradsher for their careful review of the manuscript.

At DeVry Woodbridge:

We wish to express our sincere thanks to President R. Bochino; Tom Kist, Dean of Academic Affairs; and Dean Neil Towey for their support and encouragement. Special thanks to Professor Jim Stewart for many helpful suggestions and discussions. Many thanks to EET students C. Ashley, R. Sienrukos, P. Cashman, W. Mohr, and many,

many others for valuable assistance in the development of various projects, and to K. Brockmeir for fine photography.

Most of all we wish to thank our wives, Helen and Pratibha, and our daughters, Larissa, Leena, and Neeta, for their patience and understanding in the course of the preparation of this book.

RAMAKANT GAYAKWAD

LEONARD SOKOLOFF

FOREWORD

The constantly improving cost/benefit ratio of electronic components and systems has facilitated the application of electronic control systems in an almost unlimited array of products. Closed-loop control systems, once found only in expensive, precision servo-mechanisms, are today found under the hood of the family automobile, in the heating/cooling systems in homes and businesses, and even at the toy store in robots and remote-control vehicles.

Although the basic theory of control systems remains unchanged, the microprocessor, together with components such as stepper motors, integrated-circuit analog-to-digital and digital-to-analog converters, and solid-state sensors, expands qualitative and quantitative analytical techniques beyond those employed with more traditional linear systems. This book is a significant contribution to technology education in that it provides an appropriate balance between qualitative and quantitative analysis of control systems, with analog and digital systems receiving appropriate treatment. Of critical importance is the fact that this work clearly illustrates the practical application of theory. Integrated laboratory exercises challenge students to put theory to practice, not in a ''cookbook'' fashion, but in a way that strengthens the students' analytical and problem-solving ability, a key ingredient for career growth.

This book can serve as the foundation for a strong electronic control systems sequence in an electronics engineering technology program. It can also serve students in related areas of study as a source of material for professional development.

George R. Dean

Educational Research
& Development
Vice President
DeVry Inc.

ANALOG AND DIGITAL
CONTROL SYSTEMS

INTRODUCTION TO CONTROL SYSTEMS

chapter 1 _____

1-1 INTRODUCTION

As the name suggests, a *control system* is a system in which some physical quantity is controlled by regulating an energy input. A *system* is a group of physical components assembled to perform a specific function. A system may be electrical, mechanical, hydraulic, pneumatic, thermal, biomedical, or a combination of any of these systems. The physical quantity may be any physical variable, such as temperature, pressure, liquid level, electrical voltage, or mechanical position. We may use analog, digital, or both analog and digital techniques to control a desired physical quantity. Generally, the type of technique employed is used to classify the control system as either analog or digital. More specifically, a physical quantity that is to be controlled in a given system is used to label that system. For example, temperature is controlled in a temperature control system, whereas in a liquid-level system a preassigned liquid level is maintained.

In the early days of control systems, most were analog systems employing analog techniques. These systems were relatively bulky, complex, and cumbersome both to design and to maintain. However, with the development of digital technology and the invention of integrated circuits (ICs), the design of control systems became easier as well as more economical. Presently, most control systems are microprocessor controlled, primarily because of the availability of control ICs and cheaper memories and tremendous advancements in data-handling capabilities.

An *ideal* control system is one in which an output is a direct function of an

1

input. However, in actual practice, disturbances affect the output being controlled and cause it to deviate from the desired value. The nature of these disturbances varies from system to system. For example, in an electrical control system, noise, variation in power supply voltage, component tolerances, and environmental conditions (including temperature and humidity) are some of the disturbances that adversely affect the desired output. Therefore, in a given control system it is of utmost importance to identify the disturbances. Once they are identified, the system should be modified or compensated to minimize or, if possible, eliminate the effect of the disturbances on its output. In short, even though all the disturbances for a given system may not be removed, the effect of the disturbances can certainly be minimized.

In this chapter we study various types of control systems and their characteristics. We will also learn important basic terms and conventions used to represent these control systems.

1–2 DEFINITION OF A CONTROL SYSTEM

Although a control system may be defined in a variety of ways, the most basic definition is as follows: A *control system* is a group of components assembled in such a way as to regulate an energy input to achieve the desired output. For example, in a temperature control system, thermal energy is regulated to maintain temperature at a predetermined value or within a certain range (Fig. 1–1). Control systems are classified based on the following characteristics:

1. The use of feedback
2. The type of technique(s) used in driving the output to a desired value
3. The nature of the components used for the system under study
4. The intended application of the system

Sometimes control systems are classified as "follow-up" or "regulator" systems, depending on the relationship between the output and the input. In a *follow-up control system*, the output follows the input. If for some reason the input changes, the output will change accordingly. A home heating system is an example of a follow-up system. A change in the thermostat setting will change the temperature in a given room. On

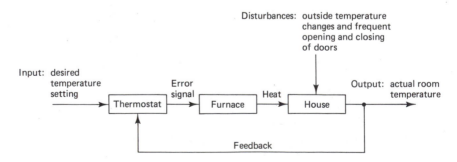

Figure 1–1. Block diagram of temperature control system.

the other hand, in a *regulator system* the output is regulated and changes in a certain way but remains within a predetermined range regardless of changes in the input. An example of a regulator system is a regulated dc power supply.

1–3 TYPES OF CONTROL SYSTEMS

Closed-Loop and Open-Loop Control Systems

The simplest and most common criterion used to classify control systems is *feedback*. In a given control system, if an output or part of an output is fed back so that it can be compared with an input, the system is said to use feedback. This arrangement forms a closed loop moving from input to output and back to input, hence the name *closed-loop system*. In short, a system that uses feedback is a closed-loop system or a system with feedback, and one that does not use a feedback is an *open-loop system*. In a closed-loop system the difference between the feedback signal and the input is called an *error signal*. The error signal is used to drive the output toward the desired value. A block diagram of a closed-loop control system is shown in Fig. 1–2a. A closed-loop system may use one of the two types of feedback:

1. Positive or regenerative feedback
2. Negative or degenerative feedback

If the feedback signal aids an input signal, it is said to be *positive feedback*. When positive feedback is used, eventually input loses control over output. In other words, output is generated without input. Hence positive or *regenerative feedback* is seldom used in practice except in a few special cases. On the other hand, if in a given system the feedback signal opposes the input signal, the system is said to use *negative* or *degenerative feedback*. The use of negative feedback generally results in many advantages; hence it is most commonly used. Accuracy and stability are the most important advantages of using negative feedback. A closed-loop control system can be represented as in the block diagram of Fig. 1–2a. In this block diagram the direction of signal flow is shown by an arrowhead. The block between the input and the output is labeled as a *forward block*. The forward block represents all the circuitry and networks that are required to condition, process, or modulate an input signal before it comes out as an output signal. On the other hand, the function of the feedback block is to convert

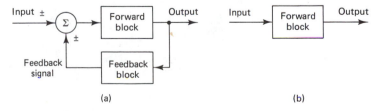

Figure 1–2. Block diagrams: (a) closed-loop control system; (b) open-loop control system.

an output signal into an appropriate form and proper size so that it can be compared with an input signal. Sometimes an output signal may not require processing and may be used as is as a feedback signal. If this happens, the closed-loop system is said to use *unity feedback* and is called a *unity feedback control system*. On the other hand, the system in which the output is not compared with the input is said to be an *open-loop system*. In this case there is no loop between an input and output or the loop is open, and hence the system is open-loop (Fig. 1–2b).

The major advantages of an open-loop control system are that it is relatively simple and economical and generally easy to maintain. However, an open-loop system is less accurate than other systems and is very susceptible to environmental changes and disturbances. Most open-loop systems are precalibrated. For example, a washer, a dryer, a dishwasher, a microwave oven, and a toaster use open-loop control systems (Fig. 1–3). These open-loop systems use precalibrated dials. On the other hand, closed-

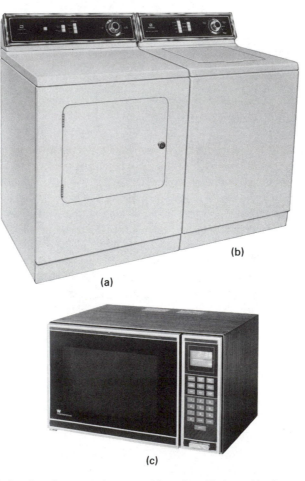

Figure 1–3. Open-loop control systems: (a) washer; (b) dryer; (c) microwave oven. [(a) and (b) Courtesy of The Maytag Company; (c) courtesy of White-Westinghouse.]

loop systems are more accurate, stable, and less sensitive to outside disturbances. However, closed-loop systems are relatively expensive and complex and not as easy to maintain.

Analog and Digital Control Systems

Control systems may be classified based on their operating techniques. In an *analog control system*, analog techniques are used to process the input signal and control the output signal, whereas in a *digital control system*, digital techniques are employed to control the output (Fig. 1–4). For a given system, the choice of techniques used depends on a number of factors; reliability, accuracy, simplicity, and economy are some of the most important. Once the system is designed, its performance is evaluated using certain techniques. For example, linear systems are analyzed using such graphical techniques as the polar or Nyquist plot, the root locus plot, and Bode plots. Bode plots or the Nyquist plot are used to establish the steady-state performance of the system, whereas the root locus plot is used if one's primary interest is in the transient behavior of the system. Generally, the sampled data technique and the z transform are used to analyze the performance of a given digital system. Note that this does not mean that analog techniques such as Bode and root locus plots cannot be used to analyze a digital system.

Linear and Nonlinear Control Systems

The nature of the components utilized in the fabrication of a given system generally determines whether the system is linear. A system is said to be *linear* if it satisfies the amplitude proportionality property and the principle of superposition. According to the *amplitude proportionality property*, if the system output is $c(t)$ for a given input (excitation) $r(t)$, an input of $Kr(t)$ must produce an output of $Kc(t)$, where K is a constant referred to as the *proportionality constant*. Similarly, according to the *superposition principle*, if an input $r_1(t)$ produces an output $c_1(t)$ and an input $r_2(t)$ produces an output $c_2(t)$, an input $[r_1(t) + r_2(t)]$ must produce an output $[c_1(t) + c_2(t)]$, where $r_1(t)$ and $r_2(t)$ are any arbitrary inputs, such as step, sine, ramp, and exponential.

Obviously, a *nonlinear system* does not follow amplitude proportionality and the superposition principle. Moreover, a linear system is of the first degree but can be of any order, whereas a nonlinear system can be of any order and any degree. The best example of a nonlinear system appears in airplanes (Fig. 1–5a). Nonlinear systems are generally analyzed by using techniques such as sampled data and the describing function. These techniques are not only complex but also cumbersome. Therefore, a nonlinear system is usually compensated so that it behaves like a linear system. Often, the nonlinear components' operation is restricted to a specific range so that their behavior can be approximated as linear. As this book is devoted to the analysis and representation of linear systems in their entirety, we will not examine various compensating techniques that are used to linearize nonlinear control systems.

An example of a linear control system is a direct-current (dc) motor speed control system (e.g., a cruise control system) (Fig. 1–5b). In the figure, the larger the input setting, the higher the speed, or vice versa.

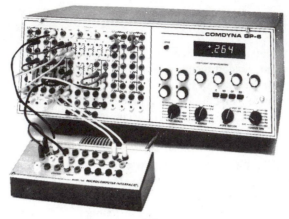

(a)

(b)

Figure 1–4. (a) Analog and (b) digital computers. [(a) Courtesy of Comdyna, Inc.; (b) courtesy of Systolic Systems.]

(a)

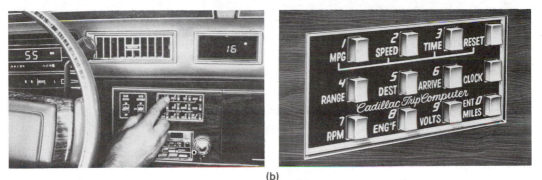

(b)

Figure 1–5. (a) Airplane, an example of a nonlinear system; (b) automobile cruise control, an example of a linear system. [(a) Courtesy of The Port of New York Authority; (b) courtesy of General Motors Corporation.]

Classification of Control Systems According to Application

Control systems are also classified according to their application. For example, servomechanism, sequential control, numerical control, and process control are some of the most important application classifications of control systems.

Servomechanisms. A *servomechanism* is a control system in which the output or the controlled variable is a mechanical position or the rate of change of mechanical position (a motion). A dc motor speed control system and a stepper motor position control system are the two most common examples of servomechanisms (Fig. 1–6).

Sequential control systems. A system in which a prescribed set of operations performed is referred to as a *sequential control system*. The automatic dishwasher and washing machine provide simple examples of sequential control systems. The control system in the automatic washing machine performs the following prescribed operations

Figure 1–6. Robot arm, an example of servomechanisms. (Courtesy of Cincinnati Milacron.)

in sequence: fill the tub, wash the clothes, drain the tub, rinse the clothes, and spin-dry the clothes (Fig. 1–7). Often, a sequential control system is an electromechanical system that uses switches, relays, or solenoids.

Numerical control systems. These days, mainly because of the advent of microprocessors, many manufacturing operations are performed by *numerical control systems*. Numerical control systems act on "numerical information" stored on a "control medium." The numerical information includes such controlled variables as position, direction, velocity, and speed and is normally coded in the form of instructions. The control medium is simply a storage medium such as punched cards, paper tape, magnetic tape, or stored memory which contains all the instructions necessary to accomplish a desired manufacturing operation. Such manufacturing operations include the following: milling, grinding, boring, welding, punching, drilling, and so on. The major advantage of a numerical control system is the flexibility of its control medium. In other words, the same machine or setup may be used to manufacture a number of different parts simply by using different control media. Figure 1–8 shows a drilling machine that can be used to drill different-size parts.

Process control systems. In a process control system, the variables in a manufacturing process are controlled. Some of the most commonly controlled variables include temperature, pressure, liquid level, conductivity, pH content, and composition. An automobile assembly plant, a dairy, an electric power plant, a cloth mill, and a

Figure 1-7. Automatic washing machine, an example of a sequential control system. (Courtesy of the Maytag Company.)

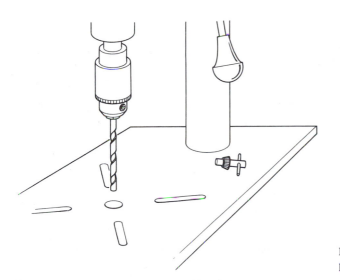

Figure 1-8. Drilling machine, an example of a numerical control system.

Figure 1–9. Auto assembly plant, an example of a process control system. (Courtesy of Chrysler Corporation.)

refinery are typical examples of manufacturing processes. Although process control systems may be either open-loop or closed-loop, most practical process control systems are closed-loop because of the inherent advantages of negative feedback. Figure 1–9 shows an auto assembly plant, which is an example of a process control system.

1–4 CHARACTERISTICS OF CONTROL SYSTEMS

Although different systems are designed to perform different functions, all of them have to meet some common requirements. The major characteristics of a typical control system include stability, accuracy, speed of response, and output sensitivity to component and environmental changes. The degree of importance between these characteristics is prioritized by the designer based on the nature of the system as well as its application. However, for a system to be of practical use, it must be stable. Often stability, accuracy, speed of response, and sensitivity are used as measures of performance to evaluate a system under consideration. In other words, stability, accuracy, speed of response, and sensitivity are the criteria for a good control system. Next, we examine the significance of each of these characteristics.

Stability

A system is said to be *stable* if its output attains a certain value in a finite time after an input is applied. A stable system attains a steady-state value at time $t = \infty$ after an input is initially applied at time $t = 0$. When the output of a system remains constant

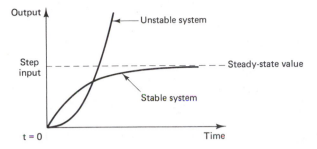

Figure 1–10. Responses of first-order stable and unstable systems to a step input.

and does not change as a function of time, the output is said to attain a *steady-state value* (Fig. 1–10). On the contrary, an unstable system never attains a steady-state value. The output of an unstable system increases with an increase in time until the system breaks down. In short, a practical system must be stable. An unstable system may be made stable by using certain techniques, of which the most common is the use of compensating networks. Thus stability is a criterion of good control and must be achieved by all practical control systems. A variety of graphical and analytical techniques are used to determine the stability of control systems. The Routh–Hurwitz criterion is an analytical technique, and Bode plots and the Nyquist plot are the graphical techniques most commonly used to determine the stability of a given system.

Most practical systems having time delays or dead times tend to be unstable. Therefore, extra precautions must be exercised in designing such systems. A *dead time* or *time delay* is the difference between the time when input to the system is applied and the time when output is produced. Dead time in control systems is the same as propagation time in ICs. Often, an unstable system is made stable simply by using negative feedback.

Accuracy

Another important parameter that is used to evaluate a given control system is accuracy. *Accuracy* indicates deviation of the actual output from its desired value and is a relative measure of system performance. In most practical systems, stability and accuracy interact with each other. In other words, if we are not careful, we may lose the stability of a system in trying to improve its accuracy, or in trying to improve the stability of a system we may decrease its accuracy. Generally, the accuracy of a control system is improved by using such control modes as integration or integration plus proportional. In practice, we seldom expect systems to be totally accurate. For example, in a follow-up system such as a home heating system, the actual temperature in a given room is never exactly the same as the thermostat setting. Usually, systems using feedback, particularly negative feedback, are more accurate than those that do not use feedback (open-loop systems). Thus, accuracy is a relative term and is defined by the user based on the nature and application of the system under consideration. For example, in an open-loop system such as a dishwasher, the cleanliness of utensils is a measure of accuracy. However, cleanliness is a relative term and may vary from user to user.

Speed of Response

In addition to stability and accuracy, the speed of response is an important factor to consider in designing control systems. *Speed of response* is a measure of how quickly an output attains a steady-state value after an input is applied. In a time domain, the response (an output) of a given system is composed of a transient portion and a steady-state portion. Figure 1–11 shows different responses of a second-order system to a step input (a dc input). In practice, it is very difficult to analyze systems that are higher than second order. Hence for the sake of simplicity, systems that are higher than second order are approximated to be of second order and analyzed with a step input. Again the step input is used because it makes analysis of system performance easier. In addition, if a system is stable to a step input, it must be stable to any other input. Remember that before the system is fabricated it is modeled by using mathematical (differential) equations and analyzed by using a variety of graphical and/or analytical techniques. Once the analytical results are satisfactory, the system may actually be built. A practical system must have a finite response time. In Fig. 1–11 the response time is finite for all outputs except the oscillatory response.

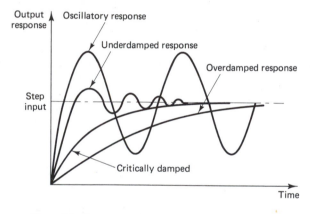

Figure 1–11. Varying responses of a second-order system to a step input.

Sensitivity

The *sensitivity* of a system is a measure of how sensitive its output is to changes in the values of physical components as well as environmental conditions. In a well-designed system the output depends primarily on the input and not on undesired signals, referred to as *disturbances*. The dependence of output on disturbances can be minimized by using certain compensating networks. Obviously, there are mathematical equations that can be used to determine the sensitivity of a given system as a function of variation in a specific system component. This information is then used to improve the efficiency of the system under consideration.

1–5 REPRESENTING A CONTROL SYSTEM

Graphical methods are most commonly used to represent control systems in order to improve communication between design engineers and users. Block diagrams and signal

Introduction to Control Systems   Chap. 1

flow graphs are two such widely used methods. They help us to visualize the system under consideration at a glance. These graphical methods also save time, yet convey all the important information about a system.

Block Diagram

As its name suggests, the *block diagram* of a system consists of blocks, directed line segments joining these blocks, and the summing junctions or error detectors that are used to add the signals algebraically. Figure 1–12 shows two different types of block diagrams, open-loop and closed-loop, which are used to represent all control systems. Any open-loop control system may be represented by the block diagram of Fig. 1–12a, and any system using feedback may be represented by the block diagram of Fig. 1–12b. In Fig. 1–12 the direction of signal flow is indicated by the arrowheads. Although an open-loop control system may be composed of more than one block, it can always be represented in a single block by combining all the blocks into one. Obviously, there are certain block diagram simplification rules that must be followed in reducing a given block diagram to a single block. (More information on block diagrams is presented in Chapter 4.) Similarly, any control system with feedback can always be represented by the block diagram of Fig. 1–12b simply by using two blocks: the forward block and the feedback block. Generally, a mathematical equation relating output to input, referred to as a *transfer function*, may be enclosed inside each block. The transfer function represents the amplitude and time relationship between an input and an output. In Fig. 1–12b the function of an error detector is to detect the difference between the input and primary feedback in generating an error signal. *Primary feedback* is a signal that is compared with the input and is derived from the output. A given closed-loop system may use more than one feedback; however, the primary feedback is the most important.

Besides its simplicity, another major advantage of block diagram is that the overall transfer function of the system under consideration can be obtained very easily. The overall transfer function is then used to analyze the system performance through the use of analytical and/or graphical techniques.

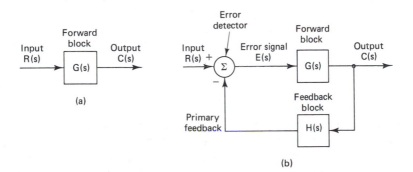

Figure 1–12. Block diagrams: (a) open-loop and (b) closed-loop control systems.

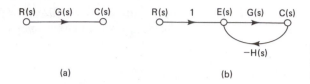

Figure 1-13. Signal flow graphs for (a) open-loop and (b) closed-loop control systems of Fig. 1-12.

Signal Flow Graph

Another graphical method widely used to represent control systems is the signal flow graph. A *signal flow graph* is a diagram that indicates the manner in which the signal flows in a given system. The signal flow graph is a one-line diagram that uses directed segments, as shown in Fig. 1-13. The block diagrams of Fig. 1-12 are represented by using signal flow graphs in Fig. 1-13. Note that in Fig. 1-13 each variable is denoted by a small circle called a *node*, and the directed line segment is denoted by its corresponding transfer function (gain), including its designated sign. A 1 above the directed line segment indicates that its transfer function is unity. The overall transfer function of a given system whose signal flow graph has been drawn can be obtained by using Mason's rule. The signal flow graph method is simpler than the block diagram method, especially when it is used to obtain an overall transfer function of a very complex system. In other words, the signal flow graph method can be used to obtain the transfer function for any system, regardless of how complicated it is. The transfer function is then used to evaluate the performance of that system either by using analytical methods such as the Laplace inverse transform or graphical methods such as Bode plots and the root locus plot. (More information on the signal flow graph is presented in Chapter 4.)

1-6 TERMS AND CONVENTIONS

Before we get involved any further in control system characteristics, design, and performance evaluation techniques, it is important to know some extremely basic terms and conventions that are almost always used with control systems. Unfortunately, there are no absolutely standard terms and conventions used by everyone all the time; however, some terms and conventions are commonly used and therefore will be introduced. In addition, we will define some new terms and conventions that will be used in later chapters. Let us consider the block diagram of a closed-loop system as shown in Fig. 1-14. In this diagram, blocks are described by their functions, so that it will be easier to define certain pertinent terms as follows:

1. *Input* [$r(t)$]. An *input* is an excitation applied to a system as an input from an external source. Based on the nature of the control system, an input to the system may also be referred to as a *reference input* or *set point*. For example, in a follow-up system such as a temperature control system, the actual temperature tries to equal the set input temperature (the thermostat setting). In this temperature control system the input may commonly be referred to as a set point. In a traditional control system an input is commonly referred to as a reference input since it is used as a reference for the output.

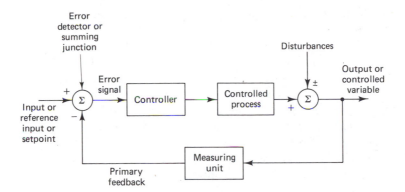

Figure 1–14. Block diagram of a typical closed-loop control system.

2. *Driving function.* Another name for input excitation is the *driving function*, a term used more commonly with electromechanical systems, in which an input excitation drives the system or input excitation is needed for the system to function.

3. *Summing junction or error detector.* A *summing junction* is a physical device that algebraically sums the incoming signals to produce an output signal. Normally, the summing junction in a closed-loop system has two input signals, one a reference input and the other a primary feedback. The primary feedback is always a function of the desired output. Hence if the feedback is negative, the output signal of the summing junction will be the difference between the input signal and the feedback signal (Fig. 1–14). Sometimes the summing junction is referred to as an *error detector*, especially in follow-up systems. This is because, in a closed-loop follow-up system, the output of the summing junction is the difference between the input and the measured output. As long as the actual output is not almost the same as the set input, the system is activated. However, when the actual output is approximately equal to the input, the output of the summing junction is almost zero, and hence the system is disabled. In other words, as long as the error signal (the difference between set input and actual output) is finite, a follow-up system is activated; otherwise, it is disabled. In electrical systems an operational amplifier (op amp) is normally used as a summing junction. Note that in an open-loop system there is no need for a summing junction because there is no feedback. However, if a summing junction is used in an open-loop system, the amplitude of its output will be the same as that of the input signal because the feedback signal is zero.

4. *Controller.* In almost all control systems an input to the controller is an error signal, which is the output of an error detector. The *controller* acts on the input error signal to produce an output signal, which in turn drives the controlled process, which tries to reduce the error signal to zero (Fig. 1–14). The controller is activated unless the error signal is zero. Thus the controller is a device that controls the process in a given system. For example, in a home heating system, the furnace is the controller; it controls the heat inside the house based on the thermostat setting. Similarly, in a toaster, the heating element is the controller and the toast darkness level setting is an input.

5. *Controlled process.* Any physical quantity such as temperature, pressure, or liquid level can be controlled through a process which includes everything that affects

the physical variables. In other words, the *controlled process* represents everything that the equipment performs to control a physical quantity.

6. *Controlled variable* [c(t)]. The *controlled variable* is a physical quantity such as temperature, pressure, or liquid level which is to be controlled in a given system. For example, temperature is the controlled variable in a temperature control system. The controlled variable is normally referred to as an *output*.

7. *Feedback*. A system is said to use *feedback* if the output or part of the output is fed back so that it can be compared with an input. The use of feedback normally improves the stability and accuracy of a system. A given system may use multiple feedbacks. However, primary feedback is a feedback signal representing an output which is compared with an input with the help of an error detector.

8. *Disturbance* [d(t)]. A *disturbance* is any unwanted signal that adversely affects the desired output. For example, changes in outside temperature and frequent closing and opening of doors are disturbances to a home heating system trying to maintain a particular temperature. Similarly, the tolerance of components, variation in component values as a function of temperature and humidity, and electrical noise are disturbances to a voltage regulator unit in regulating voltage to a specific value. Although the effect of disturbances on the desired output cannot be completely eliminated, it can be reduced significantly by using certain networks called compensating networks.

9. *Compensating network*. A network that is used to compensate a system is called a *compensating network*. Normally, a system is said to be compensated when it is modified to include networks that make its actual response approach its desired response. There are a variety of compensating networks; lag, lead, lag–lead, and lead–lag are the most important. The type of compensating network used for a given system depends on the difference between the actual response and the desired response. Sometimes a control system may also be compensated simply by adjusting its component values and their characteristics.

10. *Control mode*. The *control mode* determines the way a controller acts on an input signal in producing an output signal or control action. The most important control modes are integral, derivative, proportional plus integral, proportional plus derivative, optimum, predictive, and on–off. The choice of control mode depends primarily on the desired accuracy of the system and the speed of response of the controller used.

11. *Steady-state value* [$c_{ss}(t)$]. The *steady-state value* is the value that an output of a system attains at time $t = \infty$ after an input is applied. Once attained, the steady-state value does not change unless the input is changed, intentionally or unintentionally. Generally, most practical systems attain steady-state value in a finite time. Often, the steady-state value of a system is referred to as the *steady-state response*.

12. *Settling time* (T_s). The *settling time* is the time it takes the output of a system to reach the steady-state value after an input is applied or changed.

13. *Steady-state deviation* [$\Delta c_{ss}(t)$]. The *steady-state deviation* is the value of an output of a system at time $t = \infty$ due to a disturbance alone. We will denote steady-state deviation by $\Delta c_{ss}(t)$. In an ideal control system the steady-state deviation is zero. In other words, a system with steady-state deviation equal to zero is 100% accurate. Most practical control systems have some finite steady-state deviation. However, with the use of an appropriate control mode, the steady-state deviation can be significantly

reduced. Stated differently, the steady-state deviation and the accuracy are inversely proportional to each other.

14. *Steady-state error* [$e_{ss}(t)$]. The output of an error detector or a summing junction is referred to as an *error signal* (Fig. 1–14). More specifically, the error signal is the difference between the input signal and the feedback signal(s), assuming that negative feedback is used. The value of the error signal at time $t = \infty$ is called the *steady-state error* and is denoted by $e_{ss}(t)$. The steady-state error can be computed by using the final value theorem as shown in Chapter 5.

15. *Transfer function* [$t(t)$ or $T(s)$]. In the time domain the *transfer function* $t(t)$ of a given system or device is the ratio of output to input. However, in the complex frequency domain (the s domain) the *transfer function* $T(s)$ is defined as the Laplace transform of the output divided by the Laplace transform of the input with all the initial conditions assumed to be zero. Initial conditions are assumed to be zero because they affect only transient response, not the steady-state value of a system. Almost all practical systems are designed with a specific steady-state response in mind.

The transfer function of a given system or a device is used to evaluate its characteristics or properties by applying specific graphical and/or analytical techniques. When transformed in the time domain the transfer function reveals the time phase and amplitude relationship between the output and input. In equation form the transfer function may be expressed as follows:

$$t(t) = \frac{\text{output}}{\text{input}}$$

$$T(s) = \frac{O(s)}{I(s)} = \frac{N(s)}{D(s)}$$

(1–1)

where $O(s)$ = Laplace transform of output
$I(s)$ = Laplace transform of input
$T(s)$ = transfer function
$N(s)$ = numerator polynomial
$D(s)$ = denominator polynomial

Note that s denotes the complex plane ($\alpha + j\omega$). The transfer function may be either open-loop or closed-loop. We will use $T(s)$ for the closed-loop transfer function and $G(s)$ for the open-loop transfer function.

16. *Poles and zeros.* For a given transfer function the roots of the numerator polynomial are referred to as *zeros* and the roots of the denominator polynomial are called *poles*. For example, the system whose transfer function is

$$T(s) = \frac{s + 5}{s^2 + 3s + 2}$$

$$= \frac{s + 5}{(s + 1)(s + 2)}$$

has one zero and two poles. The zero is at $s = -5$, and the poles are at $s = -1$ and $s = -2$. For most practical systems the number of zeros is always less than or equal to the number of poles.

17. *Time constant* (τ). For a given system the *time constant* determines the settling time. A stable first-order system with a single pole has a single time constant. For such a system the settling time T_s is approximately equal to five times the time constant. In other words, $T_s = 5\tau$. However, some second-order systems may have two time constants. Obviously, the larger time constant of the two determines the settling time. In short, the time constant is the measure of system's settling time and in turn represents the response characteristic of the system under consideration.

18. *Damping constant* (α). The *damping constant* is an inverse of time constant τ, that is, $\alpha = 1/\tau$. Like the time constant, the damping constant is a measure of a system's settling time. Specifically, the smaller the damping constant, the longer the settling time, and vice versa. The damping constant indicates the degree of damping associated with the response of a given system. For example, as shown in Fig. 1–11, the response of the second-order system to a set input could be:

a. Critically damped
b. Overdamped
c. Underdamped
d. Oscillatory

The characteristic of a *critically damped* response is that it has two real and equal damping constants. An *overdamped response* has two real and simple damping constants, and an *underdamped response* has simply a single damping constant. An *oscillatory response* has a damping constant of zero. In other words, a system with a zero damping constant never attains a finite steady-state value; instead, its response oscillates between two fixed limits.

The following parameters are applicable to second- and higher-order systems.

19. *Natural resonant angular frequency* (ω_n). This parameter is applicable to a special type of second-order systems. Depending on their composition, these second-order systems oscillate at a certain angular frequency called a *natural resonant frequency*. For example, a parallel *RLC* network is said to be in resonance when its inductive and capacitive reactances are equal. The angular frequency at which this happens is given by $\omega_n = 1/\sqrt{LC}$ (with zero circuit resistance). In short, in natural resonance, the form of an oscillation is produced by the circuit itself, regardless of the type of excitation. Besides electrical systems, the concept of natural resonance is also applicable to other physical systems including mechanical, hydraulic, and pneumatic.

20. *Damping ratio* (ζ). The most convenient way to identify the response of second-order systems is by the damping ratio. The *damping ratio* is defined as the ratio between the actual damping in the system and that damping which would produce critical damping in the system. The damping ratio is denoted by the symbol ζ (lowercase Greek zeta) and is given by

$$\zeta = \frac{\alpha}{\omega_n} = \frac{1}{\tau\omega_n} \tag{1-2}$$

In other words, the damping ratio ζ is a function of the time constant τ (or the damping constant α) and the natural resonant angular frequency ω_n. More specifically, by definition, a second-order system with (Fig. 1–15)

Introduction to Control Systems   Chap. 1

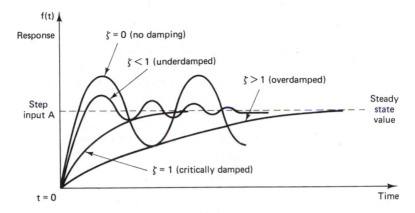

Figure 1–15. Responses of a second order system to a step input as a function of damping ratio ζ.

a. $\zeta < 1$ will have an underdamped response.
b. $\zeta = 1$ will be critically damped.
c. $\zeta > 1$ will have an overdamped response.
d. $\zeta = 0$ will have an oscillatory response.

An important characteristic of these responses is that a critically damped system responds in the least amount of time. An underdamped system gets close to the final value more rapidly but takes longer to reach the final steady-state or equilibrium value. On the other hand, an overdamped system is sluggish in responding, and an oscillatory system responds with persistent sinusoidal oscillations at the system's natural frequency. Note that the responses to a step input of all second-order systems, whether electrical, mechanical, hydraulic, or biomedical, are the same: underdamped, critically damped, overdamped, or oscillatory.

21. *Damped natural resonant angular frequency* (ω_d). For a second-order system with an underdamped response as shown in Fig. 1–11, the decaying or damped sine wave does not oscillate at the natural frequency ω_n but at the damped natural frequency ω_d. The decay or damping of the response is caused by the energy dissipating elements. For example, in a parallel *RLC* circuit, if R is nonzero, the circuit oscillates at a frequency ω_d, which is described by

$$\omega_d = \frac{1}{\sqrt{LC}} \sqrt{1 - \frac{R^2 C}{L}} \qquad (1-3)$$

The damped natural resonant angular frequency ω_d is smaller in magnitude than the natural resonant angular frequency ω_n since

$$\omega_d = \omega_n \sqrt{1 - \zeta^2} \qquad (1-4)$$

Note that the parameters ω_d, ω_n, and ζ are used in describing a decaying oscillation in both the time domain and the s domain (Laplace transform).

SUMMARY

1. A control system is a group of components assembled so that they regulate an energy input that drives an output to a desired value.

2. Control systems are classified based on the following characteristics:

   a. The use of feedback
   b. The type of technique used to drive the output to the desired value
   c. The types of components used
   d. The intended application

3. According to the classification above, control systems may be defined as open-loop or closed-loop, analog or digital, and linear or nonlinear. Control systems are also classified according to the type of application they are used for: servomechanism, sequential control, numerical control, or process control.

4. The major characteristics of control systems are stability, accuracy, speed of response, and sensitivity to component and environmental changes.

5. Block diagrams and signal flow graphs are the most commonly used graphical methods for representing control systems. The block diagram is composed of blocks and directed line segments, whereas the signal flow graph consists of nodes and directed line segments. The block diagram is easier to understand and conveys more information at first glance than does its equivalent signal flow graph.

6. Some of the most important terms commonly used in conjunction with control systems include input, summing junction or error detector, controller, controlled variable, feedback, disturbance, compensating network, control mode, steady-state value, steady-state deviation, settling time, and steady-state error.

QUESTIONS

1–1. What is a control system?

1–2. List four factors that are used to classify control systems.

1–3. Classify the following systems as either open-loop or closed-loop and explain your answers briefly: **(a)** window air conditioner; **(b)** automobile cruise control; **(c)** portable fan; **(d)** hand dryer; **(e)** automatic car wash machine; **(f)** vending machine; **(g)** automatic water sprinkler system; **(h)** jukebox; **(i)** electric iron.

1–4. What is the difference between positive and negative feedbacks? Which is in more common use and why?

1–5. Compare and contrast the usefulness of open-loop and closed-loop control systems.

1–6. What is the difference between analog and digital control systems? List the advantages and disadvantages of each.

1–7. List the criteria that are used to classify control systems as either linear or nonlinear.

1–8. Is a home heating system a linear control system? Explain your answer.

1-9. What are the four different types of applications of control systems? Briefly explain each type and give at least one example.

1-10. List the major characteristics of a control system, and explain the significance of each.

1-11. Evaluate the following systems in terms of stability, accuracy, speed of response, and sensitivity: **(a)** washing machine; **(b)** home heating system; **(c)** automobile cruise control system.

1-12. List two graphical methods used to represent control systems.

1-13. What are the parts of a typical block diagram? Briefly explain each.

1-14. What is the composition of a typical signal flow graph?

1-15. List the advantages and disadvantages of the block diagram method compared to the signal flow graph method.

1-16. Define the following terms: **(a)** input; **(b)** summing junction; **(c)** controlled variable; **(d)** feedback; **(e)** steady-state error; **(f)** damping ratio.

CONTROL SYSTEM COMPONENTS

chapter 2 _____

2-1 INTRODUCTION

In later chapters control systems will often be represented by block diagrams. Each block represents a control system component which performs a specific function in the overall operation of that system. Associated with the characterization of control system components by linear blocks is the concept of the *transfer function*, which is a mathematical relationship between the output and the input of a given linear block. The transfer function may simply be a constant or it may take on a rational form in terms of the Laplace variable *s*.

Many different types of components may be used in the design of a control system. A broad category of components, generally termed *transducers*, play a particularly important role in control systems. The transducer, which is generally defined as a component that converts energy from one form to another, may be placed into one of several categories. An electromechanical transducer such as a motor converts energy from electrical to mechanical form, and a generator does exactly the opposite. There are other transducers which convert physical quantities such as position, velocity, temperature, and others into electrical form. Any control system whose operation is based on electrical components relies heavily on transducers which convert physical quantities to an electrical form.

The use of linear blocks considerably simplifies the configuration of a control system, making it easier to analyze a system and to explain its operation. At the same time, a linear block model obscures the physical structure and the characteristics of the component that it represents. Some components are radically different, yet they have the same type of transfer function. The transfer function for the dc motor, for example, is identical in form to that of a simple *RC* low-pass filter. It is the purpose of this chapter to present the characteristics and operation of commonly used transducers and of other electrical, mechanical, and electromechanical components. Motors fit these categories, of course; however, due to their extensive nature, they are covered separately in Chapter 3.

2–2 TEMPERATURE SENSORS

Thermistor

The *thermistor* (*ther*mal re*sistor*) is a transducer that changes its resistance with temperature. A ceramic-based mixture may be used as a base material from which the thermistor is manufactured. Thermistors are also made from oxide-based semiconductor materials, including oxides of iron, copper, manganese, nickel, and others. Thermistors are commercially available in a variety of packages, such as disks, rods, beads, and chips.

One of the more important characteristics of a thermistor is its *temperature coefficient* (TC), which expresses a change in the thermistor's resistance over a given temperature range. The TC is negative and may be expressed mathematically as

$$TC = \frac{\Delta R/R}{\Delta T} \qquad (2-1)$$

where the change in resistance $\Delta R/R$ is usually expressed as a percent or in parts per million (ppm). The current used in the measurement of resistance must be sufficiently small so as not to cause any significant power dissipation, which might otherwise contribute to errors.

Sometimes a manufacturer may specify the value of material constant β, which is related to the resistance temperature curve, such as the one shown in Fig. 2–1. If the value of β is known, the equation

$$\frac{R_2}{R_1} = e^{\beta[(1/T_1)-(1/T_2)]} \qquad (2-2)$$

is used to calculate the value of R_2 at any temperature T_2. The manufacturer generally specifies in the data sheet the value of R_1 at some reference temperature T_1 (T_1 is typically 25°C).

When a thermistor is used to sense the temperature of a given environment, it is generally necessary to convert the change in its resistance to an equivalent analog voltage. This is commonly accomplished with a Wheatstone bridge and an operational amplifier (op amp), as shown in Fig. 2–2. Often, due to application constraints, the thermistor is located some distance away from the bridge. It is connected to the arm of the bridge between points *A* and *D* by means of a two-wire cable l_1–l_2. The third lead, l_3, which is used for the purposes of temperature compensation, is also brought to the bridge from the remote location and is connected to the adjacent arm of the bridge in series

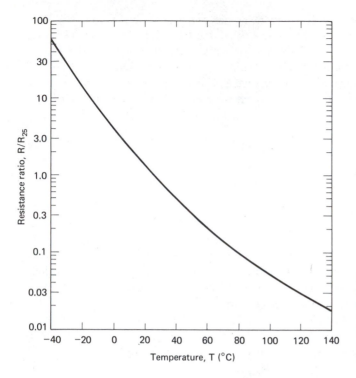

Figure 2–1. Variation of the resistance ratio with temperature of a ceramic rod thermistor type H rods, beta = 4400. R/R_{25} = resistance at temperature T/resistance at 25°F. (Courtesy of Sohio Engineered Materials Co./Carborandum, Niagara Falls, N.Y.)

with the calibration resistor R_2. Any change in the wire resistance of l_1 due to temperature changes of the surroundings is canceled by exactly the same change in the resistance of l_3 because l_1 and l_3 are in the adjacent arms of the bridge, l_2 being common to both arms. Consequently, the bridge is unbalanced only by the change in resistance R_x of the thermistor.

The bridge is initially balanced by adjusting R_2 so that the output of the bridge $V_i = 0$ V, and therefore $V_o = 0$ V. When the thermistor R_x experiences a change in

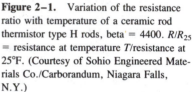

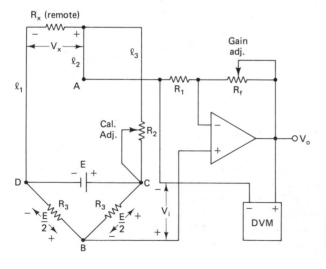

Figure 2–2. Measurement of an unknown temperature with a remote thermistor.

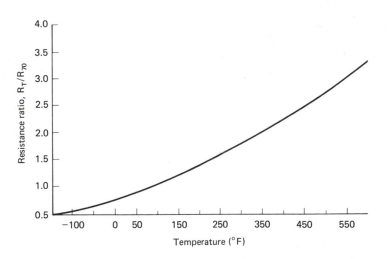

Figure 2–3. Resistance variation with temperature of the nickel resistive temperature sensor. R/R_{70} = resistance at temperature T/resistance at 70°F. (Courtesy of BLH Electronics, Canton, Mass.)

resistance $X = \Delta R_x/R_x$ due to the temperature change of the remote environment, the bridge output may be expressed as

$$V_i = E\left(\frac{1+X}{2+X} - \frac{1}{2}\right) = \frac{E}{2}\frac{X}{2+X} \qquad (2\text{–}3)$$

and the output of the op amp may be expressed as

$$V_o = \frac{E}{2}\frac{X}{X+2}\left(1 + \frac{R_f}{R_1}\right) \qquad (2\text{–}4)$$

Resistive Sensor

A *resistive temperature sensor* resembles a thermistor in its basic behavior. However, the resistive temperature sensor has a positive temperature coefficient and is capable of operation over a much wider temperature range. This is due to the physical characteristics of the materials from which these devices are made. All resistive materials exhibit a variation in resistance with temperature. Not all materials, however, possess a stable and repeatable temperature characteristic which may be accurately predicted over a wide temperature range.

One such material is nickel. Fine wires of high-purity nickel, when combined with a suitable insulating material. make accurate resistive temperature sensors (Fig. 2–3). Nickel phenolic and nickel[*] Teflon® temperature sensors are commercially manufactured for surface temperature measurement in the range −100 to +600°F. The sensor grid is typically attached to a thin backing, and the entire structure is bonded to the surface whose temperature is to be measured. The Wheatstone bridge configuration

[*] Teflon® is a registered trademark of Dupont.

shown in Fig. 2–2 may be used in the measurement process to convert the resistance variation to an analog voltage signal.

Platinum is another material that is used as the base material for resistive temperature sensors. It may be operated continuously from −320 to +1500°F without degradation in its characteristics. Platinum sensors are available commercially in a weldable series and in a free filament series. In the weldable series, the platinum sensing grid is bonded to a 6-mil stainless steel substrate with refractory oxide. The steel substrate is then welded to the test surface. Free-filament platinum sensors, on the other hand, are mounted on the carrier, which is then bonded to the test surface with bonding materials such as EPY150, CER1200, Rokide®, or Silastic® 140.

Thermocouple

In contrast to the thermistor and resistive temperature sensors, a *thermocouple* is a transducer that converts temperature change directly to a voltage signal. A thermocouple is made by welding two dissimilar materials. The voltage at the junction of two dissimilar materials exhibits a predictable variation with temperature.

Some of the materials that may be used in the manufacture of a thermocouple are shown in Table 2–1. As can be seen, the materials exhibit a broad range of operating temperatures, some as high as 5200°F. The thermocouple response times, shown in Table 2–1, are all in the millisecond range. This is an advantage from an application standpoint, as the thermocouple is capable of making temperature measurements directly and quickly.

Figure 2–4 illustrates a commercial thermocouple configuration. This is a grid structure which is available in a copper–constantan junction which is mounted on Teflon–fiberglass tape suitable for direct surface mounting.

As shown in Fig. 2–5, the thermocouple may also be packaged in a probe configuration. It is small, designed to withstand the temperature of the hot junction, and is available in three options: insulated, flattened, and exposed. The exposed junction is recommended for measurement of gas temperatures, and the flattened junction, which may be spot welded to the surface, is suitable for surface temperature measurements. The insulated junction is designed to protect the thermocouple wires and to isolate the thermocouple from the surface to avoid electrical pickup. The spring-loaded probe is suitable for surface temperature measurements where direct attachment of the probe to the surface is undesirable. The spring loading maintains good thermal contact between the probe and the surface.

The *hot junction* is an active thermocouple junction whose voltage is proportional to the measured temperature. A *cold junction* results when the leads of the measurement circuit are connected to the thermocouple. Because the potential due to the cold junction also varies with temperature and appears in series with the hot junction, its effects cannot be simply nulled out. Consequently, the effect of the cold junction, which adds to the measurement error, must be compensated.

One approach to compensation may be to immerse the cold junction in an ice-water bath and thus maintain the cold junction at a constant temperature, and therefore

---

Rokide® is a registered trademark of Norton Co. Silastic® is a registered trademark of Dow Corning Corp.

TABLE 2–1 THERMOCOUPLE MATERIALS, OPERATING TEMPERATURES AND TRANSIENT RESPONSE OF THERMOCOUPLES

a. Temperature Ranges (°F)

Material	Atmosphere	Exposed junction[*]		Shielded junction	
		Short time	Long time	Short time	Long time
Chromel–alumel[†]	Oxidizing	1800	900	2000	1000
Copper–constantan	Oxidizing or reducing	600	300	750	400
Chromel–constantan[†]	Oxidizing	1700	1100	1800	1200
2.1% gold, cobalt–copper	Oxidizing	160	125	160	125
Platinum 10% rhodium–platinum	Oxidizing	3200	2500	3200	2500
Platinum 13% rhodium–platinum	Oxidizing	3100	2500	3100	2500
Tungsten–tungsten, 26% rhenium	Inert	5000	3200	5200	3200

b. Typical Response Times

Room temperature to 350°F in hot moving air	Exposed junction: $\tau = 0.013$ s
Air velocity = 45 ft/s	Insulated junction: $\tau = 0.32$ s
Time constant for 63.2% of a step change	Flattened junction: $\tau = 0.21$ s
in temperature	Welded junction and sheath: $\tau = 0.020$ s in water

[*] Short-time exposure should be considered in terms of minutes, while long-time exposure in hours is representative. These reduced temperature limits result from the use of 0.001-in.-diameter wires.

[†] Hoskins Manufacturing Co., Detroit, Mich.

Source: BLH Electronics, Canton, Mass.

at a constant potential, which can be nulled out. The use of ice water, as shown in Fig. 2–6a, is impractical. An alternative to the ice-water bath is a temperature sensor which is placed in close proximity with the cold junction, as shown in Fig. 2–6b. The output of the sensor varies in the same way as does the potential of the cold junction. By properly scaling the output of the sensor and combining it with the amplified signal from the thermocouple in the 2B56 cold junction compensator, the effects of the cold junction may be completely nulled out through calibration. Figure 2–6c shows the

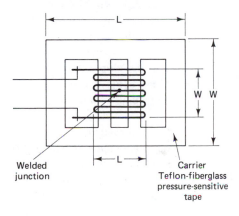

Welded junction

Carrier Teflon-fiberglass pressure-sensitive tape

Figure 2–4. Grid-type thermocouple construction for bonding with conventional cements on flame spray techniques. (Courtesy of BLH Electronics, Canton, Mass.)

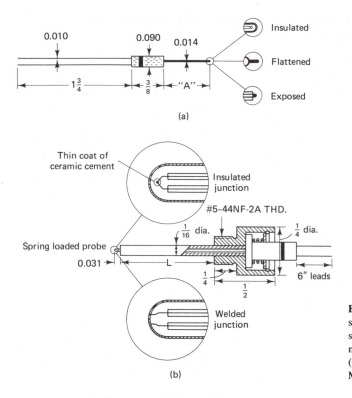

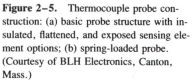

Figure 2–5. Thermocouple probe construction: (a) basic probe structure with insulated, flattened, and exposed sensing element options; (b) spring-loaded probe. (Courtesy of BLH Electronics, Canton, Mass.)

overall diagram, which includes a thermocouple, temperature sensor, and major components of the 2B56.

IC Temperature Sensor

There are also temperature sensors made of semiconductor material in chip form. An example of this type of device is the AD590, which is a monolithic IC temperature sensor. The AD590 circuit and its $I-V$ characteristics are shown in Fig. 2–7a and b. The AD590 is a two-terminal, calibrated, temperature-dependent current source whose current varies linearly with temperature at the rate of 1 μA/°K over the temperature range −55 to +150°C. The current TC may be deduced from the $I-V$ characteristics as follows. Assuming that the supply voltage is greater than 3 V, the change in current $\Delta I = 80$ μA (218 to 298), corresponding to the temperature change $\Delta T = 80$°C = 80°K (−55 to +25). The resulting temperature coefficient TC = $\Delta I/\Delta T = 1$ μA/°K. By applying the laser trimming technique to the chip's thin film resistors at the wafer level, the manufacturer is able to control and achieve accurately the desired value of the temperature coefficient. AD590 is laser trimmed for an output current of 298.2 μA at 298.2°K.

The AD590 is basically a PTAT (proportional to absolute temperature) current regulator i.e., its output current is equal to a scale factor times the absolute temperature. The scale factor is trimmed at the factory by adjusting the output current (which represents the temperature) to agree with the actual temperature. The difference between the indicated temperature (the value of the output current) and the actual temperature represents

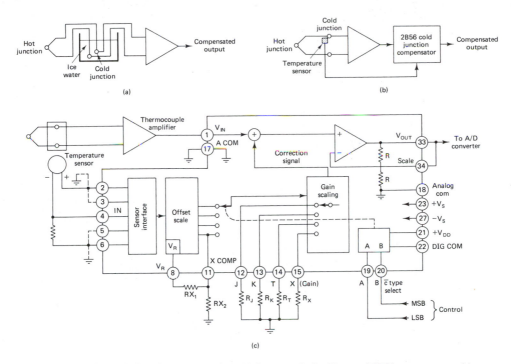

Figure 2–6. Cold junction compensation: (a) ice-water bath; (b) use of 2B56 compensator; (c) thermocouple and 2B56 functional block diagram. [(b), (c) Courtesy of Analog Devices, Norwood, Mass.]

the calibration error. The calibration error at 25°C for AD590 family ranges from ±0.5°C for AD590M to ±10°C for AD590I.

The calibration error is the primary contributor to maximum total error. Since the calibration error is a scale factor error, it may be trimmed in a circuit where AD590 is connected in series with a 5 V reference supply and a potentiometer. The potentiometer is adjusted until the voltage across the potentiometer divided by the measured temperature is 1 mV/°K. When the calibration error is trimmed out at one temperature, its effect is zero over the entire temperature range. With the calibration error trimmed out at one temperature, the absolute error over the operating temperature range is not negligible. It is due to the nonlinear thermal characteristics of the device and its range for the AD590 family extends from ±1°C (AD590L) to ±5.8°C (AD590I).

The AD590 may be packaged in the form of a probe, as shown in Fig. 2–7c. The probe package is particularly useful in applications where the sensor is to be immersed in liquid or gaseous media.

2–3 STRAIN GAGE

The *strain gage transducer* is a grid of fine wire attached to a flexible backing, a carrier sheet. In a typical application, the strain gage is bonded to a mechanical member that is stressed by the applied forces. The stress causes the mechanical member as well

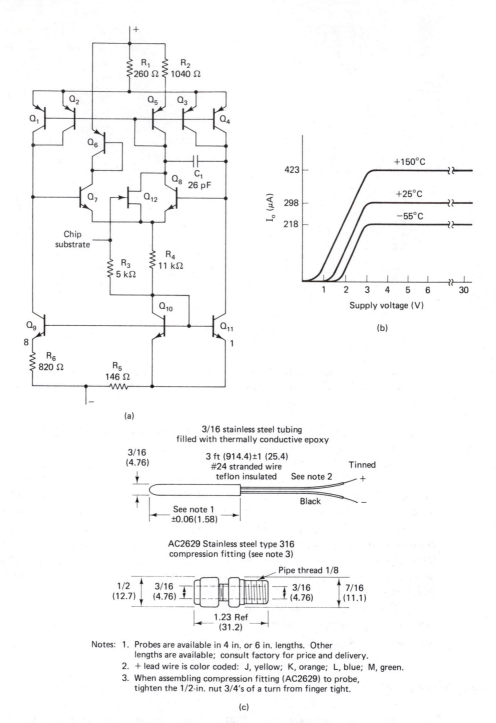

Figure 2–7. AD590 Monolithic IC temperature sensor: (a) circuit; (b) I–V characteristics; (c) AD590 in the probe package AC2626. (Courtesy of Analog Devices, Norwood, Mass.)

Control System Components   Chap. 2

as the strain gage to undergo a dimensional change known as *strain*. The change in resistance of the strain gage wire due to stretching or compression, as the case may be, is converted to an analog voltage by the additional electronics. It is assumed that the dimensional changes of the wire are sufficiently small so as not to deform the wire permanently, and consequently, the stress-produced strain occurs within the elastic limits of the wire.

Figure 2–8 shows a typical flat grid strain gage mounted on a suitable carrier, which may be bonded to the test surface. The resistance of any wire may be expressed by

$$R = \rho \frac{L}{A} \quad \Omega \tag{2-5}$$

where L = length
A = cross-sectional area
ρ = resistivity of the wire

It can be seen from (2–5) that stretching produces an increase, and compression produces a decrease in R. One important characteristic of the strain gage is the gage factor G, which is defined as follows:

$$G = \frac{\Delta R/R}{\Delta L/L} \tag{2-6}$$

where $\Delta L/L$ is the change in the length of gage wire divided by its original length. Both R and L increase when the gage is subjected to tension, and decrease when the gage is in compression.

As mentioned earlier, additional circuitry is required to convert the gage resistance change to an analog signal. The bridge circuit with the operational amplifier shown in Fig. 2–2 can perform this function. If four strain gages are connected, one in each arm of the bridge as shown in Fig. 2–9, then by standard theory the output of the bridge is expressed by

$$\frac{V_o}{E} = \frac{R_1}{R_1 + R_2} - \frac{R_4}{R_3 + R_4} \tag{2-7}$$

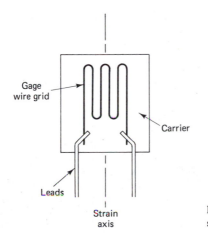

Gage wire grid

Carrier

Leads

Strain axis

Figure 2–8. Flat grid strain gage construction.

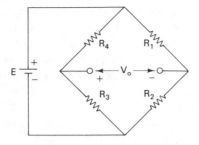

Figure 2–9. Wheatstone bridge in strain gage measurement application.

When the condition $R_1R_3 = R_2R_4$ is satisfied, the bridge is in balance and $V_o = 0$ V in Eq. (2–7). Suppose that the bridge is initially balanced, and then each of the four strain gages is subjected to stress. The resulting change in bridge output due to change in the resistance of the gages is deduced from Eq. (2–7) by the total increment approach [i.e., assuming that $\Delta V_o = f(R_1, R_2, R_3, R_4)$, then $\Delta V_o = f(R_1 + \Delta R_1, R_2 + \Delta R_2, R_3 + \Delta R_3, R_4 + \Delta R_4) - f(R_1, R_2, R_3, R_4)$], with the following result:

$$\frac{\Delta V_o}{E} = \frac{R_1 + \Delta R_1}{R_1 + R_2 + \Delta R_1 + \Delta R_2} - \frac{R_4 + \Delta R_4}{R_3 + R_4 + \Delta R_3 + \Delta R_4} \qquad (2\text{–}8)$$

Not included in Eq. (2–8) is the term $R_1/(R_1 + R_2) - R_4/(R_3 + R_4)$, which is equal to zero because the bridge is initially balanced, and at balance $R_1R_3 = R_2R_4$.

Apparent strain occurs when the remote active gage produces a change in resistance due to a change in temperature. In an actual application where many gages are involved, the measurement may take some time. If the temperature in the proximity of the gage changes while the gage is being stressed, we must expect that a part of the bridge output signal is temperature related, not stress related. Consequently, the apparent strain can produce significant errors in measurement, and must therefore be compensated.

A typical compensation scheme is shown in Fig. 2–10, where a dummy gage is placed next to an active gage. Whereas the active gage is subjected to stress, the dummy gage is not. The result is that the temperature-related resistance change produced by the dummy gage and the active gage are exactly the same because the two gages are

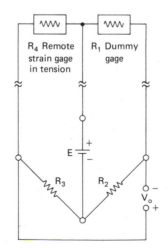

Figure 2–10. Temperature compensation is provided by the dummy gage, which is not subjected to stress but is exposed to the same temperature as the active gage.

Control System Components   Chap. 2

electrically identical; because the two gages appear in the adjacent arms of the bridge, the effects of the apparent strain are canceled out.

A variety of strain gage bridge configurations are possible. A bridge containing one strain gage is called a $\frac{1}{4}$ *bridge*; a $\frac{1}{2}$ *bridge* includes two strain gages (they generally appear in the adjacent arms of the bridge, with one gage in compression while the other is in tension). When the bridge is configured as a *full bridge*, it produces the maximum output signal. The full-bridge configuration uses four strain gages arranged so that the gages in the two opposite arms of the bridge are in tension while the remaining two gages are in compression.

The bridge output equation (2−8) is general, and applies to all the bridge configurations noted above. Readers may wish to test their algebraic skills and show that in the case of the $\frac{1}{4}$ bridge with $R_3 = R_4$, $R_1 = R_2$, and R_1 being the active gage, the bridge output may be expressed as

$$\frac{\Delta V_0}{E} = \left(\frac{1}{2}\right)\frac{X}{2 + X} \qquad (2-9)$$

where $X = \Delta R_1/R_1$, and X is positive when R_1 is in tension and negative when R_1 is in compression. Also, in the case of the $\frac{1}{2}$ bridge, where $R_1 = R_2 = R_3 = R_4 = R$ and R_1 and R_4 are the active strain gages, R_1 in tension and R_4 in compression, the bridge output may be expressed as

$$\frac{\Delta V_o}{E} = \frac{2X}{4 - X^2} \qquad (2-10)$$

where $X = \Delta R/R$. X assumes a negative value when R_1 is in compression and R_4 is in tension. Finally, consider the full-bridge circuit of Fig. 2−9, where four strain gages are used, one in each arm of the bridge. Assuming that R_1 and R_3 are subjected to tension while R_2 and R_4 are in compression, and also assuming that all four gages are electronically identical, the bridge output may be expressed as

$$\frac{\Delta V_o}{E} = X = G\epsilon \qquad (2-11)$$

where $X = \Delta R/R$ (R being the nominal gage resistance at reference temperature) and $\epsilon = \Delta L/L$ is the strain, which typically is expressed in μin./in. Each of the expressions (2−9) to (2−11) is derivable from Eq. (2−8). Apparent strain compensation through the use of dummy gages is generally not necessary in the case of $\frac{1}{2}$ or full bridges where the active gages are remote and exposed to the same temperature environment.

The G factor for all wire or foil (foil strain gages use metallic foil instead of wire) strain gages is typically 5 or less. Hence the bridge output for these types of gages is small, and additional amplification of the bridge output is generally required, as shown in Fig. 2−2. The G factor for *semiconductor strain gages*, on the other hand, is 100 or better. Consequently, the bridge output due to a semiconductor strain gage may be of the order of several volts, which is sufficient for most applications to be used directly without additional amplification.

The operation of the semiconductor strain gage depends on the piezoresistive effect, which occurs in semiconductors under stress. From the physics standpoint, stress-related strain (i.e., dimensional change in the semiconductor material occurring within its elastic limits) has an effect on the carrier transport phenomenon in that it affects the conduction electron energy, carrier mobility, and carrier lifetime. From an electrical point of view, strain causes the semiconductor to change its conductivity, and therefore its resistance, in a predictable and repeatable manner. The manufacturing techniques used in today's chip production may also be applied to the manufacture of strain gages, thus making the production of semiconductor strain gages cost-effective.

The sensitivity versus strain relationship is shown in Fig. 2–11 for selected semiconductor strain gages. As shown, the sensitivity (normalized change in gage resistance

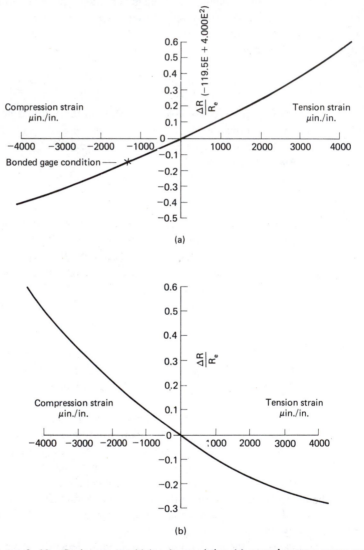

(a)

(b)

Figure 2–11. Strain gage sensitivity characteristics: (a) general-purpose gages; constant-current circuitry; (b) temperature-compensated *n*-type semiconductor gages. (Courtesy of BLH Electronics, Canton, Mass.)

$\Delta R/R$) is along the ordinate, and strain measured in μin./in. is along the abscissa. The slope of the curve represents the value of the gage factor. In the case of the general-purpose gage, the slope is positive, and therefore the gage factor is also positive (its value from the graph is approximately 100, although manufacturer-specified nominal values are somewhat higher), which means that the gage resistance increases when the gage is in tension and decreases when the gage is subjected to compression.

In some applications the semiconductor strain gage bonded to a test surface may experience a large temperature change. If the thermal expansion coefficients of the test surface and of the gage are unequal, the gage will experience an apparent strain which is unrelated to the applied stress. In addition, the gage resistance will change, due to the change in temperature. The latter effect may be compensated, as discussed earlier, through the use of a dummy gage in an adjacent arm of the bridge. However, the apparent stress due to unequal coefficients of expansion poses a rather serious problem for the user in cases where a strain gage is exposed to large temperature variations. For applications such as this, the manufacturers produce temperature-compensated strain gages. To achieve temperature compensation, the manufacturer specifies the material with an appropriate temperature coefficient to which the strain gage is to be bonded. The change in the strain gage resistance due to temperature is canceled by an opposite resistance change which is caused by expansion of the material to which the gage is bonded. The temperature versus strain curve for a typical temperature-compensated strain gage is shown in Fig. 2–11b. The slope of the curve is negative; hence the gage factor in this case is negative.

Because temperature-compensated gages are expensive, their use is recommended in applications where large temperature fluctuations occur or where high accuracy of measurement is required. In most other applications general-purpose gages are satisfactory; they are cheaper and more linear than temperature-compensated gages. All semiconductor gages are generally small; most are under $\frac{1}{4}$ in. and some are as small as 0.06 in. in active gage length.

2–4 PHOTOTRANSISTOR

The operation of the *phototransistor* depends on the photoeffect, which occurs in semiconductors and other solids when light of the proper wavelength strikes the material surface. In the case of an *npn* transistor, which is typically packaged in a TO-18 can with a window provided for the incident light, as shown in Fig. 2–12, the incident light causes holes generated in the collector to be injected into the base and to diffuse across the base toward the emitter. Upon reaching the emitter, the holes are injected into the emitter region, and at the same time electrons are injected from the emitter into the base. Due to the higher emitter injection efficiency, more electrons are injected into the base. These electrons diffuse across the base region, are injected into the collector, and become the photocollector current which flows in the external circuit. The collector–base region is made larger in a phototransistor, as compared to an ordinary transistor, to enhance the number of photo-generated carriers.

The phototransistor may be operated as either a two-terminal or three-terminal device. When an electrical source is applied to the base, the transistor operates in the normal manner with or without the incident light. However, when the base is open

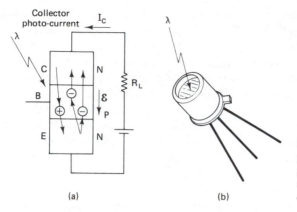

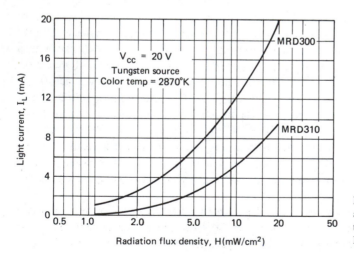

Figure 2–12. Phototransistor: (a) light of proper wavelength generates hole–electron pairs in an *npn* transistor; (b) phototransistor packaged in TO-18 can with the window provided for the incident light.

and the phototransistor operates as a two-terminal device, the light must be applied to produce a collector current.

A typical variation of a phototransistor light current with irradiance is shown in Fig. 2–13. The irradiance, which expresses light flux density incident on the phototransistor in the *radiometric* or physical system, is measured in W/cm^2. In the *photometric* system, which is based on the visual effect (the response of the human eye, which extends from 400 to 700 nm, with the peak at 555 nm), the equivalent of irradience is illuminance measured in lumens/ft^2 (footcandles). The two systems are related at 555 nm, where 1 W of radient flux is equal to 680 lumens of luminous flux.

A tungsten lamp is often used as a source of radiation for testing phototransistors and other photodetectors. As shown in Fig. 2–14, it emits little radiation in the visible spectrum at low color temperature. As the color temperature is increased, its response shifts closer to the visible and the MRD phototransistor series spectral responses. *Color temperature* is defined as the temperature that the ideal blackbody must assume to produce the same illuminance as the lamp.

Tungsten lamp is often used as a source of radiation for a phototransistor. The response of the MRD300 phototransistor when the lamp operates at a color temperature

Figure 2–13. Photo current versus irradiance for a typical phototransistor. (Courtesy of Motorola Semiconductor Products, Inc., Phoenix, Ariz.)

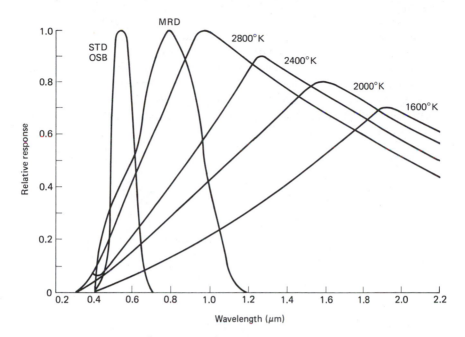

Figure 2–14. Spectral response of a tungsten lamp at selected color temperatures. Also shown for comparison are the visual and the MRD series phototransistor responses. (Courtesy of Motorola Semiconductor Products, Inc., Phoenix, Ariz.)

of 2870 °K is shown in Fig. 2–13. This response changes if the color temperature is changed. Figure 2–15 shows the variation in relative value of the response with color temperature; the 100% point is at 2870 °K. From Fig. 2–15, the response falls off to 80% at 2550 °K; hence the value of light current, $H = 10$ mW/cm^2 in Fig. 2–13, for example, is reduced for MRD300 from 12 mA at 2870 °K to 0.8(12) = 9.6 mA at 2550 °K.

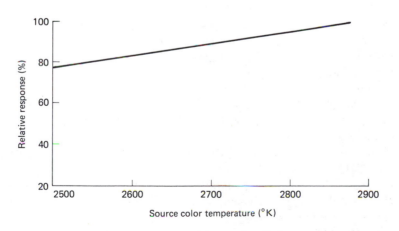

Figure 2–15. Relative response of MRD300 as a function of color temperature. (Courtesy of Motorola Semiconductor Products, Inc., Phoenix, Ariz.)

The response of the phototransistor also falls off, due to differences in the angular alignment of the phototransistor and the radiation source. This type of a response is shown in Fig. 2–16. As shown, the response is narrower when the radiation is focused onto the phototransistor through a lens, and wider when a flat glass is used. In the case of the wider response curve, the peak of the response occurs when the normal to the photodetector plane and the radiation axis coincide (0°). The radiation falls off to one-half of its peak value when the normal and radiation axes are misaligned by approximately 45°.

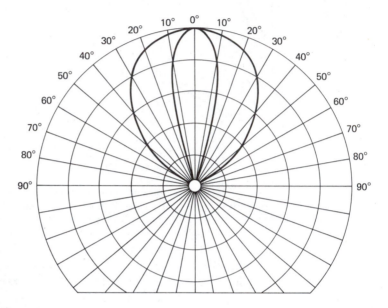

Figure 2–16. Polar response of MRD300. Inner curve with lens, outer curve with flat glass. (Courtesy of Motorola Semiconductor Products, Inc., Phoenix, Ariz.)

The collector characteristics of a typical phototransistor (i.e., MRD300), shown in Fig. 2–17a, are very similar to those of an ordinary transistor except that the input parameter is the base current in the case of an ordinary transistor, and irradiance H in the case of a phototransistor. The I_L versus H curve shown in Fig. 2–13 may be derived from the collector characteristics.

The open-base sensitivity of the phototransistor to the incident radiation is shown in Fig. 2–17b to be a linear function of the irradiance H. This curve may also be deduced either from the collector characteristics or from Fig. 2–13; for example, at $H = 5$ mW/cm^2, the sensitivity is 1.5 mA/(mW/cm^2) which may be obtained from Fig. 2–13 or from Fig. 2–17a. The sensitivity is increased by connecting a series base resistance. The effect of the base resistance on sensitivity is shown in Fig. 2–17c for $H = 5$ mW/cm^2. For example, the sensitivity is increased by 40% when a 200-kΩ resistor is added to the base.

The frequency response of a typical phototransistor (i.e., MRD300) is characterized, as shown in Fig. 2–18, by the upper 3-dB break frequency and its dependence on the

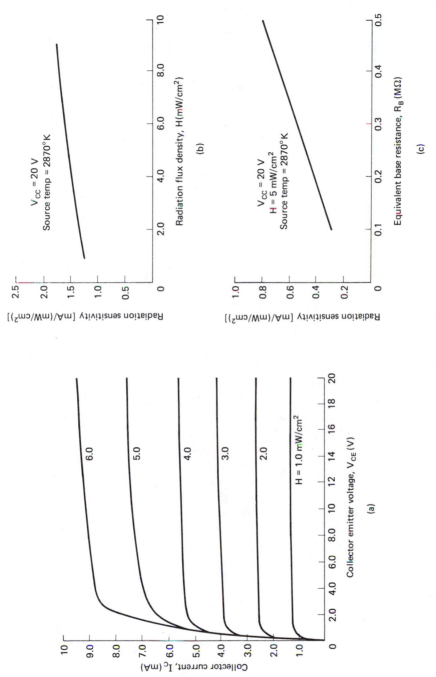

Figure 2–17. Phototransistor characteristics (MRD300): (a) collector characteristics, $V_{cc} = 20$ V, source temperature = 2870 K; (b) radiation sensitivity (open base); (c) effect of base resistance on radiation sensitivity. (Courtesy of Motorola Semiconductor Products, Inc., Phoenix, Ariz.)

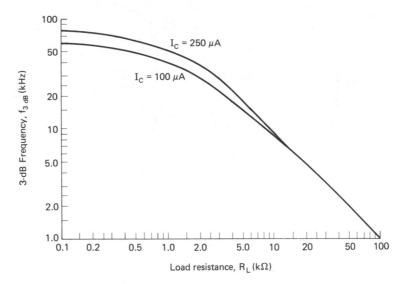

Figure 2–18. Upper 3-dB break frequency versus R_L for the MRD series phototransistors. (Courtesy of Motorola Semiconductor Products, Inc., Phoenix, Ariz.)

value of the load resistor R_L. As can be seen from the graph, the upper 3-dB frequency, and therefore the bandwidth of the phototransistor frequency response, decrease with increasing R_L.

The switching speed of the phototransistor is described by its response to an optical pulse input shown in Fig. 2–19. Suppose that the response waveform represents the collector photocurrent. The approximate model of the phototransistor shown in Fig. 2–20a may be helpful in drawing various conclusions regarding its switching mechanism. There is a delay time t_d at the beginning of the response during which the photo-generated current charges the collector-to-base capacitance C_c and the base-to-emitter capacitance C_e. As the voltage V_{be} exceeds the threshold value, the collector-to-emitter current $g_m V_{be}$ begins to flow and rises exponentially toward the steady-state value. This results in the rise-time delay t_r. Figure 2–20b shows the variations in t_d and t_r as a function of the collector current for two values of R_L. As can be seen from the graph, t_d, being device dependent, does not depend on the value of R_L, and t_r increases with R_L. Both t_d and t_r decrease with an increase in the collector current.

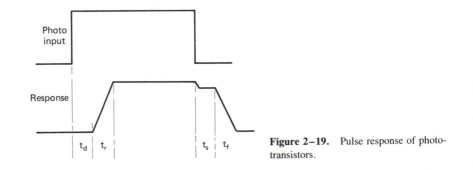

Figure 2–19. Pulse response of phototransistors.

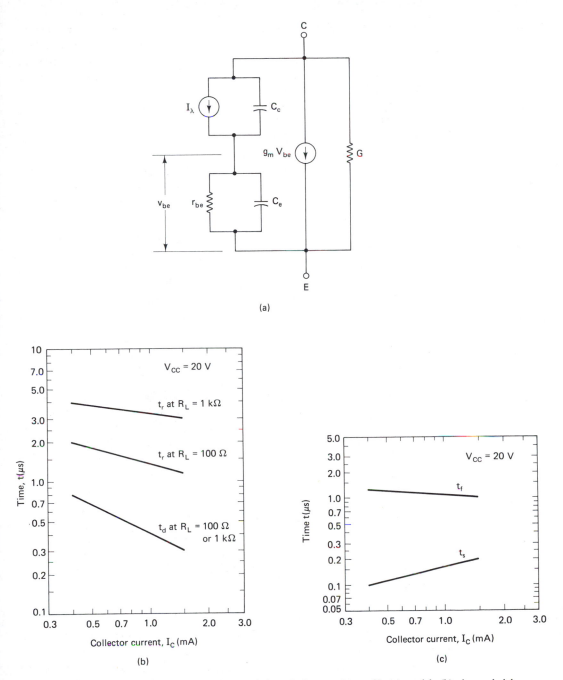

Figure 2–20. Switching characteristics of phototransistors, V: (a) model; (b) rise and delay times (MRD300); (c) storage and fall times (MRD300). (Courtesy of Motorola Semiconductor Products, Inc., Phoenix, Ariz.)

As the input is turned off at the trailing edge, the collector current continues to flow for the time duration t_s. This is the storage-time delay caused by the minority-carrier stored charge in the base region. The collector-current decay begins only after the stored charge leaves the base region. The effects of stored charge are most severe when the transistor is saturated, resulting in a large t_s.

As soon as the base region is free of stored charge, C_c and C_e begin to discharge, and only as V_{be} begins to decline does the collector current fall exponentially to zero, resulting in the t_f time delay, the fall time. As shown in Fig. 2–20c, the fall time decreases, and as expected, the storage time increases, with increasing collector current.

The rise time is generally the largest contributor to the total switching time. One way to reduce the rise time is to use a low value of R_L, which results in a wider bandwidth and a lower rise time, as can be verified from Figs. 2–18 and 2–20b. This is consistent with the general amplifier theory, which states that the product of bandwidth ($B = f_2$, the upper 3-dB frequency, which is $R_L C$ dependent) and rise time is equal to 0.35:

$$t_r f_2 = 0.35$$

Also, the amplifier's gain–bandwidth product is constant and equal to f_T, that is,

$$f_2 A_v = f_T$$

Eliminating the bandwidth from the two equations results in

$$A_v = 2.86 f_T t_r$$

Thus a reduction in the phototransistor's rise time must be accompanied by a reduction in its gain. A_v represents the voltage gain of a transistor amplifier, but in the case of a phototransistor it has a slightly different meaning. In the case of a phototransistor, its transfer characteristics are described by the sensitivity $S = I_c/H$; that is, the sensitivity is equal to the ratio of collector current and the input irradiance and is measured in mA/(mW/cm^2). The phototransistor's gain may thus be expressed by the product $A_v = SR_L$, which represents the output voltage per unit input irradiance.

In any case, it is apparent from the discussion above that in general a compromise must be made between the speed and the output voltage level unless compensation is used. One possible compensation approach is shown in Fig. 2–21a, where a common-base transistor stage is included in the emitter circuit of the phototransistor. The very low resistance that the common-base stage produces in the emitter circuit of the phototransistor is responsible for the considerable improvement in rise time, fall time, and bandwidth, as shown in Fig. 2–21b, c, and d.

The current that the phototransistor generates also appears in the collector circuit of Q_2 and flows through R_L. Because R_L is in the collector of the common-base stage, its value may be varied over a wide range without affecting the emitter resistance of the phototransistor. It is thus possible to achieve a large output level and a low switching speed.

Typical applications of the phototransistor are shown in Fig. 2–22. In the case of the voltage regulator for the projection lamp, the unijunction transistor (UJT) oscillator controls the firing angle of the silicon-controlled rectifier (SCR), which supplies the power to the projection lamp. The timing capacitor in the emitter of UJT is charged by

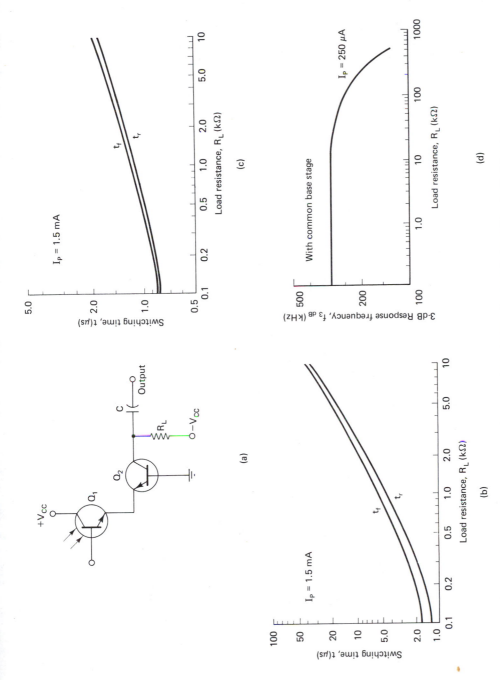

Figure 2–21. (a) Grounded-base transistor compensating stage in the emitter circuit of the photo-transistor; (b) t_r, t_f without compensation, (c) t_r, t_f with compensation, (d) bandwidth versus R_L with compensation, common-base stage. (Courtesy of Motorola Semiconductor Products, Inc., Phoenix, Ariz.)

43

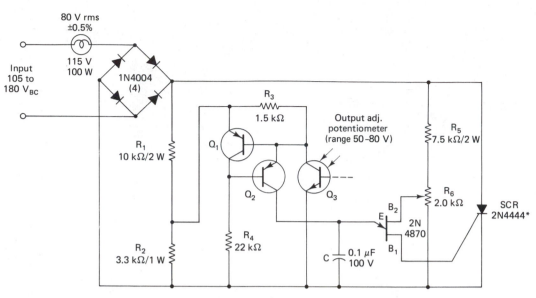

(a)

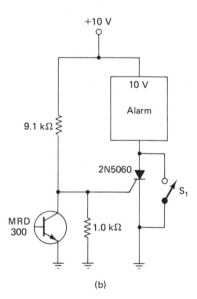

(b)

Figure 2–22. Phototransistor applications: (a) voltage regulator for a projection lamp (Q_1 and Q_2, MPS6516, Q_3, MRD300); (b) alarm system. (Courtesy of Motorola Semiconductor Products, Inc., Phoenix, Ariz.)

a constant-current source Q_1, Q_2, and the phototransistor Q_3. Q_3 receives its light input from the projection lamp and provides a shunt path for the current, thus reducing the value of the charging current. The values of the charging current, C, and the setting of R_6 determine the firing point of the UJT, which in turn controls the conduction angle of SCR.

If the projection lamp intensity should increase, Q_3 conducts more current, reducing the current that charges C. Since the rate of change of voltage across C is I/C (where I is the charging current), a reduction in I results in the UJT being fired at a later time, and consequently, a reduction in the SCR conduction angle. The power delivered to the lamp is reduced and its brightness decreases. The opposite process occurs if the lamp brightness is reduced: the charging current increases. The lamp thus operates at the desired brightness level set by R_6.

In the second application, the alarm circuit remains OFF due to the conduction of the phototransistor, which holds the SCR gate at the ground potential. However, when the light beam is broken, causing the phototransistor to be turned OFF, +10 V is applied to the gate of SCR. The SCR, being thus triggered, turns ON and activates the alarm circuit. S_1 serves as the reset switch. Its momentary closure turns the SCR OFF, thus resetting the alarm.

In contrast to the tungsten light source, which is broadband, some of the LED sources of radiation shown in Fig. 2–23 are narrow-band, closely approximating a monochromatic source. The MRD series phototransistors, for instance, are much more sensitive to the GaAs source than they are to the tungsten source at 2870 °K. This is to be expected since the degree of overlap between the spectral response of the MRD and that of GaAs is greater than the overlap with other sources.

Other types of photosensors which are commercially available are shown in Fig.

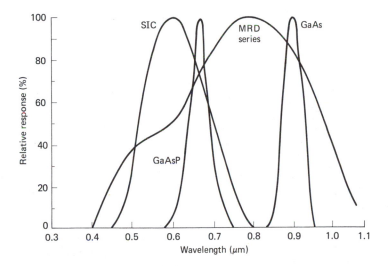

Figure 2–23. Other radiation sources and spectral characteristics for several LEDs compared with MRD series. (Courtesy of Motorola Semiconductor Products, Inc., Phoeniz, Ariz.)

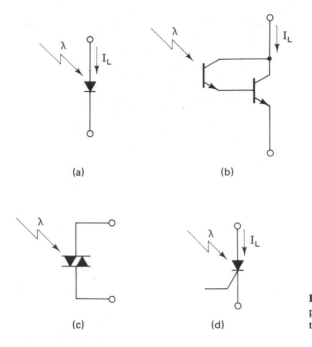

(a)   (b)

(c)   (d)

Figure 2–24. Other photosensors: (a) photodiode; (b) photo Darlington; (c) phototriac; (d) photo SCR.

2–24. The photo triac and photo SCR are used in power-switching applications, and a photo Darlington is suitable for applications requiring large current gains.

2–5 PHOTOCONDUCTIVE CELL

The *photoconductive cell* depends, as does the phototransistor, on the photo effect for its operation. Electrically, however, they differ. The phototransistor exhibits the photo effect as discussed before, and at the same time behaves electrically in the same way as does the conventional bipolar transistor. On the other hand, the photoconductive cell, as shown in Fig. 2–25a, consists of photosensitive material deposited on the substrate and encapsulated in a moisture-resistant, hermetically sealed package. The light-sensitive material used in the illustration is cadmium sulfide (CdS). Its spectral response is very similar to that of the human eye. Light that passes through the package window and is incident on the photo element causes it to change its conductive properties. Specifically, the photoconductive cell undergoes a large change in resistance between low and high values of illuminance. As shown in Fig. 2–25b, the resistance of the NSL4440 device, for example, is 120 kΩ at 1 fc (footcandle) illuminance. It decreases to 2.5 kΩ at 100 fc. The dark resistance is 50 MΩ minimum for a hermetically sealed package, and the maximum operating current is 100 mA. The photoconductive cell is a rugged device and is often used in street-lamp and oil-burner applications, where high voltages and high powers are encountered and a long operational life is required.

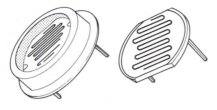

(a)

Silonex part no.	Sensitive material	Resistance at 1 ftc (Ω)	Typical resistance at 100 ftc (Ω)	Minimum dark resistance (Ω)		Maximum current (mA)
				Hermetic	Plastic	
NSL-4440		120 k	2500	50 M	10 M	100
NSL-4450	Type 4 peak at 530 nm	26 k	550	15 M	1.5 M	100
NSL-4460		11.4 k	200	5 M	500 k	100
NSL-4470		5.5 k	140	500 k	200 k	100

(b)

Figure 2–25. (a) Photoconductive cell structure; (b) typical photoconductive cell characteristics. (Courtesy of Silonex, Inc., Plattsburgh, N.Y.)

2–6 OPTOCOUPLER

An *optocoupler* (also called an *optoisolator*) is a device that provides electrical isolation between a source and a load. As shown in Fig. 2–26a, it is enclosed in a miniature six-pin plastic package (other package configurations are also possible). The optocoupler consists of a light-emitting diode (LED) on the input and a photosensor at the output, with no electrical connection between them.

Figure 2–26 shows several optocoupler configurations. They differ only in the type of the photosensor being used. The SCR or triac outputs are suitable for ac power-switching applications. The Darlington output, on the other hand, is recommended for those applications that require large current gains and where the phototransistor output is used in many general-purpose applications.

The triac and SCR optocouplers are particularly useful in applications that require interfacing of low-power IC chips to loads, such as in teletypes, motors, solenoids, and general appliances.

The operation of the optocoupler depends on forward biasing the input light-emitting

(a)

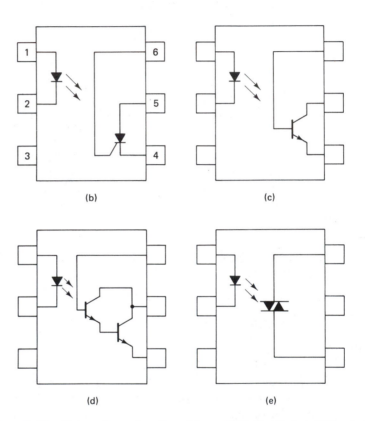

(b) (c)

(d) (e)

Figure 2–26. Optocoupler configurations: (a) case 730A-01; (b) photo SCR output; (c) phototransistor output; (d) photo Darlington output; (e) phototriac output. Some commercially available types: (b) MOC3002; (c) 4N26; (d) MOC119; (e) MOC633A.

diode from an external source. Forward current through the LED causes it to emit radiation [GaAs diode emits in the infrared (IR) region]. This radiation is detected by the photosensor, which generates current to drive the external load.

One of the most important characteristics of the optocoupler is the *current transfer ratio*, which is the ratio of the output current to the input forward LED current, or $N = I_L/I_F$. Figure 2–27 shows the current transfer characteristics for the 4N26 optocoupler, where I_L is the collector current of the output phototransistor. The current transfer

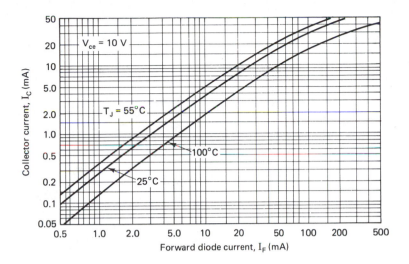

Figure 2–27. Current transfer characteristics for 4N26 optocoupler with phototransistor output, $V_{ce} = 10$ V. (Courtesy of Motorola Semiconductor Products, Inc., Phoenix, Ariz.)

ratio N is the slope of the tangent line to the characteristic curve at the selected value of I_F.

Optocoupler devices are generally operated in one of two modes: the pulsed mode or the linear mode. These modes of operation are illustrated in Fig. 2–28. The linear mode is illustrated by the amplifier circuit in Fig. 2–28a. In this mode of operation the LED is first biased by a dc current. The applied ac current is superimposed over the dc current, thus causing the LED current to vary above and below the dc value by the amount of the peak ac value. The dc biasing is necessary to avoid distortion of the ac signal that is to be amplified. The reader may recall that the conventional class A transistor amplifier must also be properly biased before applying an ac signal to its input. This is done for the same reason—to avoid clipping either the positive or the negative peak of the amplified signal. When multiplied by the current transfer ratio N, the LED current becomes the phototransistor collector current. In Fig. 2–28 this current is dropped across the 100-Ω resistor, and the resulting ac voltage (the dc component is blocked by the 1-μF capacitor) is amplified before appearing as V_o. The value of V_o is calculated in the following example.

EXAMPLE 2–1

Calculate V_o in the optoamplifier configuration shown in Fig. 2–28a.

SOLUTION   The total LED current is 10 ± 5 mA. Thus it varies from 5 to 15 mA. The corresponding change in the phototransistor collector current is obtained from Fig. 2–27 as 1.8 to 6 mA or 4.2 mA peak to peak. The peak-to-peak voltage across the 100-Ω resistor, which is the input to the inverting op amp with a gain of 10, is therefore $4.2(100) \times 10^{-3} = 0.42$ V p-p. This voltage is given a gain of 10; hence $V_o = 4.2$ V p-p.

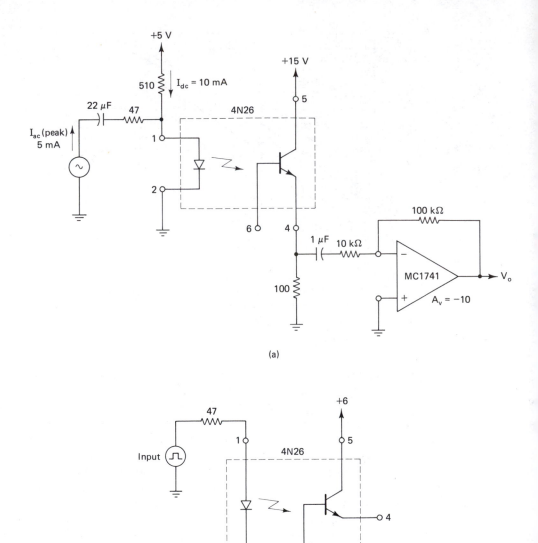

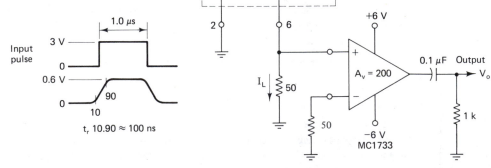

Figure 2–28. Optocoupler modes of operation: (a) linear mode; (b) pulsed mode. (Courtesy of Motorola Semiconductor Products, Inc., Phoenix, Ariz.)

The pulsed mode, which is illustrated in Fig. 2–28b, is another mode of operation for the optocoupler. In contrast to the linear mode, where much attention is given to the linear operation of the optocoupler to assure minimum distortion of the input signal, in the pulsed mode the optocoupler behaves as a simple switch. In this mode of operation, the emitter remains open, and the output is taken across the load resistor which is connected between the base and ground. The phototransistor is thus converted to the photodiode (base–collector *p-n* junction). One of the most apparent disadvantages in doing this is that the load current, which normally would have been the collector current, is now the base current, which is much smaller than the collector current; $I_B = I_C/h_{FE}$. The current transfer characteristics in Fig. 2–27 still apply if the I_C scale is divided by h_{FE}. The most important advantage, however, in open emitter use of the optocoupler is a significant improvement in the switching speed. The rise time, which would have been 1 or 2 μs for the phototransistor, is in the neighborhood of 100 ns in this application. The low value of load resistor, which matches the 50 Ω connected to the inverting terminal of the op amp, is used to minimize the offset current of the op amp, and the 0.1-μF capacitor blocks the dc component from appearing at the output. The low value of voltage across the load requires the use of a high-gain amplifier. In the following example V_o is calculated for the pulsed mode.

EXAMPLE 2–2

Calculate V_o for the circuit shown in Fig. 2–28b.

SOLUTION The LED current $I_F = (3 - 1)/47 = 42.6$ mA assuming a 1-V drop across LED. Using the 25°C curve in Fig. 2–27, the collector current for the calculated I_F is 15 mA. Using $h_{FE} = 325$ (from the 4N26 specifications), the value of base current $I_B = 15/325 = 46.2$ μA and the voltage across the 50-Ω resistor is $46.2(50) \times 10^{-6} = 2.3$ mV. This voltage is amplified by the op amp, whose voltage gain is 200, resulting in an output voltage of

$$V_o = 2.3(200) \times 10^{-3} = 0.46 \text{ V}$$

Some of the applications of the optocoupler are shown in Fig. 2–29. The emphasis here is clearly on the isolation characteristic of the optocoupler. In both cases the load, which is electrically isolated from the source, requires large dc power. As can be seen, the required control source current is relatively small.

2–7 OPTOINTERRUPTER DEVICE

The *optointerrupter device* (OID) is based on the same principle as that of the optocoupler. The main difference lies in the package configuration. As shown in Fig. 2–30a, the OID package has a slot that provides a means of blocking or interrupting the light emitted by the LED.

The typical OID circuit shown in Fig. 2–30b is basically the same as that of an optocoupler. It consists of the optical link formed by the LED (typically, GaAs, which

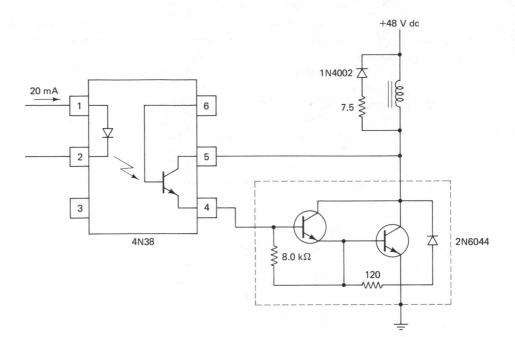

(a)

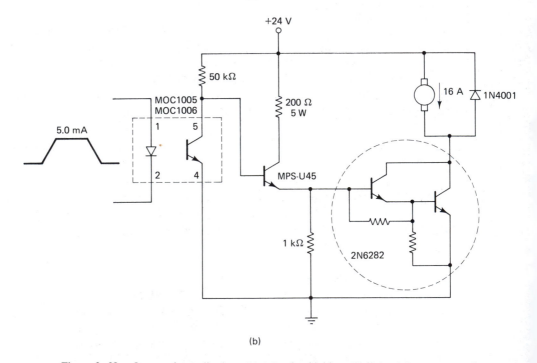

(b)

Figure 2–29. Optocoupler applications: (a) 4-A solenoid driver; (b) isolated dc motor controller. (Courtesy of Motorola Semiconductor Products, Inc., Phoenix, Ariz.)

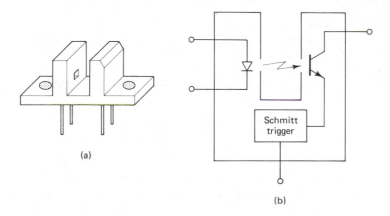

(a)

(b)

Figure 2–30. Optointerrupter: (a) typical case; (b) typical circuit.

emits light in the IR region) and the photosensor, which may be any one of those mentioned earlier. The Schmitt trigger shown in Fig. 2–30b is included by some manufacturers (e.g., the GP-1A01, manufactured by Sharp Electronics, includes a Schmitt trigger). The purpose of the Schmitt trigger is to provide wave shaping to the output waveform and thus achieve a waveform that is as close as possible to the ideal square wave. As mentioned earlier, there may be some variation in the shape of the package and also in the circuit; some circuits may include a Schmitt trigger, while others include an SCR or triac output for power-driving applications. The characteristics of the Motorola's M0C7811 family of OIDs are included in Appendix C.

One important application of the OID is in sensing the speed of the motor. As shown in Fig. 2–31, the disk that is attached to the shaft of the motor rotates partially inside the slot of the OID. The construction of the disk includes n slots or holes cut at

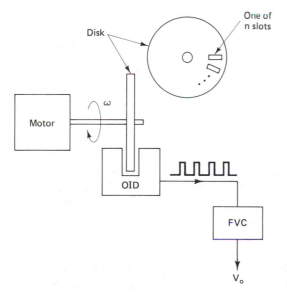

Figure 2–31. Use of OID In motor speed-sensing application.

regular angular intervals near the edge. As the disk rotates, the light from the LED passes through the slots in the disk and is blocked by the solid portion of the disk. The OID therefore outputs a pulse train whose duty cycle depends on size and the spacing of the slots and whose frequency depends on the speed of the motor ω. Since the disk has n slots, the OID must output n pulses for each revolution of the disk. If the motor's speed is ω rpm or $\omega/60$ rps, the frequency of the pulses may be expressed as

$$f = \frac{n\omega}{60} \qquad \text{Hz} \qquad\qquad (2-12)$$

The pulse train is then applied to the frequency-to-voltage converter (FVC), whose output is a dc voltage proportional to the frequency of pulses. This speed-sensing configuration may be used to provide the feedback voltage (which is proportional to the speed of the motor) in the closed-loop speed control system, replacing the commonly used tacho generator.

The OID may also be used in sensing the angular position of the motor shaft. If the disk attached to the motor shaft has n slots, the angular interval between any two slots must be $360/n$ degrees; in the case of 120 slots, the slots are spaced at $3°$ intervals. As the disk moves, the OID outputs one pulse for each slot in the disk. Consequently, the count of the total number of pulses must be related to the angular displacement of the motor shaft. If k pulses are outputted by the OID, for example, the motor shaft must have moved through an angle

$$\theta_m = \frac{360k}{n} \qquad\qquad (2-13)$$

As shown in Fig. 2–32, the pulses from OID B are counted by the 8-bit binary up/down counter, and the resulting 8-bit word at its output (the binary equivalent of θ_m) is converted to the dc voltage level V_o (the electrical equivalent of θ_m) by the digital-to-analog converter (DAC).

There are instances where the shaft changes the direction of rotation many times before reaching its equilibrium position. This occurs, for instance, in the case of the closed-loop, underdamped position control system (analyzed in Chapter 8), whose step response undergoes damped oscillations before reaching the steady state. The counter, in this case, will continue its count regardless of the direction of shaft rotation, resulting in an incorrect response at the output of the DAC.

To correct this condition, two OIDs are used. OID A output is used as a clock input to the negative edge-triggered D flip-flop, and OID B output is used as the input to the counter and also to the D flip-flop. The two OIDs are adjusted so that the A output leads the B output by approximately $90°$ for clockwise (CW) rotation of the disk, and for counterclockwise (CCW) rotation, the A output lags the B output by approximately $90°$. Everytime the flip-flop is triggered on the $1 \rightarrow 0$ transition of waveform A, the input to the D flip-flop, and therefore its output, is a 1 for CW rotation of the motor shaft, and 0 for CCW rotation. The output of the D flip-flop thus serves as the control for the up/down input of the counter, reversing the direction of its count whenever the disk reverses its direction of rotation. This configuration may be used to provide the position feedback signal in the closed-loop position control system, thus

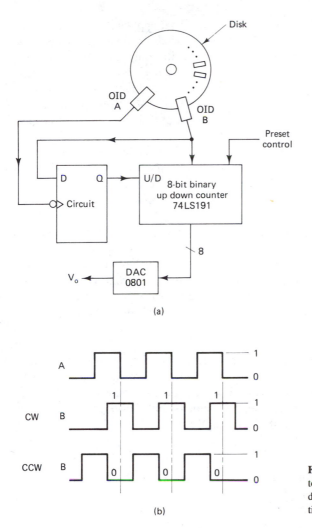

(a)

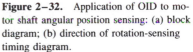

(b)

Figure 2–32. Application of OID to motor shaft angular position sensing: (a) block diagram; (b) direction of rotation-sensing timing diagram.

replacing the commonly used potentiometer. It has a significant advantage over a potentiometer, in that it provides the angular position of the motor shaft in digital form, which may be used directly by a computer or microprocessor.

2–8 HALL EFFECT DEVICE

The *Hall effect device* (HED) was discovered by Edwin Hall in 1879 when he was experimenting with conductivity in metals. Hall discovered that when he placed a magnet at a right angle to the conductor and therefore at a right angle to the flow of current, a voltage appeared across the two edges of the conductor which are at right angles to both the current flow and the field. This voltage is the *Hall voltage*, named after its discoverer.

The Hall effect device was not used in practical applications for almost a century. The reason for this is that the Hall voltage in metals is very small (in the microvolt range). In the days when the vacuum tube was the only amplifier available, it was generally felt that Hall voltage was much too small to be accurately measured and used in a practical application. With the birth of semiconductor devices in the 1950s, a semiconductor-based HED was developed with a Hall voltage several magnitudes higher than that of its metal-based counterpart. However, at that time it was too expensive for practical applications.

Today, semiconductor HEDs are commercially available and generally inexpensive (although there are some special-purpose HEDs on the market which are in excess of $100 each). The availability and low price of the HED opened the door to many applications, including current and magnetic field sensing, power monitoring, position and speed sensing, modulation, and many others.

The operation of the Hall effect device depends on the presence of a magnetic field, which is at right angle to the flow of current. As shown in Fig. 2–33a, the externally applied current is in the $-x$ direction with respect to the coordinate system as shown; the electrons therefore flow in the $+x$ direction, and the magnetic field B_z is applied in the $+z$ direction. Consider first the case where no magnetic field is applied. As shown in Fig. 2–33b, the only force acting on the electrons is

$$F_{Ex} = -|q|E_x \tag{2–14}$$

where E_x is the electric field due to the externally applied voltage. It is acting in the $-x$ direction; hence the force F_{Ex} acting on the electrons is in the $+x$ direction.

When the magnetic field B_z is applied, the electrons experience another force due to B_z. This force depends on the electron velocity and the magnetic field and may be expressed as a cross product in the following form:

$$\mathbf{F} = -|q| \, \mathbf{V} \times \mathbf{B} \tag{2–15}$$

$\mathbf{F}$ is a vector that is perpendicular to the plane of $\mathbf{V}$ and $\mathbf{B}$. Applying this equation to the present case, we obtain

$$F_{+y} = F_{Bz} = |q|V_x B_z \tag{2–16}$$

This is the force on the electrons which acts in the $+y$ direction, causing the electrons to deflect toward the lower edge, C. This results in a lateral charge distribution, with edge C acquiring negative charge and edge D becoming positively charged. This type of charge unbalance (due to the presence of B_z) is responsible for the presence of the electric field E_y, which creates another force,

$$F_{-y} = F_{Ey} = -|q|E_y \tag{2–17}$$

where q is the electron charge. F_{Ey} acts on the electrons in $-y$ direction. As the charge unbalance increases, E_y is strengthened, causing an increase in F_{Ey}. Equilibrium condition is finally reached when the two forces in the y direction are equal (i.e., $F_{Ey} = F_{Bz}$). As the electrons experience no net force in the y direction at this time, their average velocity is again in the $+x$ direction. Under equilibrium conditions, the net charge distribution along the C and D edges gives rise to the difference in potential known as

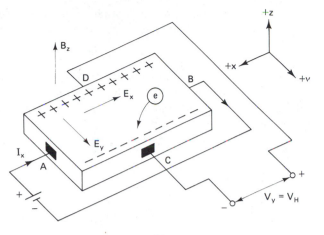

(a)

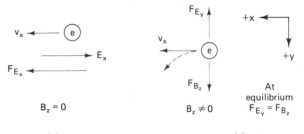

(b)

F_{E_y}

v_x

e

$+x$

$+y$

F_{B_z}

At equilibrium
$F_{E_y} = F_{B_z}$

$B_z \neq 0$

(c)

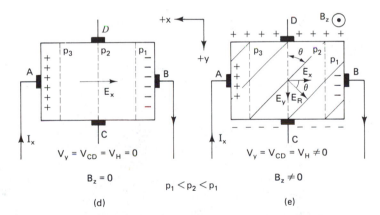

(d)

$p_1 < p_2 < p_1$

(e)

Figure 2–33. (a) Hall effect device configuration with applied magnetic field; (b) electron flow with no magnetic field; (c) in the presence of the magnetic field, the electrons are deflected toward the lower edge due to the force F_{Bz} until the equilibrium is reached when $F_{Bz} = F_{Ey}$; (d) equipotential lines with no magnetic field, $V_H = 0$; (e) magnetic field causes the equipotential lines to be rotated by an angle θ resulting in $V_H \neq 0$.

the Hall voltage. The Hall voltage, which depends on the magnitudes of the current I_x and the magnetic field B, may be expressed as

$$V_H = kI_x B \cos \phi \qquad (2-18)$$

where k is a sensitivity constant expressed in mV/(mA-kG), and ϕ is the angle between the magnetic field vector and the z axis. If the magnetic field is in the $x - y$ plane, $\phi = 90°$ and $V_H = 0$. If, on the other hand, B is perpendicular to the $x - y$ plane, $\phi = 0°$ and Eq. (2–18) reduces to

$$V_H = kI_x B_z \qquad (2-19)$$

Hall voltage may also be explained by considering the equipotential lines, which are perpendicular to edges C and D in the case of zero magnetic field. As shown in Fig. 2–33d, equipotential p_2 is between C and D (as an example); hence there is no difference in potential between C and D, and the Hall voltage must therefore be zero. On the other hand, when the magnetic field is applied in the $+z$ direction, the equipotential lines are rotated by an angle θ, as shown in Fig. 2–33e. Since $p_3 > p_1$, there must be a potential rise from C to D, and consequently, $V_H \neq 0$.

As can readily be deduced from the geometry in Fig. 2–33e, the angle between E_x and E_R is also θ. Hence at equilibrium when $F_{Bz} = F_{Ey}$, the following result is obtained using Eqs. (2–16) and (2–17):

$$\tan \theta = \frac{E_y}{E_x} = -\frac{V_x B_z}{E_x} = -u_e B_z \qquad (2-20)$$

where u_e is the electron mobility, defined as the electron drift velocity per unit electric field. Since θ is generally less than 1 rad, Eq. (2–20) reduces to

$$\theta \doteq -u_e B_z$$

showing that equipotential line rotation is directly proportional to the magnetic field. It also follows from Fig. 2–33 that the Hall voltage should also increase with increased θ. The negative sign is due to the negative electron charge.

Manufacturers of Hall devices usually specify the HED magnetic sensitivity as $K_s = kI_x = V_H/B_z$. Equation (2–19) may then also be expressed as

$$V_H = K_s B_z \qquad (2-21)$$

The plot of V_H versus B_z is, as expected, a straight line whose slope is K_s. The magnetic sensitivity for semiconductor-type HEDs ranges from 10 to approximately 50 mV/kG. A typical magnetic sensitivity curve for a semiconductor HED is shown in Fig. 2–34.

As noted earlier, a metal HED offers a very small output voltage, much too small for practical applications. A semiconductor HED, on the other hand, has a much higher magnetic sensitivity. There are two types of semiconductor HEDs: bulk and thin film. A bulk HED is made by cutting Hall plates of required dimension from a semiconductor slice and then bonding them to a substrate. After attaching the four contact leads, the device is usually packaged in epoxy.

The thin-film HED is made by the process used in the IC chip fabrication. The semiconductor patterns and the ohmic conducting paths are deposited on a substrate using the photolithographic method. In contrast to the bulk type, thin-film HEDs use

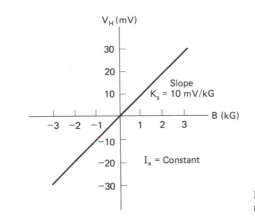

Figure 2–34. Typical output characteristics of a semiconductor type HED.

less current and exhibit better sensitivity, although their linearity is not as good. Two thin-film HEDs, which differ only in lead configuration, are illustrated in Fig. 2–35b.

The use of a magnetic flux concentrator helps to improve HED sensitivity. This is illustrated in Fig. 2–35c, where a ferrite core is used to concentrate the magnetic flux lines in the current-sensing application. The HED is placed in the gap of the core, where the magnetic field is stronger due to concentrator action. The use of a concentrator is particularly important in applications where the current to be sensed is small and is not able to generate a field strong enough for the HED to sense. Another concentrator arrangement, and a graph that shows the improvement in HED sensitivity as a function of the concentrator length, appear in Fig. 2–35a.

An HED may also be used in speed and shaft position-sensing applications. As shown in Figs. 2–31 and 2–32, tiny magnets may be attached to the disk in place of the slots and the OID is replaced by an HED. The resulting configuration may be used in speed-sensing or position-sensing applications as described in Section 2–7.

2–9 DIGITAL-TO-ANALOG CONVERTER

The *digital-to-analog converter* (DAC) performs the function of transforming an n-bit digital input to an analog voltage level. Thus an 8-bit DAC accepts an 8-bit digital input and generates at its output a unique analog voltage level.

The essential part of any DAC is the resistive network. The most commonly used resistive network is the *R-2R ladder*, so called because it uses only two values of precision resistors. The combination of an R-$2R$ ladder and an op amp, shown in Fig. 2–36a, constitutes the basic DAC structure. Shown here is an inverted ladder structure, in which the binary-weighted currents in the $2R$ branches remain constant regardless of the state of the analog switches.

As shown, the resistive ladder switches binary-weighted currents between the grounded line and the line that is connected to the inverting input of the op amp. The switching is done by the analog switches S_0 to S_7, whose state is controlled by external signals. Suppose that the most significant bit (MSB) alone is applied; that is, switch S_7 is connected to the inverting input of the op amp and the remaining switches are grounded.

Using flux concentrators to increase sensitivity:
Flux concentrators can be used to increase Hall generator sensitivity. The ferrite-embedded BH-702 is particularly well suited for cencentrator modifications, because of the small gap possible. The sensitivity curve on the right illustrates how BH-702 magnetic sensitivity varies with concentrator length.

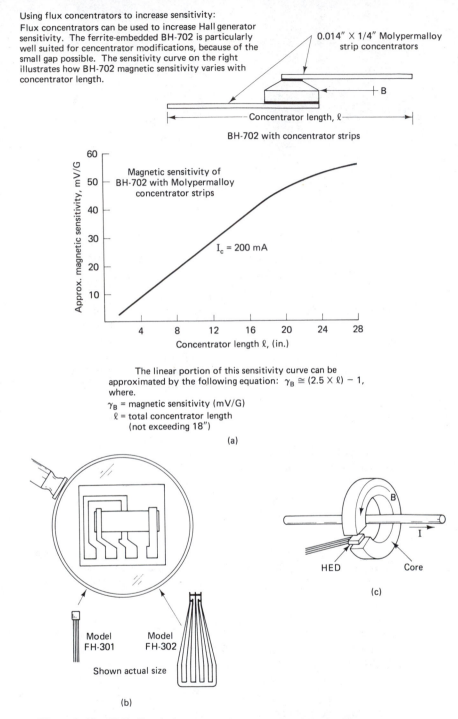

0.014″ × 1/4″ Molypermalloy strip concentrators

⊢ B

⊢ Concentrator length, ℓ ⊣

BH-702 with concentrator strips

Magnetic sensitivity of BH-702 with Molypermalloy concentrator strips

I_c = 200 mA

Approx. magnetic sensitivity, mV/G

Concentrator length ℓ, (in.)

The linear portion of this sensitivity curve can be approximated by the following equation: $\gamma_B \cong (2.5 \times \ell) - 1$, where.

γ_B = magnetic sensitivity (mV/G)
ℓ = total concentrator length (not exceeding 18″)

(a)

Model FH-301   Model FH-302

Shown actual size

(b)

B

HED   Core

I

(c)

Figure 2–35. Hall effect device: (a) use of concentrator strips to improve the magnetic sensitivity of HED and the graph which shows such improvement; (b) example of thin-film HED made by F. W. Bell; (c) use of a core and an HED in a current measurement application. [(a), (b) Courtesy of F. W. Bell, Orlando, Fla.]

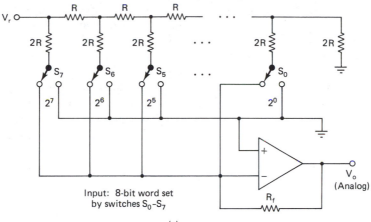

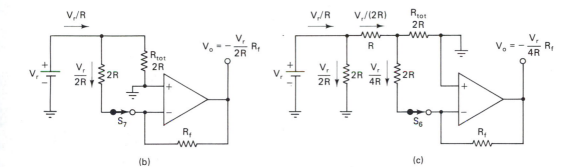

$R_f = R$

2^7	2^6	2^5	2^4	2^3	2^2	2^1	2^0	V_o/V_r
1	0	0	0	0	0	0	0	$-\frac{1}{2}$
0	1	0	0	0	0	0	0	$-\frac{1}{4}$
0	0	1	0	0	0	0	0	$-\frac{1}{8}$
0	0	0	1	0	0	0	0	$-\frac{1}{16}$
0	0	0	0	1	0	0	0	$-\frac{1}{32}$
0	0	0	0	0	1	0	0	$-\frac{1}{64}$
0	0	0	0	0	0	1	0	$-\frac{1}{128}$
0	0	0	0	0	0	0	1	$-\frac{1}{256}$
0	0	0	0	0	0	0	0	0

(d)

Figure 2–36. (a) Basic DAC structure using the inverted R-$2R$ resistive ladder. Simplified DAC circuit when the input is: (b) 10000000; (c) 01000000; (d) tabulation of the normalized DAC output corresponding to the digital inputs which are applied by means of analog switches, one at a time beginning with MSB.

In the equivalent circuit shown in Fig. 2–36b, R_{tot}, which is equal to $2R$, represents the combination of all resistors to the right of the $2R$ branch in series with switch S_7. The current through this $2R$ branch, as can easily be shown, must be equal to $V_r/2R$. This current also flows through R_f and develops the output voltage $V = -(V_r/2R)R_f$. If $R = R_f$, this voltage, as shown also in the table of Fig. 2–36d, is $-V_r/2$.

Consider next the 01000000 input and corresponding equivalent circuit shown in Fig. 2–36c, where R_{tot} represents the combination of all resistors to the right of the $2R$ branch, which is in series with switch S_6. It may be noted that the current through the $2R$ branch, which is in series with S_7, has not changed, despite the fact that S_7 changed state from a 1 to a 0. It also follows that since the branch current always remains constant, the total current drawn from the reference source V_r also remains constant for any state of the switches—an important feature of the inverted R-$2R$ ladder structure. It is left as an exercise for the reader to verify all currents shown in Fig. 2–36c. The current in series with switch S_6 also flows through R_f and developes $V_o = -(V_r/4R)R_f$, which is equal to $-V_r/4$ if $R = R_f$. This type of analysis may be extended to the remaining switches and the resulting outputs tabulated as shown in Fig. 2–36d.

By the superposition principle the output V_o due to any combination of switch states may be expressed in the general form for an n-bit DAC as follows:

$$V_o = -\frac{V_r R_f}{R} \frac{b_0 2^0 + b_1 2^1 + b_2 2^2 + \cdots + b_{n-1} 2^{n-1}}{2^n} \qquad (2-22)$$

where b_0 through b_{n-1} are the bits whose value is 0 or 1 as determined by the digital input.

EXAMPLE 2–3

Calculate the value of the output for an 8-bit DAC. The reference source is -10 V dc and $R_f = R$. The input is 10110011.

SOLUTION   Substituting the given values in Eq. (2–22) yields

$$V_o = +10 \left(\frac{1 \times 2^0 + 1 \times 2^1 + 1 \times 2^4 + 1 \times 2^5 + 1 \times 2^7}{2^8} \right)$$

$$= \frac{10(167)}{256} = +6.52 \text{ V dc}$$

An 8-bit DAC fabricated using the complementary metal-oxide semiconductor (CMOS) process is illustrated in Fig. 2–37. It is available in a 16-pin ceramic dual-in-line package (DIP) or in a higher-power-dissipating plastic package. As shown in Fig. 2–37c, the AD7524 uses the inverted resistive ladder discussed earlier (Fig. 2–36a) with $R = R_f = 10$ kΩ. Since the output depends on the ratio R_f/R, the feedback resistor R_f required by the op amp circuit is included on the chip to ensure a predictable R_f/R ratio.

The AD7524 DAC is compatible with and may be interfaced to 8-bit microprocessors such as the 6800, 8085, Z80, and so on, through the use of the $\overline{CS}$ and $\overline{WR}$

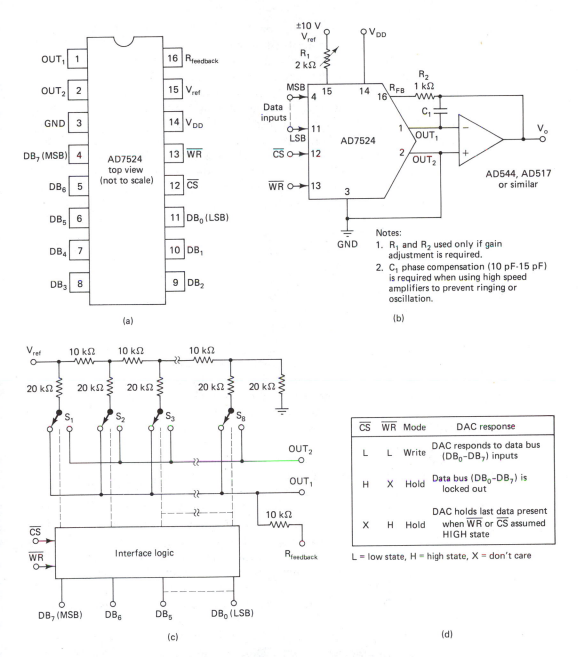

Figure 2–37. AD7524 multiplying DAC made by Analog Devices: (a) case outline; (b) DAC configuration is completed when the external components shown are connected to the AD7524 chip for the unipolar binary operation (two-quadrant multiplication); (c) AD7524 functional diagram; (d) mode selection table for AD7524. (Courtesy of Analog Devices, Norwood, Mass.)

control signals. Several operation modes of AD7524 which are dictated by the states of $\overline{CS}$ and $\overline{WR}$ control signals are tabulated in Fig. 2–37d.

A *multiplying DAC* is a DAC designed to operate not only with a fixed dc reference V_r but also with an ac analog input in place of the fixed V_r. A AD7524 may be configured to operate as a multiplying DAC by changing the external circuit in Fig. 2–37b. The output of the multiplying DAC may be represented by $V_o = V_i X$, where V_i is the ac analog signal and X is the fraction associated with the digital input. Since both V_i and X may assume both polarities, this type of operation is called *bipolar four-quadrant* operation. By contrast, when V_r assumes a fixed value, *two-quadrant* operation results.

The *resolution* of the n-bit DAC is defined as the smallest change in the output that can be resolved by the n-bit DAC. Since the smallest change in the output is determined by the least significant bit (LSB), the resolution may be expressed as a percent of the full-scale output:

$$R = \frac{100}{2^n} \qquad \% \qquad\qquad (2\text{--}23)$$

or it may also be expressed in terms of volts as

$$R = \frac{V_r}{2^n} \qquad V \qquad\qquad (2\text{--}24)$$

The analog output of an n-bit DAC is divided into 2^n discrete voltage ranges. All voltage values within any such range are represented by the same digital code, which is generally assigned the midrange value. Since each of the 2^n-ranges has a span of 1 LSB, the digital code must represent all analog values $\frac{1}{2}$LSB above and $\frac{1}{2}$LSB below the midrange point. This is a consequence of the DAC's resolution limitation, which results in an error known as the *quantizing error* of $\pm\frac{1}{2}$ LSB.

The *DAC accuracy* is generally expressed in terms of its error, which is defined as the difference between the actual analog output and the output as predicted by Eq. (2–22). The sources of error include system nonlinearity, noise, calibration, zeroing error, gain error, and possibly op amp offset drift. In theory the accuracy and resolution for a given DAC must be compatible (i.e., approximately equal), as it makes little sense to have a system with high resolution but poor accuracy, or vice versa. In practice, however, manufacturers design DACs whose accuracy is somewhat better than their resolution. Analog Devices, for example, specifies the accuracy of their AD7524 (an 8-bit DAC) as ± 0.002 of full-scale reading (FSR), the resolution being equal to 2^{-8} or 0.0039 of FSR. Thus the maximum error in this case is almost one-half that of the resolution.

A monotonic mathematical function is one that always increases or always decreases. A ramp is an example of a monotonic function. The *monotonicity* waveform is obtained at the output of the DAC by driving the DAC with a counter that goes through its full-scale count. The monotonicity waveform is a staircase type of a waveform with step height equal to $V_r/2^n$, being limited by the resolution of the DAC, and step width equal to $1/f_c$, the period of the clock. Ideally, the height of all steps must be the same; in other words, the DAC output must be exactly equal to the voltage corresponding to 1 LSB for any two adjacent codes on the input to the DAC. Any deviation from this ideal case represents a measure of the DAC's *differential linearity*, or more specifically, its differential nonlinearity. Frequently, manufacturers substitute "monotonic" for the differential nonlinearity spec to indicate that a given DAC's differential nonlinearity is less than 1 LSB, which is generally acceptable. Differential linearity errors greater than 1 LSB are generally used to characterize a nonmonotonic response. Such errors may be the reason why codes are missed in an analog-to-digital converter that uses a DAC with a differential linearity error in excess of 1 LSB.

2-10 ANALOG-TO-DIGITAL CONVERTERS

The *analog-to-digital converter* (ADC) is a component that converts the applied analog voltage to the digital form. A variety of ADC configurations are commercially available in chip form. Each ADC type features special characteristics which are suitable for a specific application. Generally, the speed of the converter dictates its price; the faster the converter is, the more expensive it is. Although the ADC design is not limited to one structure, there are configurations that are more popular than others. Some of these are presented next.

Simultaneous ADC

The *simultaneous ADC* is also known as the *flash converter* and the *parallel converter*. A 2-bit simultaneous ADC configuration is illustrated in Fig. 2-38. The analog input V_i to be converted to digital form is applied simultaneously to the + side of each of the three comparators, and the input to the − side depends on the converter bit size. For the 2-bit ADC considered here, the reference supply is divided into 2^2 or 4 ranges, as shown in Fig. 2-38b. Consequently, $V_r/4$, $V_r/2$, and $3V_r/4$ are applied to the − input of comparators 1, 2, and 3, respectively. Whenever V_i exceeds the voltage on the − input of a comparator, the comparator's output is driven HIGH; otherwise, it remains LOW. As shown in Fig. 2-38b, the value of V_i (which must not exceed V_r) determines in a unique manner the output states of the three comparators. The diode and the potentiometer at the output of each comparator provide TTL compatibility.

The encoder network converts the comparator outputs to a 2-bit word. Each 2-bit word represents one of the four ranges, and therefore the analog input being converted must be within one of four ranges. The input V_i may vary by as much as $V_r/4$ within a given range with no change in the digital output. This limits the system resolution, which may be improved by increasing the number of bits.

In general, 2^n ranges are associated with an n-bit converter, and the value of each range is $V_r/2^n$ volts. The converter also requires $2^n - 1$ comparators. A 12-bit simultaneous ADC, for example, has 4096 ranges, and each range is equal to 2.4 mV. The converter's resolution in this case is 2.4 mV, which is excellent; however, the 4095 comparators and a large encoding network add to its complexity and make this converter very expensive.

The most important advantage of the simultaneous ADC is its speed, which is limited only by the propagation delay of the individual components. It is probably the fastest converter on the market today. In applications where fast conversion time is

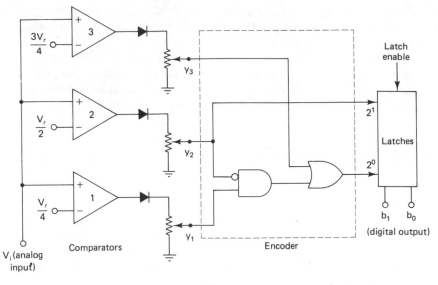

(a)

x = V_i/V_r	y_3	y_2	y_1	2^1	2^0
$0 \leqslant x < 1/4$	0	0	0	0	0
$1/4 \leqslant x < 1/2$	0	0	1	0	1
$1/2 \leqslant x < 3/4$	0	1	1	1	0
$3/4 \leqslant x < 1$	1	1	1	1	1

$$b_0 = y_3 + \bar{y}_2 y_1$$
$$b_1 = y_2$$

(b)

Figure 2–38. (a) Simultaneous ADC (2-bit) circuit diagram; (b) encoder truth table.

required, the additional expense associated with the simultaneous ADC may well be justified.

Monotonic ADC

The *monotonic ADC* configuration is illustrated in Fig. 2–39a. At the beginning of the conversion cycle the counter is reset to all 0s, resulting in $V_a = 0$ V. The analog input V_i (assumed to be positive) drives the comparator's output HIGH, which enables the AND gate and allows the clock pulses to be applied to the counter. As each clock pulse advances the binary state of the counter by 1, the output of the DAC makes a step equal to $V_r/2^n$. The DAC's output is a staircase-type waveform (also called a *monotonicity waveform*—hence, *monotonic ADC*). At *coincidence*, which occurs when $V_a = V_i$, the comparator's output is driven LOW, the AND gate is disabled, and the clock pulses are cut off from the counter. The counter stops its count, and its digital state

Control System Components   Chap. 2

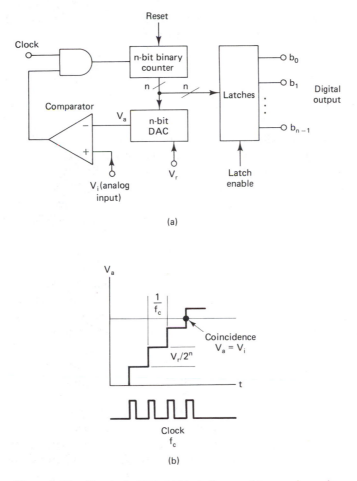

Figure 2-39. Monotonic ADC: (a) block diagram; (b) conversion cycle.

represents the analog input V_i. With the latch enable pulse, the contents of the counter may be transferred to the latches and thus made available for processing.

The resolution for this converter is expressed by Eqs. (2–23) and (2–24) and illustrated in Fig. 2–39b. The time that it takes the converter to complete one conversion is expressed by the conversion time T_c. If it takes k clock pulses to reach coincidence, the conversion time may be expressed as

$$T_c = \frac{k}{f_c} \tag{2–25}$$

where $1/f_c$ is the period of the clock pulses. The conversion time varies as the analog input fluctuates. The conversion rate, which is equal to $1/T_c$, expresses the number of conversions that the converter makes per second. The minimum conversion rate occurs when the counter advances through the full-scale count. It is expressed as

$$R_{cm} = \frac{f_c}{2^n} \qquad \text{conversions/second} \tag{2–26}$$

The average conversion time is $[(0 + 2^n)/2]/f_c = 2^{n-1}/f_c$, and the reciprocal of the average conversion time expresses the average conversion rate as

$$R_{ca} = \frac{f_c}{2^{n-1}} = 2R_{cm} \tag{2-27}$$

Tracking ADC

One of the most objectionable properties of the monotonic ADC is the resetting of the counter to 0 on each conversion. The counter thus wastes a considerable amount of time attempting to reach coincidence. This is particularly true for large values of the analog input.

This problem is resolved by the *tracking ADC*, shown in Fig. 2–40. At the beginning of the operation it does exactly what the monotonic ADC does on each conversion, that is, after being reset to 0, the counter chases the analog input, and as shown in the illustration, reaches coincidence on the fourth clock pulse. The similarity between the two converters ends here because after this, the tracking ADC follows or tracks the analog input. The tracking feature is made possible by the use of an up/down binary counter. Whenever the input V_i exceeds the DAC output V_a ($V_i > V_a$), the HIGH output of the comparator enables the up count control, and when V_a exceeds V_i ($V_i < V_a$), the LOW output of the comparator after being inverted, enables the down count control. In this manner the tracking ADC remains most of the time within 1 LSB of the coincidence (unless V_i makes an extremely sharp transition in a fraction of the clock period's time).

The latch enable strobe transfers the contents of the counter to the latches for external processing. This need not be done on each clock pulse, particularly if V_i remains constant for any length of time. A latch control circuit may be added to the given configuration. The control circuit can compare the state of the latches with that of the counter and output a latch enable strobe whenever the two differ by a predetermined count. The latch enable strobe thus updates the latches whenever the state of the counter changes.

Single-Slope ADC

In contrast to the converters discussed thus far, the *single-slope ADC* does not use a DAC. Instead, as shown in Fig. 2–41a, it uses a ramp generator and a control unit which provides the necessary timing pulses. At the start of conversion cycle (SOC), the control unit resets the counter and the ramp generator to 0. As the ramp generator begins to ramp up, $V_i > X_5$ and the HIGH output of the comparator enables ($X_1 = 1$) the AND gate, thus allowing the clock pulses to be applied to the counter. The counter advances its count with each clock pulse as long as $V_i > X_5$. At the instant when $X_5 = V_i$ (or slightly greater than V_i), the comparator output is driven LOW, the AND gate is disabled, the clock pulses are therefore cut off from the counter, and the counter stops. At this time, the state of the counter represents the analog input V_i.

The 1-to-0 transition of X_1 (the comparator output) marks the end-of-conversion cycle (EOC). It is detected by the control unit, which in turn transfers the counter contents to the latches by means of the latch enable pulse X_3, and prepares the system

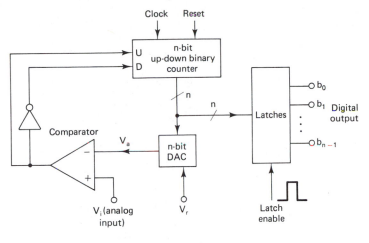

(a)

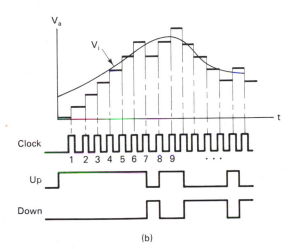

(b)

Figure 2–40. Tracking ADC: (a) block diagram; (b) DAC output V_a tracks the analog input V_i.

for the next conversion cycle by resetting the counter and the ramp generator to 0. The conversion cycle then repeats.

If at the end of conversion the counter has counted k clock pulses and the clock frequency is f_c, the conversion time may be expressed as

$$T_c = \frac{k}{f_c} = \frac{V_i}{m} \qquad (2\text{–}28)$$

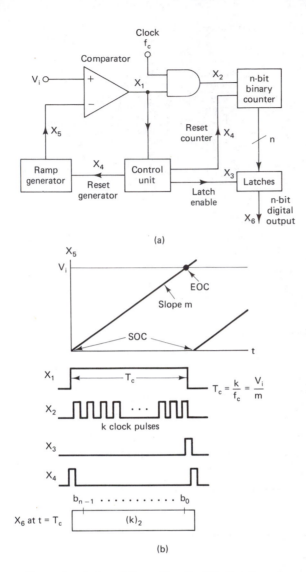

(a)

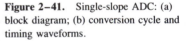

(b)

Figure 2–41. Single-slope ADC: (a) block diagram; (b) conversion cycle and timing waveforms.

It can be deduced from Fig. 2–41b that the conversion time is also equal to the analog input V_i divided by the slope m of the ramp. Suppose that $V_i = 5$ V dc, $f_c = 1$ MHz, and $m = 100$ mV/μs; then the value of k from Eq. (2–28) is 50, and an 8-bit counter stores 00110010_2 at the end of conversion. If the output of the latches is decoded (binary-to-seven segment) and applied to the seven-segment display, the display will show 5.0 after the decimal point is correctly placed. This example illustrates, of course, the use of the single-slope ADC as a digital voltmeter.

The conversion time and rates as expressed by Eqs. (2–25) to (2–27) apply here also. The slope of the ramp generator depends on the values of electrical components, which may be affected by temperature, aging, drift, power supply fluctuations, and so on. Consequently, any variations in the value of the slope will affect the conversion accuracy.

Dual-Slope ADC

A typical *dual-slope ADC* configuration is shown in Fig. 2–42a. At the beginning of the conversion cycle the control unit resets the counter to all 0s and applies V_i by means of the analog input switch to the integrator. For a negative value of V_i, the output of the integrator is a positive-going ramp which may be expressed as $V_a = V_i/RC$, where V_i/RC is the slope of the ramp (the reader should verify the expression for V_a). The positive-going ramp drives the comparator output HIGH and enables the AND gate, allowing the clock pulses to be applied to the counter. After a fixed time interval T_1 (T_1 remains constant for all conversions), the counter has counted k_1 clock pulses, and thus stores the binary equivalent of k_1. The integrator output at this time is $V_a(T_1) = V_i T_1/RC$.

At $t = T_1$, the control unit detects the state of the counter [i.e., $(k_1)_2$], resets the counter to all 0s, and applies the positive reference supply V_r to the integrator input. The positive input to the integrator reverses the polarity of the slope to $-V_r/RC$, causing the integrator to ramp down from $+V_i T_1/RC$ volts toward 0 V. The counter stops its count at $t = T_1 + T_2$ when the integrator output $V_a = 0$ V, which disables the AND gate, cutting off the clock pulses from being applied to the counter. During the time interval T_2 the counter has counted k_2 clock pulses, and hence stores the binary equivalent of k_2 [i.e., $(k_2)_2$].

Since the change in the integrator output voltage is exactly the same during the time intervals T_1 and T_2, it follows that

$$\frac{V_i T_1}{RC} = \frac{V_r T_2}{RC} \qquad (2\text{--}29)$$

Solving for k_2 after substituting $T_1 = k_1/f_c$ and $T_2 = k_2/f_c$, the following result is obtained:

$$k_2 = \frac{k_1}{V_r V_i} \qquad (2\text{--}30)$$

which shows that the contents of the counter k_2 at the end of the conversion cycle is independent of RC constant.

Upon detecting the 1-to-0 transition from the comparator at $t = T_1 + T_2$, the control unit transfers the contents of the counter to the latches, resets the counter to all 0s, and returns the input analog switch to V_i. The conversion cycle now repeats. In the case of a lower input V_i, the values of T_1 and k_1 remain unchanged; however, the values of T_2 and k_2 are also lower. This is illustrated in Fig. 2–42b by the dashed-line response and is consistent with Eq. (2–30).

The conversion time

$$T_c = T_1 + T_2 = \frac{k_1 + k_2}{f_c} \qquad (2\text{--}31)$$

which is a function of k_2, assumes its minimum value for $k_2 = 0$ ($V_i = 0$ V) and its maximum value for $k_2 = 2^n$ when the n-bit counter advances through its full-scale

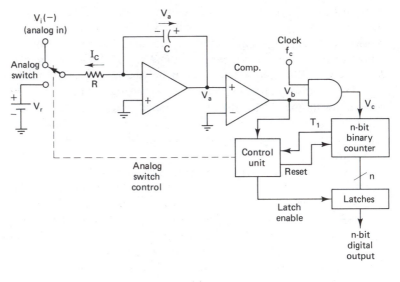

(a)

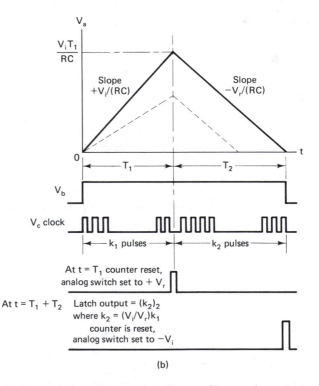

(b)

Figure 2–42. Dual-slope ADC: (a) block diagram; (b) conversion cycle, timing, and control signals.

Control System Components   Chap. 2

count during the time interval T_2. The conversion rates, being the reciprocal of the conversion times, are expressed as

$$R_{c(min)} = \frac{1}{T_{c(max)}} = \frac{f_c}{k_1 + 2^n} \tag{2-32}$$

$$R_{c(max)} = \frac{1}{T_{c(min)}} = \frac{f_c}{k_1} \tag{2-33}$$

It is possible to convert both polarities of V_i by adding a polarity sensing circuit on the input to the converter. One of the most inportant features of the dual-slope ADC is its stability. The conversion time is independent of the RC constant; therefore, as long as f_c and V_r are stable, the conversion time is also stable. If k_1/V_r in Eq. (2-30) is a power of 10, then with proper decoding, the value of k_2 may be displayed digitally as being equal to V_i after an adjustment in the position of the decimal point.

Successive Approximation ADC

A typical *successive approximation ADC* block diagram which uses a two-phase clock is shown in Fig. 2-43. At the beginning of the conversion cycle, the ring counter stores a 1 in the LSB position with 0s in the remaining bit positions, and the successive approximation register (SAR) stores 0s in all bit positions. The SAR is an n-bit register with parallel-in/parallel-out capability. The first ϕ_1 clock pulse (ϕ_1), occurring at t_1, rotates the ring counter 1 bit CW, thus shifting the 1 from the LSB to the MSB position. Assuming that the ring counter responds to the leading edge of the clock, the ϕ_1 clock pulse at t_1 and the 1 from the MSB stage of the ring counter are both applied to AND gate 1, causing the MSB stage of SAR to be set to a 1 with 0s in the remaining bit positions.

As the conversion process continues, the SAR output is applied to the DAC, whose output $V_a = V_r/2$ is compared to V_i in the comparator stage. The HIGH from the comparator ($V_a > V_i$), together with the HIGH from the MSB stage of the ring counter, enable AND gate 2, and the ϕ_2 clock pulse, at t_2, resets the MSB of SAR to 0. When $V_a < V_i$, the LOW output of the comparator disables AND gate 2, thus preventing the MSB of SAR from being reset by the ϕ_2 clock pulse at t_2. Similarly, the next ϕ_1 clock pulse at t_3 stores a 1 in the 2^{n-2} stage of the SAR, and if the resulting V_a exceeds V_i, the 2^{n-2} stage is reset to 0 by the ϕ_2 clock pulse at t_4, but if $V_a < V_i$, the 2^{n-2} stage retains the 1 state.

This process continues whereby the ϕ_1 clock pulse inserts a 1 into the next SAR stage, the resulting state of the SAR is compared against V_i, and a decision is made by the comparator at the next ϕ_2 clock pulse time whether to leave the 1 in or to take it out. After n periods of the ϕ_1 clock, the conversion is complete, and SAR stores an n-bit word that represents V_i. The branching of SAR states is illustrated for a 3-bit converter in Fig. 2-43b. It is clear that the conversion time is always the same regardless of the value of V_i and is expressed as

$$T_c = \frac{n}{f_c} \tag{2-34}$$

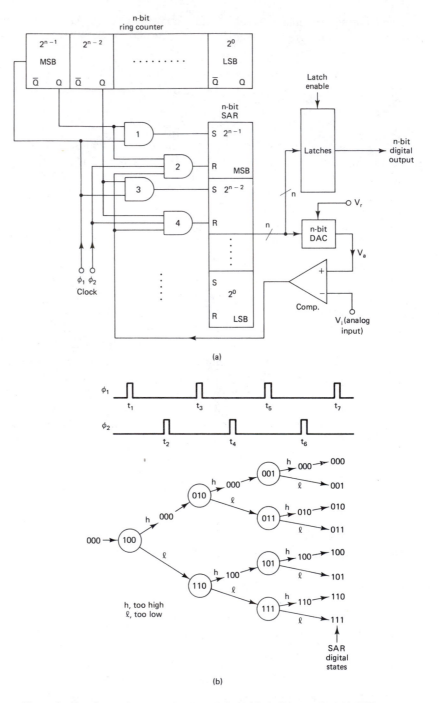

Figure 2–43. Successive approximation ADC: (a) block diagram; (b) 3-bit SAR state-transition diagram; (c) illustrative example.

       Control System Components   Chap. 2

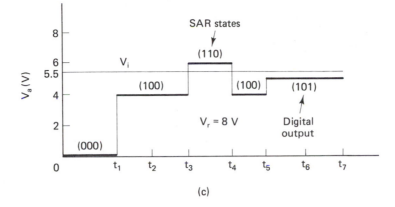

Figure 2–43. (*Continued*)

where n represents the number of bits associated with the converter. Next to the simultaneous converter, the successive approximation converter is the fastest, and for that reason it is probably used more than others. Figure 2–43c illustrates a step-by-step conversion process as a 3-bit successive approximation converter with $V_r = 8$ V converts the 5.5-V input. The reader may wish to follow the conversion path for this example on the SAR state diagram of Fig. 2–43b as the states progress in the sequence 000–100–110–100–101, the final state being 101 after three periods of the ϕ_1 clock.

A commercial 8-bit successive approximation ADC made by Analog Devices is shown in Fig. 2–44. It is fabricated on a single chip and is available in a 800-mW, 18-pin DIP plastic or ceramic package. It can be configured for bipolar -5 to $+5$ V operation, or 0 to 10 V unipolar operation.

The tri-state output buffers, the BLANK-CONVERT (B & $\overline{\text{C}}$) input control, and the DATA READY ($\overline{\text{DR}}$) output control, are special features that permit interfacing the AD570 to a microprocessor. The conversion is initiated by bringing the B & $\overline{\text{C}}$ line LOW and is completed after about 25 μs, at which time the $\overline{\text{DR}}$ line goes LOW, and 500 ns later the output lines become active with new data.

Another commercial ADC is shown in Fig. 2–45. The TSC7116 is fabricated on a single chip using CMOS technology and is available in a 40-pin DIP plastic package. The seven-segment decoders, polarity and digit drivers, voltage reference, clock circuits, and the backplane driver for the LCD display are among the functions that are integrated into the chip.

The conversion time of the TSC7116 remains constant at 4000 counts of the clock: 1000 counts is used for input integration, 0 to 2000 counts for reference integration, and what remains of the 4000 counts, which may be anywhere from 1000 to 3000 counts is used for the auto zero (A/Z) phase. In the A/Z phase, the analog input is replaced by a short to simulate the 0-V input condition, a closed loop is formed around the integrator and the comparator, thus allowing the C_{AZ} capacitor to charge to a voltage which is used to compensate the comparator offset error voltage, and thus guarantee a 0-V display for a 0-V input. In control system applications the TSC7116 offers a simple

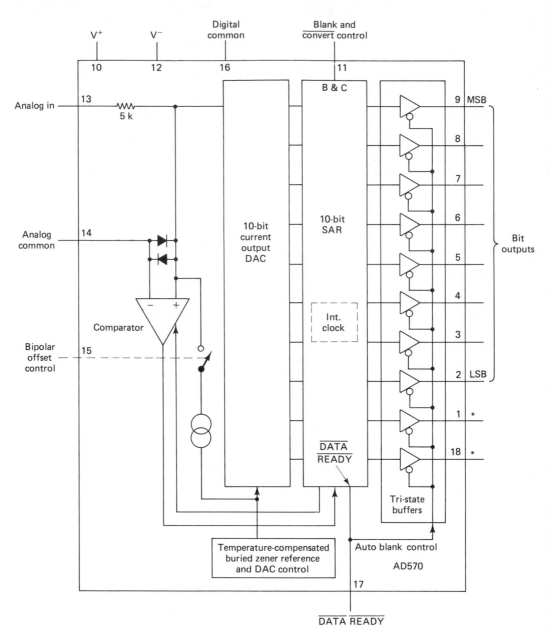

Figure 2–44. AD570, an 8-bit successive approximation ADC. (Courtesy of Analog Devices, Norwood, Mass.)

and relatively inexpensive means of digitally displaying control system data, such as motor rpm, armature current, analog input, and so on.

Included below are some definitions that will help the reader to better understand the characteristics and the performance of ADCs.

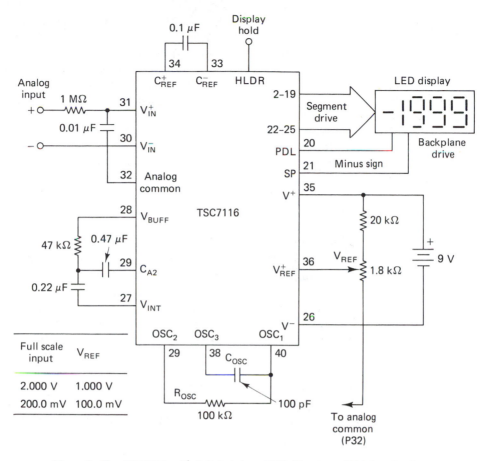

Figure 2–45. TSC7116, a 3½-digit dual-slope ADC. (Courtesy of Teledyne Semiconductor, Mountain View, Calif.)

Quantizing error. This is the error associated with the partitioning of the analog voltage range by an *n*-bit ADC into 2^n discrete voltage intervals or quantums. Any voltage value within one such interval produces the same output code. Suppose, for example, that an 8-bit DAC is used in Fig. 2–40, and $V_r = 10$ V. The analog range is partitioned in this case into 256 discrete voltage intervals, each interval being equal to $10/2^8$ or approximately 39 mV. This means that the analog input may vary almost 39 mV, with the output code remaining unchanged (39 mV is also the voltage level associated with 1 LSB). Generally, a code is assigned the midpoint of the interval value, and for that reason the quantizing error is equal to $\pm \frac{1}{2}$LSB. In the example above, the quantizing error is ± 19.5 mV.

Absolute accuracy. This error is expressed as the difference between the theoretical input and the actual input voltages required to produce a given code at the output of an ADC. The quantization of the analog input into discrete voltage intervals, as discussed above, results in many or the entire interval of voltage levels which produce the same code. Due to the quantizing uncertainty, the analog input is therefore assigned the midpoint value of the interval of voltage levels which produce the same code. Suppose, for example, that 10 V ($\pm$ 19.5 mV) produces a theoretically expected full-scale 8-bit code (11111111). Then an 8-bit ADC which produces a full scale-code for the input range 9.88 to 9.98 V (the midpoint value is 9.93 V) has an absolute error of 9.93 $-$ 10 $=$ -70 mV. The sources of absolute error include zero error, gain, nonlinearity, and noise. Standard test conditions and very accurate equipment must be used in absolute accuracy measurement.

Relative accuracy. This expresses the error due to the deviation of the analog voltage at any code from the theoretical value. This deviation is generally expressed in percent, ppm (parts per million), or as a fraction of the LSB of the full-scale analog range (FSR) after the full-scale analog range has been properly calibrated. The theoretical transfer characteristic of the ADC (graph of the output codes versus analog input values) is a straight line drawn through the midpoints of the quantization intervals at each code.

Linearity error: differential and integral. The *differential linearity error* expresses the deviation of the measured step width from the ideal step width (the ideal or theoretical step width must be exactly 1 LSB wide or 2^{-n} of the full-scale analog range for an *n*-bit converter). The differential linearity error in excess of 1 LSB is responsible for the missed codes by the ADC. Manufacturers often specify an acceptable differential linearity error as "no missed codes," implying that it is less than 1 LSB. The *integral linearity error* is the deviation of the overall shape of the conversion response from the straight-line response. The zero-and-gain adjustment is a calibration procedure that must be performed on the ADC. The zero adjustment is done so that the output code makes a transition from all 0s to the 1 in the LSB position when an input equal to $\frac{1}{2}(2^{-n})$FSR $= 2^{-(n+1)}$ (FSR) is applied. The gain adjustment is made so that the transition in the output code to all 1s occurs when the input is $[1 - (3/2)2^{-n}]$FSR. These adjustments apply to the unipolar ADC.

2–11 VOLTAGE-TO-FREQUENCY AND FREQUENCY-TO-VOLTAGE CONVERTERS

The *voltage-to-frequency converter* (VFC) operation is similar to that of a VCO; it converts the voltage level of the input signal to a unique frequency. The output wave shape of the VFC is typically a square wave. The VFC is a linear device whose transfer characteristic is a straight line, the slope of the straight line representing the value of the VFC's transfer function in Hz/V. As such, the VFC may be represented by a linear block.

The *frequency-to-voltage converter* (FVC) is also a linear device, which converts the frequency of the input signal to a unique voltage level. It, too, may be modeled by a linear block whose transfer function value is the slope of FVC's straight-line transfer characteristics.

The circuit configurations for the VFC and FVC are very similar, and for that reason the manufacturers offer both functions on the same chip. The operation of the chip to provide either the VFC or the FVC function is determined only by the external to the chip's passive components.

The AD650 chip made by Analog Devices has been selected for the purpose of analysis. Figure 2–46 shows the AD650 connection diagram for VFC operation, and the diagrams shown in Fig. 2–47 are used for analysis. When the AD650 operates as a VFC, there are two modes of operation: the integrate mode and the reset mode. In the integrate mode shown in Fig. 2–47c, the internal 1-mA current source is connected to the output of the integrator. In this mode the analog input voltage V_i, shown in the simplified diagram in Fig. 2–47a, develops the input current $I_i = V_i/R_i$, which is applied to the integrator capacitor C_{int}. The capacitor current may be expressed in general as

$$I_c = C\frac{dV_c}{dt} \qquad (2-35)$$

Substituting I_i for I_c and V_o, the output voltage of the integrator, for V_c in Eq. (2–35), the following result is obtained:

$$\Delta V_o = -\frac{I_i}{C_{int}}\Delta t = -\frac{V_i}{R_iC_{int}}\Delta t \qquad (2-36)$$

Since V_i is constant, V_o is related linearly to t. The slope of the straight line is negative and its magnitude is $V_i/(R_iC_{int})$, being a function of the analog input V_i. The negative slope is due to the manner in which C_{int} is being charged by I_i. As shown in Fig. 2–47c, the output of the integrator V_o, which is same as the voltage across C_{int}, is negative.

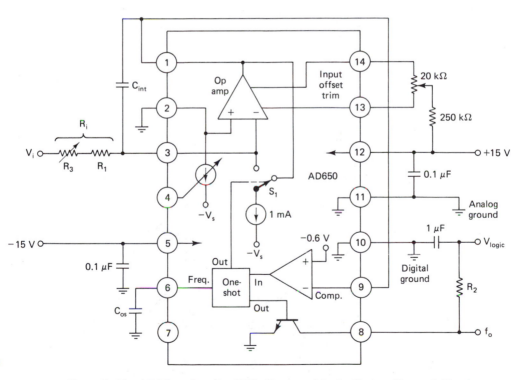

Figure 2–46. AD650 configured as VFC. (Courtesy of Analog Devices, Norwood, Mass.)

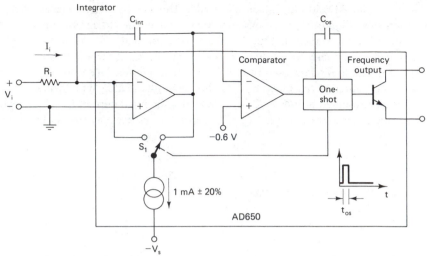

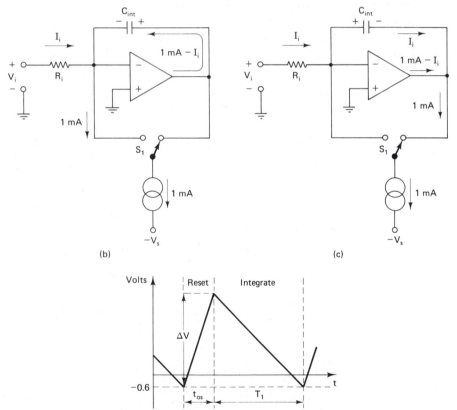

Figure 2–47. AD650 VFC: (a) simplified block diagram; (b) reset mode; (c) integrate mode; (d) integrator output. (Courtesy of Analog Devices, Norwood, Mass.)

It may therefore be concluded that in the integrate mode V_i causes the output of the integrator to ramp down. When V_o reaches -0.6 V, the comparator outputs a negative step which triggers the one-shot (a monostable multivibrator), thus initiating the reset mode. At the beginning of the reset mode, the one-shot switches the 1-mA current source to the integrator input as shown in Fig. 2–47b. As the C_{int} charging current is equal to 1 mA, I_i reverses direction because $I_i < 1$ mA, and C_{int} charges as shown in Fig. 2–47b, with polarity so as to cause V_o to ramp in the positive direction. The duration of the reset period is constant and is determined by the one-shot timing, which consists of a propagation delay equal to approximately 300 ns and an additional fixed time interval associated with the charging of C_{os}. C_{os} is normally shorted to the analog ground, but at the beginning of the reset period, an internal 0.5-mA current source is connected to pin 6, causing the current to flow through C_{os} into pin 6, charging C_{os} negative. When the voltage across C_{os} reaches -3.4 V, the one-shot is reset, the reset period ends, and the integrate mode as described above, begins. Equation (2–35) may be used to describe that part of the reset period t_{os} which is related to the charging of C_{os}, and t_{os} may be expressed as

$$t_{os} = \frac{\Delta V C_{os}}{I_{os}} + t_{pd} = \frac{(-3.4)C_{os}}{-5 \times 10^{-4}} + 3 \times 10^{-7}\,\text{s} \qquad (2\text{--}37)$$

The change in the integrator output voltage during the reset period may be calculated using Eq. (2–35). Since the slope of the positive ramping of V_o during the reset period is the current through C_{int} divided by C_{int}, the increment in V_o during t_{os} must be equal to that slope times t_{os}, or expressed mathematically,

$$\Delta V_o = \frac{t_{os}}{C_{int}}(1\text{ mA} - I_i) \qquad \text{V} \qquad (2\text{--}38)$$

where $I_i = V_i/R_i$ and t_{os} is defined in Eq. (2–37). Since the increment ΔV_o made by the integrator during the reset mode and during the integrate mode is the same, the duration of the integrate mode T_I may be obtained by dividing ΔV from Eq. (2–38) by the slope of the ramp during the integrate mode, which is equal to I_i/C_{int}, resulting in

$$T_I = \frac{(t_{os}/C_{int})\,(10^{-3} - I_i)}{I_i/C_i} = t_{os}\left(\frac{10^{-3}}{I_i} - 1\right) \qquad (2\text{--}39)$$

The frequency of the square wave at pin 8 is the reciprocal of the total time associated with the reset and the integrate modes, or

$$f_o = (t_{os} + T_I)^{-1} = \frac{V_i/R_i}{6.8C_{os} + 3 \times 10^{-10}} \qquad \text{Hz} \qquad (2\text{--}40)$$

Equation (2–40) expresses the transfer characteristics of the AD650 chip when it is configured as VFC. It shows that a linear relationship exists between the frequency of the output signal and the analog input voltage. It may be noted that C_{int}, which determines the amplitude of the integrator output, does not affect the output frequency. The values of R_i and C_{os} dictate the full-scale frequency range for a given V_i range. Better linearity is achieved when larger values are used for R_i and C_{os}. For example, Analog Devices data show that for the input range 0 to 10 V, the typical range of R_i is 20 to 100 kΩ, and the typical range of C_{os} is 50 to 1000 pF, for full-scale frequency range values of 10 kHz to 1 MHz. The typical nonlinearity for the ranges above falls into the range 20 to 1000 ppm (20 ppm corresponds to 100 kΩ and 1000 pF).

The connection diagram for the operation of the AD650 as an FVC is shown in Fig. 2–48. The input signal whose frequency is to be converted to a voltage level is applied through the RC differentiating circuit to pin 9 and the output is taken across the parallel combination of C_{int} and $R_1 + R_3$.

Each time the input signal drops below -0.6 V (the comparator threshold), the comparator outputs a positive step that triggers the one-shot. The one-shot, in turn, switches the 1-mA current source to the integrator input for a time period as determined by the quasi-stable state of the one-shot. As shown by Eq. (2–37), the time duration t_{os} of the quasi-stable state is determined by the value of C_{os}. If the input waveform is always more positive than -0.6 V and is applied directly to pin 9, the one-shot will never be triggered and the circuit will not work. For that reason the input differentiating circuit is used to assure that regardless of the average value of the input waveform, a negative-going spike will be generated by the differentiating circuit, which will bring the $-$ input of the comparator below -0.6 V. It is important that this time duration [when the $-$ input of the comparator is less than -0.6 V] not exceed the time of the quasi-stable state of the one-shot; otherwise, the one-shot triggered more than once per cycle of the input waveform. To meet this requirement, the value of the RC time constant

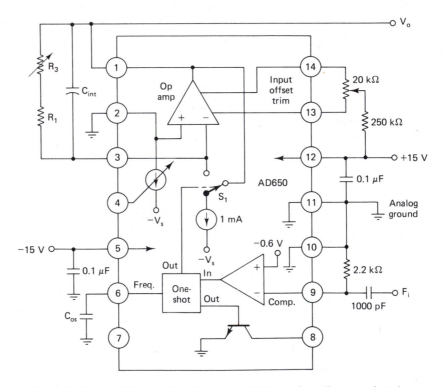

Figure 2–48. AD650 connection diagram for FVC operation. (Courtesy of Analog Devices, Norwood, Mass.)

Control System Components   Chap. 2

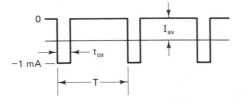

Figure 2–49. Current waveform applied to the integrator of the FVC in Fig. 2–48.

of the differentiating circuit (it is 2.2 μs in the given configuration) must be carefully selected. Analog Devices suggests that the duration of the input pulse should be greater than 100 ns and should not exceed $0.3t_{os}$.

As stated earlier, the one-shot switches the 1-mA current source to the integrator input for the time duration t_{os} each time the comparator is triggered. The resulting current waveform is shown in Fig. 2–49. During this time, charge is delivered to C_{int} with a corresponding increase in voltage. During the remainder of the cycle, the 1-mA current source is connected to the output of the integrator and the capacitor loses some charge. If the frequency of the current pulses, which corresponds to the frequency of the input, is constant, the average value of current pulses (the dc component) I_{av}, shown in Fig. 2–49, is proportional to the duty cycle and may be expressed as

$$I_{av} = I_{pk} \text{(duty cycle)} = (10^{-3}) \frac{t_{os}}{T} \qquad \text{A}$$

$$= t_{os}f \qquad \text{mA}$$

(2–41)

where f is the frequency of the input waveform. Thus, as the input frequency increases, I_{av} increases also. I_{av} flows through the series combination of R_1 and R_3, producing a positive output V_o.

2–12 SHAFT POSITION ENCODER

Shaft position encoders are electromechanical devices that convert the angular position of the shaft to a pulse train or a coded n-bit word. A disk that represents the mechanical portion is used to accomplish the conversion. The wide range of practical applications where angular position encoding is used includes the numerical control of machining processes, printers, *X-Y* plotters, computer disk position controls, servomechanisms, radar antennas, digital displays, remote position controls, radar antenna position controls, and many others. There are two types of encoders: the incremental type and the absolute type. These will be discussed presently.

There are two types of contacting incremental encoders: the contacting type and the optical type. A typical structure for a *contacting incremental encoder* is shown in Fig. 2–50a. The metallic strips (electrical conductors) are deposited along tracks *A* and *B* on the surface (made of electrically nonconductive material) of the disk with a single index strip in the outer track. The strips are electrically connected to the metallic back of the disk, which is maintained at ground potential. The contacting spring-loaded pins (one for each track) make a pressure contact with the disk and ride along the tracks as the disk rotates. Since the pin unit is electrically a short circuit, the output

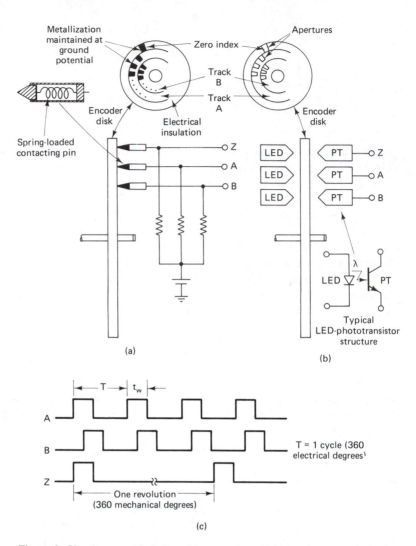

Figure 2–50. Incremental shaft position encoders: (a) contacting type; (b) optical type; (c) encoder output waveforms.

becomes 0 V every time the pin makes contact with the strip, and during the times when the pin is between the strips, the output rises to the full supply voltage.

Typical output waveforms are shown in Fig. 2–50c. To provide maximum flexibility for the user, manufacturers generally design the encoding disks so that waveforms A and B are in quadrature (90° phase difference). As shown, the zero index track generates one pulse per revolution of the disk, thus serving as a convenient marker for synchronization purposes.

If the disk contains N strips on a given track, one revolution of the disk produces N pulses whose duty cycle dc $= t_w/T$. Furthermore, if angular velocity of the disk is ω (rpm), the frequency of pulses of output A or B may be deduced from

$$\omega(\text{rev/min})N(\text{pulses/rev})1(\text{min/60 s}) = f\,(\text{pulses/s})$$

thus

$$f = \frac{\omega N}{60} = \frac{1}{T} = \frac{dc}{t_w} \qquad (2-42)$$

where t_w represents the pulse width. The resolution that represents the maximum rotation of the shaft before the next pulse is produced depends on N, and may be expressed as

$$R = \frac{360°}{N} \qquad (2-43)$$

Thus, for a 1° resolution, 360 strips per track are required. To determine the angular displacement of the shaft, one simply needs to count the pulses as described in Section 2–7. The accuracy of such a measurement clearly depends on the resolution.

The major disadvantage of the contacting incremental encoder is the contact wear, which limits the useful life of the encoder. An encoder that does not depend on contact for monitoring the rotation of the shaft is the optical incremental encoder shown in Fig. 2–50b. As shown, the apertures or holes in the disk replace the strips and the sensing unit consists of the LED–phototransistor combination. As the disk rotates, the light from LED is detected by the phototransistor when the aperture is aligned with the light beam from the LED, causing the output to become LOW (depending on the circuit design). At times when the disk blocks the light, the output rises to a HIGH. The resulting output is also a pulse train, as shown in Fig. 2–50c. Equations (2–42) and (2–43), where N this time represents the number of apertures per track, apply as well to the optical incremental encoder.

The *optical incremental encoder* does not have the contact wear of the contacting encoder, and in that respect it is more reliable. On problem which applies to both of these encoders is that in the event that the power is interrupted, the encoding information is lost. The incremental encoder thus behaves in the same way as the volatile random access memory (RAM) chip. The absolute encoder structure, which solves this problem, is treated next.

As shown in Fig. 2–51a, the 5-bit encoding disk is divided into five tracks and 2^5 or 32 sectors. An n-bit encoder that uses the binary code will require n tracks and 2^n sectors. The dark areas on the disk represent a binary 1 and the light areas represent a 0. LSB (2^0 weight) is assigned to the outer track and the MSB is assigned to the inner track. Consequently, the 5-bit code associated with a sector represents all angular positions of the shaft within that sector. In the illustration, the 5-bit code represents all angular values of the shaft position within the 11.25° sector. The code assignment truth table for the 5-bit binary disk is shown in Fig. 2–51b.

Codes other than the natural binary may also be used to encode the disk. Binary-coded-decimal (BCD) code, for example, may use n bits to encode a decimal number. Some of the commonly used 4-bit BCD codes include the 8421 weighted code, the 2421 weighted code, the unweighted excess-3 code, the Gray code, and others. The Gray code has the disadvantage of being incompatible with most computer systems that require the Gray code converter. The Gray code does have the distinct advantage of changing only one bit at a time as the codes change in sequence during the rotation

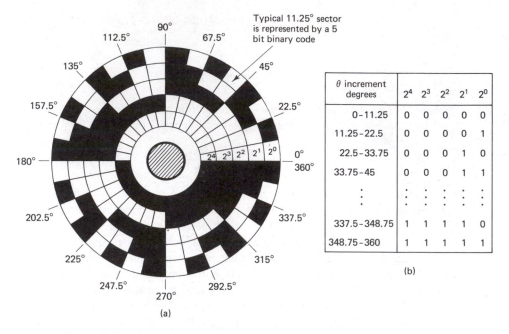

Figure 2–51. Absolute shaft position encoder: (a) encoding disk (5-bit); (b) angular code assignment truth table.

of the disk. The 2421 and the XS-3 codes are self-complementing, and that may be important in some mathematical applications. Five and more bits may also be used to encode a decimal number.

If the disk is encoded with a BCD code, the number of decades decides the track/sector encoding pattern. For example, if the BCD code uses 8421 weighting and two decades are required, eight tracks will be needed; four outer tracks may be used to represent the 10^0 weighted decimals 0 to 9, and the four inner tracks may be used to represent the 10^1 weighted decimal numbers. With two decades the decimal numbers 0 to 99 are represented by 100 sectors, and each sector will output an 8-bit BCD code. With optical sensing, eight LEDs and eight phototransistors will be required for this application. The resolution in this case is 3°, obtained by dividing the 360° (one rotation) of the disk by 100. If 5-bit BCD code was used in the application above, the number of tracks will increase to 10, but the resolution will remain unchanged at 3°. Clearly, the resolution is a function of the number of decades used and may be expressed for an m-decade, BCD-encoded disk as

$$R = \frac{360}{10^m} \qquad (2-44)$$

The resolution for the disk encoded with an n-bit natural binary code is

$$R = \frac{360}{2^n} \qquad (2-45)$$

where 2^n represents the number of sectors in the disk. If the encoded disk is rotating, the waveform from the LSB (the outer track) sensor will resemble waveform A or B in

Fig. 2–50c, with a 50% duty cycle, meaning that the pulse width t_w is equal to one-half the period [i.e., $t_w = T/2 = 1/(2f_{LSB})$, where f_{LSB} is the frequency of pulses from the LSB track sensor]. Then

$$\frac{T\,(\text{seconds})}{2(\text{sectors})} \times \frac{2^n(\text{sectors})}{\text{rev}} \times \frac{1(\text{minute})}{60(\text{seconds})} = \frac{T2^n}{120} = \frac{2^n}{120f_{LSB}} = \frac{1}{\omega} \qquad \text{min/rev}$$

interrelates the speed ω in rpm, the number of sectors 2^n, and the frequency of LSB track output f_{LSB}. This equation may be expressed in a more compact form as

$$2^n\omega = 120f_{LSB}$$

If the disk was encoded using a BCD code and m decades, the equation above will apply if 2^n is replaced by 10^m. A more general equation which encompases both the binary and the BCD codes may be expressed as follows:

$$X\omega = 120f_{LSB} \qquad (2-46)$$

where $X = 2^n$ for an n-bit binary code or $X = 10^m$ for an m-decade BCD code.

In some applications it may be necessary to encode multiple rotations of the shaft. To meet such requirements, manufacturers combine several encoder disks with gears between the disks in one housing. Suppose, for example, that two identical disks (each disk is same as that shown in Fig. 2–51a) are geared so that disk 1 and the primary gear are coupled to the driving shaft whose angle is to be encoded, while disk 2 is coupled to the secondary gear. The teeth ratio of the secondary to the primary gears is $360/11.25 = 32$, where $11.25°$ is the angular width of the sector. The gears provide the angular step down so that disk 2 moves one sector ($11.25°$) for one full rotation of disk 1. In this arrangement, which requires eight optical sensors, the output of disk 2 may be used as a counter of rotations of disk 1. Since disk 1 outputs 32 counts per rotation, a maximum count of $32^2 = 1024$ will result when disk 1 completes 32 revolutions. Multiturn optical absolute encoders with a maximum count of 500,000 or better using BCD or binary codes are available. A commercial incremental encoder is illustrated in Fig. 2–52.

Figure 2–52. Commercial incremental modular encoder. (Courtesy of Litton Encoder Division, Chatsworth, Calif.)

EXAMPLE 2–4

The disk is encoded using the BCD (8421) code and 10 tracks to produce the maximum possible count from this encoding pattern. Calculate (a) the resolution; (b) shaft rotation to produce a 1001100111 code; (c) ω in rpm if the pulses from the LSB track sensor have a pulse width of 20 μs.

SOLUTION (a) The outer four tracks represent the units, the next four tracks represent the 10s, and the two innermost tracks are used for the 100s, with the two MSB bits in the 100s decade always being 0. The count therefore extends from 0 to the maximum count of 399. The full rotation of the disk is divided into 400 parts or sectors. Therefore,

$$R = \frac{360}{400} = 0.9°$$

(b) The given BCD code may also be written as 001001100111, which is equal to 267_{10}. Assuming that the disk started its rotation from the 0 sector, by advancing 267 sectors, it rotated through an angle

$$\theta = 267(0.9) = 240.3°$$

(c) $$f_{LSB} = \frac{1}{2t_\omega} = \frac{1}{40 \times 10^{-6}} = 25 \text{ kHz}$$

using Eq. (2–46) and 400 in place of X yields

$$\omega = \frac{1202.5 \times 10^4}{400} = 7500 \text{ rpm}$$

2–13 SOLENOID

A *solenoid* is an electromechanical component that converts electrical energy to a mechanical form. The typical solenoid construction shown in Fig. 2–53 includes a plunger made of iron-based (magnetic) material surrounded by the coil and an iron frame which houses the solenoid structure. The plunger is free to move within the nonmagnetic form, which also provides a mechanical support for the turns of the coil.

The current I through the coil generates the magnetic field or flux which loops around the coil. The iron frame, with its high magnetic permeability, helps to concentrate the magnetic flux and establish the magnetic circuit. In a manner similar to the closed-loop electric circuit, which provides the path for the flow of electric current, the magnetic circuit, shown by the dashed lines, provides the path for the magnetic flux, which extends through the frame, the plunger, the stop, and the air gap, which is between the end of the plunger and the stop.

The presence of the magnetic field causes the iron-based frame and the plunger to be magnetized. The magnetization of the iron-based material occurs when the field induces into the material the magnetic domains or the magnetic dipoles, which become

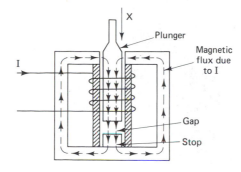

Figure 2–53. Solenoid construction.

oriented or aligned in the same direction, thus providing a low-reluctance path for the magnetic flux. Unlike the electric field vector, which begins and ends on a charge distribution, the magnetic field must take the form of closed loops, as there are no known magnetic charges. Thus in the illustration the magnetic circuit must be a closed loop. The magnetic continuity across the air gap is produced when the bottom of the plunger becomes, in effect, polarized as the north pole and the top of the stop as the south pole. This type of magnetic polarization creates a force of attraction between the stop and the plunger, causing the plunger, which is free to move axially, to accelerate toward the stop. If a load is attached to the end of the plunger, it would be displaced a distance X as shown in the illustration. Mechanical work is thus performed for a given value of applied electric current.

Some solenoids may be operated from an ac or a dc source, yet others are designed strictly for dc operation. All solenoids, however, depend on the size of the coil (number of turns) and the value of applied current for the force that must be generated to drive a given load. A typical force–stroke relationship for a general-purpose commercial solenoid is illustrated in Fig. 2–54b. As one would expect, the solenoids must have a power rating. The size of the wire and the temperature characteristics of the wire insulation set the limit for the maximum applied current. In the illustration the C9 solenoid is rated at 8.6 W of continuous applied dc power. It is based on the maximum coil temperature of 105°C, which is equal to 80°C coil heat rise above the 25°C ambient temperature. If the solenoid is operated in an ambient temperature above 25°C, the 8.6 W must be derated accordingly so as not to exceed the maximum coil temperature of 105°C. Linear derating can be used, so that, for example, the maximum rating of 8.6 W reduces to 2.15 W at 85°C ambient temperature. Of course, if the power was applied on intermittent basis, say 3 minutes maximum ON and 9 minutes minimum OFF, allowing the coil to cool in the OFF-time, the solenoid above is rated at 15.5 W at 25°C ambient. In pulsed applications with 100 ms ON and 900 ms OFF, the C9 solenoid is rated at 86 W. The efficiency with which the solenoid converts the applied electrical energy to mechanical energy or mechanical work depends on the materials used for the frame, the plunger, and the stop.

Some solenoids are of the "push" type, others are of the "pull" type. The solenoid in the illustration in Fig. 2–53 is of the pull type. However, if the stop is constructed on the opposite end of the plunger, this solenoid would become the push type.

Solenoids are used in home appliances, vending machines, and in various industrial

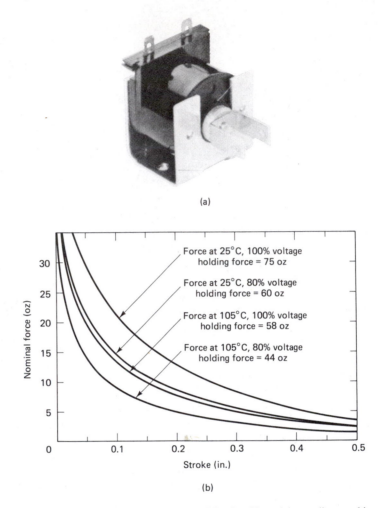

(a)

Force at 25°C, 100% voltage
holding force = 75 oz

Force at 25°C, 80% voltage
holding force = 60 oz

Force at 105°C, 100% voltage
holding force = 58 oz

Force at 105°C, 80% voltage
holding force = 44 oz

(b)

Figure 2–54. (a) General-purpose commercial solenoid used in vending machines, appliances, valve operators, etc.; (b) force–stroke characteristic of the C-9 solenoid shown, dc continuous duty, 8.6 W. [Courtesy of Deltrol Controls, Milwaukee, Wis. (Solenoids Engineering Manual).]

controls. In home heating systems, solenoids are used as zone control valves to control the flow of hot water.

2-14 GEARS

Gears are mechanical components used in the transmission of mechanical power and in the transformation of speed or torque. In control system applications, for example, some motors cannot operate smoothly at low speeds (they cog). Through the use of gears, the motor may operate at a higher speed, which is then transformed to a lower value as required by the load. Gears may also be used in other control applications to produce very small angular motion of the load.

Gears are available commercially in a variety of sizes and shapes. The material from which gears are made may be metal for heavy-load applications or plastic for light loads. The *spur gear* construction shown in Fig. 2–55 is probably simplest and is used here for the purposes of analysis.

The shafts of two spur gears are always parallel, and the shaft of the driver is typically coupled to the shaft of the motor. When the driver gear rotates, its teeth engage the teeth of the driven gear. The force applied to the tooth of the driven gear results in a torque that causes the driven gear, and therefore the load applied to the shaft of the driven gear, to move. The mechanical power is thus transmitted by means of two gears, from the motor to the load.

When the driver gear engages the driven gear during the motion, their teeth at any given instant make contact in at most two points. Whether contact is made at one or two points depends largely on the amount of backlash. *Backlash* is the play between the teeth of the two gears due largely to wear in the teeth. Then locus of all such contact points forms the *pitch circle*, which passes approximately through the midpoints of gear's teeth. The *circular pitch* p_c, on the other hand, is the distance measured along the pitch circle between the corresponding points on two consecutive teeth. It follows, then, that the length of the pitch circle, whose diameter is D, must be equal to the product of the number of teeth N and the circular pitch. Hence

$$P_cN = \pi D \tag{2-47}$$

or

$$P_cP_d = \pi$$

where $P_d = N/D$ is defined as the *diametral pitch*.

To produce proper mashing of the teeth during motion, the teeth of the driver and those of the driven gears must be identical in shape, regardless of the relative size of the two gears. To satisfy this requirement, the circular pitches of the driver and the driven gears must be equal. As defined earlier, the circular pitch is equal to the arc

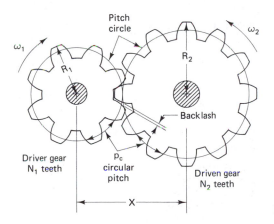

Figure 2–55. Spur gear construction.

length $R\theta$, where R is the radius of the pitch circle and θ is the angle subtended by the arc. Using the subscript 1 for the parameters on the driver's side, and the subscript 2 for the driven-side parameters, it follows that

$$R_1\theta_1 = R_2\theta_2 \tag{2-48}$$

Using Eq. (2–47), Eq. (2–48) may be reduced to the form

$$N_1\theta_1 = N_2\theta_2 \tag{2-49}$$

Differentiating Eq. (2–49) twice leads to the following velocity and acceleration relationships:

$$N_1\dot{\theta}_1 = N_2\dot{\theta}_2$$
$$N_1\ddot{\theta}_1 = N_2\ddot{\theta}_2 \tag{2-50}$$

Defining the gear teeth ratio as $N_G = N_2/N_1$, Eqs. (2–48) and (2–49) may be rewritten as

$$N_G = \frac{\theta_1}{\theta_2} = \frac{\dot{\theta}_1}{\dot{\theta}_2} = \frac{\ddot{\theta}_1}{\ddot{\theta}_2} = \frac{N_2}{N_1} \tag{2-51}$$

In a well-designed gear train, the losses are minimized, and therefore the mechanical power applied to the input gear is approximately the same as the output power. Since the mechanical power is the product of torque and velocity,

$$P = T_1\omega_1 = T_2\omega_2 \tag{2-52}$$

Using Eq. (2–51) in Eq. (2–52) yields

$$\frac{\omega_1}{\omega_2} = \frac{T_2}{T_1} \tag{2-53}$$

The constancy of P in Eq. (2–52) requires that whenever speed is stepped down, the torque must necessarily be stepped up, and vice versa. The expression for power in terms of horsepower may be deduced from Eq. (2–52) by using some of the following conversions:

$$1 \text{ hp} = 550 \text{ ft-lb/s} = 746 \text{ W} = 2545 \text{ Btu/h}$$
$$1 \text{ Btu} = 778 \text{ ft-lb} = 1055 \text{ W-s} \tag{2-54}$$

as

$$P = \frac{T\omega}{5252} \qquad \text{hp} \tag{2-55}$$

where T is in ft-lb and ω is in rpm (the reader should verify this expression), and if T is in oz-in., the value of the denominator in Eq. (2–55) is 1,008,405.7.

If the load torque is purely inertial, $T_2 = \ddot{\theta}_2 J_L$, the reflected or equivalent inertia on the motor side is determined from Eqs. (2–51) and (2–53) as follows:

$$T_1 = \frac{T_2}{N_G} = \frac{\ddot{\theta}_2 J_L}{N_G} = \frac{J_L}{N_G^2}\ddot{\theta}$$

Therefore, the reflected inertia on the motor side is

$$J_R = \frac{J_L}{N_G^2} \qquad (2-56)$$

By following a similar procedure, the effects of load viscous friction $B_L \dot{\theta}_2$ on the motor side may be expressed by the reflected viscous friction coefficient

$$B_r = \frac{B_L}{N_G^2} \qquad (2-57)$$

It can be seen from these results that the load inertia and load viscous friction are attenuated by N_G^2 in speed step-down configurations and are multiplied by N_G^2 in speed step-down arrangements.

Whenever a particular design specifies the distance X between the gear centers and N_G, the following expression, which the reader should derive from Fig. 2–55, may prove to be useful:

$$R_1 = \frac{X}{1 + N_G} \qquad \text{and} \qquad R_2 = \frac{X N_G}{1 + N_G} \qquad (2-58)$$

Thus if $X = 20$ in., $N_G = 4$, the pinion has 25 teeth, and the gear diameters from Eq. (2–58) are determined as $D_1 = 8$ in. and $D_2 = 32$ in., the larger gear must have 100 teeth and the circular pitch must be approximately 1 in. for each gear.

The *rack-and-pinion gear* configuration shown in Fig. 2–56 is useful in applications requiring a linear motion of the load. The pinion, which is generally the smaller of the two gears, is in this case driven by the motor, with the result that the rotational motion of the motor's shaft is converted to the translational motion of the load, which is on top of the rack. The rack may be regarded as a gear with infinite radius. All equations that were developed for gears apply in this case also.

In contrast to spur gears, whose shafts are parallel, the shafts of *bevel gears*, if extended by imaginary lines, intersect in space at point O, as shown in Fig. 2–57. Because the shafts of bevel gears can be at any angle—right angle (90°), acute angle (<90°), or obtuse angle (>90°)—it is possible to couple the load to motor's shaft at any angle, which is an important feature of bevel gears. The speeds of two bevel gears are inversely related to the sines of their respective gear center angles, hence $\omega_g/\omega_p = \sin\theta_p/\sin\theta_g$. Equation (2–47), which was defined for the spur gears, also applies to the bevel gears. The teeth of bevel gears may be straight, spiral, or skewed. Straight teeth are simplest in form and are used at speeds below 1000 rpm. Spiral teeth are

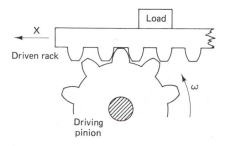

Figure 2–56. Rack-and-pinion configuration converts rotational motion to translation motion.

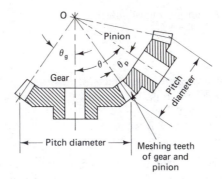

Figure 2–57. Bevel gear construction. Shaft axes of the gear and the pinion make angle θ and intersect at O.

curved and oblique in form, allowing for greater contact area between the teeth, and therefore providing smoother motion and greater load-handling capability. Bevel gears are often used in motion-picture equipment and sewing machine applications, where smoothness of motion is the primary requirement.

Worm gears are used to connect nonintersecting, nonparallel shafts. A typical worm gear construction is shown in Fig. 2–58a. The efficiency of a worm drive depends on the helix angle α such that $\tan \alpha = d/\pi D_w$, where D_w is the worm diameter and d is the distance advanced by the worm in one revolution. To achieve optimum efficiency, the value of α should be close to 30° in speed-reduction drives, and almost double that value in speed-step-up applications. In both of these cases, the value of the angle β (which measures the steepness of the tooth side) should be approximately 30°.

One important application of a worm gear is in speed-reduction drives; a typical

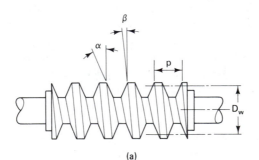

(a)

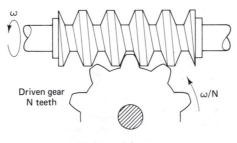

(b)

Figure 2–58. Worm gear: (a) construction; (b) single-threaded worm driving a pinion.

Control System Components   Chap. 2

configuration is shown in Fig. 2–58b. As shown, the worm rotates with a speed ω driving a pinion with N teeth. The number of teeth T in the worm is defined as

$$T = \frac{N}{N_G} = \frac{d}{p} \qquad (2\text{–}59)$$

where N_G = speed ratio (speed of the worm ω_w divided by the speed of the pinion ω_p
        p = axial pitch of the worm
        d = distance advanced by the worm in one revolution.
Large speed reduction is thus possible. For instance, in the case of a single threaded worm ($T = 1$) driving a pinion with 50 teeth, the speed of the pinion is 1/50 that of the worm.

In addition to the gears already discussed, there are helical gears, planetary gears, and others. *Helical gears* are used to couple parallel shafts, or nonintersecting shafts which are at right angles (a worm gear is a special case of a helical gear), or to couple nonintersecting shafts at any angle. In the *planetary gear* structure, the central gear (the sun gear) drives several pinions (the planets). This structure permits the simultaneous drive of multiple loads.

Gears are subject to wear, which results in backlash, and backlash creates positional errors. When gears are made of metal, they may present a considerable inertial load to the motor. Gears are also expensive. These are some of the negative features associated with gears. On the positive side, gears offer an alternative for transmission of mechanical power as well as a means of transforming speed and torque. Gears are particularly useful in driving loads with unusual axial orientation in relation to the motor shaft.

2–15 LINEAR VARIABLE DIFFERENTIAL TRANSFORMER

The *linear variable differential transformer* (LVDT) is a component that converts the translational motion to an equivalent electrical signal. As shown in Fig. 2–59b, it has linear transfer characteristics, where k is the slope of the straight line. The value of k then becomes the constant transfer function of the linear block model of LVDT.

The operation of LVDT may be described from its construction diagram, shown in Fig. 2–59a. There are three windings, which are on a common form: the primary winding, and the two secondary windings W_1, W_2 connected in series. Since the sense of W_1 is opposite to that of W_2, the voltage V_1 induced in W_1 is 180° out of phase with that of V_2. The magnetic core, which is free to move inside the form, increases the permeability of the region inside the form where the core is present; that is, it concentrates or increases the number of magnetic flux lines that cut the secondary winding. It is therefore possible to vary the coupling between the primary and secondary windings by adjusting the position of the core. As shown in the diagram, the core is centered to provide equal coupling to both secondary windings. Since W_1 and W_2 have the same number of turns, the voltages induced into the secondary windings, V_1 and V_2, are equal in magnitude and differing in phase by 180°, resulting in the output voltage $V_o = 0$ V.

By moving the core in the $+X$ direction reduces the coupling to W_2, resulting in $V_2 < V_1$ and a positive output. The coupling is reduced to W_1, on the other hand,

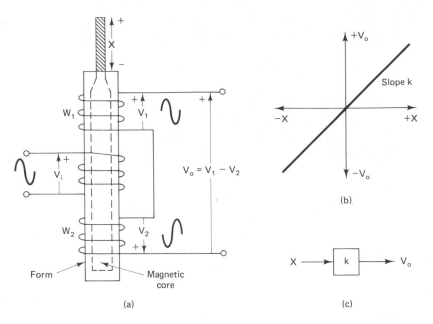

Figure 2–59. LVDT: (a) construction; (b) transfer characteristics; (c) linear block model.

when the core is moved in the $-X$ direction, resulting in $V_1 < V_2$ and a negative output.

As stated before, the LVDT converts the translational motion of the core to an analog voltage in a linear manner. The motion of the core is generally somewhat limited to approximately $\pm\frac{1}{2}$ in. The LVDT is a very sensitive device, being capable of detecting positional changes smaller than 0.01 in.

2–16 ACCELEROMETER

An *accelerometer* is a configuration of several components which convert acceleration to an electrical analog voltage. From an overall viewpoint, an accelerometer may be regarded as a transducer that converts the mechanical energy associated with motion to the electrical form. It may be used, as an instrument, to monitor changes in velocity (i.e., acceleration) due to shock, vibration, or impact.

An accelerometer may be configured in several ways. A typical structure that uses an LVDT is shown in Fig. 2–60a. It includes a cantilever spring, fixed to the support on one end; a weight attached to the free end; and an LVDT, whose magnetic core is positioned directly below the free end of the spring. The accelerometer housing is bonded to the mechanical member being accelerated. The upward acceleration, as shown, results in the downward reaction force due to the mass W/g. The cantilever spring, which is being deflected in the downward direction a distance X (the magnitude of the displacement being proportional to the spring constant k_s) produces an upward

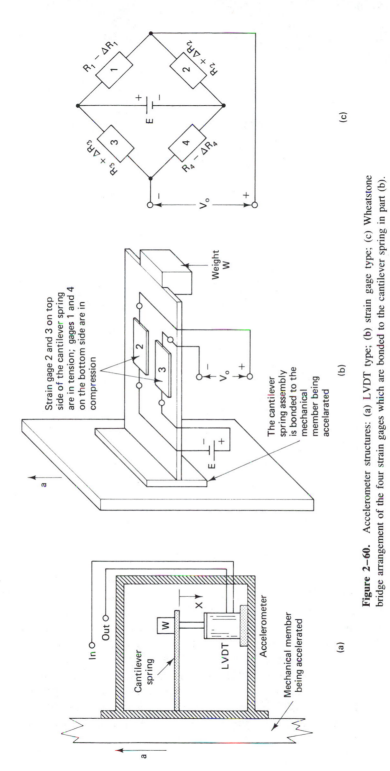

Figure 2–60. Accelerometer structures: (a) LVDT type; (b) strain gage type; (c) Wheatstone bridge arrangement of the four strain gages which are bonded to the cantilever spring in part (b).

97

reaction force equal to k_sX. In the absence of other external forces, these two forces must balance each other, that is,

$$\frac{W}{g} a = k_s X \qquad (2-60)$$

and the acceleration when normalized to g may be expressed as

$$a_n = \frac{a}{g} = \frac{k_s}{W} X \qquad (2-61)$$

since the electrical output of LVDT may be expressed as $k_L X = V_o$, Eq. (2–61) may be expressed in terms of the LVDT parameters as

$$a_n = \frac{k_s}{W k_L} V_o \qquad (2-62)$$

Equation (2–62) shows that the acceleration is linearly related to the output voltage V_o. The coefficient $k_s/W k_L$ determines is the slope of the straight line that results when a_n is plotted versus V_o.

Figure 2–60b shows another approach to the accelerometer configuration. This structure uses four strain gages; two are bonded to the top side of the cantilever beam, and two are bonded to the bottom side. As the unit is given an upward acceleration, gages 2 and 3 on the top side are subjected to tension as the cantilever beam is deflected downward, and their resistance increases due to the stretching of the gage wire. At the same time, gages 1 and 4 are subjected to compression, their resistance decreasing because the compression causes the gage wire diameter to increase. As shown in Fig. 2–60c, the four gages are connected in the Wheatstone bridge configuration. To get the maximum output from the bridge, the opposite arms of the bridge contain gages which are either in tension or in compression. If all gages are identical (i.e., their nominal resistance is R and their resistance change is ΔR), then, by Eq. (2–11), the bridge output is

$$V_o = \frac{E \Delta R}{R} = \epsilon E G \qquad (2-63)$$

where G is the gage factor and ϵ is the gage strain. The bridge output is thus proportional to the change in the gage resistance, which is, in turn, proportional to the cantilever spring deflection, and the deflection of the spring depends on the value of the acceleration. Consequently, the output of the bridge may be used to represent the acceleration.

The accelerometer structure, based on discrete strain gages which are bonded to the cantilever spring, may be extended to the solid-state monolithic beam structure, which contains piezoresistive strain-sensing elements. A structure of this type is shown in Fig. 2–61a, where the two dark patterns represent the diffused strain gages. Each strain-sensing element (318 mils long and 1 mil wide) is achieved by diffusing boron, p-type material, through the use of the photolithographic process into n-type silicon

Control System Components   Chap. 2

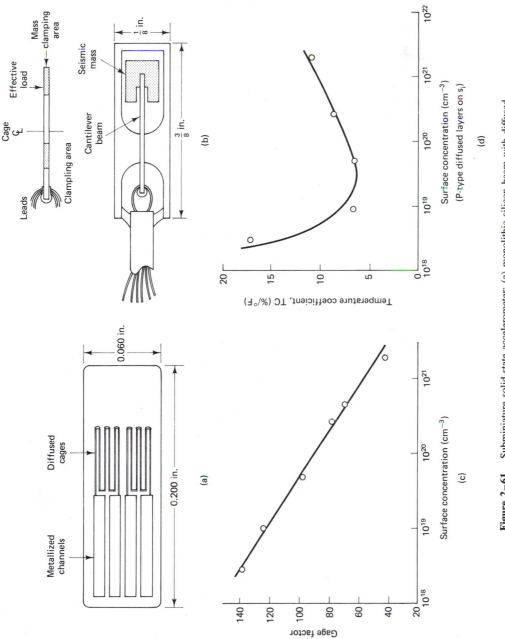

Figure 2–61. Subminiature solid-state accelerometer: (a) monolithic silicon beam with diffused piezoresistive strain gages; (b) accelerometer components; (c) gage factor variation with surface concentration of p-type diffused layers on silicon; (d) strain gage TC variation (percent per 100°F) with surface concentration of p-type diffused layers on silicon. (Courtesy of Joseph R. Mellon, Jr., Kulite Semiconductor Products, Inc., Ridgefield, N.J.)

substrate, which serves the function of the cantilever beam. This type of diffusion is done routinely today in the fabrication of the IC chips. There are 318 squares in each diffused resistive pattern to achieve a target resistance of 500 Ω, resulting in the sheet resistance of approximately 1.6 Ω/square (500/318).

The illustration shows two strain gages diffused into the *n*-type silicon beam. Two more are also diffused into the back side of the beam, and the four gages are interconnected in a Wheatstone bridge configuration. The metallization is used to provide electrical connections. The four diffused strain gages are physically an integral part of the beam structure, but electrically they are isolated from each other and from the beam by the *p-n* junction which may be biased to form the reverse-biased diode. The reverse-biased diode is a common approach used in IC technology to isolate components on the same chip. The solid-state accelerometer configuration and its components are shown in Fig. 2–61b.

The operation of the solid-state accelerometer depends on the piezoresistive effect. According to this effect, when a crystalline material is stressed, it exhibits a change in resistance. A doped silicon material also exhibits this type of resistance change under stress. In a typical measurement application the solid-state accelerometer shown in Fig. 2–61 is clamped at the point shown to the mechanical member being accelerated. The acceleration causes the mass attached to the free end of the silicon beam to bend the beam slightly and thus produce the stress in the strain-sensitive elements. The strain gages change resistance, unbalancing the bridge, thus producing an output signal that is proportional to the beam deflection, which, in turn, depends on the value of the acceleration.

The bridge output depends on the value of the gage factor G, which is a strong function of the doping level or the surface concentration, as shown in Fig. 2–61c. For a doping level of 10^{21} cm^{-3}, for example, the value of G is about 56, resulting in the bridge output of 20 mV for a 5-V input and a 100-g load. The beam deflection under these conditions is less than 0.6 mil (0.0006 in.). The surface concentration also affects the TC of the diffused strain gage as shown in Fig. 2–61d. The equation for the bridge output as stated in Eq. (2–63) also applies in this case.

There are several advantages associated with the solid-state accelerometer described here. First, it is very small. The housing used for the accelerometer shown in Fig. 2–61 occupies a volume of 10^{-2} cm^3 and weighs less than 0.5 g without the leads and mounting. Small accelerometers are generally used in dynamic and static testing of mechanical members which themselves are very small. In such applications it is therefore important that the size and weight of the accelerometer has a minimal effect on the apparent measurement. Second, since the diffused sensors are an integral part of the silicon beam structure, a better thermal stability can be realized. Very good thermal sensitivity is also obtained through adjustments in the doping level. The diffusion of the strain-sensitive elements into the silicon beam eliminates the need for the cements and adhesives that would typically be used to bond the strain gage to the force-sensing member, such as the cantilever spring in Fig. 2–60b. This eliminates the source of hysteresis and creep, and also the apparent strain, which is due to the mismatch of the thermal coefficients that may exist between the strain gages and the surface to which they are bonded.

A *potentiometer* can function as a transducer of mechanical motion. Shown in Fig. 2–62a is a rotational-type potentiometer. The angular displacement θ of its wiper arm is related to the output voltage V_o through the potentiometer constant K_p determined as follows:

$$K_p = \frac{E}{\theta_a} \qquad \text{V/rad or V/deg} \qquad (2-64)$$

where θ_a is the total active angular displacement of the wiper arm; it is equal to 360° minus the dead zone, which is the angular separation between outer terminals of the pot. As shown in Fig. (2–62), when the potentiometer is modeled by a linear block, K_p represents its transfer function such that $V_o = K_p\theta$.

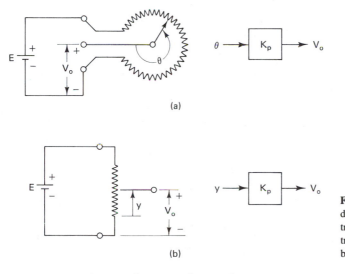

(a)

(b)

Figure 2–62. Potentiometer as the transducer of motion: (a) rotational motion transducer and its linear block model; (b) translation motion transducer and its linear block model.

A potentiometer whose wiper arm executes motion in the straight line is shown in Fig. 2–62b. This configuration may serve as a transducer of translational mechanical motion, in contrast to the potentiometer just described, which serves as a transducer of rotational motion. The potentiometer constant, in this case, may be determined from

$$K_p = \frac{E}{L} \qquad \text{V/unit length} \qquad (2-65)$$

where L represents the total active translational motion of the wiper arm.

Linearity and accuracy are among some of the important characteristics of interest and concern to both the manufacturer and the user. The ideal potentiometer transfer characteristics are shown as a straight line in Fig. 2–63b. Load is not included. The actual transfer characteristic without the load deviates from the straight line. The deviation or difference between the theoretical and actual responses expresses the degree of linearity. The same deviation may also be used to express the error. In Fig. 2–63b the

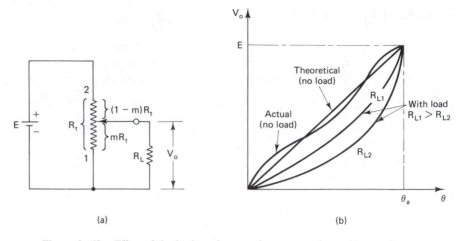

Figure 2–63. Effect of the load on the potentiometer transducer: (a) potentiometer circuit with the load; (b) effect of the load on the potentiometer transfer characteristics.

straight line is fixed at both ends and passes through the origin, which along the abscissa marks the beginning of wiper arm travel. In this case the *absolute linearity* is the maximum deviation of the actual output from the theoretical straight line. It is expressed as a percent of the applied voltage E and may be measured at any angle θ within the range $0 < \theta < \theta_a$.

The unloaded output of a potentiometer may easily be shown to be equal to mE, where m is the ratio of the resistance between the wiper arm and end 1 to the pot resistance R_t. When the load R_L is applied then from Fig. 2–63a, the loaded output may be expressed as

$$V_{oL} = \frac{mE[R_t R_L/(mR_t + R_L)]}{mR_t R_L/(mR_t + R_L) + (1 - m)R_t}$$

and reduced to

$$V_{oL} = \frac{mE}{1 + m(1 - m)(R_t/R_L)} \tag{2–66}$$

As shown in Fig. 2–63b, the loaded response is nonlinear, and its deviation from the ideal straight-line response increases with decreasing R_L. It may also be noted in Eq. (2–66) that for $R_L \gg R_t$, the second term in the denominator may be ignored in relation to the unity resulting in the unloaded output. The deviation of the loaded response may be used to express the degree of linearity, and it may also be used to express the accuracy or the error due to loading. Hence the error may be expressed as the difference between the unloaded and loaded responses:

$$e = V_{oU} - V_{oL} = mE\left[1 - \frac{1}{1 + m(1 - m)(R_t/R_L)}\right]$$

This expression reduces to

$$e = E \frac{m^2(1 - m)(R_t/R_L)}{1 + m(1 - m)(R_t/R_L)} \qquad (2-67)$$

In the case when $R_L > R_t$, Eq. (2–67) reduces to the approximate form

$$e \doteq Em^2 (1 - m) \frac{R_t}{R_L} \qquad (2-68)$$

The maximum error occurs when $de/dm = 0$, and the value of $m = \frac{2}{3}$ under those conditions. Thus the maximum error occurs when the wiper arm is set to approximately 67% from end 1. Equation (2–67) must be used to obtain the exact value of the maximum error e.

The resolution of a potentiometer is defined as the smallest change in the output voltage that corresponds to the smallest increment in the wiper arm motion. For example, a wire-wound pot with 1000 turns and 10 V dc across it has 10-mV resolution, which corresponds to the voltage increment per turn. The resolution of potentiometers, which use a resistive film instead of wire, is infinite.

Figure 2–64a shows a 10-turn potentiometer. In this configuration the wiper arm moves along a helix of the warm gearlike structure. With this configuration it is possible to advance through an electrical angle of 3600°. Potentiometers of this type with up to 100 turns are available. Other commercial potentiometers are also shown in Fig. 2–64.

(a)　　　　　　　　　　　　　　　　(b)

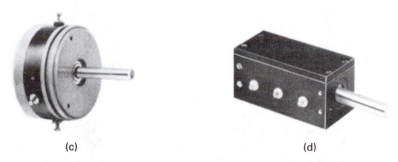

(c)　　　　　　　　　　　　　　　　(d)

Figure 2–64. Commercial potentiometers: (a) model 20010, 10-turn infinite resolution, 5-500 kΩ, 4 W; (b) model 7813, 10-turn, infinite resolution, 5 to 500 kΩ, 2 W; (c) model 105, single-turn, infinite resolution, 1 to 250 kΩ, 2 W; (d) model 111, linear motion, infinite resolution, 500 Ω/in. to 125 kΩ/in., electrical stroke to 48 in., 1 W/in. (Courtesy of Vernitech, Division of Vernitron Corp., Deer Park, N.Y.)

SUMMARY

1. Temperature-sensing components include the thermistor, resistive sensor, thermocouple, and IC chip. The thermistor (negative TC) and the resistive sensor (positive TC) produce a change in their resistance due to a change in temperature. The change in resistance may be converted to an analog voltage by the Wheatstone bridge and amplifier configuration shown in Fig. 2–2. The thermocouple is a junction of two dissimilar materials. The junction potential thus produced varies linearly with temperature; it may be amplified and displayed to represent an unknown temperature. The cold junction that results when test leads are connected to the thermocouple must be compensated to avoid measurement errors. IC-chip sensors, whose operation is based on a temperature-dependent current source (typical configuration shown in Fig. 2–7), offer a relatively inexpensive and accurate temperature-sensing alternative.

2. A strain gage converts a stress-induced strain to a change in resistance, which in turn is converted to the electrical form by means of a Wheatstone bridge. Strain gages may be made of wire or a semiconductor material. Semiconductor strain gages depend on piezoresistive effect for their operation.

3. Light may be sensed by a phototransistor, photodiode, photo Darlington, photo SCR, photo triac, or photoconductive cell. In all cases, the photosensor produces a current that is proportional to the frequency and the intensity of incident irradiation. The photodiode and the phototransistor are used in low-power applications, whereas the photo SCR and photo triac are intended for high-power applications. The photo Darlington, also used in low-power applications, produces a large current gain, as do Darlingtons in general. The photoconductive cell differs from other photosensors in that it changes its resistance in response to incident light.

4. An optocoupler includes an LED radiation source and a photosensor. It is generally packaged in a DIP case and its main feature is the electrical isolation between the source and the load. An optointerruptor device (OID) is electrically the same as the optocoupler, differing only in the package design. The OID package contains a slot intended for the optical wheel (used in speed- and position-sensing applications), which rotates partially inside the slot, interrupts the LED light beam, and produces pulses at the output of the OID for processing.

5. A Hall effect device (HED) is an effect that occurs in all metals but is more pronounced in semiconductors. This effect depends on the current through the device and the externally applied magnetic field. The Hall voltage, which is in quadrature with the current flow, depends on the values of the current and the magnetic field.

6. A digital-to-analog converter (DAC) transforms a digital signal to analog form, and an analog-to-digital converter (ADC) transforms an applied analog voltage to digital form. In contrast to the DAC, which is based on the R-$2R$ ladder, there are many ADC configurations and many of them contain a DAC. ADCs vary in complexity and performance, ranging from the simultaneous ADC, which contains an extremely complex coding section and is the fastest, to the successive approximation ADC, which is not as complex and is probably the most popular.

7. A frequency-to-voltage converter (FVC) produces an analog signal proportional

to frequency, and a voltage-to-frequency converter (VFC) produces an output whose frequency varies linearly with the amplitude of the applied voltage.

8. Shaft position encoders produce a unique digital signal corresponding to the position of the disk, and therefore to the position of the motor shaft that holds the disk. The code deposited on the surface of the disk may be sensed by a physical contact or optically by an LED–photosensor arrangement.

9. A solenoid is an electromechanical device that converts the applied electrical current to the motion of the plunger, which performs mechanical work.

10. Gears are used to transmit mechanical power and to transform speed and torque. The spur gear is the simplest in construction, but others, including bevel, worm, helical, planetary, and so on, are considerably more complex in their design.

11. A linear variable differential transformer (LVDT) does the reverse of what the solenoid does—it converts the translational motion of its core to an electrical signal.

12. An accelerometer is a component that converts translational acceleration to an equivalent electrical signal. The accelerometer configuration may be based on the use of LVDTs or on the use of a strain gage–Wheatstone bridge arrangement. The solid-state accelerometers are manufactured on a chip using the IC monolithic technology. The operation of the IC accelerometers depends on the piezoresistive effect.

13. A potentiometer converts mechanical rotational or translational motion to an electrical signal. The potentiometer thus offers an alternative in position-sensing applications.

QUESTIONS

2–1. A temperature change may be responsible for measurement errors when a remote sensor is used in conjunction with the Wheatstone bridge. Explain how temperature compensation is achieved in the case of the remote thermistor and a strain gage. How is the thermocouple cold junction compensation achieved?

2–2. Compare and contrast the electrical characteristics and the performance, advantages, and the disadvantages of thermistors, resistive sensors, thermocouples, and IC sensors.

2–3. Explain the piezoresistive effect and relate it to strain gage applications.

2–4. Explain the operation of a phototransistor using concepts from physics, optics, and electronics.

2–5. How does the value of R_L affect the bandwidth and gain of a phototransistor? Is it possible to achieve a large bandwidth and a large value of the output simultaneously? Explain.

2–6. Describe the operation of an OID and its application in speed and position sensing. Describe the operation of the circuit that senses the direction of shaft rotation, and explain why this circuit plays an indispensible role in position control that uses optical sensing of shaft position.

2–7. Describe all fields and the sources of such fields, and all forces acting on the current that take part in the operation of an HED.

2–8. Compare and contrast the advantages and disadvantages, from the standpoint of operation, applications, and circuit complexity, of the following ADCs: simultaneous, monotonic, tracking, single- and dual-slope, and successive approximation.

2–9. Compare and contrast the operation and applications of incremental (two types) and absolute shaft position encoders. Include in your discussion the use of various codes, and their advantages and disadvantages.

2–10. Compare and contrast the operation, construction, characteristics, applications, advantages and the disadvantages of strain gage, solid-state, and LVDT accelerometers.

2–11. Discuss the effect of loading on potentiometer linearity and accuracy.

2–12. Why does the force generated by the solenoid decrease as the length of stroke increases?

PROBLEMS

2–1. The thermistor is subjected to a temperature change from 75°C to 25°C. Its resistance at 25°C is 510 Ω and its TC = −400 ppm/°C. Calculate the thermistor resistance at 75°C.

2–2. In the circuit shown in Fig. P2–2, R_a represents the active strain gage, whose unstressed resistance is 500 Ω and the gage factor is 10. The bridge is first balanced, then the active gage is subjected to a compression resulting in a 1% length reduction of its wire.
  (a) Calculate the value of R if the magnitude of V_{o1} is to be 1 V. Determine the polarity of V_{o1}.
  (b) Repeat part (a) for V_{o2}. Ignore V_{o1}.

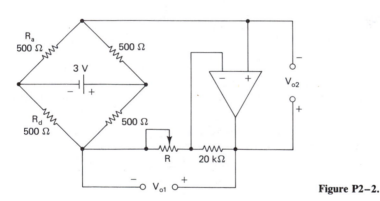

Figure P2–2.

2–3. Using the semiconductor strain gage sensitivity curve shown in Fig. 2–11a, determine the approximate value of the gage factor.

2–4. If V_o = 6 Vp-p in Fig. 2–28a, calculate the peak value of the ac input current. Assume the dc value of diode current remains at 10 mA

2–5. Calculate the new value of the input pulse voltage in Fig. 2–28b so that the amplitude of the output pulse is 1 V.

2–6. In the optical motor speed-sensing system shown in Fig. P2–6a, calculate **(a)** ω_m. **(b)** the value of the transfer function G shown in Fig. P2–6b, which represents the system in Fig. P2–6a.

106   Control System Components   Chap. 2

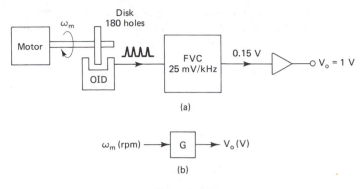

(a)

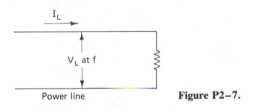

(b)

Figure P2–6.

2–7. Given the power line shown in Fig. P2–7, where I_L is the line current and V_L is the line voltage at frequency f, determine the block diagram configuration which shows how an HED may be used to monitor the power on the line.

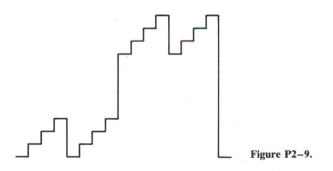

Power line   **Figure P2–7.**

2–8. Given an n-bit DAC using the inverting R-$2R$ resistive ladder, and $V_r = 10$ V, if the resolution must be less than 1 mV, how many bits must the ladder have?

2–9. When a binary n-bit counter is applied to an n-bit DAC and the counter advances through its full-scale count, the DAC output drawn to scale is as shown in Fig. P2–9.
 (a) Determine the value of n.
 (b) Assuming that the counter works, which DAC bits are not properly operating?

Figure P2–9.

2–10. Determine the coding section logic diagram for a 3-bit simultaneous ADC.

2–11. The output V_a in Fig. 2–39a rises monotonically at the rate of 20.48 V/ms; $f_c = 2048$ kHz, and the average conversion rate is 4000 s^{-1}. At coincidance the counter stores 750_{10}. Calculate (a) V_i; (b) quantizing error in mV; (c) V_r; (d) number of bits.

2–12. In the dual-slope converter shown in Fig. 2–42, $C = 0.005$ µF, $V_r = 2.5$ V, $f_c = 500$ kHz, and the counter stores 0101101000_2 at the end of conversion (Fig. P2–12).
  (a) Calculate the value of R.
  (b) Calculate the value of V_i.
  (c) Sketch I_c over the time interval $T_1 + T_2$ and show all important values.

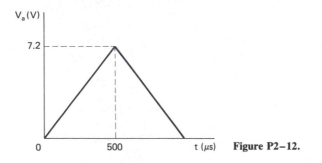

Figure P2–12.

2–13. In the configuration shown in Fig. P2–13, $P_m = 0.1$ hp. Calculate **(a)** ω_m; **(b)** load torque; **(c)** motor torque.

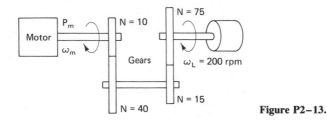

Figure P2–13.

2–14. A 10-vdc supply is across a 100-kΩ single-turn rotary potentiometer with 2000 turns, and the value of R_L is also 100 kΩ. Calculate **(a)** the potentiometer setting m so that $V_{oL} = 5$ V dc; **(b)** the potentiometer resolution in mV; **(c)** the maximum error in mV; **(d)** the angle advanced by the potentiometer rotor if $V_{oL} = 4.2$ vdc.

LAB EXPERIMENT 2–1
OPTICAL SENSING OF MOTOR SPEED

Objective. To determine experimentally the transfer characteristics of components used in motor speed sensing by optical means. These components include the optical disk, the OID, and the FVC.

Equipment

Frequency counter
Digital voltmeter

Materials. As shown in the test circuit.

Test system. The circuit shown in Fig. 8–11 is used in this experiment and operated as an open-loop system. To configure this system as an open-loop system, switch S_1 is disconnected, the output of OA$_4$ is fed to the input of OA$_1$, and the 1-kΩ resistor between the $-$ input of OA$_2$ and S_1 is grounded.

Procedure

1. Set the gains of OA$_1$ and OA$_3$ to 5 by adjusting the respective feedback pots. Apply a positive dc voltage to the input V_i for the motor speed of 500 rpm as measured on the frequency counter at the output of OA$_6$. Measure and record V_i, f, W_m, and V_o. Note that W_m is calculated (in rpm) from frequency measurements, and V_o is measured at the output of OA$_1$. All dc voltage measurements are to be done on a digital voltmeter.

2. Repeat step 2 for motor speeds of 750 to 2500 rpm in steps of 250 rpm. Record all data as required.

Test data. The format for the test data and the calculated results, as well as the report format, must comply with and meet the requirements of your lab instructor.

Evaluation

1. Plot V_o versus W_m using the appropriate graph paper.
2. Plot V_o versus f using the appropriate graph paper.
3. Determine the values of the slopes in steps 1 and 2 above by the least-squares method of fitting data to a straight line.
4. Explain the significance of the slopes obtained in step 3.
5. Is the value of the slope for the graph obtained in step 2 within the FVC manufacturer's spec limit? If not, explain any discrepencies through additional experimentation.
6. Draw the linear block models and include the values of the transfer functions for the following: (a) the combined effect of the optical disk and the OID; (b) FVC; (c) the combined effect of the optical disk, OID, FVC, OA$_4$, and OA$_1$.
7. By means of a block diagram, suggest a practical application for the system in this experiment.

LAB EXPERIMENT 2–2
OPTICAL SENSING OF ANGULAR POSITION

Objective. To investigate the operational characteristics of a circuit that senses optically the angular position of motor's shaft. The sensing circuit includes an optical disk, OID, binary counter, DAC, and the direction-sensing circuit.

Equipment

Digital voltmeter
Dual-trace oscilloscope

Materials. As shown in the test circuit.

Test system. The position control system shown in Fig. 8–12 is used in this experiment and operated as an open-loop system. To configure this system as an open-loop system, the feedback path at the output of OA_4 must be broken.

Procedure

1. Adjust the gain of OA_4 to 1 by the 20-kΩ feedback gain-adjust pot.
2. Apply a dc voltage to V_i, causing the motor to rotate continuously. Display the two OID outputs on a dual-trace oscilloscope. If necessary, adjust the position of one of the OIDs so that two OID waveforms are approximately in quadrature phasing.
3. Reset the binary counter to all 0's and make any necessary adjustments to ensure that the output of the DAC at pin 4 is 0 vdc, and that V_f is also equal to 0 vdc.
4. Construct a circuit using buffers or latches and LED miniature bulbs which will provide a visual indication of the binary state of the counter. This circuit will be helpful in the test steps that follow.
5. With the motor not energized electrically, advance the optical disk by hand in the CW direction without missing steps from the counter state 0_2 to 50_2. Measure and record at each counter state the value of OA_4's output V_f (1% or better measurement accuracy).
6. Advance the optical disk manually in the CCW direction without missing steps from counter state 50_2 and 30_2. Measure and record the value of V_f (with 1% or better measurement accuracy) at each state of the counter. The output of the D flip-flop should be complemented in this step, thus reversing the direction of count.
7. Advance the optical disk manually in the CW direction without missing steps from 30_2 to 100_2. Measure and record the value of V_f (with 1% or better measurement accuracy) at each state of the counter.

Test data. The format for the test data and the calculated results, as well as the report format, must comply with and meet the requirements of your lab instructor.

Evaluation

1. Plot the data obtained in steps 5 to 7 of the Procedure on a graph as suggested in Fig. LE2–2. All steps must have equal time duration, but the duration used for each step is not critical.

Control System Components

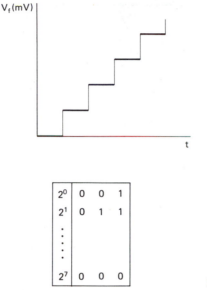

2^0	0	0	1
2^1	0	1	1
.			
.			
.			
.			
2^7	0	0	0

Figure LE2–2.

2. Using theory, calculate the value of system resolution.

3. Write software for IBM PC in BASIC, and obtain hard copies of the programs and graphs for the following: (a) system accuracy for all data points in steps 5 to 7 of the Procedure as a percent of the full-scale output V_f; (b) differential linearity as a percent of step height for the three data groups (i.e., monotonically increasing data in step 5, decreasing in step 6, and increasing again in step 7). (c) *Optional*: The graph required in step 1 of this Evaluation may be generated on the computer.

4. From the software results in step 3, determine (a) maximum error; (b) maximum differential nonlinearity.

5. Are the results of step 4 within the DAC manufacturer's spec limits? Any discrepencies must be explained and justified through additional experimentation.

6. Represent by means of a single linear block the effect of all components between the motor's shaft and OA_4 (the input to the linear block is θ_o and its output is V_f). Include in the linear block the value of its transfer function.

7. By means of a block diagram, suggest an application for the system in this experiment (the feedback path may be reconnected for purposes of illustration).

8. Explain the meaning and the difference between the accuracy and the resolution.

9. What is the value of the ratio of resolution in mV to maximum error obtained in this experiment?

10. Should the ratio in step 9 be < 1, $=$, 1, or > 1? Explain. What do manufacturers of DACs generally like this ratio to be?

MOTORS

chapter 3

3-1 INTRODUCTION

Motors are transducers that convert energy from electrical form to mechanical form. The electrical current supplied to the motor causes the motor shaft to rotate. Depending on the coupling arrangement between the load and the motor shaft, the load will undergo a prescribed motion which may be rotational or translational or perhaps some other type of irregular motion, especially if cams are used as part of the coupling arrangement. Regardless of the type of motion, the torque and the speed supplied to the load may be converted to work and to mechanical power.

As a source of mechanical power, motors are used in a broad range of applications, too numerous to be described here. The examples of motor applications that follow provide a very modest sample indeed. Large power motors, for instance, are used in heavy industrial machinery, in electrical trains, or in general to position or move heavy loads. Appliance manufacturers use motors in refrigerators, air conditioners, washing machines, and fans, and tool manufacturers use motors in a variety of power tools, such as saws and drills. The automobile industry uses dc motors to operate power equipment, such as power windows. Small dc motors are also used in consumer applications such as VCR or casette tape drives, 35-mm film advances, movie projectors, and many others. The computer industry uses motors in applications such as read/write head positioning in disk drives, printers, and other peripherals. Motors also play a dominant role in automatic control systems such as speed or position control servos.

Motors vary greatly in the design of the rotor and the stator; and in the form of electrical power that must be supplied—ac or dc. The generation of magnetic field is a common feature that all motors share and depend on for their operation. Some motors use a permanent magnet in their construction and others use an electromagnet. An electromagnet requires a separate winding, generally referred to as the field winding, and a power supply, which provides the field current for the production of the magnetic field. In some designs the same power supply is used to supply current to both the armature and the field winding.

The electrical characteristics and performance also vary from one motor type to another. The type of motor used depends mainly on the application. An ac motor, for instance, may be quite satisfactory for driving an electric fan at constant speed. It may not perform as well, however, in a control system, where it may operate at a low speed and smooth control of its speed is required. In that case a dc motor is more suitable. In the past ac motors have been used predominantly in industrial and consumer applications. Since World War II, however, when control engineering has become a formal discipline, more and more control systems have been designed using dc motors, and today the dc motor represents one of the most important control system components. Over the years, extensive research on magnetic materials and the design of low-inertia armatures have been aimed at maximizing the torque-to-inertia ratio and minimizing the motor response. As a result, current designs of permanent-magnet dc torque motors (some are brushless) offer optimum output torques per unit motor volume, with response times in the low-millisecond range.

Stepper motors, another class of motors, operate in contrast to all other types of motors, in discrete angular steps of the shaft in response to the input pulses, one step per input pulse. Stepper motors can provide very accurate positioning of the load since there is generally a one-to-one correspondence between the angular displacement of the shaft and the applied pulses. Stepper motor technology has been available for decades; however, the use of stepper motors has been limited by a lack of extensive applications. The enormous growth of the computer industry over the past couple of decades has revived interest in the stepper motor as a control component in various peripherals, such as printers and disk drives. Today, there are several types of stepper motor commercially available, and they are used predominantly in digital and computer-based systems.

In this chapter we present the characteristics and operation of the permanent-magnet dc servomotor and several types of stepper motors. The first component to be described is a dc generator. It will serve as a good introduction to the dc motor.

3–2 DC GENERATOR

All motors and generators of electrical power are based on a discovery by an Englishman, Michael Faraday (1791–1867), who found that it is possible to generate electricity by passing a conductor in a magnetic field. His law of electromagnetic induction is possibly one of the greatest discoveries of all time. Ironically, an American, Joseph Henry (1797–1878), made the discovery first, but it was Faraday who published it first and hence was given the credit.

Faraday found that the EMF induced into a conductor that moves in a magnetic field is proportional to the rate at which the magnetic flux changes; hence

$$e = -\frac{d\phi}{dt} \tag{3-1}$$

where the magnetic flux is defined as the integral of the dot product as follows:

$$\phi = \int \mathbf{B} \cdot d\mathbf{A} = \int |B| \cos \theta \, dA \tag{3-2}$$

where B is the magnetic flux density expressed in Wb/m^2. B may be taken over any desired area A, and θ represents the angle between B and the normal to the surface area A. Using the results of Eq. (3-2) in Eq. (3-1) yields

$$e = -\frac{d}{dt} \int \mathbf{B} \cdot d\mathbf{A} \tag{3-3}$$

From Eq. (3-3) it is clear that the induced EMF may be achieved in one of several ways. One way would be to apply an alternating magnetic field to a stationary conductor. A practical application of this is a transformer, where the applied ac current to the primary produces an alternating magnetic flux which induces voltage into the stationary secondary winding. EMF may also be induced by placing a conductor in a constant magnetic field and then changing the area of the conductor or its orientation. Visualize a loop of wire suddenly stretched—its area undergoes a sudden change—or simply a loop of wire that is rotated in a constant magnetic field. In all cases described above, EMF is induced due to a change in the magnetic flux which is linked by the conductor.

The minus sign in Eq. (3-1) gives us a clue to the polarity of the induced EMF. The equation states basically that the induced EMF opposes the flux change that produced the EMF. This is a consequence of the Lenz's law and is in agreement with the conservation of energy. Actually, it is not the induced EMF but rather, the effect of the induced EMF which opposes the original flux change. This is illustrated in Fig. 3-1a. When the conductor is moving in the upward direction (in the direction of increasing flux), the polarity of the induced voltage which gives rise to the current I must be as shown. The reason for this is that the current I, whose direction is as shown, produces the magnetic field, which encircles the conductor in the clockwise direction. As shown in the adjacent diagram, above the conductor the magnetic field due to I and the original magnetic field B are in the same direction, giving rise to the *repulsive force*, and below the conductor the two fields are in opposite direction, giving rise to the *attractive force*. The net result is a force in the downward direction which opposes the upward motion of the conductor, Lenz's law clearly being satisfied. The same result may also be obtained from the current I, field B, and force F_B (which the magnetic field exerts on the moving electron charge); the vector triad representation is as shown. Since $\mathbf{F}_B = -|q_e| (\mathbf{V}_e \times \mathbf{B})$, where the electron velocity V_e is opposite in direction to that of I, the direction of F_B must be downward. In Fig. 3-1b the conductor is moving downward (in the direction of decreasing flux); the polarity of the induced voltage and the current direction is now reversed to produce the field interaction as shown, and the net upward force opposes the downward motion of the conductor, Lenz's law being satisfied once again.

The magnitude of the induced EMF for the conductor following a straight-line path in a uniform magnetic field may be determined from Fig. 3-1 using Eq. (3-3) or (3-1). The magnitude of the induced voltage may then be expressed as

$$e = \frac{d}{dt} (BA \cos \theta) = B \frac{dA}{dt}$$

where B is taken out of the differentiation operation because it is independent of time and $\theta = 0°$ because the conductor moves in a plane perpendicular to B, tracing out an area whose normal is aligned with B. If the length of the conductor is L (this is the part of the conductor that interacts with the field B) and its velocity is V, in time dt it moves a distance $V\,dt$, which is perpendicular to the field, resulting in $dA = LV\,dt$. Substituting this expression for dA in the equation above yields

$$e = \frac{BLV\,dt}{dt} = BLV \qquad (3{-}4)$$

Two conductors may be connected in series in the form of a loop, and then rotated in the magnetic field as shown in Fig. 3–2. Not shown in the simplified diagram

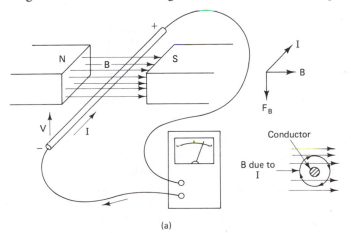

(a)

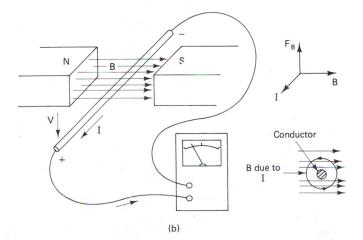

(b)

Figure 3–1. Voltage induced into a moving conductor. The conductor moves in the direction of (a) increasing magnetic flux and (b) decreasing magnetic flux.

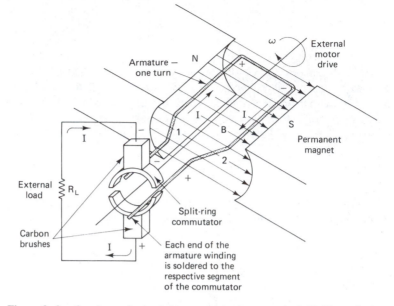

Figure 3–2. One loop of wire being rotated in the magnetic field. The split-ring, carbon brush commutator provides the electrical path for the induced voltage to the load.

is the armature construction, which is typically cylindrical in form, its shaft supported at both ends by ball bearings or bushings, which allow it to rotate freely. The conductors are embedded in the slots, which are cut along the axial direction to keep the conductors from moving around due to forces that act on the conductors during rotation of the armature. The electrical connection to the external load is provided by the commutator, consisting of metallic segments made of copper which are bonded to the tapered end of the armature shaft, and carbon brushes. Spring loading causes the brushes to make pressure contact with the copper segments, thus providing an electrical connection to the load. As the armature rotates, the brushes contact in sequence, the coils of the armature.

The arrangement in Fig. 3–2 uses a one-turn coil which consists of conductors 1 and 2. The two ends of the coil are connected to the two-segment or split-ring commutator. As the armature rotates, the conductors in this case follow a circular path of radius R, in contrast to the straight-line path followed by the conductors in Fig. 3–1.

The voltages induced into conductors 1 and 2 are in series aiding. The polarities of these voltages are consistent with Lenz law requirements since the current through conductor 1 reacts with the magnetic field B, producing a downward force that opposes the motion. Similarly, the current in conductor 2 reacts with B to produce an upward force that also opposes the motion. The sum of the two induced voltages minus the drops due to the armature resistance and brush contact resistance appears across the load.

The two-segment commutator produces a full-wave (FW) rectified waveform as shown in Fig. 3–3b. This can be understood by considering the motion of the loop. When the loop is rotated 180° clockwise from its position as shown in Fig. 3–2, the

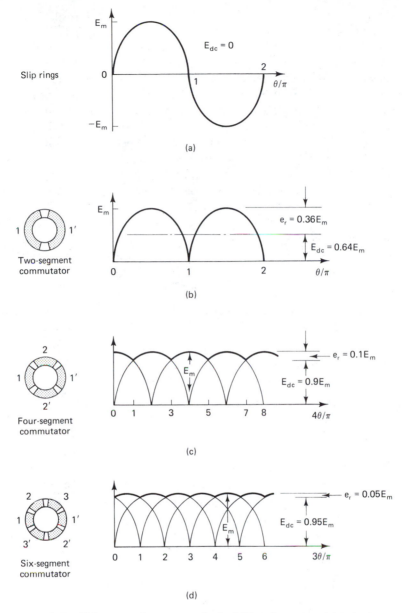

Figure 3–3. Voltage waveforms across the load for various commutator structures: (1) slip rings provide ac output; (b) split-ring or two-segment commutators provide FW rectified waveform; (c) two armature coils wound at 90 mechanical degrees use a four-segment commutator; (d) three armature coils wound at 60 mechanical degrees use a six-segment commutator.

positions of conductors 1 and 2 are reversed. At this time the polarity of the induced voltages, as well as the direction of currents in each conductor, must be reversed to satisfy Lenz's law, which once again demands that the currents due to the induced voltages, in reacting with the magnetic field B, must produce forces that oppose the motion of the loop. As a consequence of this, as the loop continues to rotate, the polarities of the induced voltages and the directions of the currents must be reversed after each 180° of loop's rotation. Therefore, the current through the load R_L never changes direction, resulting in a FW-rectified type of waveform. The expression for the voltage induced in each conductor may be obtained from Eq. (3–4) and Fig. 3–4. As shown in Fig. 3–4, the conductor which is attached to the armature moves in a circular path with tangential velocity $V_t = \omega R$. The component of V_t which is normal to the magnetic field is $V_n = \omega R \sin \theta$, where $\theta = \omega t$ is the angle that the velocity vector V_t makes with the field B. B in Eq. (3–4) is equal to Φ/A, where Φ is the magnetic flux per pole and A is the area over which the magnetic flux is incident. The rotation of the conductor traces out a cylindrical surface whose area must be equal to $2\pi RL$. Substituting these results in Eq. (3–4) yields

$$e_o(t) = \frac{\Phi}{A} LV = \frac{\Phi N_p}{2\pi RL} (L) (\omega R \sin \theta) = \frac{\Phi N_p \omega \sin \omega t}{2\pi} \qquad \text{V/conductor} \qquad (3–5)$$

where N_p represents the number of magnetic poles. This is the instantaneous voltage induced in one conductor. If the armature winding contains N_c conductors and if these conductors are distributed over N_{pp} parallel paths, the instantaneous generator output voltage may be expressed as

$$e_o(t) = \frac{\Phi N_c N_p \omega \sin \omega t}{2\pi N_{pp}} \qquad \text{V/generator} \qquad (3–6)$$

In Eq. (3–6), N_c may be replaced by $2N_t$, where N_t represents the number of turns in the armature coil, as there must always be two conductors per turn. It is important to recognize the units that must be used in Eq. (3–6). If the SI system is used, Φ is in webers and ω is in rad/s. When ω is expressed in rpm, as it often is, Eq. (3–6) must be multiplied by $2\pi/60$. In the case of a simple two-pole generator where the armature winding consists of N_t turns and is distributed over two parallel paths, and ω is in rpm, Eq. (3–6) reduces to

$$e_o(t) = \frac{N_t \Phi \omega \sin \omega t}{30} \qquad \text{V/generator} \qquad (3–7)$$

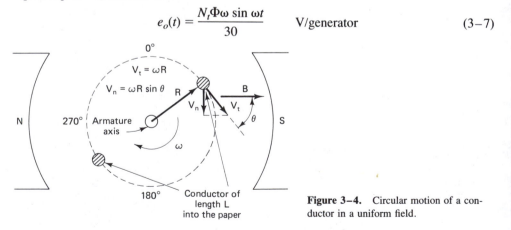

Figure 3–4. Circular motion of a conductor in a uniform field.

If in place of the segmented commutator, slip rings are used which allow the armature coil to be connected to the load throughout the complete rotation of the armature, without switching the conductors of the armature coil from one side of the load to the other as does the two-segment commutator (the construction of the typical slip rings is shown in Fig. 3–5), the output is ac and may be represented by Eq. (3–7). Its graphical representation is shown in Fig. 3–3a, where the peak voltage E_m is the coefficient of $\sin \theta$ in Eq. (3–7). The period in this case is $60/\omega$, where ω is in rpm.

If a two-segment or split-ring commutator is used, Eq. (3–7) may still be used to describe the output waveform, except that $\sin \theta$ must be replaced by the absolute value $|\sin \theta|$ since the current and output voltage never change polarity. In this case the average or dc value of the waveform may be calculated by the standard methods as follows:

$$E_{dc} = \frac{1}{\pi} \int_0^{\pi} E_m \sin \theta \, d\theta = 0.64 E_m \tag{3–8}$$

The peak value of the ripple in this case is $0.36E_m$ or 36% of the peak value E_m. The ripple is defined as the deviation of a given waveform from its dc value.

Suppose that another coil is added to the armature in Fig. 3–2. If it is wound in quadrature (i.e., at 90° in relation to the existing coil), a four-segment commutator may be used with coil 1 connected as shown in the simplified diagram in Fig. 3–3c to two opposite segments, and coil 2 connected to the two quadrature segments. As shown by the waveform in Fig. 3–3c, each phase is alternately connected to the load for approximately 90° of armature rotation. The period of the resulting waveform is one-fourth that of the waveform shown in Fig. 3–3a, and one-half that of waveform shown in Fig. 3–3b. Its dc value is

$$E_{dc} = \frac{2}{\pi} \int_{\pi/4}^{3\pi/4} \sin \theta \, d\theta = 0.9 E_m \tag{3–9}$$

and the ripple in this case is 10% of E_m.

Extending this discussion to a six-segment commutator which accommodates three armature coils, each coil using two parallel path and the angular displacement between two consecutive coils being 60°. The period of the resulting waveform, which is shown by the heavy line in Fig. 3–3d, is one-sixth that of the waveform in Fig. 3–3a, one-third that in Fig. 3–3b, and two-thirds that in Fig. 3–3c. Its dc value

$$E_{dc} = \frac{3}{\pi} \int_{\pi/3}^{2\pi/3} \sin \theta \, d\theta = 0.95 E_m \tag{3–10}$$

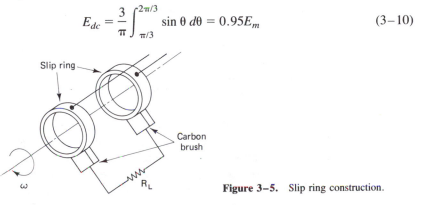

Figure 3–5. Slip ring construction.

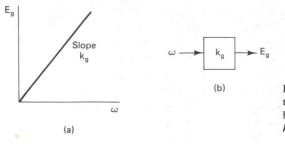

(b)

(a)

Figure 3–6. Dc generator characterization: (a) transfer characteristics; (b) linear block model, where the generator constant k_g is also the transfer function.

and the peak value of the ripple is now 5% of E_m. Thus as the number of coils and the number of segments on the commutator are increased, the output approaches the pure dc value, with the value of the ripple decreasing and the frequency of the ripple increasing. It must be understood that the actual output waveform is not a pure sinusoid; it may be distorted considerably. Consequently, the evaluations of E_{dc} above are at best only approximations.

The peak value of the generator voltage may be expressed from Eq. (3–6) as

$$E_m = \frac{\Phi N_c N_p}{2\pi N_{pp}}\omega \qquad (3–11)$$

where all terms in the coefficient of ω are constant for a selected generator. They may, therefore, be lumped into a single constant k_g, the generator constant; the units of k_g are either V/rpm or V(rad/s). The generator output may then be expressed in general as

$$E_g = k_g\omega$$

The graph of this equation in Fig. 3–6 shows that the generator has linear transfer characteristics, making it possible for the generator to be modeled by a linear block, which is also shown in Fig. 3–6.

3–3 DC MOTOR

We have thus far considered the effects of electromagnetic induction, where a voltage is induced into a conductor that moves in a magnetic field. This is clearly illustrated in Fig. 3–2, where the armature of the generator may, for example, be rotated by a motor, and the induced voltage is applied to the load through the split ring and the brushes of the commutator.

Let us now consider the configuration shown in Fig. 3–7. It does resemble in some respects the generator structure of Fig. 3–2. There are significant differences, however. Among these is the absence of the motor that drives the armature shaft. The armature shaft in this case is free to rotate on its own. Also, the load resistor is replaced by the dc power supply V_a, which supplies the current through the commutator to the armature winding, as shown in the diagram. Clearly, we are dealing here with a different concept, which must first be closely examined.

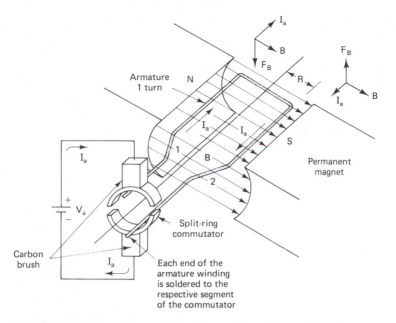

Figure 3–7. Basic dc motor construction. Forces acting on the current-carrying conductors in the magnetic field produce rotational motion of the armature.

According to the Lorentz force expression, a charge that moves in a magnetic field experiences a force

$$\mathbf{F}_B = q\mathbf{V} \times \mathbf{B} \qquad (3-12)$$

where $\mathbf{V}$ is the velocity of the charge and $\mathbf{B}$ is the magnetic field. This is a cross product; hence the force vector $\mathbf{F}_B$ is perpendicular to the plane formed by the $\mathbf{V}$ and $\mathbf{B}$ vectors. The magnitude of the force is given by

$$F_B = |\mathbf{F}_B| = q|\mathbf{V}|\,|\mathbf{B}|\sin\theta \qquad (3-13)$$

where θ is the angle between $\mathbf{V}$ and $\mathbf{B}$. The direction of the force vector $\mathbf{F}_B$ is determined by the right-hand-screw rule. In the configuration of Fig. 3–7 the direction of the force acting on conductor 1 is downward for the direction of current, as shown. The current I_a is due to the flow of electrons in the opposite direction. Therefore, the velocity vector (for the electron flow) is opposite to the direction of I through conductor 1, and the cross product between $\mathbf{V}$ and $\mathbf{B}$ is in the upward direction. When this cross product is multiplied by the negative electron charge, the correct downward direction of the force vector is obtained. The same result is obtained for the direction of the force vector when the I vector is crossed into B. It may easily be verified that the force acting on conductor 2 in Fig. 3–7 is in the upward direction. As these forces act in the direction perpendicular to the plane of the loop, they are responsible for creating *torque*, equal to the product of the force F_B and the perpendicular distance R (R is shown in the diagram as the distance from the conductor to the axis of the armature). The torques acting on conductors 1 and 2 cause the armature to rotate in the CCW direction.

As the armature rotates through 180°, the positions of conductors 1 and 2 are reversed (i.e., conductor 2 is now near the north pole, and conductor 1 is near the south pole). Not only are the conductor positions reversed, but the action of the commutator reverses the polarity of the externally applied voltage from + to − for conductor 1 and from − to + for conductor 2. The switching by the commutator is responsible for reversing the direction of currents in each conductor, and therefore for reversing the directions of forces acting on each conductor. Now the force acting on conductor 2 is in the downward direction and the force on conductor 1 is in the upward direction. The result of all this is that the loop is now able to complete the remaining 180° of the rotation and return to its original position, where the commutator switches the directions of currents once again. The switching action of the commutator plays an important role in the operations of the motor, making possible continuous rotation of the motor armature. If slip rings of the type shown in Fig. 3–5 were used instead of the split-ring commutator, the armature would simply rock back and forth, never being able to complete one revolution.

The expressions for force and torque may be determined from Eq. (3–12). Substituting dX/dt for $\mathbf{V}$, where X represents the displacement along the conductor,

$$d\mathbf{F}_B = q_e \frac{d\mathbf{X}}{dt} \times \mathbf{B} = \frac{dq_e}{dt}(d\mathbf{X} \times \mathbf{B}) = I_a d\mathbf{X} \times \mathbf{B} \tag{3–14}$$

where $I_a = dq_e/dt$ is the fundamental definition of current as the flow of electrons. The force expression may be obtained by integrating Eq. (3–14):

$$\mathbf{F}_B = I_a \left(\int_{x_1}^{x_2} d\mathbf{X} \right) \times \mathbf{B} = I_a(\mathbf{L} \times \mathbf{B}) = L(\mathbf{I}_a \times \mathbf{B}) \tag{3–15}$$

where I_a and B are taken outside the integration operation because they are constant, and the result of integrating dX is $x_2 - x_1 = L$, the length of the conductor. The magnitude of the force on the conductor may be expressed from Eqs. (3–14) and (3–15) as

$$|\mathbf{F}_B| = BLI_a \tag{3–16}$$

In Eq. (3–16), $\sin \theta$ is replaced by 1 because the conductor makes a 90° angle with the field B. The direction of the force may be obtained by crossing $\mathbf{I}_a$ into $\mathbf{B}$ (the direction of I_a is along L).

In determining the expression for the torque, Eq. (3–16) is used and the mechanics of substituting for various quantities may be duplicated from eqs. (3–5) and (3–6). That is, $B = \Phi/A$, where Φ is the magnetic flux per pole and A is the cylindrical surface area of the armature (i.e., the area over which the flux is incident), and also the path of the conductor during its rotation. The resultant torque expression, which is the product of the force F_B and the perpendicular distance R may then be expressed as

$$T_m = R|\mathbf{F}_B| = RBLI_a = RL \frac{\Phi N_p N_c}{2\pi RLN_{\text{pp}}} I_a \tag{3–17}$$

$$= \frac{\Phi N_p N_c}{2\pi N_{\text{pp}}} I_a$$

where the flux Φ is in Wb (webers), the torque in N·m (newton-meters), and current in amperes, which is consistent with the SI system of units. In this book the English system of units is used to express the motor-developed torque T_m in oz-in. Some adjustment of units is therefore necessary in Eq. (3–17). As 1 ft-lb = 192 oz-in. = 1.356 N·m, then 1 N·m = (192/1.356) = 141.6 oz-in. In the English system flux is expressed in lines, 1 Wb = 10^8 lines. Combining these conversion constants, the multiplying factor in Eq. (3–17) must be $(141.6/2\pi)10^{-8} = 2.25 \times 10^{-7}$, resulting in

$$T_m = \left[(2.25 \times 10^{-7}) \frac{\Phi N_p N_c}{N_{pp}} \right] I_a \tag{3–18}$$

where torque is in oz-in., flux in lines, and current in amperes. If the expression in brackets is replaced by the motor torque constant k_i, the motor developed torque may be expressed by

$$T_m = k_i I_a \tag{3–19}$$

where T_m is in oz-in. if k_i is in oz-in./A and I is in amperes. As shown in Fig. 3–8, the motor has linear torque transfer characteristics. The torque developed by the motor depends on the value of k_i, which is a constant for a selected motor, and the value of the armature current, which of course is externally adjustable. In this respect the motor may be regarded as the current-to-torque transducer.

The similarity between Eqs. (3–6) and (3–17) is actually not surprising since the coefficients k_g and k_i are based on the physical parameters of the armature and the stator. This also suggests that any motor can behave as a generator if its shaft is rotated by another motor, or as a motor if its armature is driven by an external electrical power source. This is indeed the case, as all motors and generators obey this type of a reciprocal law. Suppose that the peak generator voltage, the coefficient of sin θ in Eq. (3–6), is divided by the motor torque as expressed by Eq. (3–18). First, some adjustments are necessary to equalize the units in both equations. If flux is to be in lines in Eq. (3–6), and if we wish ω to be 1000 rpm, Eq. (3–6) must be multiplied by $10^{-8}(2\pi/60)(10^3) = (2\pi/6)10^{-6}$. Taking the ratio

$$\frac{E_g}{T_m} = \frac{\Phi N_c N_p \omega}{2\pi N_{pp}} \left(\frac{2\pi}{6} \times 10^{-6} \right) \frac{N_{pp}}{(2.25 \times 10^{-7})\Phi N_c N_p I_a}$$

$$= \frac{\omega}{1.35 I_a}$$

or

$$\frac{T_m}{I_a} = 1.35 \frac{E_g}{\omega} \tag{3–19a}$$

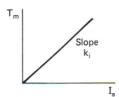

Figure 3–8. Motor torque characteristic.

Substituting for the ratios in Eq. (3–19a), defined previously as k_i and k_g, respectively, the following relationship is obtained:

$$k_i = 1.35k_g \qquad (3-19b)$$

where k_i is in oz-in./A if k_g is in V/1000 rpm. The value of the motor torque constant may then be measured by driving the shaft of the motor under test with another motor at 1000 rpm, and measuring the voltage V_0 across the input terminals of the motor under test. The value of k_i is then calculated as

$$k_i = 1.35V_0 \qquad \text{oz-in./A} \qquad (3-19c)$$

The symbolic representation of the motor as a control system component is shown in Fig. 3–9. As shown, the armature rotates with the speed ω driving the load. The armature circuit consists of the armature resistance R_a and the armature inductance L_a; both are due to the total armature winding, which makes an electrical contact with the brushes of the commutator. The voltage V_b is due to the generator property of the motor. As the armature rotates in the magnetic field, a voltage is induced into the armature winding as described earlier for the dc generator. This voltage, however, is induced in the direction such that it opposes the externally applied voltage. Note the polarities of V_b and V_a in the diagram. V_b, generally referred to as the *back EMF*, is a consequence of Lenz's law. It is proportional to the motor speed ω through the back EMF constant k_b as

$$V_b = k_b\omega \qquad (3-20)$$

where k_b is in V/rpm if ω is in rpm or V/(rad/s) if ω is in rad/s.

The magnetic field may be generated by a permanent magnet, as was the case for both the dc generator and the dc motor discussed earlier. It may also be generated by an electromagnet. As shown in Fig. 3–10, the external dc source supplies field current I_f to the turns of the field winding, which is wound around the poles of the electromagnet (made of iron-based material). I_f generates the magnetic field in the direction shown and magnetizes the poles of the electromagnet north, south, as shown. If the direction of I_f is reversed, the direction of magnetization would be reversed, reversing the magnetic polarization, and consequently the direction of induced field B. In contrast to the permanent magnet, the electromagnet loses its magnetism when I_f is shut off; however, the strength of the magnetic field B may be controlled through I_f, something that is not possible to do with a permanent magnet.

It becomes necessary to redefine the motor-developed torque equation (3–19) if

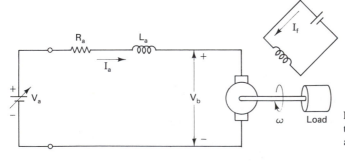

Figure 3–9. Symbolic representation of the motor as a control system component and the armature circuit.

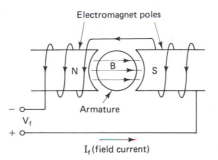

Electromagnet poles

N   B   S

Armature

V_f

I_f (field current)

Figure 3–10. Basic electromagnet motor field structure.

the permanent magnet is replaced by an electromagnet. Since the value of the magnetic flux within the brackets of Eq. (3–18) is a function of the field current, I_f may be factored out together with I_a as follows:

$$T_m = kI_fI_a \qquad (3\text{–}21)$$

From the resulting equation it is evident that the motor-generated torque may be controlled by the armature current or by the field current. Generally, one of these is varied to control the torque while the other is held constant. If the permanent magnet is used as a source of magnetic field, the field is fixed and the armature current is used to control the torque. Similarly, if an electromagnet is used and the field current I_f is held constant, the armature current may again be varied to control the motor-developed torque. In both cases we achieve what is generally referred to as the *armature-controlled* mode of operation of the motor.

On the other hand, with the electromagnet, the field current may be varied to control the motor torque while the armature current is held constant. This is known as the *field-controlled* mode of operation of the motor. Dc motors that depend on the electromagnet for magnetic field production are called *field-wound* dc motors. Field-wound motors are further subdivided in accordance with the manner in which power is applied to the field winding as well as the design of the field winding. Some of the more popular field winding designs are illustrated in Fig. 3–11.

In the *series-field-winding* configuration, the field winding is connected in series with the armature winding. The same current therefore flows through both windings. Since the field winding is subjected to the large armature current, the field winding may contain fewer turns to produce the required magnetic flux, but lower-gage wire must be used to dissipate the higher power. This configuration is shown in Fig. 3–11a. In Fig. 3–11b the *split series field winding* actually consists of two field windings, but only one of these may be used at any given time in series with the armature winding. This configuration is useful in applications where quick reversal in direction of motor rotation is required. By switching to the second winding, which is wound in the opposite sense with respect to the first winding (the direction of the magnetic field is reversed), it takes less time to bring the motor to a momentary stop and then reverse its rotation. With this configuration, the direction of the input current does not have to be reversed to reverse the direction of motor rotation.

Figure 3–11c shows the *shunt field winding* structure, where the armature and the field windings are in parallel. Additional control of the field current in this configuration may be achieved with a rheostat in series with the field winding. The *compound*

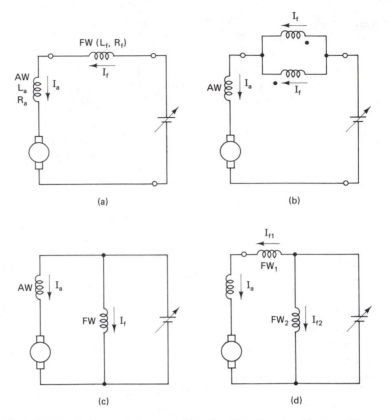

Figure 3–11. Field wound dc motors: (a) series; (b) split series; (c) shunt; (d) compound. AW, armature winding; FW, field winding.

field winding configuration shown in Fig. 3–11d uses two field windings, one in series with the armature winding and the other in shunt. The two field windings may be wound with the same sense or with the opposite sense. In the former case, the two field windings produce magnetic fields which aid each other, and in the latter case the two magnetic fields oppose each other. The latter field structure is often referred to as the *compound differential field winding.*

The field winding may also be *separately excited.* In this case a separate power supply is used to supply current to the field winding, which is completely isolated from the armature winding.

One common characteristic that the four field-wound motor structures shown in Fig. 3–11 have in common is a nonlinear speed versus torque response. This may create a problem in the analysis, where the motor is treated as a linear component. In contrast to the field-wound motor, the speed–torque response for a permanent-magnet motor is much more linear. The motor speed–torque characteristics will be discussed in greater detail later.

As mentioned earlier and also expressed in Eq. (3–21), a dc motor may operate in the field control or in the armature control modes. A permanent-magnet motor, of course, can operate only in the armature control mode. Generally, field control offers less torque and a slower response, but it uses less power from the power supply. In contrast, armature control requires much more electrical input power, but in return we gain greater output torque and quicker response.

The design of the armature can have a significant effect on the speed of response and on the ability of the motor to operate smoothly at low speeds without cogging. We shall discover later that the mechanical time constant of the motor is directly proportional to the armature inertia. The moment of inertia for a solid cylinder is $\frac{1}{2} MR^2$. It is then possible to reduce the inertia of the cylindrical armature by reducing its diameter or by reducing its weight, or both. By lowering the inertia, the mechanical time constant would decrease thus allowing the motor to respond faster to sudden changes in input.

The conductors that make up the armature winding may be placed in the slots of the solid armature core, or they may be bonded to the armature surface as shown in Fig. 3–12. When the conductors are in the slots, the forces which they experience during the motion cause them to move inside the slots. This does not create any problems at high armature speeds, but at lower speeds, unevenness of motion or cogging may be observed. This is not desirable in control applications, where the motor is often required to operate at low speeds. To reduce cogging, some armature designs (as shown in Fig. 3–12b) use the surface-wound armature approach, where the conductors are permanently bonded to the surface of the armature to prevent them from moving around during armature rotation. In this illustration the armature is hollow and made of nonmagnetic material to reduce its weight, resulting in lower inertia and an improved speed of response.

The problem with placing the conductors on the surface of the armature is that they occupy space, forcing the gap between the armature and the stator poles to be larger. The consequence of this is that to produce an equivalent magnetic field in the gap, a stronger magnet (which is larger and probably more expensive) must be used. In the case of a solid-core armature with conductors in the slots, the gap is smaller and the stator magnet may be smaller, to produce an equivalent field.

The commercially available *dc permanent-magnet* (PM) *torque motor* includes most of the important features discussed above. It has a high torque-to-inertia ratio, allowing it to start and stop quickly with high initial torque; it can also operate at low

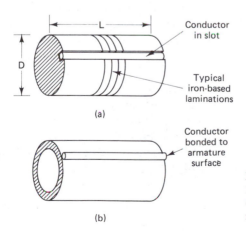

(a)

(b)

Figure 3–12. Armature construction: (a) laminated, with conductors placed in slots; (b) hollow, with conductors bonded to the nonmagnetic surface.

speeds with minimal cogging. These features make the torque motor suitable for control applications. Obviously, the magnet plays an important role in the design of the motor. Years of research in magnetic materials to reduce the magnet size in motor applications has resulted in the smaller, better quality magnets available today, which make possible smaller-volume motors with greater output torques (larger torque-to-motor volume ratio). The armature core may be made of solid magnetic material, or as shown in Fig. 3–12a, it may be laminated. The laminated core reduces the eddy-current losses.

A simple two-segment commutator was used in Fig. 3–7 to illustrate the switching of the two-conductor coil during its rotation. As suggested by Eq. (3–18), the torque developed by the motor would be increased if the number of current-carrying conductors in the magnetic field is increased. To optimize the torque production by a given motor, many more current-carrying conductors must be included in the armature circuit. To accomplish this, many coils are wound on the armature, each coil consisting of many turns wound over two parallel paths, and the coils are connected in series. The commutator provides the connection between coils, and switches the current through each coil in succession after it has rotated 180°.

Figure 3–13 shows an eight-segment commutator with eight armature coils. Also shown for reference is the position of the magnet and the direction of the magnetic field in relation to the position of the brushes, which must be along the axis perpendicular to the magnetic field. This requirement has not been explicitly emphasized in our previous discussion of the motor operation. It is, however, an important one. It has been mentioned earlier that if the motor armature is to rotate continuously in the given direction, the spatial orientation of currents must be held fixed. In Fig. 3–7, the spatial orientation of current between the armature axis and the north pole of the stator must be in the direction away from the commutator and toward the commutator on the other side of the armature axis to assure, in this case, CCW rotation of the motor. To ensure that the spatial orientation of currents is held fixed, the commutator must switch the conductor currents at the instant when the loop is in the plane perpendicular to the magnetic field. This means that the brushes must be positioned along the axis perpendicular to the magnetic field. The position of the brushes is clearly illustrated in Figs. 3–7 and 3–13a.

The operation of the commutator in Fig. 3–13 is similar to that of Fig. 3–7, except that in this case switching is done more frequently because there are more coils. Each coil must rotate 180° before its current is reversed, but because there are eight coils in this illustration, the commutator performs the switching every 45°. The rolled-out representation of the commutator helps to show the direction of conductor currents at the instant when brush 1 contacts segment 1 and brush 2 makes contact with segment 5. Note that the external current I_a divides between conductors 1 and 8′ upon entering segment 1, and leaves segment 5 as a sum of currents from conductors 4′ and 5.

If the armature in Fig. 3–13 rotates 45° CCW, segment 2 makes contact with brush 1 and segment 6 makes contact with brush 2. As can be seen from Fig. 3–13b, the current in conductors 1′ and 5′ must reverse its direction; consequently, the current in coils 1–1′ and 5–5′ must reverse its direction. The switching of currents by the commutator will continue in sequence as the successive commutator segments make contact with the two brushes.

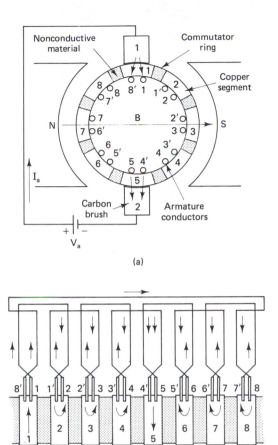

(a)

(b)

Figure 3–13. Eight-segment commutator: (a) basic configuration; (b) two-dimensional (rolled-out) representation of the cylindrical commutator ring, showing the directions of all conductor currents at the instant the brushes make contact with segments 1 and 5.

Motor Speed-Torque Characteristics

Let us consider the dc PM motor and the armature circuit shown in Fig. 3–9. Assuming that the motor has attained the steady-state speed for an externally applied input V_a, the voltage drop across L_a is zero. The armature current must then be equal to the difference in potential across R_a divided by R_a. This may be expressed mathematically as

$$I_a = \frac{V_a - V_b}{R_a} \qquad (3-22)$$

Substituting this expression for I_a together with Eq. (3–20) in Eq. (3–19), the following result is obtained:

$$T_m = \frac{k_i}{R_a}(V_a - k_b\omega) \qquad \text{oz-in.} \qquad (3-23)$$

Making the substitutions

$$k_t = \frac{k_i}{R_a} \qquad \text{oz-in./V} \qquad\qquad (3-24)$$

and

$$B_m = k_t k_b \qquad \text{oz-in.-s or oz-in./rpm} \qquad (3-25)$$

Eq. (3–23) may be rearranged in the form

$$T_m = (-B_m)\omega + k_t V_a \qquad\qquad (3-26)$$

In Eq. (3–26), T_m represents the motor-developed torque. This is the torque that the motor generates to drive the load T_L, to overcome disturbances and to overcome friction which is internal to the motor. Stated mathematically,

$$T_m = T_L + T_d + T_f = k_i I_a \qquad\qquad (3-27)$$

We shall assume that the motor has no internal losses (i.e., $T_f = 0$) and that there are no disturbances, thus reducing Eq. (3–27) to $T_m = T_L$. The torque developed by the motor must then be exactly equal to the torque demanded by the load. If T_m is plotted along the ordinate and ω along the abscissa, the form of Eq. (3–26) is that of a straight line with the slope $-B_m$ as shown in Fig. 3–14a.

The *stall* condition occurs when the motor stops. Substituting $\omega = 0$ in Eq. (3–26), the stall torque

$$T_s = k_t V_a \qquad\qquad (3-28)$$

occurs at the beginning of motion, and also in cases when the applied load is greater than the stall torque.

The no-load point on the speed–torque characteristics in Fig. 3–14a is defined when $T_L = 0$. The no-load speed is determined from Eq. (3–26) for $T_L = T_m = 0$ as

$$\omega_{NL} = \frac{k_t V_a}{B_m} = \frac{V_a}{k_b} \qquad\qquad (3-29)$$

If the motor speed is ω_{NL}, corresponding to V_a as shown in Fig. 3–14a, and the load T_x is applied to the motor shaft, the operating point of the motor shifts from ω_{NL} to the point defined by coordinates (ω_x, T_x), as shown. Thus the applied load torque reduces the motor speed, and the reduction in motor speed results in a lower back EMF. It can be deduced from Eq. (3–22) that for a constant armature input V_a, a reduction in V_b causes an increase in the armature current. This is consistent with Eq. (3–27), from which $\Delta T_m = k_i \Delta I_a$; that is, the increased armature current is responsible for an increase in the motor-developed torque to accommodate the applied load $\Delta T_L = T_x$. If the new operating speed ω_x is too low and the original speed must be restored, the armature voltage must be increased. This is illustrated in Fig. 3–14b, where the armature voltage must be increased from V_1 to V_2 to restore the original no-load speed ω_1 with the load T_x applied. The power rating of the motor must always be observed, and the applied

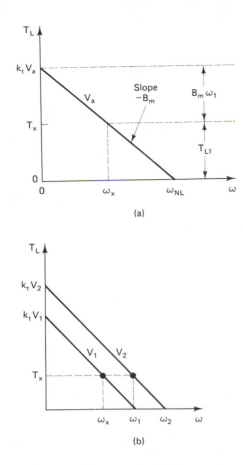

(a)

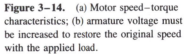

(b)

Figure 3–14. (a) Motor speed–torque characteristics; (b) armature voltage must be increased to restore the original speed with the applied load.

armature current must never exceed the maximum rated current as suggested by the manufacturer.

Although a single line is shown in Fig. 3–14b, a family of parallel lines is possible if V_a in Eq. (3–26) is used as a parameter. The slope B_m dictates the rate at which the motor speed changes with the applied load. This is a property of the individual motor. As the value of B_m increases, the change in speed decreases with increased load. If in a hypothetical case the value of B_m is infinite (the line is vertical), there is no change in speed for any change in load.

EXAMPLE 3–1

The given motor has the speed–torque characteristics shown in Fig. E3–1. Assume that $R_a = 2\ \Omega$.

(a) If $V_a = 5$ V, calculate the motor no-load speed.
(b) Calculate the speed of the motor for $V_a = 5$ V and $T_L = 7$ oz.-in.
(c) What must be the input V_a to restore the motor speed calculated in part (a) assuming that the 7-oz-in. load is applied?

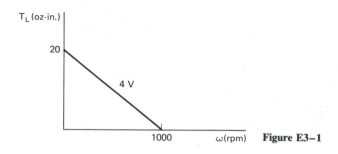

Figure E3-1

SOLUTION (a) From the speed–torque diagram $k_b = 4$ V/krpm. From Eq. (3–29), $\omega = 5/4 = 1250$ rpm.

(b) The slope of the given speed–torque characteristics is 20 oz/krpm $= B_m$ and $k_t = T_s/V_a = 20/4 = 5$ oz-in./V. Using Eq. (3–26), we obtain

$$7 = -20\omega + 5(5)$$

Solving for ω yields

$$\omega = \frac{18}{20} = 0.9 \text{ krpm} = 900 \text{ rpm}$$

(c) Using Eq. (3–26) gives us

$$7 = -20(1.25) + 5V_a$$

Solving for V_a yields

$$V_a = \frac{32}{5} = 6.4 \text{ V}$$

Motor Models

Before proceeding with the derivation of motor models, it is necessary to characterize the motor load torque T_L in terms of inertial and frictional effects. Let us consider first the general forms of frictional torque. As shown in Fig. 3–15, T_s represents the static friction torque that occurs only just before the motion starts. The Coulomb friction torque T_c occurs during motion; however, it is independent of speed, as it remains constant during the motion. The viscous friction torque T_B is proportional to speed through the viscous friction coefficient B. The total friction during the motion may then be expressed as

$$T_f = T_c + T_B = B\omega + T_c \tag{3–30}$$

A motor driving a cylinder immersed in high-viscosity oil is an example of a viscous friction torque. The effect of viscous friction is negligible when the same cylinder rotates in air unless its speed is extremely high or unless a fan is mounted on its shaft, which would increase the viscous friction.

There are other load configurations that the motor may be required to drive. Some of these are illustrated in Fig. 3–16. In the case of the weight and pulley, the linear and angular displacements are related by

$$X = R\theta \tag{3–31}$$

Motors   Chap. 3

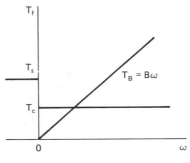

Figure 3–15. Types of friction. Static friction T_s occurs before the beginning of motion, Coulomb friction T_c remains constant during motion, and viscous friction T_B is proportional to speed.

Differentiating both sides of Eq. (3–31) yields

$$\dot{X} = R\dot{\theta} \qquad\qquad (3\text{–}31)$$

or

$$V = \omega R \qquad\qquad (3\text{–}32)$$

where ω is the angular velocity of the pulley and V is the linear velocity of the weight W. The torque that the motor must develop at the shaft to drive the weight must be RW. This type of load torque is independent of speed, and thus behaves as does a Coulomb torque.

In the case of the rack-and-pinion drive, Eqs. (3–31) and (3–32) apply also. The load torque, however, may not be expressed as simply as above because it depends

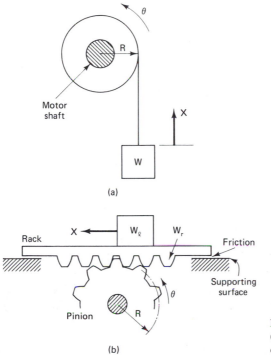

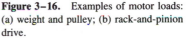

Figure 3–16. Examples of motor loads: (a) weight and pulley; (b) rack-and-pinion drive.

on the type of mechanical support used for the rack. The rack may slide on a supporting surface or roll on ball bearings. The friction between the rack and the supporting surface results in a force $F_f = \mu(W_l + W_r)$ (where μ is the coefficient of dynamic friction) and consequently in a torque RF_f which acts as a load on the motor. The motor must, therefore, develop at least this much torque in order to drive the rack.

Inertial effects present to the motor another type of load torque. In general, to position a load or to bring the load to its final or steady-state speed, the load must be accelerated. To accelerate a load whose inertia is J_L to some final speed, the motor must develop a torque

$$T_J = J_L \dot{\omega} \tag{3-33}$$

where $\dot{\omega}$ is the angular acceleration of the load. When the load attains the steady-state speed, $\dot{\omega} = 0$, and therefore the inertial torque must also be zero. Sometimes the load inertia is not given in an explicit form, and an equivalent inertia must be determined. This applies to a system such as the weight and pulley shown in Fig. 3–16a. The acceleration $\dot{\omega} R$ is obtained by differentiating Eq. (3–32). The reaction force of the mass W/g due to acceleration may be expressed as

$$F = Ma = \frac{W}{g}(\dot{\omega}R)$$

and the torque

$$T = FR = \left[\frac{(W)R^2}{g}\right]\dot{\omega} = J_e\dot{\omega} \tag{3-34}$$

where the expression in brackets must represent the equivalent inertia J_e due to the acceleration of the weight,

$$J_e = \frac{W}{g}R^2 = \frac{W}{g}\left(\frac{X}{\theta}\right)^2 \tag{3-35}$$

where R was replaced by X/θ from Eq. (3–31). J_e is the equivalent inertia of the weight when it is accelerated. In the case of the weight and pulley, the motor senses other inertias as well: the inertia of its own armature J_m and the inertia of the pulley. All these inertias must be added, and when multiplied by the acceleration, the product represents the total torque that the motor must develop to drive the inertia.

Equation (3–35), which was developed for the weight and pulley load, also applies to the rack-and-pinion configuration as the equivalent inertia of the rack with the load. The additional inertias which must be included in that system are the inertias of the motor armature and that of the pinion.

From the preceding discussion, J_L in Eq. (3–33) must represent the total inertia of the load, including the motor armature. The total load torque may then be expressed as the sum of the total friction and inertial torques as

$$T_L = T_f + T_J \tag{3-36}$$

Motors   Chap. 3

Returning now to the development of the motor model, Eq. (3–26) may be expressed as follows after assigning $T_m = T_L$ and using Eqs. (3–36), (3–30), and (3–33):

$$-B_m\omega + k_t V_a = J_t\dot{\omega} + B_L\omega + T_c$$

Rearranging terms gives us

$$J_t\dot{\omega} + (B_m + B_L)\omega = k_{te}V_a \tag{3–37}$$

where the equivalent k_t is defined as

$$k_{te} = \frac{k_t V_a - T_c}{V_a} \tag{3–38}$$

The practical significance of k_{te} may be realized by referring to the speed–torque curves shown in Fig. 3–14. Since $k_t V_a$ represents the stall torque, which is the maximum torque that the motor develops at $\omega = 0$ for a given V_a, the effect of the Coulomb friction load torque is to lower the stall torque, by the amount T_c, to some lower value. In the new position, the straight line is parallel to its original position, and for that reason the definition of B_m as expressed by Eq. (3–25) is still valid. If $T_c = 0$, clearly $k_{te} = k_t$, but if $T_c \neq 0$, the speed–torque straight line with the lowered stall torque must be used. Dividing both sides of Eq. (3–37) by $(B_L + B_m)$ gives us

$$\tau_m\dot{\omega} + \omega(t) = k_m V_a(t) \tag{3–39}$$

where $\tau_m = J_t/(B_L + B_m)$ (seconds) = motor time constant, load included
  $k_m = k_{te}/(B_L + B_m)$ (rad/s/V or rpm/V) = motor speed constant, load included
  J_t = total load inertia, including the motor armature (oz-in.-s^2)
  B_L = total load viscous friction coefficient (oz-in.s or oz-in./rpm)
B_m and k_{te} are as defined in eqs. (3–25) and (3–38), respectively. Equation (3–39) is the differential equation mathematical model of the motor. It will be used elsewhere to analyze motor performance.

Taking the Laplace transform of both sides of Eq. (3–39) results in

$$\tau_m[s\Omega(s) + \omega(0)] + \Omega(s) = k_m V_a(s)$$

Equating the initial speed of the motor to zero and solving for the ratio $\Omega(s)/V_a(s)$, we have

$$\frac{\Omega(s)}{V_a(s)} = \frac{k_m}{(\tau_m)s + 1} \tag{3–40}$$

(see Fig. 3–17). $\omega(0)$ was equated to 0 only because it would not be possible otherwise to solve algebraically for the ratio of output to input. This is true not only here but for all transfer functions. All transfer functions are defined when the system is at rest (i.e., all initial conditions must be set to zero).

Equation (3–40) is another model of the motor, the transfer function model. As

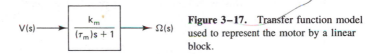

Figure 3–17. Transfer function model used to represent the motor by a linear block.

noted above, it is valid only when all initial conditions are zero. The differential equation model, on the other hand, admits the use of initial conditions, and in that respect it is more general.

EXAMPLE 3–2

Suppose that the motor drives the weight-and-pulley arrangement shown in Fig. 3–16a. The 1-lb weight is attached to the pulley, whose weight is 0.5 lb, by a string of negligible weight. The inertia of the motor armature, including its shaft, is 0.1 oz-in.-s^2. The motor accelerates the load W from rest to 60 ft/s in 400 ms and in doing so delivers 0.7 hp to the load. There is no viscous friction during the motion; however, the Coulomb friction may be approximated by 50 oz-in. Calculate the radius R of the pulley in centimeters assuming that it is cylindrical and made of solid material. Assume a constant acceleration of the load.

SOLUTION   The ω of the motor may be calculated from $V = \omega R$ in terms of R as

$$\omega = \frac{V}{R} = \frac{60}{R} \text{ rad/s} = \frac{60}{R}\left(\frac{60}{2\pi}\right) = \frac{1800}{\pi R} \text{ rpm} \qquad R \text{ in ft}$$

Using the ω above in Eq. (2–63) gives us

$$P = \frac{T\omega}{5252}$$

$$0.7 = \frac{1800T_m}{192(5252)\pi R} \qquad T_m \text{ in oz-in.}$$

Solving for T_m yields

$$T_m = 1232R = T_J + 50$$

$$T_J = 1232R - 50 \text{ oz-in.}$$

Calculating all inertias, we have

$$J_e = \frac{W}{g}R^2 = (1/32.2)R^2(192 \text{ oz-in./ft-lb}) = 5.96R^2 \qquad R \text{ in ft}$$

$$J_{\text{pul}} = \tfrac{1}{2}MR^2 = \tfrac{1}{2}\left(\frac{1}{2\cdot32.2}\right)R^2(192) \qquad = 1.49R^2$$

$$J_{\text{arm}} \qquad\qquad\qquad\qquad = \underline{\;0.1\;}$$
$$J_t = 7.45R^2 + 0.1$$

The acceleration may be calculated using the ω calculated earlier as

$$\dot{\omega} = \frac{\Delta\omega}{\Delta t} = \frac{60}{R}\left(\frac{1}{0.4}\right) = \frac{150}{R} \qquad \text{rad/s}^2$$

The inertial torque

$$T_J = J_t \dot{\omega}$$

Substituting for T_J, J_t, and $\dot{\omega}$ we obtain

$$1232R - 50 = (7.45R^2 + 0.1)\frac{150}{R}$$

Rearranging terms gives us

$$114.5R^2 - 50R - 15 = 0$$

Solving the quadratic equation for R yields

$$R = 0.64 \text{ ft} = 7.7 \text{ cm}$$

Stepper Motors

In contrast to dc motors, which operate in a continuous manner, the shaft of a *stepper motor* moves in discrete angular increments or steps. Instead of the dc voltage which is applied as the input to a dc motor, pulses of a specific frequency are applied to the input of a stepper motor. In general, a stepper motor makes a step for each applied pulse. The size of the step depends on the stepper motor design. Several commercial designs are described in this section.

The operation of a stepper motor depends on the application of pulses to its phase windings in a proper sequence. In addition to the sequence requirement, the applied pulses must provide the phase windings with sufficient current, which can be quite large. Thus to operate a stepper motor, the user must first design the sequencer logic, whose output provides the required sequence of energizing the phase windings of the selected stepper motor, and the driver, which provides the current as required by the phase windings. Fortunately, most commercial stepper motors are provided with a controller specifically designed for that stepper motor by the manufacturer of the stepper motor or by an independent company. The controller, simplifies considerably for the user the task of operating the stepper motor. As shown in Fig. 3–18, the user needs to apply the clock pulses and the direction of rotation input to operate the given motor. Simple controllers are available in IC chip form while the more elaborate controllers which include a power supply and other special features are constructed on the board.

In contrast to a dc motor, a stepper motor requires neither brushes nor a commutator for its operation. What is even more significant is that the stepper motor–controller structure may be directly interfaced to a microprocessor or computer. Interfacing the dc motor to a computer system involves additional circuitry and expense. It is also clear that a stepper motor can position a load very accurately, as each input pulse causes the stepper motor to advance precisely one angular increment. It is not surprising, therefore, that stepper motors find numerous applications in computer and microprocessor systems.

Before proceeding with a description of the first stepper motor selection, the PM type, let us consider briefly some basic principles used by all stepper motors which are essential and necessary for the reader to understand.

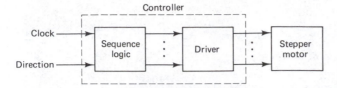

Clock ——

Direction ——

Figure 3–18. A controller simplifies the task of operating the stepper motor.

The magnetic flux, magnetic lines of force, of the permanent-magnet loop from the north pole to the south pole is shown in Fig. 3–19a. The dipolar property of the permanent magnet may be induced into an electromagnet by the current I through the winding, as shown in Fig. 3–19b. The magnetic polarity of the electromagnet may be reversed by reversing the direction of current I or, as shown in Fig. 3–19c, by reversing the sense of the winding, leaving the direction of current unchanged.

In Fig. 3–19d the permanent magnet is on the shaft and is free to rotate as an armature does. There are two electromagnet poles, which are part of the stator metallic

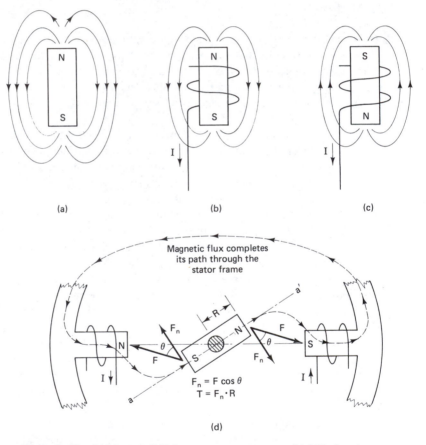

(a) (b) (c)

(d)

Figure 3–19. (a) Magnetic field due to a permanent magnet; (b) field of an electromagnet due to current I; (c) magnetic polarity reverses when the winding sense is reversed and the direction of I remains unchanged; (d) the force of attraction produces a torque that rotates the armature to its equilibrium position, where the axes of the armature and the stator poles are aligned, $F_n = F \cos \theta$, $T = F_n R$.

frame. In the position shown, axis $a-a'$ of the magnet armature is slightly displaced from the pole axis of the electromagnets. The magnetic force due to attraction of unlike magnetic poles produces the normal component $F_n = F \cos \theta$, which is perpendicular to axis $a-a'$. The resulting torque $F_n \cdot R$, rotates the armature in the CW direction until the armature axis $a-a'$ is aligned with the stator pole axis. This represents the equilibrium position of the armature, as no net torque acts on the armature in this position. If the current I is reduced to zero, thus deenergizing the electromagnet poles shown in the diagram, and if another pair of nearby poles is energized, the torque due to the force of attraction of the newly energized poles rotates the armature to the new equilibrium position, where the armature axis $a-a'$ is aligned with these poles. If there are many electromagnet pole pairs around the stator, and if these poles are activated by current pulses in sequence, the armature will execute a stepping motion of rotation following the rotating magnetic field, resulting from the sequential switching of the stator pole windings.

3–4 PERMANENT-MAGNET STEPPER MOTOR

A typical construction of the PM stepper motor is shown in Fig. 3–20. This is a four-phase motor, and each phase is wound on two stator poles. The stator in this design must therefore have eight poles. The rotor (armature, which is made of permanent-magnet material, is aligned with stator poles 1 and 1'. It is held in this position by the phase 1 current I_1, which magnetizes 1 as the south pole and 1' as the north pole. Note the winding sense for windings 1 and 1' required to produce this magnetization state, where the flux must enter pole 1 (because the rotor north pole is next to it) and leave pole 1'. As the phase windings $\phi_1, \phi_4, \phi_3, \phi_2$ (1–4–3–2 sequence) are energized by the corresponding current pulses I_1, I_4, I_3, I_2 (one at a time in the direction as

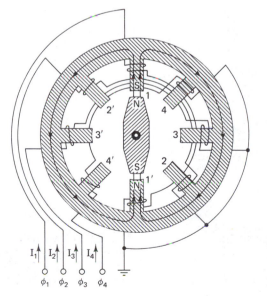

Figure 3–20. Permanent-magnet stepper motor construction. There are four phases and each phase is wound on two stator poles. The step angle is 45°.

shown in the diagram), the rotor advances clockwise making 45° with each step (360/8). When the north end of rotor magnet reaches stator pole 2, the excitation sequence 1–4–3–2 is repeated except the direction of currents I_1, I_4, I_3, I_2 must be reversed for the remaining 180° of rotation in order to produce the induced south poles at stator poles 1′, 4′, 3′, and 2′ and thus insure a continuous motion. Thus, the excitation sequence 1–4–3–2 will produce clockwise rotation of the rotor magnet assuming the currents are reversed in direction after each 180° of rotation. Clearly, to achieve counterclockwise rotation the sequence 1–2–3–4 must be followed with negative currents applied for the first 180° rotation and positive currents (direction as shown in Fig. 3–20) applied for the remaining 180° rotation.

3–5 SINGLE-STACK VARIABLE-RELUCTANCE STEPPER MOTOR

A typical construction of the single-stack variable-reluctance (VR) stepper motor is shown in Fig. 3–21. The single-stack VR stepper motor uses one rotor, in contrast to the multistack VR stepper motor, which uses several rotors. The rotor and the stator are both made of magnetic material. There are no permanent magnets here. There are three phases and each phase winding is wound on four poles or teeth of the stator. Phase 1, for example, is wound on poles 1, 4, 7, and 10 of the stator. There are a total of 16 rotor teeth, and of course, 12 stator teeth, as each of the three phases requires four rotor teeth. The opposite poles are wound in the opposing sense, so that there is a balance between the flux entering and leaving the rotor. Suppose that the current I_1 is applied to phase 1 as shown in the illustration and four teeth of the rotor are aligned with teeth 1, 4, 7, and 10 of the stator. The flux enters the rotor from stator teeth 4 and 10, and leaves the rotor by way of stator teeth 1 and 7, completing its closed path (as it always must) through the stator frame. It may be observed that the tip of stator tooth 4 is the induced north pole (as flux is leaving tooth 4), and the tip of rotor tooth which is aligned with stator tooth 4, and through which the flux

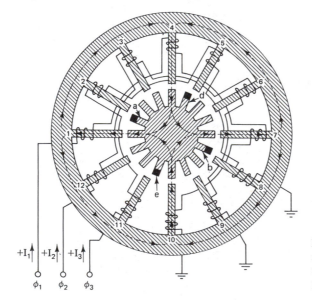

Figure 3–21. Variable-reluctance stepper motor: three-phase single-stack construction; $N_r = 16$, $N_s = 12$, $X = 4$ poles/phase, $\theta_s = 7.5°$, $R_s = 48$ steps/rev.

Motors   Chap. 3

enters the rotor, is effectively an induced south pole. This magnetic polarization must exist to allow the flux continuity through the gap between the two aligned teeth. The same is true for the remaining three pairs of aligned stator-rotor teeth.

To advance the rotor one step in the CW direction, phase 3, which is wound on stator teeth 2, 5, 8, and 11, must be energized by applying current I_3 in the direction shown after the current I_1 is removed. The flux must now find a different path to complete the magnetic circuit. Just as the current in an electrical circuit seeks the path of lowest resistance, so does the flux in a magnetic circuit seek the path of lowest reluctance. The air gap between the teeth presents reluctance to the magnetic flux; the larger the gap, the larger the reluctance. Consequently, the flux in leaving stator poles 2 and 8, which are induced north poles, will jump the gap to the nearest rotor teeth. Rotor teeth a and b, which are closest to the rotor teeth, become the induced south poles. The flux leaves rotor teeth d and e through the air gap, enters stator teeth 5 and 11, and thus completes the remaining part of the magnetic circuit through the stator frame. In the meantime, the magnetic force or attraction exists between stator tooth 2 (the induced north pole) and the rotor tooth a (the induced south pole), the force also exists between pole pairs (11, e), (8, b), and (5, d). As described in Fig. 3–19d, this results in a torque acting on the rotor, causing it to advance until rotor teeth a, d, b, and e align with stator teeth 2, 5, 8, and 11, respectively, at which time the gap between the respective teeth is minimal, resulting in minimum reluctance and maximum flux through the magnetic circuit. This represents the equilibrium position for the newly activated phase 3, the position in which the rotor experiences no net torque acting on it. In the process the rotor advances in the CW direction one step angle, which is 7.5° in this case. The complete sequence is illustrated in Fig. 3–22, where the initial positions of four rotor teeth are darkened so that the reader can clearly follow how the rotor advances in the CW direction one rotor tooth pitch of 22.5° in three steps as the phases are activated in the sequence 1–3–2–1. This sequence is repeated for continuous rotation. For CCW rotation the reverse sequence, 1–2–3–1, must be followed.

Before defining and deriving the applicable expressions pertaining to the operation of the single-stack VRSM, the following list of symbols will be helpful:

N_r = number of rotor teeth P_s = stator tooth pitch (deg)
N_s = number of stator teeth θ_s = step angle (deg)
N_p = number of phases R_s = stepping rate (steps/rev)   (3–41)
P_r = rotor tooth pitch (deg) $X = N_s/N_p$ = number of stator teeth per phase

The tooth pitch is defined as the angular separation between two corresponding points of adjacent teeth. The rotor and the stator tooth pitches may then be defined as

$$P_r = \frac{360}{N_r} \quad \text{and} \quad P_s = \frac{360}{N_s} \qquad (3\text{–}42)$$

If N_p phases of the single-stack VRSM are excited in sequence, it takes N_p steps before the stator teeth which were initially aligned with some rotor teeth are aligned once again with another set of rotor teeth. The rotor thus advances, as illustrated in Fig. 3–22, an angle P_r in N_p steps. The step angle must then be

$$\theta_s = \frac{P_r}{N_p} = \frac{360}{N_r N_p} \qquad \text{deg/step}$$

Sequence	Rotor position and flux orientation

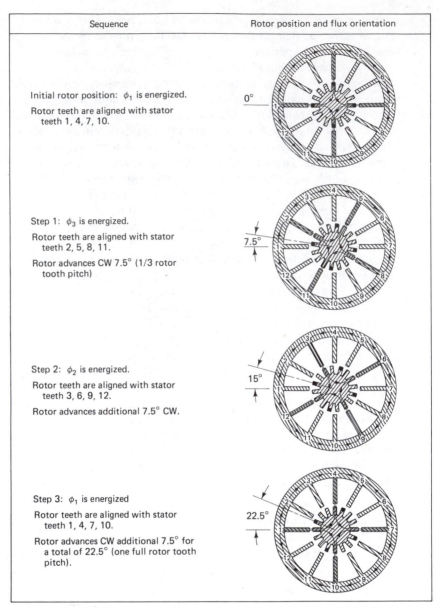

Initial rotor position: ϕ_1 is energized.

Rotor teeth are aligned with stator teeth 1, 4, 7, 10.

0°

Step 1: ϕ_3 is energized.

Rotor teeth are aligned with stator teeth 2, 5, 8, 11.

Rotor advances CW 7.5° (1/3 rotor tooth pitch)

7.5°

Step 2: ϕ_2 is energized.

Rotor teeth are aligned with stator teeth 3, 6, 9, 12.

Rotor advances additional 7.5° CW.

15°

Step 3: ϕ_1 is energized

Rotor teeth are aligned with stator teeth 1, 4, 7, 10.

Rotor advances CW additional 7.5° for a total of 22.5° (one full rotor tooth pitch).

22.5°

Figure 3–22. Three-step switching sequence of the single stack VRSM. The rotor position and flux orientation are shown for each step of the sequence.

Another way of obtaining the step angle is to realize that just before a particular phase is excited, some of the rotor and stator teeth must be separated by exactly θ_s because these teeth become aligned upon excitation of that phase. The step angle must then be also equal to the difference between the rotor and the stator tooth pitches. Combining all of the results above, the step angle is defined as

$$\theta_s = \frac{P_r}{N_p} = \frac{360}{N_r N_p} = |P_r - P_s| \quad \text{deg/step} \tag{3–43}$$

The stepping rate is expressed as

$$R_s = \frac{360}{\theta_s} = N_r N_p \quad \text{steps/rev} \tag{3–44}$$

If the frequency of applied pulses is f, and assuming 1 step per pulse,

$$\frac{1}{R_s} \frac{\text{rev}}{\text{step}} \times f \frac{\text{step}}{\text{s}} \times 60 \frac{\text{s}}{\text{min}}$$

represents the speed of the motor in rpm:

$$w = \frac{60f}{R_s} = \frac{60f}{N_p N_r} = \frac{\theta_s f}{6} \quad \text{rpm} \tag{3–45}$$

For the motor in Fig. 3–21, $P_r = 360/16 = 22.5°$, $P_s = 360/12 = 30°$, and the difference $30 - 22.5 = 7.5°$ is the step angle, which may also be calculated as $360/[16(3)]$, $R_s = 360/7.5 = 48$ steps/rev.

In the illustrative motor configuration of Fig. 3–21 each of the three phases was wound on four stator poles. The number of phases and the number of stator poles used per phase may not be randomly chosen, as they are related. Consider Eq. (3–43):

$$P_r - P_s = \pm\theta_s \quad \text{(the } \pm \text{ results from the absolute value)}$$

Substituting from Eq. (3–42) yields

$$\frac{360}{N_r} - \frac{360}{N_s} = \frac{\pm 360}{N_r N_p}$$

then $1/N_r - 1/N_s = \pm 1/(N_r N_p)$, and combining the left-hand side under a common denominator giving us

$$\frac{N_s - N_r}{N_s N_r} = \frac{\pm 1}{N_r N_p}$$

$$N_s - N_r = \pm \frac{N_s}{N_p}$$

Substituting $N_s = X N_p$ from Eq. (3–41), and $N_r = R_s/N_p$ from Eq. (3–44) in the equation above, and solving for X, we have

$$X = \frac{R_s}{N_p(N_p \pm 1)} = \frac{N_r}{N_p \pm 1} \tag{3–46}$$

Thus the number of stator poles per phase and the stepping rate or the number of rotor teeth are interrelated. Some of the parameter choices for the motor in Fig. 3–21 are summarized in Table 3–1. If it is required that the stepping rate be 48 steps/rev, three or four phases may be used. For three phases the rotor must have 16 teeth, and the value of X from Eq. (3–46) may be 4 or 8: the stator must have 12 teeth if four stator

TABLE 3–1 PARAMETER OPTIONS

N_p	R_s	N_r	X	N_s
3	48	16	4	12
			8	24
4	48	12	4	16
4	64	16	?	?

teeth are used by each phase, and if eight stator teeth are used per phase, the stator must have 24 teeth. If four phases are used, still assuming that $R_s = 48$ is required, the value of X must be 4, and the stator must use 16 teeth. Suppose, for example, that we insist on four phases to be used and a rotor with 16 teeth. The stepping rate can no longer remain at 48; its new value from Eq. (3–44) is 64. Under such conditions there is no solution for X and therefore no solution for N_s.

Suppose, for example, that the required step angle be 9°. The 9° step angle restricts the stepping rate to 40 steps/rev. Under these conditions four or five phases may be used, with two stator poles per phase. For four phases $N_r = 10$, $N_s = 8$, and for five phases $N_r = 8$, $N_s = 10$. No other motor parameter values will satisfy the foregoing conditions.

3–6 MULTISTACK VARIABLE-RELUCTANCE STEPPER MOTOR

The variable-reluctance stepping motor (VRSM) may have more than one stack. Typical VRSM structures may include two, three, four, or more stacks. A stack, also referred to as a *phase*, typically includes a toothed rotor and the surrounding stator structure. A three-stack (three-phase) VRSM has been chosen to illustrate the basic principles of its operation. Figure 3–23 shows the construction of this type of a motor. Shown in the top diagram is the view normal to the motor shaft, which reveals the three sections. Each section includes the rotor and the stator with the windings. The bottom diagram exposes the details of the rotor and the stator structure of the three stacks. In this design the stator of each stack consists of four poles, and each pole has teeth, in contrast to the single-stack VRSM, where each pole is basically a single tooth. It may also be noticed that the number of stator and rotor teeth is the same in each stack, as opposed to the single-stack VRSM, where the number of rotor teeth must not be the same as the number of stator teeth, or else the step angle would be zero. One wonders, then, how the structure where the number of stator and the rotor teeth is the same can ever work. The answer to this lies in the fact that there is more than one stack, and the stacks are indexed. As shown in the bottom diagram of Fig. 3–23, the teeth of the rotor and the stator of stack 1 are aligned, the teeth of the stator and the rotor of stack 2 are misaligned by 10° due to indexing (i.e., rotating the stator of stack 2 by 10° in relation to the position of the rotor), and the stator of stack 3 is rotated or indexed by 20° so that its teeth are misaligned by 10° with respect to the stator teeth of stack 2, and by 20° with respect to the stator teeth of stack 1. The teeth of the three rotors,

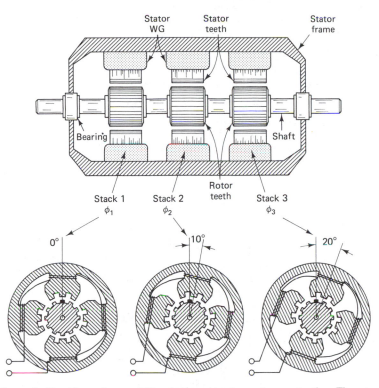

Figure 3–23. Three-phase variable-reluctance stepping motor construction. The rotor and the stator each have 12 teeth and $\theta_s = 10°$. The successive stator sections are indexed by $\frac{1}{3}$ rotor tooth pitch ($10°$), whereas the teeth of the three rotor sections are aligned.

which are on the same shaft, are in perfect alignment. In general, for N_p phases or stacks, the value of the index angle is

$$\theta_i = \frac{P_r}{N_p} = \theta_s \tag{3–47}$$

where P_r is the rotor tooth pitch, as defined by Eq. (3–42). In the present case, $N_r = N_p = 12$ (number of teeth on rotor and stator) and $P_r = 360/12 = 30°$; therefore, the index angle is $10°$.

The mechanism described for the single-stack VRSM and illustrated in Fig. 3–19d, where the stator teeth of the excited phase are brought into alignment with the nearest rotor teeth (the rotor is forced to rotate), thus reaching the equilibrium condition, applies also in the case of the three-stack (or multistack) VRSM since both the rotor and the stator are made of magnetic material, and even though the three rotors are fixed to the same shaft, the three stacks are more or less magnetically isolated. Accordingly, if phase 1 is initially excited and its rotor-stator teeth are aligned, the rotor-stator teeth of stack 2 are at this time misaligned by $10°$, and those of stack 3, by $20°$. Removing the excitation (armature winding current) from stack 1 and applying it to stack 2 will rotate the rotor by 10° and thus align the rotor-stator teeth of stack 2. At this time the rotor-stator teeth of stack 3 are misaligned by 10°. Removing the excitation from stack 2 and applying it to stack 3 will result in an additional 10° rotation of the motor shaft, where the teeth of stack 3 are now aligned, and those of stack 1 are misaligned by 10°. The switching sequence just described is illustrated in Fig. 3–24, where the motor shaft is shown to advance one rotor tooth pitch (30°) in three steps.

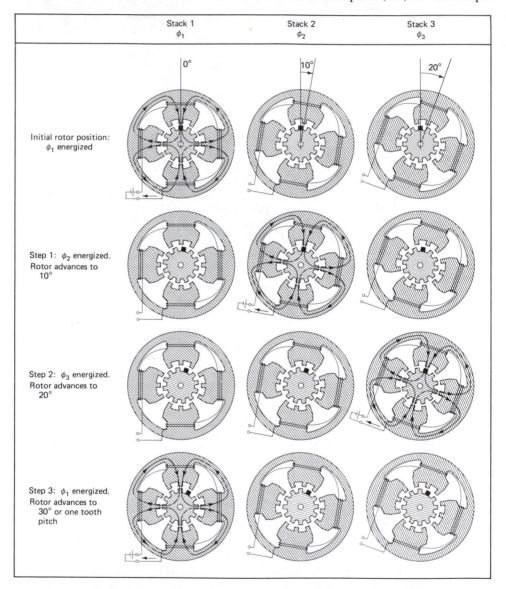

Figure 3–24. Stepping sequence of the three-phase VRSM. $N_r = N_s = 12$, $P_r = 30°$, and $\theta_s = 10°$. As shown, the black rotor tooth advances CW 10° with each step for a total of 30° at the end of the three-step sequence. Sequence 1–2–3–1 is used for CW rotation, and the reverse, 1–3–2–1, for CCW rotation.

Motors   Chap. 3

In general, the motor shaft will advance one rotor tooth pitch in N_p steps, where N_p is the number of stacks used. It is clear at this time that the index angle defined earlier must be the step angle θ_s. Repeating the $1-2-3-1$ sequence illustrated in Fig. 3–24 results in continuous rotation of the motor shaft. Reversing the sequence to $1-3-2-1$ results in CCW rotation.

The switching sequence shown in Fig. 3–24 is also expressed in tabular form in Fig. 3–25. The typical switching circuit includes the three-phase VRSM represented by its symbol, and the phase excitation is symbolically represented by switches and dc sources.

Exciting phases 1 and 2 in sequence causes the motor to make one step. However, if phases 1 and 2 were excited simultaneously, the motor shaft would advance one-half step. Exciting phases 2 and 3 next causes the motor to advance one full step. As shown in Fig. 3–25, two-phase excitation may be done sequentially: $1-2$, $2-3$, and so on. Exciting two phases or one phase at a time produces the same step angle, and

Step	S_1	S_2	S_3
1	X		
2		X	
3			X
1	X		

(a)

Step	S_1	S_2	S_3
1	X	X	
2		X	X
3	X		X
1	X	X	

(b)

Step	S_1	S_2	S_3
1	X	X	
2		X	
3		X	X
4			X
5	X		X
6	X		
1	X	X	

(c)

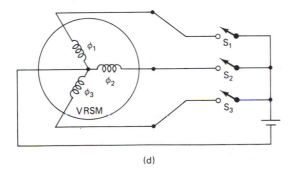

(d)

Figure 3–25. Three-phase VRSM: (a) single-phase excitation CW sequence; (b) dual-phase excitation CW sequence; (c) half-stepping CW sequence; (d) circuit switching diagram. For CCW rotation, reverse the sequence, reading the tables (a), (b), and (c) from the bottom in the upward direction. (X, switch closed.)

the only difference between the two methods is that the latter method produces a rotor motion which leads the former by one-half step angle. Two-phase excitation requires twice as much current from the driving source.

Alternating the two-phase excitation with one-phase excitation is referred to as the *half-stepping* mode of operation. In the half-stepping mode the step angle is half its normal value, and as shown in Fig. 3–25c, it takes twice as many steps (six steps versus three steps in normal mode to complete the sequence).

As there is no need to reverse the current direction in operating the multistack VRSM, *unipolar drive* (single polarity or simply one power supply) is generally used, in contrast to the bipolar drive required by some stepping motors.

As we have seen, two phases may be excited simultaneously in the multistack VRSM. This is generally not done with the single-stack VRSM, due to the close proximity of its phase windings, which results in a large mutual inductance and excessive coupling between phases.

3–7 HYBRID STEPPER MOTOR

The hybrid stepper motor (HSM) combines the features of the PM and the VR motors. As shown in Fig. 3–26, the construction of a typical HSM includes two sections (more than two sections are possible) with an axial magnet between the two sections. Each section includes a toothed rotor and stator poles (also toothed) with windings. The detailed stator-rotor construction for each section is shown in the bottom diagram.

The number of teeth on the rotor and the stator are different, in contrast to the multistack VRSM, where they are the same. Sections A and B are identical in construction; however, the stator teeth of the two sections are perfectly aligned, and the teeth of the two rotors are misaligned by $\frac{1}{2}P_r$. (Recall that in the multistack VRSM the rotor teeth are aligned and it is the stator teeth that are indexed.) For this design $P_r = 360/30 = 12°$; therefore, the rotors are indexed by 6°.

The stator phase windings are distributed between the poles of the two sections. As shown in the illustration, phase 1 is wound on stator poles 1, 3, 5, and 7 of section A and on poles 1, 3, 5, and 7 of section B, and phase 2 is distributed among poles 2, 4, 6, and 8 in each section.

The axial permanent magnet magnetizes the rotor of section A as the north pole and the rotor of section B as the south pole. The added complexity due to the sharing of the phase windings by the two sections complicates the magnetic circuit, resulting in a flux path which is radically different from all others. The direction of flux through the stator poles is determined by the direction of magnetization of these poles, which is dictated, in turn, by the applied phase currents. The direction of stator pole magnetizations, which correspond to the directions of the applied phase currents as shown, is indicated in Fig. 3–26 by arrows.

Consider the instant when phase 1 (ϕ_1) is energized by I_1 applied in the direction shown. The rotor teeth of section A are aligned with the stator teeth of poles 1 and 5, and those of section B are aligned with the teeth of poles 3 and 7. As illustrated in Fig. 3–27, the magnetic circuit takes the following path: flux from the north pole of the permanent magnet enters the rotor of section A and leaves through poles 1 and 5

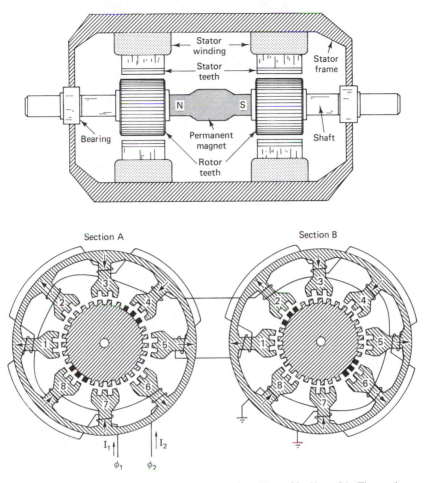

Figure 3–26. Hybrid stepper motor construction: $N_r = 30$, $N_s = 24$. The teeth of the stator sections are aligned, and the teeth of the two rotors are misaligned by $\frac{1}{2}P_r$ (= 6°), $\theta_s = 3°$. The direction of the stator pole magnetization corresponds to the direction of applied currents as shown.

(poles 1 and 5 are magnetized by I_1 in the direction to provide the low-reluctance path for the flux at this time). The flux continues its path along the stator frame and enters the rotor of section B through stator poles 3 and 7 (note again that I_1 magnetizes poles 3 and 7 in the direction to provide a low-reluctance path for the flux to enter the rotor of section B), thus completing its closed path at the south pole of the permanent magnet. It may be added that the permanent magnet flux is reinforced by the flux due to the phase windings.

To advance the motor shaft one step in the CW direction, I_1 must be removed and I_2 applied to phase 2 (ϕ_2). To understand why ϕ_2 must be energized next to produce CW motion, observe the position of the stator-rotor teeth in both sections in Fig. 3–26. Observe the teeth, which are blackened for reference. The black teeth are the closest to being aligned (they are exactly one step away from being aligned) with the

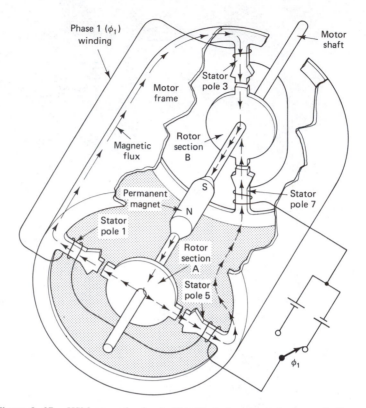

Figure 3–27. HSM magnetic circuit. This illustration shows the magnetic flux path when phase 1 is energized, and flux exits section A via poles 1 and 5, and enters section B rotor via poles 3 and 7. Most of the rotor-stator teeth and other details have been omitted to focus attention on the magnetic flux path.

teeth of stator poles 4 and 8 in section *A* and poles 2 and 6 in section *B*. There is one problem, however. Poles 4 and 8 and 2 and 6 are magnetized in the wrong direction by I_1, whose direction is as shown. To solve this problem, the direction of I_1 must be reversed. The complete four-step switching sequence for CW rotation is illustrated in Fig. 3–28, which shows the rotor position and stator pole magnetization direction in each section for the given polarities of phase currents. For CW rotation (as shown in the illustration) the sequence is $1^+-2^--1^--2^+-1^+$. This sequence is reversed for CCW rotation.

As the motor shaft advances one rotor tooth pitch in four steps, the step angle must be $\frac{1}{4}P_r$, or it may be expressed as the difference in the rotor-stator tooth pitches. Hence

$$\theta_s = \frac{P_r}{4} = \frac{360}{4N_r} = \frac{90}{N_r}$$

$$= |P_s - P_r| \qquad (3\text{–}48)$$

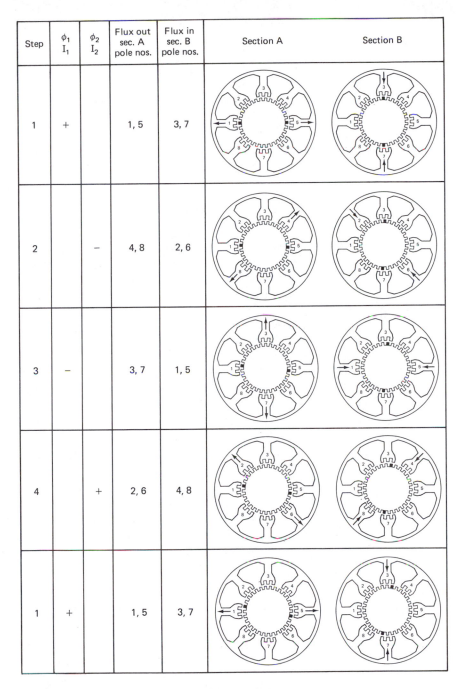

Step	ϕ_1 I_1	ϕ_2 I_2	Flux out sec. A pole nos.	Flux in sec. B pole nos.	Section A	Section B
1	+		1, 5	3, 7		
2		−	4, 8	2, 6		
3	−		3, 7	1, 5		
4		+	2, 6	4, 8		
1	+		1, 5	3, 7		

Figure 3–28. Four-step sequence of the two-phase HSM. The rotor position and the flux orientation are shown at each step of the sequence. $N_r = 30$, $N_s = 24$, $\theta_s = 3°$. As shown, the black rotor tooth advances CW 3° with each step for a total of 12° (one rotor tooth pitch) at the end of the sequence. For CW rotation the sequence is $1^+-2^--1^--2^+-1^+$. This sequence is reversed for CCW rotation.

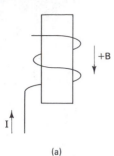

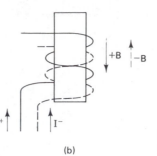

Figure 3–29. Windings: (a) unifilar; (b) bifilar.

(a)

(b)

In Fig. 3–28, $N_r = 30$ and $N_s = 24$; therefore, $\theta_s = 90/30 = 3°$, or θ_s may be calculated as $360/24 - 360/30 = 3°$.

As shown earlier, two polarities of phase currents must be provided to operate the HSM shown in the illustration, and thus impose a requirement that two power supplies (commonly referred to as a *bipolar drive*) are needed. The bipolar-drive requirement is unattractive from an economic standpoint; a *unipolar drive* (one power supply) is generally preferred.

The difference between a motor that uses a bipolar drive and one that works with a unipolar drive lies in the phase-winding structure. A stator phase winding that is *unifilar* (i.e., wound in a single file) requires a bipolar drive, whereas a unipolar drive can be used with a *bifilar* type of winding. With a unifilar winding, which was used in Fig. 3–26 and is also shown in Fig. 3–29a, the direction of current must be reversed to reverse the direction of B.

The bifilar winding shown in Fig. 3–29b includes two windings which are wound one over the other but with opposing sense. Consequently, applying current I^+ to the solid-line winding magnetizes the core with the north pole at the bottom and produces the field $+B$ in the downward direction, but when current I^- of the same magnitude and from the same power supply is applied to the dashed-line winding, the result is that the direction of magnetization and the direction of the field B are both reversed.

Imagine now that the HSM in Fig. 3–26 is bifilar wound. The original ϕ_1 winding is replaced by two windings ϕ_1^+ and ϕ_1^-, which are wound with opposing sense, and ϕ_2 is replaced with the opposing sense windings ϕ_2^+ and ϕ_2^-. We now have in effect four phases, and each of the phases may be excited by one polarity current. The superscripts $+$ and $-$ are used here to designate the direction of stator pole magnetization. Consequently, if ϕ_1^+ and ϕ_1^- are wound on poles 1, 3, 5, and 7 of both sections in Fig. 3–26, then if $+$ refers to the magnetization as shown, $-$ produces magnetization in opposite direction.

Figure 3–30 illustrates two switching circuits: one for the two-phase HSM using bipolar drive, and the other for the four-phase HSM with unipolar drive. The four-step switching sequence to produce CW rotation with one phase or two phases excited simultaneously applies to both configurations and is shown in Fig. 3–30c and d. The half-stepping eight-step sequence which is a combination of the sequences in Fig. 3–30c and d also applies to both configurations and is shown in Fig. 3–30e. For CCW rotation, the sequence must be reversed.

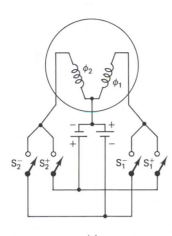
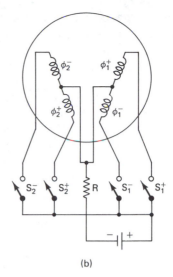

(a)   (b)

Step	S_1^+	S_1^-	S_2^+	S_2^-
1	X			
2				X
3		X		
4			X	
1	X			

(c)

Step	S_1^+	S_1^-	S_2^+	S_2^-
1	X			X
2		X		X
3		X	X	
4	X		X	
1	X			X

(d)

Step	S_1^+	S_1^-	S_2^+	S_2^-
1	X			X
2				X
3		X		X
4		X		
5		X	X	
6			X	
7	X		X	
8	X			
1	X			X

(e)

Figure 3–30. Hybrid stepper motor: (a) two-phase unifilar HSM using bipolar drive; (b) four-phase bifilar HSM using unipolar drive. The four-step CW sequence with phases excited: (c) one at a time; (d) two at a time; (e) eight-step half-stepping CW sequence. For CCW rotation the sequence in all cases is reversed.

3-8 STATIC TORQUE CHARACTERISTICS OF THE STEPPER MOTOR

When some of the rotor teeth are brought into alignment with the corresponding stator teeth, the rotor reaches an equilibrium position. It will remain in this position indefinitely unless the phase receives another excitation input. One such rotor equilibrium position is indicated by 0 in Fig. 3–31. In this position the motor develops no torque. If the rotor is displaced in the CCW direction an angle θ_1 by an external load T_L, the motor develops positive torque in the opposite direction (CW), to just balance the load. Had the rotor been displaced in the CW direction, the motor would then develop negative torque in the CCW direction. Thus the motor develops the restoring torque, which is always in the direction toward the original rotor equilibrium position. The resulting torque versus rotor displacement sine-wave-like characteristic (the actual shape depends on the rotor-stator structure), which is periodic with the period of the rotor tooth pitch P_r, is shown in Fig. 3–31.

The peak torque is maximum motor torque corresponding to the applied phase current. This torque is shown as T_H, the *holding torque* (some manufacturers refer to this torque as the *static torque*). The value of the peak torque as given in the commercial data sheets usually corresponds to the rated current. Although the motor-developed torque increases in some cases almost linearly with the applied phase current, it reaches saturation at the rated current where the rotor-stator magnetic structure saturates, and no appreciable increase in torque is realized for an increase in phase current.

The motor-developed torque is zero when the rotor is displaced by $\frac{1}{2}P_r$, and if displaced any further the rotor will slip to the next equilibrium position or perhaps to another position which is at a distance equal to some multiple of the rotor tooth pitch. When this happens, the synchronization between the applied pulses and motor steps, which is so critical in positioning applications, is lost. To assure that this does not happen, the stepper motor whose holding torque is much larger than the load torque must be selected.

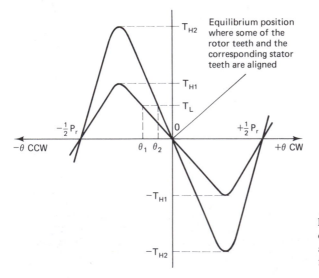

Figure 3–31. Static stepper motor torque characteristics. The motor-developed torque as a function of the rotor displacement from the equilibrium position.

Motors   Chap. 3

As the external load is removed, the rotor will advance (CW in the illustration) toward the original equilibrium position, designated by 0. As it does so, the motor-developed torque gradually decreases, reaching a point where the internal friction of the motor exceeds the motor-developed torque. The rotor must stop at this point, unable to reach the true equilibrium position. The uncertainty in the rotor position near equilibrium is generally referred to as the position error (with no applied load), which is typically less than 1°. This error is noncumulative; it is same for 1 step or 1000 steps.

As shown in the illustration, the motor with holding torque T_{H1} produces the positional error θ_1, and the motor with T_{H2} produces an error θ_2. As $\theta_1 > \theta_2$, it is clear that to reduce the error, either the load must be reduced or a motor with a larger holding torque must be used. The positional error with the external load may be estimated by assuming that the torque curve is sinusoidal (i.e., by the representation $T_m = -T_H \sin \theta$). As the period of the sine wave is 2π and the period of the torque curve is P_r, θ must be replaced by $(2\pi/P_r)\theta_d = N_r\theta_d$, where θ_d is the angular displacement of the rotor from the equilibrium position and N_r is the number of rotor teeth. Equating $T_m = T_L$ and $\theta_d = \theta_e$, the positional error may be calculated from

$$T_L = -T_H \sin N_r\theta_e \qquad (3-49)$$

Suppose that $T_H = 40$ oz-in. for a four-phase HSM, $N_r = 30$, and $T_L = 12$ oz-in. applied in the CW direction. The positional error may be calculated from Eq. (3-49) as

$$\theta_e = \frac{\sin^{-1}(-12/-40)}{30} = 0.58° = 34.9' \text{ (minutes of arc)}$$

The positional error is approximately 19% of the step angle, which is 3° (and 7% of the rotor tooth pitch, which is 12°). The positional error increases to 1.77° for a motor with $T_H = 15$ oz-in.

The positional error may also be estimated from the straight-line approximation of that part of the torque curve which is linear between the two peaks. Then

$$T_L = mT_H \qquad (3-50)$$

where the slope m is generally referred to as the *stiffness*. For a given load, the positional error is smaller for a motor with greater stiffness.

The choice of phase excitation method can affect significantly the torque production capability of the stepper motor. As discussed earlier, a stepper motor may be excited one phase at a time or two phases at a time. In general (and there are some exceptions), motor output torque is greater when more than one phase is excited at a time. This is illustrated in Fig. 3-32, where the four-phase HSM torque variation with rotor position is shown for both single-phase and dual-phase excitation. Although dual-phase excitation offers approximately 50% more torque over the single-phase excitation method, the power taken by the motor from the power supply is twice as much. This is the case in general; we get less additional torque for the additional expanded power. This type of uneven exchange may be acceptable, however, in a number of practical applications. Dual-phase excitation, as shown, does not change the step angle, and the rotor equilibrium position is merely shifted by half a step angle.

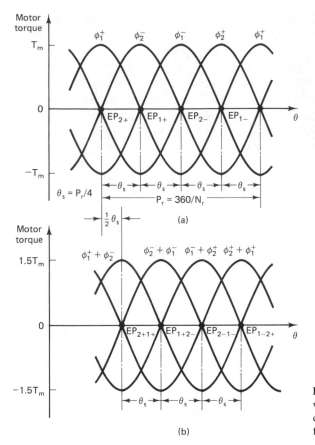

Figure 3–32. Motor torque variation with rotor position for: (a) single-phase excitation; (b) dual-phase excitation of the four-phase HSM.

3–9 DYNAMIC TORQUE CHARACTERISTICS OF THE STEPPER MOTOR

In Section 3–8 we considered a static case of the stepper motor, one where the rotor is in the equilibrium position and for all practical purposes the rotor does not move. In a practical application, however, the rotor must move so as to impart the required motion to a given load. In a load-positioning application, for instance, the motor would typically start from rest and accelerate the load to the desired position. To provide this type of motion, the motor must develop sufficient torque to overcome friction and to accelerate the total inertia. In accelerating the inertia, the motor may be required to develop a large amount of torque, particularly if the acceleration must be completed in a short time so as to position the load quickly. Inability of the motor to develop sufficient torque during motion may cause the motor stall, resulting in a loss of synchronization between the motor steps and phase excitation and consequently, resulting in incorrect positioning of the load.

The torque-producing ability of the stepper motor as a function of the operating speed is usually presented by the manufacturer in graphical form. One such representation

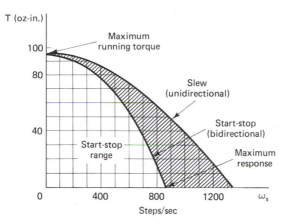

T (oz-in.)

100 — Maximum running torque

80

Slew (unidirectional)

40

Start-stop (bidirectional)

Start-stop range

Maximum response

0   400   800   1200 ω_s

Steps/sec

Figure 3–33. Stepper motor speed–torque characteristic.

is shown in Fig. 3–33. Closer inspection of this graph reveals two distinct regions of operation: the start-stop range and the slew range.

The *start–stop range* is between the coordinate axes and the start–stop curve. When operating within this range, the stepper motor may stop and start again or reverse the direction of motion without losing a step. The coordinates of the maximum torque developed by the motor at the operating speed within the start–stop range lie on the start–stop curve. The point of intersection between the start–stop curve and the vertical axis is defined as the *maximum running torque* (it is 95 oz-in. in this case), and the *maximum response* is defined at the point where the start–stop curve intersects the horizontal axis. The maximum response in the illustration is 850 steps/s. Thus the motor with no load can operate at the maximum speed of 850 steps/s. If the load torque is 50 oz-in., then according to the graph, the motor must not exceed the speed of 600 steps/s if it is to remain within the start–stop operating range.

Suppose that the given speed–torque characteristics apply to the motor in Fig. 3–26 and we wish to know the least time that this motor would require to advance a purely frictional load of 75 oz-in. through an angle of 240° without acceleration, assuming that the load must stop abruptly at the end of motion. The sudden stop requires the motor to operate in the start–stop range. From the graph, the maximum speed corresponding to 75 oz-in. in the start–stop range is 400 steps/s. Since the step angle of the given motor is 3°, the motor must make 240/3 = 80 steps, and the time required to advance 80 steps at 400 steps/s must be 80/400 = 0.2 s.

The *slew range* is shown in the illustration as the shaded area between the start–stop and slew curves. If the motor is to operate within the slew range, it must not stop or start or change direction. This is the unidirectional range of operation, where no steps will be missed by the motor during the motion, provided that the motor does not stop, start, or reverse direction in that range. The operation above the slew range is not permitted under any circumstances.

To establish operation within the slew range, the motor must first establish operation in the start–stop range, and then using the controlled acceleration ramp as shown in Fig. 3–34, make the transition to the slew range. To make such a transition, the motor must develop additional torque to accelerate the total inertia, including the rotor and

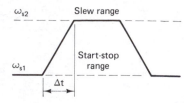

Figure 3–34. Ramping from start–stop range to slew range.

the load from ω_{s1}, as shown, to ω_{s2} in the time interval Δt using, as an approximation, a perfectly linear acceleration ramp. Equation (3–33) may be used to calculate the inertial torque, where $\dot{\omega}_s = d\omega_s/dt$ may be replaced by $\Delta\omega_s/\Delta t$; hence

$$T_J = J_T \frac{\Delta\omega_s}{\Delta t} = J_T \frac{\Delta\omega_s}{\Delta t} \frac{1}{R_s} 2\pi \qquad \text{oz-in.} \qquad (3\text{–}51)$$

where J_T = total inertia, including rotor and load in oz-in.-s^2

$\Delta\omega_s = \omega_{s2} - \omega_{s1}$ in steps/s

R_s = stepping rate as defined by Eq. (3–44) in steps/rev and

2π rad/rev converts ω_s from steps/s to rad/s

The transition must be made under controlled-acceleration conditions from the slew range to the start–stop range if the motor must be stopped without loss of steps. It may be assumed that it takes the same time to ramp down as to ramp up (actually it takes slightly less time). During the ramping-down-time interval, the inertial reaction torque is negative; consequently it aids the motor developed torque.

Suppose that the motor shown in Fig. 3–26 drives a frictional load of 20 oz-in. at 500 steps/s. Assuming that the speed–torque characteristics if Fig. 3–33 apply, the combined inertia of the rotor and the load is 0.15 oz-in.-s^2. It is required to determine the shortest time to make the operational transition to 1000 steps/s. From the speed–torque characteristics, the motor develops the maximum torque of 40 oz-in. at 1000 steps/s and 80 oz-in. at 500 steps/s. Since the actual net torque available to accelerate the inertial load into the slew range is between (40–20) and (80–20), it is a common practice to use the average of the two resulting in $T_{\text{avail}} = [(40 - 20) + (80 - 20)]/2 = 40$ oz in. Using this value in Eq. (3–51) gives us

$$40 = \frac{0.15(1000 - 500)2\pi}{120 \, \Delta t}$$

Solving for Δt yields

$$\Delta t = 98 \text{ ms}$$

Since the area under the speed versus time curve represents the distance travelled, it follows that it takes $\frac{1}{2}(1000-500) (0.98) = 24.5$ steps to advance the rotor speed at a constant rate from 500 steps/s to 1000 steps/s.

The motor-developed torque in Fig. 3–33 is often called the *pull-out torque*. This terminology applies particularly to the stepper motor, whose rotor is required to make a step for each applied excitation pulse to the phase winding. The pull-out torque is the maximum motor-developed torque. When the load torque exceeds the pull-out torque, the one-to-one relationship between the applied pulses and rotor steps, is lost.

As shown in Fig. 3–33, the pull-out torque decreases with increasing speed. The degradation in torque is due primarily to the reduction in the phase-winding current at high operating speeds. A typical phase winding may be represented by a series RL circuit. The response of this circuit to step input of amplitude E is, from basic theory, $I_{ph} = (E/R)(1 - e^{t/\tau})$, where E/R is the final value of phase current, which in this case must be the rated phase current, and $\tau = L/R$ is the time constant of the phase winding. The phase excitation may be taken as a 50% duty cycle square wave of frequency ω_s pulses/s; hence the duration of one pulse $T = \dfrac{2\pi}{\omega_s}$. If $T/\tau = 5$ (the duration of phase excitation is five time constants) is used in the equation above, then I_{ph} is approximately equal to the rated value of E/R at the end of the excitation, as illustrated in Fig. 3–35a for 100 steps/s. In Fig. 3–35b the duration of phase excitation is shorter for the operating speed of 500 steps/s and the phase current reaches 63% of the rated value at the end of the excitation period. At 1000 steps/s the phase current is at 39% of the rated value when the excitation pulse terminates (Fig. 3–35c). The phase current is thus unable to establish its rated value as the operating speed is increased. As the value of phase current has a direct bearing on the pull-out torque, the torque must decline with the increase in the operating speed.

This problem may be corrected by inserting an external resistor in series with the winding, as shown in Fig. 3–35d. When a resistor whose value is four times the phase winding resistance is connected in series, the time constant is reduced to one-fifth of its original value, resulting in a dramatic improvement in phase current, which is able in the case of 1000-steps/s input to establish 92% of its rated value at the end of the excitation period in contrast to its value of 39% before the $4R$ external resistor was added. Although there is not much change in torque at low speeds, at high speeds, as illustrated in Fig. 3–36, the start–stop curve in Fig. 3–35e with the maximum response of 850 may be pulled up considerably with the added $4R$ series resistor. The price that

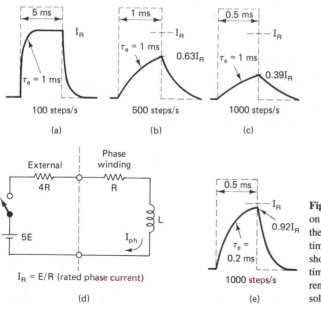

Figure 3–35. Effect of the stepping rate on phase current. At high stepping rates the phase current does not have sufficient time to establish its rated value. Part (e) shows that an external resistor reduces the time constant, thus raising the phase current and torque. Dashed lines, excitation; solid lines, phase current.

Sec. 3–9   Dynamic Torque Characteristics of the Stepper Motor

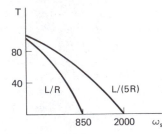

Figure 3–36. The added external resistor reduces the phase winding time constant and improves the motor pull-out torque.

one pays for all this is a bigger power supply. In the case of the $4R$ added resistor, the power supply must output five times as much voltage and five times as much power if the phase-winding rated current is to be maintained at its original value.

Consider next the dynamic situation illustrated in Fig. 3–37, where the load torque T_1 is driven by a three-phase motor. This is a graph of the instantaneous motor-developed torque as a function of rotor position. Three (one for each phase) motor torque versus rotor displacement curves of the type shown in Fig. 3–31 are combined to form the composite three-phase motor dynamic torque characteristic, where the peaks are spaced by the rotor step angle θ_s.

In a load-driving application such as this, one important condition must almost always be satisfied: The instantaneous motor-developed torque should be greater than the load torque to provide a net torque used to accelerate the rotor to the next step position. Suppose that ϕ_1 is excited and the rotor is advancing toward the ϕ_1 equilibrium position EP_1 (as it should). However, before reaching EP_1, ϕ_2 is fired at point A, resulting in the torque jump to point B (the torque at B is the torque due to ϕ_2 at the time when switching from ϕ_1 to ϕ_2 occurs). At this time ϕ_2 takes over and accelerates the load to the next step position at point F, where ϕ_3 is fired and the cycle is repeated.

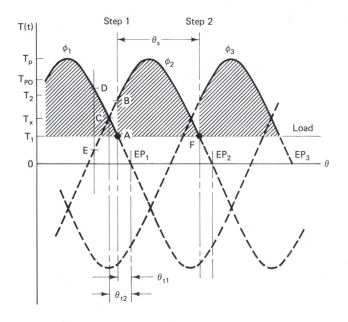

Figure 3–37. The instantaneous torque produced by the motor must exceed the load torque to accelerate to the next step position. The shaded area represents the net torque available for load acceleration.

Motors   Chap. 3

The shaded areas represent the net torque available to accelerate the load to the next step position.

If the $\phi_1 \rightarrow \phi_2$ transition was not done at point A (where the ϕ_1 torque at A is equal to the load torque), then shortly thereafter the ϕ_1 torque would have dropped below the load torque T_1 and the motor could have stalled. The firing of the phases in succession at correct rotor position where phase switching takes place is necessary to optimize the net torque production (the shaded area) which is required to accelerate the load to the next step position. For that reason the phases under loaded conditions are never fired at their individual equilibrium positions (EPs), which are along the zero torque axis.

The *torque angle* θ_t (as it is sometimes called) can be used to mark the rotor position where the firing of the given phase occurs in relation to the equilibrium position of the preceding phase. The torque angle thus specifies the angle by which the firing of phase is advanced in relation to the static equilibrium position of the rotor. In Fig. 3–37, θ_{t1} is the torque angle corresponding to the load T_1. The firing of ϕ_2 can also be done at point C, the crossover torque T_x, with the corresponding torque angle θ_{t2}. In practice, a firing point greater than θ_{t2} is seldom, if ever, used because if ϕ_2 was fired, for example at point D, the instantaneous torque would drop to point E which is below the load torque, resulting in a possible stalling of the motor.

The load torque must under no circumstance be larger than the pull-out torque T_{PO}, whose value depends on the number of phases and is generally between T_x and the peak torque T_p. If the load inertia is negligible, the maximum load torque must not exceed T_x. On the other hand, a load torque such as T_2, where the load has a very large inertia, may be acceptable. The reasoning is that even if the instantaneous motor torque at times drops below T_2, there is sufficient kinetic energy in the inertia to coast the load to the next firing position. T_x increases with the number of phases, and as was observed before, the torque-producing, and therefore the load-driving capability of the motor increase.

3–10 MECHANICAL RESONANCE

Some stepper motors exhibit an oscillatory response by the rotor as it reaches a step position. This is illustrated in Fig. 3–38. There are two excitation pulse inputs that will be used in the explanation: ω_{s1} is at the resonant stepping rate and ω_{s2} is not. First, assume that ω_{s2} is applied. The first pulse at t_1 advances the rotor to position 2. Upon reaching position 2, the rotor begins to oscillate in a typical underdamped system fashion. As the rotor moves CW, it overshoots position 2 at t_3, then it reverses direction and undershoots position 2 at t_5 in moving CCW, then it overshoots again at t_7. As the peaks of the oscillation decrease in time, the rotor finally settles at position 2 before the next pulse is applied. The next pulse, at t_{13}, advances the rotor to position 3 and after oscillating about position 3 as it has done in position 2, the rotor settles in that position. The motion repeats when the rotor is driven to position 4 by the pulse at t_{16}.

The angular speed of the rotor may be expressed as $\omega = d\theta/dt$. The speed ω may therefore be viewed as the slope of rotor's angular response. At points such as t_3, t_5, t_9, and t_{11}, where the rotor position peaks and the slope is zero, the speed must be

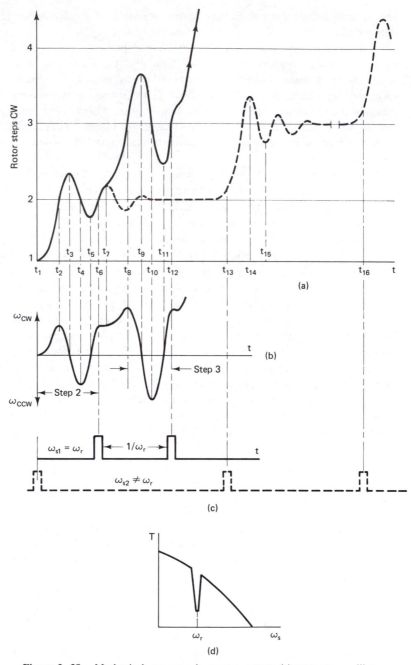

Figure 3–38. Mechanical resonance in stepper motors: (a) rotor step oscillatory response; (b) rotor velocity; (c) phase excitation at ω_r (solid line) causes loss of torque at ω_r stepping rate as shown in part (d).

zero. At points of inflection where $\ddot{\theta} = \dot{\omega} = 0$ (zero acceleration), the rotor speed goes through its relative maxima and minima: maxima (CW) at t_2, t_6, and t_8 and minima (CCW) at t_4 and t_{10}. Applying an excitation pulse at or slightly before t_6 when the rotor is moving with maximum velocity toward position 3 advances the rotor to position 3, where it oscillates with a larger amplitude than it did in position 2; compare the positive peaks at t_3 and t_9 or negative peaks at t_5 and t_{11}. Another pulse applied at t_{12} when the rotor is moving with the maximum velocity (which is greater than that at t_6) toward position 4 causes the rotor, which has acquired a sufficient kinetic energy, to shoot past position 4 in an oscillatory response whose amplitude may be several step angles. The synchronism between the steps and the excitation pulses is now lost and the value of the pull-out torque drops, as shown in Fig. 3–38d, below its normal value. For these reasons the resonant stepping rate ω_r should be avoided during normal operation.

In most PM stepper motors the high rotor kinetic energy that is associated with the resonant response is dissipated when the rotor interacts with the stator fields. The VRSM does not have a permanent magnet; hence there is no mechanism for dissipating the resonant kinetic energy, although there are some energy losses due to eddy currents and hysteresis. The VRSM is then more susceptible to the resonant response than are other stepper motors.

Damping of the resonant response may be accomplished mechanically or electronically. A viscous damper is one example of a mechanical damper. The damper unit, which contains a viscous fluid, is coupled directly to the shaft of the motor. During the fast changes in rotor speed, the viscous friction of the damper provides a means of dissipating the kinetic energy of resonant response and thus lowers or eliminates the rotor resonant oscillations. The viscous friction and the inertia of the damper act as a load on the motor and consequently may lower the torque delivered to the load and the response time of the motor. Belts, pads, and clutches are some of numerous other mechanical dampers. Electronic damping probably provides the most satisfactory type of damping, as it does not affect the normal operation of the motor. It is accomplished as follows: As the rotor moves and the phases are energized in sequence, a pulse of current in a direction opposite to the normal direction is applied to the winding which was just deenergized. This current pulse generates a magnetic field which produces a torque on the rotor (for a short time) opposite to the direction of rotor's motion, thus providing a retro torque or a breaking torque. As the precise timing of the current pulses is very critical, the electronic damper board may be complex and expensive.

3–11 PROGRAMMABLE SEQUENCER DESIGN

The sequencer is an important part of the stepper motor controller, which, as mentioned earlier, simplifies for the user the task of operating the stepper motor. The input clock pulses provide the stepping rate for the motor, and the direction input controls the direction of motor rotation. Figure 3–39 illustrates the design of a typical sequencer from the state assignment to the final hardware product. As it is assumed that the

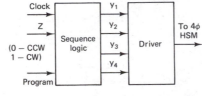

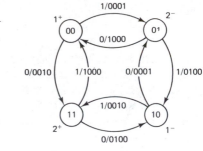

State assignment

State		y_1	y_2	y_3	y_4
1^+ or ϕ_1^+	00	1	0	0	0
2^- or ϕ_2^-	01	0	0	0	1
1^- or ϕ_1^-	10	0	1	0	0
2^+ or ϕ_2^+	11	0	0	1	0

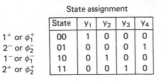

State Table

MT	Present state Q_b	Q_a	Z	Next state Q_b	Q_a	Outputs y_1	y_2	y_3	y_4	Q_b J_b	K_b	Q_a J_a	K_a
0	0	0	0	1	1	0	0	1	0	1	x	1	x
1	0	0	1	0	1	0	0	0	1	0	x	1	x
2	0	1	0	0	0	1	0	0	0	0	x	x	1
3	0	1	1	1	0	0	1	0	0	1	x	x	1
4	1	0	0	0	1	0	0	0	1	x	1	1	x
5	1	0	1	1	1	0	0	1	0	x	0	1	x
6	1	1	0	1	0	0	1	0	0	x	0	x	1
7	1	1	1	0	0	1	0	0	0	x	1	x	1

Boolean Functions

$K_a = \Sigma\,(2, 3, 6, 7)$, $\quad D = \Sigma\,(0, 1, 4, 5)$
$J_a = \Sigma\,(0, 1, 4, 5)$, $\quad D = \Sigma\,(2, 3, 6, 7)$
$J_b = \Sigma\,(0, 3)$, $\quad D = \Sigma\,(4, 5, 6, 7)$
$K_b = \Sigma\,(4, 7)$, $\quad D = \Sigma\,(0, 1, 2, 3)$
$y_1 = \Sigma\,(2, 7)$
$y_2 = \Sigma\,(3, 6)$
$y_3 = \Sigma\,(0, 5)$
$y_4 = \Sigma\,(1, 4)$

X = don't care condition

Boolean Function Minimization

Minterm designation for a 3 variable mahoney map

	$\bar{Z}$	Z	$\bar{Z}$	
$\bar{Q}_a$	0	1	5	4
Q_a	2	3	7	6
	$\bar{Q}_b$		Q_b	

$K_a = 1$

$J_a = 1$

$K_b = Z \odot Q_a$

$J_b = Z \odot Q_a$

$y_1 = Q_a(Z \odot Q_b)$

$y_2 = Q_a(Z \oplus Q_b)$

$y_3 = \bar{Q}_a(Z \odot Q_b)$

$y_4 = \bar{Q}_a(Z \oplus Q_b)$

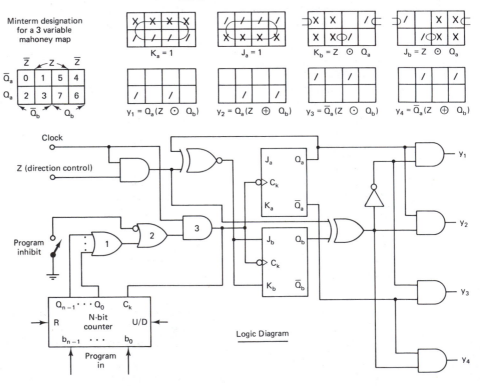

Figure 3–39. Programmable sequencer design.

reader is familiar with the sequential logic design, the discussion of the design procedure here is brief.

In this illustration the sequencer controls a four-phase stepper motor. The four phase windings to be controlled by the sequencer are designated as y_1, y_2, y_3, and y_4. The four-phase HSM with unipolar drive, as shown in Fig. 3–30b, actually has six input leads, but two of these leads are tied together and connected to one side of the power supply through R. As R is in series with the phase winding, it controls the time constant and, as shown in Fig. 3–35, improves the stepper motor operation at high speed. The remaining four leads receive their signal from the driver, which, in turn, receives its four inputs from the sequencer. The driver is a power amplifier that supplies the required value of current to the phase windings of the motor.

The first step in a design of this type is to decide on the number of memory elements required to accommodate all states. One binary state is assigned to each of the four phase windings, hence two memory elements are required to represent four states. The state assignment table (which is same as Fig. 3–30c, where one phase is excited at a time) shows 2-bit state assignments and the binary value of the corresponding outputs. This information is translated to the state diagram, which shows the state transitions induced by the clock pulse and the value of the direction input Z, together with the corresponding next-state output values for each transition.

The state table includes the information contained in the state diagram: the present state of the memory combined with Z, the next state, and the binary values of the outputs. The values shown in the Q_a and Q_b columns are based on the operation of the JK flip-flop and the given present-to-next state transitions. The resulting J, K inputs to flip-flops A and B and the four outputs are represented by Boolean functions in the first canonical form, and then minimized using the three-variable Mahoney map.

The minimized Boolean functions are finally implemented in the logic diagram with two flip-flops and gates. This is the final diagram of the sequencer, which accepts on the input the clock and the direction control Z, and provides the four-phase excitation control signals for the four-phase HSM. The programming feature is also included in this design. The n-bit counter shown is preset manually or by a microprocessor to the number of steps that the stepper motor must execute. The program inhibit switch is then open (enabling the program), the counter's down count mode enabled, and the clock pulses from the output of AND gate 3 cause the counter to count down. As long as the state of the counter is not all 0's, the output of OR gates 1 and 2 is HIGH. The HIGH, together with the clock at the input to AND gate 3, results in the clock pulses at the output of AND gate 3, which are applied to the sequencer and the counter. When the counter reaches the state of all 0's, the 0 input to AND gate 3 from OR gate 2 disables the clock from being applied to the sequencer and the counter. At that point the counter stops and the stepper motor stops, having advanced through the N steps stored in the counter at the beginning of the sequence. Closing the PROGRAM INHIBIT switch (shown as a manual control, although it can be a flip-flop receiving its input from a microprocessor) disables the program, allowing the clock pulses to be applied to the sequencer in a normal fashion. The programmable sequencer feature may prove to be particularly useful in robotics application where several programs (of the counter) must be executed in order and the sequencer outputs (y_1 to y_4) switched between several stepper motors to achieve the required motion of the robot.

The driver circuit typically includes a power switch capable of high-current operation and a low-power amplifier stage which precedes the power switch. The power darlington pair, which is available commercially in a *pnp* or *npn* configuration on an IC chip, can be used as a power switch, and an op amp can be used as a low-power amplifier. The driver circuit receives its input from the sequencer and outputs an excitation current of the appropriate level to the phase windings of the stepper motor. Figure 3–40 illustrates several typical driver configurations.

Figure 3–40a shows a driver for a three-phase stepper motor. As one power supply is used, which provides one polarity of current, this driver is unipolar. The circuit includes three identical sections, one for each phase winding. The phase windings $L_{\phi 1}$, $L_{\phi 2}$, and $L_{\phi 3}$, which are actually part of the motor's armature, are shown here for simplicity as being a part of the driver circuit. In a typical operation, the ϕ_1 output of the sequencer is amplified by the A_1 stage and its output applied to the Darlington pair Q_1–Q_2. The Darlington pair applies the required value of the excitation current (which flows into the tied collectors of Q_1 and Q_2) to $L_{\phi 1}$. When Q_1 and Q_2 are turned off, the fast decrease in current through $L_{\phi 1}$ produces a very large positive voltage ($L_{\phi 1} \, dI_1/dt$; as dI_1/dt is negative, the voltage induced into $L_{\phi 1}$ is in the same direction as the current I_1, i.e., positive on the collector side). This *inductive kick*, as it is sometimes called, is of sufficient amplitude to destroy the transistors. This is the reason the diode D_1 is connected across $L_{\phi 1}$. When Q_1 and Q_2 are ON, D_1 conducts very little, but when the Darlington pair is turned OFF and the inductive kick develops, D_1 turns ON, thus providing a discharge path for the inductive current, which at this time is in a state of decay. The added series damper resistor R_d limits the D_1 current and thus prevents possible overheating of D_1. The discharge current circulates in a series circuit consisting of D_1, R_d, the resistance, and the inductance of the ϕ_1 winding R_1 and $L_{\phi 1}$. The operation of the other two sections is the same. R_t is used to improve the high-speed performance of the stepper motor, and has the same meaning as that of the external $4R$ resistor in Fig. 3–35.

The driver shown in Fig. 3–40b is bipolar, as it uses a dual-polarity power supply. The level shifter is used to raise the bases of the ϕ_1^+ and ϕ_2^+ excitation pulses to the $+E$ level, and to lower the bases of the ϕ_1^- and ϕ_2^- excitation pulses to the $-E$ level. This is necessary because op amps A_1 and A_3 and the emitters of Darlingtons Q_1–Q_2 and Q_5–Q_6 are at $-E$ reference potential, and op amps A_2 and A_4 and the emitters of Darlingtons Q_3–Q_4 and Q_7–Q_8 are at $+E$ reference potential. The driver in this case provides two-phase control. The ϕ_1^+ and ϕ_2^+ control signals from the sequencer drive the two phase windings with the currents $+I_1$ and $+I_2$ as shown, whereas the ϕ_1^- and ϕ_2^- control signals drive the phase windings with currents $-I_1$ and $-I_2$, which are in the opposite direction. Damper diodes D_1 to D_4 provide the damping function as described for Fig. 3–40a, except in this case they are connected between the collector and emitter of the Darlington because the emitters here are not grounded as they were in Fig. 3–40a. If, for example, the Darlington Q_1–Q_2 switches from ON to OFF, the voltage induced into $L_{\phi 1}$ is positive on the collectors of Q_1 and Q_4. When this voltage exceeds $+E$, D_2

conducts and circulates current in the series circuit consisting of D_2, R_d power supply $+E$, $L_{\phi 1}$, and R_t, which consists of the $L_{\phi 1}$ winding resistance and external resistance added to improve the high-speed operation of the motor. Actually, a resistor of several kilohms may be placed across D_2, which will not significantly load the phase currents and will provide a faster discharge path for the inductive current at the point when the inductive voltage drops below $+E$ and the current would otherwise be forced to discharge through the high resistance of D_2, which is reverse biased. Such resistors should also be used across the other damper diodes. When the Q_3–Q_4 Darlington is switched OFF and its collector voltage becomes negative, D_1 turns ON and provides the discharge path for the inductive current. The operation of the circuit that controls the ϕ_2 current is exactly the same.

The driver shown in Fig. 3–40c provides two polarities of phase current with only one power supply. This is made possible by the arrangement of the Darlington switches in a Wheatstone bridge configuration; one bridge structure is used for each phase winding. As shown, the level shift is provided for the Darlingtons whose emitters are not at ground potential. The positive side of the power supply is used here as the reference potential. At any given time during the operation two Darlingtons must be conducting. A ϕ_1^- control pulse from the sequencer, for example, switches ON Darlingtons Q_5–Q_6 and Q_7–Q_8. This results in current flow through the phase winding $L_{\phi 1}$ from A to B. The current then completes its path through the Darlington Q_5–Q_6, through the power supply, and finally through the Darlington Q_7–Q_8. When the Darlingtons Q_5–Q_6 and Q_7–Q_8 are turned OFF, the voltage induced into $L_{\phi 1}$ due to a rapidly decreasing current is $-$ at A and $+$ at B. This forward biases diodes D_1 and D_4, which provide the discharge path for the current as shown in Fig. 3–41. Diodes D_2 and D_3 are reversed biased at this time. As shown, diodes D_1 and D_4 become forward biased when the winding voltage exceeds the power supply voltage E. As shown in the illustration, 10-kΩ resistors across D_1 and D_4 provide the path for the discharge current when the phase-winding voltage drops below E, and D_1 and D_3 become reversed biased.

The ϕ_1^+ control signal from the sequencer switches ON Darlingtons Q_1–Q_2 and Q_3–Q_4 and reverses the direction of current through $L_{\phi 1}$. When the Darlingtons are switched OFF, the polarity of the winding voltage is opposite to that shown in Fig. 3–41. The diodes D_1 and D_4 are now reverse biased, and D_2 and D_3, which are forward biased, provide the discharge path for the phase current (the 10-kΩ resistors may also be used across D_2 and D_3). The resistor R_t has the same meaning as discussed in the description of Fig. 3–40b. The operation of the bridge circuit used for ϕ_2 is exactly the same as that for ϕ_1. This driver provides two polarities of phase current using only one power supply, in contrast to Fig. 3–40b, which does the same but uses two power supplies. The main disadvantage of the bridge driver is that it uses twice the components of Fig. 3–40b. The cost-effectiveness, in this case, must be considered by the user, who must choose between a board with more components using one power supply or a board with fewer components which uses two power supplies.

The driver shown in Fig. 3–40d uses one power supply and provides one polarity of phase current for the stepper motor with bifilar windings. The structure and the operation of this driver are similar to the driver in Fig. 3–40a.

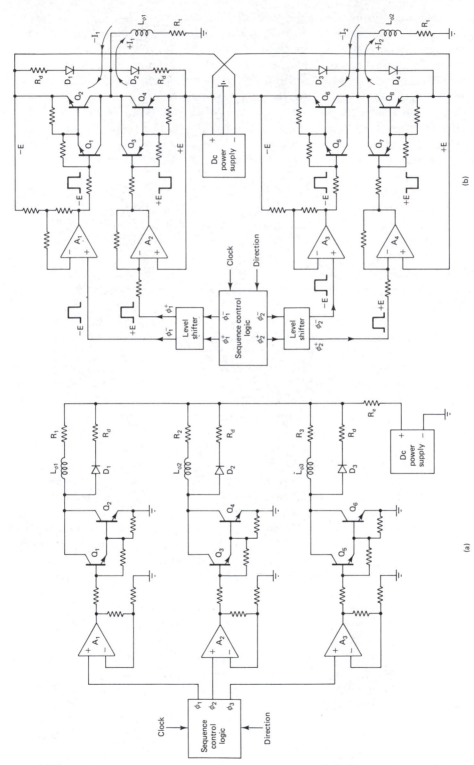

(a)

(b)

Figure 3–40. Stepper motor drivers: (a) three-phase unipolar driver that can be used to drive a three-phase VRSM; (b) two-phase bipolar driver that can be used to drive a two-phase unifilar HSM; (c) two-phase unipolar driver using a Wheatstone bridge configuration of power switches to provide the reversal of current through the phase winding using only one power supply; (d) four-phase unipolar driver that can be used to drive a four-phase bifilar HSM.

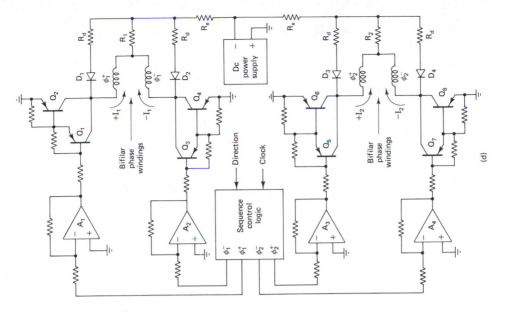

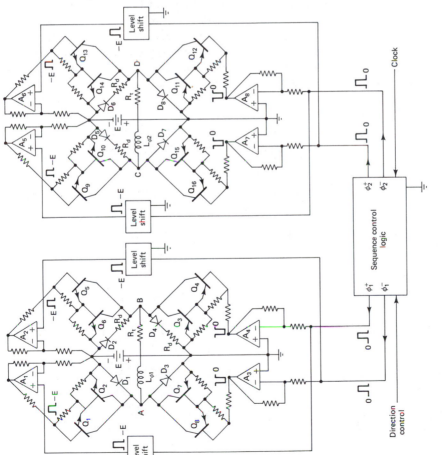

Figure 3–40. (*continued*)

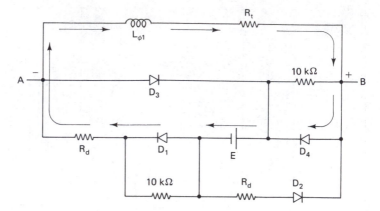

Figure 3–41. The discharge path of the ϕ_1 current at the instant Darlingtons Q_5–Q_6 and Q_7–Q_8 are switched OFF. D_2 and D_3 are reverse biased.

Bilevel Drive

One problem common to all drivers just discussed is the difficulty in establishing the rated current (required for optimum motor generated torque) in phase windings at high operating speeds. As mentioned earlier, the reason for this is the excessive value of the time constant. The added external resistor reduces the time constant and improves the high-speed operation, but at the same time it contributes to an inefficient operation, as the power supply (which must be now larger to maintain the phase current at the rated value) must provide power which is wasted in the external resistor.

The bilevel drive offers an alternative for achieving a higher motor performance with lower power consumption. A typical bilevel-drive configuration is shown in Fig. 3–42. This circuit is considerably simplified to simplify the description. The toggle

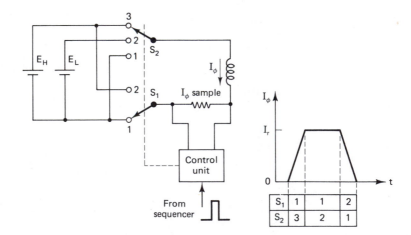

Figure 3–42. Bilevel drive.

switches are used here to represent the switching action by the control unit, and in the actual circuit, solid-state switches are used. The voltage developed across the sampling resistor provides a measure of the phase current for the control unit. The control unit also receives the phase excitation signal from the sequencer, and on the basis of these two signals generates the control signals to switch the two power supplies. At the beginning of the excitation period, the control unit applies the high voltage E_H (which is several times the rated voltage of the motor) to the phase winding, resulting in an extremely fast current rise. As soon as the current reaches the rated value I_r, the control unit disconnects E_H and applies the low voltage E_L, which is of sufficient value to maintain the rated current for the duration of the excitation period. At the end of the excitation period the control unit disconnects E_L and applies E_H after reversing its polarity. As at this time E_H opposes the phase-winding current, it speeds up its decay. Thus E_H provides a fast rise of the phase current to its rated value at the beginning of the excitation period and a fast decay to zero at the end of the excitation period, while E_L maintains the phase current at its rated value during the excitation period.

Chopper Drive

The simplified chopper drive configuration shown in Fig. 3–43 also uses a phase-current sampling resistor and a control unit that provides the control signal for switching the high-voltage power supply E_H. At the beginning of the excitation period the control unit applies E_H to the phase winding. As soon as the phase current reaches I_{max}, E_H is disconnected by the control unit, and the phase current, in the absence of drive, begins to decay. When the current falls to I_{min}, the control unit reconnects E_H, the current begins to rise, and the cycle repeats; that is, as shown in the illustration, the current rises and falls as E_H is switched in and out of the phase-winding circuit, with the average value I_{avg} maintained at the rated current I_r. The chopper drive has fast response, but the circuit is complex and expensive. Both the bilevel and the chopper drives improve the performance of the stepper motor at high stepping rates at reduced power consumption.

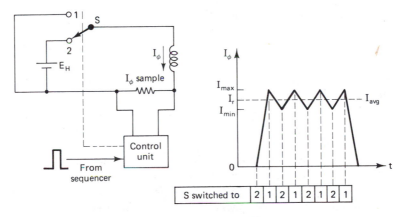

Figure 3–43. Chopper drive.

3-13 PRACTICAL STEPPER MOTOR MOTION CONTROL SYSTEM

A practical motion control system using a stepper motor is illustrated in Fig. 3–44. This system was constructed and tested in the lab with excellent performance results. The system structure consists of three general sections: the sequence logic, the power amplifier or the driver section, and of course the stepper motor itself, which may be any four-phase stepper motor (a 1.8° hybrid stepper motor and a 7.5° variable-reluctance stepper motor were tested in the lab with this system).

The input to the system may be applied in any one of several ways: as clock pulses applied to gate 3 of the sequencer together with the direction input Z, from a programmable counter, or from a microprocessor. With the PROGRAM INHIBIT switch closed, the programmable counter is inhibited (has no effect on stepper motor operation), and the applied clock pulses cause the stepper motor to operate in a constant speed mode, with the speed of the motor being dependent on the frequency of the pulses. The programmable counter may be used for position control. This is accomplished by opening the program inhibit switch, storing in the counter a number that corresponds to the number of steps that the motor is to advance (which may be done manually or by a microprocessor), and applying a clock to gate 3. As the counter down-counts to the 0 state, the motor advances the desired number of steps. At the end of the cycle the 0 state of the counter produces a 0 at the output of gate 2, thus disabling the clock and causing the motor to stop.

When the sequencer is used, switches S_1 to S_4 are set to couple the outputs of gates 7 to 10 of the sequencer to the bases of Q_1 to Q_4 of the power driver. For example, a HIGH applied to the base of Q_1 switches Q_1 ON and produces a LOW at the base of Q_5 (Q_5 to Q_8 are power transistors rated at 10 A collector current). This causes Q_5 to saturate, and its collector current (equal to the phase-winding rated current) flows through the ϕ_1 winding and through the 100-Ω resistor, which limits the collector current and also lowers the phase-winding time constant, thus improving the high-speed operation of the stepper motor. The Q_2-Q_6, Q_3-Q_7, and Q_4-Q_8 transistor combinations control the remaining phases (ϕ_2, ϕ_4, and ϕ_3, respectively) in exactly the same way. Diodes D_1 to D_4 provide their usual function, to limit the inductive kick at the instant when the power transistor turns OFF. This is the sequencer whose design is illustrated in Fig. 3–39.

When a microprocessor is used as an input device, switches S_1 to S_4 are set to connect the output port A to the input of the power driver. When configured in this mode, the microprocessor is totally dedicated to controlling the motion of the stepper motor, and the sequencer is not required.

Shown in Fig. 3–45 are two programs written for the 8085 microprocessor in Fig. 3–44. In both programs, port A is configured as the output port by the MVI A, 01H instruction, memory locations 0001 to 0004 store the phase excitation data, and memory locations 0005 and 0006 store the desired time delay data, which are used to control the motor speed. Execution of the speed control program begins by outputting a HIGH (the contents of memory location 0004 as one of 4 bits from port A and thus exciting one of the motor phases. The time delay (the contents of memory locations 0005 and 0006) is executed before the next phase is excited by the contents of 0003. In this manner the four phases are excited sequentially with a time delay between any

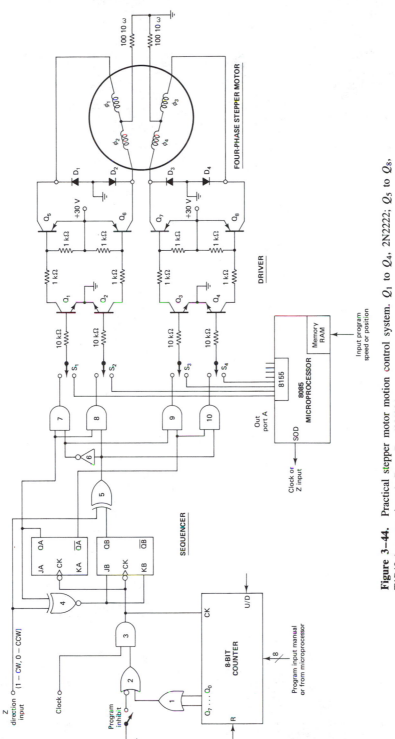

Figure 3–44. Practical stepper motor motion control system. Q_1 to Q_4, 2N2222; Q_5 to Q_8, T1P42 (power resistors); D_1 to D_4, 1N4004.

```
6 MEMORY LOCATIONS USED              0001 = 4TH PULSE
                                     0002 = 3RD PULSE
                                     0003 = 2ND PULSE
                                     0004 = 1ST PULSE
                                     0005 + 0006 = DELAY TIME

START          MVI A,01H       ;SET PORT A ON 8155 TO BE OUTPUT
               OUT X0          ;COMMAND WORD AT ANY PORT ADDRESS
REPEAT         LXI B,0004      ;4 STEPS IN TABLE
FINISH_TABLE   LXI H,0000      ;POINT TO BEGINNING OF TABLE
               DAD B           ;POINT TO STEP PULSE
               MOV A,M         ;GET STEP PULSE FROM TABLE
               OUT X1          ;APPLY STEP TO DRIVE CIRCUITRY/PORT A
               CALL DELAY      ;WAIT FOR DESIRED TIME
               DCX B           ;POINT TO NEXT STEP IN TABLE
               MOV A,C         ;
               ORA B           ;CHECK IF 4TH STEP APPLIED
               JNZ FINISH_TABLE
               JMP REPEAT
DELAY          PUSH H          ;SAVE ENVIRONMENT
               LHLD 0005       ;GET DELAY TIME FROM MEMORY
LOOP           DCX H           ;
               MOV A,L         ;
               ORA H           ;
               JNZ LOOP        ;LOOP UNTIL ZERO
               POP H           ;RESTORE ENVIRONMENT
               RET
```

(a)

```
8 MEMORY LOCATIONS USED              0001 = 4TH PULSE
                                     0002 = 3RD PULSE
                                     0003 = 2ND PULSE
                                     0004 = 1ST PULSE
                                     0005 + 0006 = DELAY TIME
                                     0007 + 0008 = # OF PULSES

START          MVI A,10H       ;SET PORT A 8155 TO BE OUTPUT
               OUT X0          ;COMMAND WORD AT ANY PORT ADDRESS
               LHLD 0007       ;GET # OF PULSES FROM MEMORY
               MOV E,L         ;
               MOV D,H         ;MOVE INTO NEW REGISTERS
REPEAT         LXI B,0004      ;4 STEPS IN TABLE
FINISH_TABLE   LXI H,0000      ;POINT TO BEGINNING OF TABLE
               DAD B           ;POINT TO STEP PULSE
               MOV A,M         ;GET STEP PULSE FROM TABLE
               OUT X1          ;APPLY STEP TO DRIVE CIRCUITRY
               CALL DELAY      ;WAIT FOR DESIRED TIME
               DCX D           ;DECREMENT # OF PULSES
               MOV A,E         ;
               ORA D           ;
               JNZ HOP_OVER    ;CONTINUE UNTIL # OF PULSES = ZERO
               HLT             ;STOP WHEN # OF PULSES = ZERO
HOP_OVER       DCX B           ;POINT TO NEXT STEP IN TABLE
               MOV A,C         ;
               ORA B           ;CHECK IF 4TH STEP APPLIED
               JNZ FINISH_TABLE
               JMP REPEAT
DELAY          PUSH H          ;SAVE ENVIRONMENT
               LHLD 0005       ;GET DELAY TIME FROM MEMORY
LOOP           DCX H           ;
               MOV A,L         ;
               ORA H           ;
               JNZ LOOP        ;LOOP UNTIL ZERO
               POP H           ;RESTORE ENVIRONMENT
               RET
```

(b)

Figure 3–45. (a) Speed control and (b) position control software written for the 8085 microprocessor used with the system in Fig. 3–44.

two excitations. Each new time delay loaded into locations 0005 and 0006 represents another motor speed. The four-step sequence is executed by the program indefinitely.

Operating the stepper motor in the slew range requires that its operating speed first be established in the start–stop range and then be ramped (accelerated in a controlled manner) into the slew range. The speed control program discussed above may be modified to accommodate the ramping requirement in the following manner. Assuming that the ramp begins at ω_i (in the start–stop range) and terminates at ω_f (in the slew range) and the duration of the ramp is t_r, the slope of the ramp (which is proportional to the acceleration of the motor during ramping) $a = (\omega_f - \omega_i)/t_r$. The ramp is subdivided into k intervals so that within the kth interval $(\Delta\omega)_k = (\omega_{k+1} - \omega_k)/t_k$ and $(\Delta t)_k = t_{k+1} - t_k$ and the constancy of ramp slope requires that for all intervals the ratio $(\Delta\omega)_k/(\Delta t)_k = a$. If the time intervals are all equal, $(\Delta\omega)_k = (\omega_f - \omega_i)/k$ and $(\Delta t)_k = t_r/k$ must be satisfied by all intervals.

Because the time delay between the successive phase excitations is inversely related to the motor stepping speed (i.e., $\omega = 1/t_d$), the choice of equal time intervals as described above produces nonlinear (actually parabolic) response in the motor speed and the constancy of the slope (as required by the ramp) is no longer satisfied. The nonlinear speed response, and consequently the nonlinear acceleration, results in higher acceleration during a part of the ramping interval. Remembering that the slope of the ramp is dictated by the maximum available torque which is required to accelerate the load into the slew range, periods of accessive acceleration may result in a loss of synchronization between the phase excitation pulses and the rotor motion.

Returning to the implementation of the ramping requirement into the speed control program of Fig. 3–45, it must be remembered that the microprocessor is limited in its capability to do extensive mathematical operations. To produce a linear increase in the stepping rate, the time interval between phase excitations must be decreased in a nonlinear manner. To achieve this, a variable-delay subroutine must be able to reduce the time delay between excitations by an amount that decreases progressively with each succeeding excitation and maintains a count of the number of intervals k during the ramping time, so that at the end of the ramping interval, the subroutine converts to a fixed-delay subroutine that maintains the motor speed at ω_f. For example, if the ramping is to be achieved from 1000 to 1500 steps/s in 100 ms, and 100 intervals ($k = 100$) are selected, each interval produces a 5-step/s speed increase, and the *change* in the time delay between the successive excitations declines from 5 μs at the beginning of the ramp to approximately 2.2 μs at the end of the ramp. The slope of the ramp may be controlled by the choice of ramp-time duration.

The position control software in Fig. 3–45 is similar in structure to that of the speed control. The additional memory locations 0007 and 0008 store the number of steps that the motor is to advance. The repetitive sequence of the motor phase excitation is done in exactly the same way as in the speed control program. The principal difference between the two programs is in the number of pulses to be counted, which is moved from memory locations 0007 and 0008 to register pair DE. Each time a phase excitation pulse is outputted from port A, the register pair DE is decremented and compared with 0. When the content of the register pair DE is reduced to 0, the program executes a Halt, thus stopping the motor after it has advanced the desired number of steps. The two programs serve here as an illustration of simple stepper motor motion. This can be extended to more complex motion where the motor executes a variety of speed changes, ramping, and precise positioning, as well as a reversal of motion.

SUMMARY

1. According to Faraday, a voltage is induced into a conductor moving in a magnetic field. This is shown in Fig. 3-4.

2. A dc generator is based on Faraday's law. As shown in Fig. 3-2, when an armature is rotated in a magnetic field of a permanent magnet, a voltage is induced into the conductors, which are a part of the armature. A generator is an electromechanical transducer, as it converts energy from mechanical to electrical form.

3. The commutator provides an electrical connection between the rotating armature circuit and the external circuit. Increasing the number of segments on the commutator reduces the ripple amplitude, as shown in Fig. 3-3.

4. The polarity of the induced voltage is based on Lenz's law, which requires that the polarity of the induced voltage be such that if it produced current through the external circuit and therefore through the armature conductors, the torque produced by the current will oppose the motion responsible for the induced voltage.

5. A motor is a transducer that converts electrical energy to the mechanical form. The externally applied armature current experiences a force due to the magnetic field, and the force that acts at right angle to the armature's radius produces rotational motion capable of driving a load. This is shown in Fig. 3-7.

6. The dc motors may derive their magnetic field from a permanent magnet or from an electromagnet. An electromagnet uses a field winding which requires a field current. Field-wound dc motors are classified according to the field-winding configuration. As shown in Fig. 3-11, field-wound dc motors fall into categories such as series, split series, shunt, compound, or separately excited. Most of these motors have nonlinear speed-torque characteristics.

7. The motor specifically designed for control applications is the dc PM torque motor. It has linear speed-torque characteristics, fast response, and is able to operate smoothly without cogging at low speeds. The commutators used by dc motors generally have many segments. A typical commutator structure and the direction of all conductor currents are shown in Fig. 3-13.

8. The speed-torque characteristics of the dc motor describes the torque-producing capability of the motor as a function of the operating speed in a family of parallel lines with the armature voltage used as a parameter.

9. There are two motor models which are used to analyze motor performance: the differential equation and the transfer function. The differential equation is more general, as the transfer function, which does not allow the use of initial conditions, is derivable from the differential equation. The typical use of the transfer function is to represent a linear block.

10. Stepper motors produce incremental motion in response to applied excitation pulses. A stepper motor controller simplifies the user's task of operating a stepper motor. A controller typically includes a sequencer and a driver. The sequencer consists of

combinational logic and memory cells, and the driver is a power stage that provides the required value of excitation currents for the motor phase windings.

11. The most common types of stepper motors include the following: permanent magnet; variable reluctance (single stack or multistack), which uses no magnets; and the hybrid, which resembles a variable-reluctance type and uses a permanent magnet. The operation of the stepper motor depends on the excitation of the phase windings in proper sequence. More than one winding may be excited at the same time, which requires more power but generally produces more torque. One winding excitation may be alternated with two winding excitations to produce finer stepping, known as half-stepping.

12. The speed–torque characteristics of the stepper motor includes the start–stop range, where the motor may start, stop, or reverse direction, and the slew range, where the motor must operate unidirectionally without starting or stopping. Ramping is used to operate the motor in the slew range.

13. An external resistor in series with the phase winding reduces the winding-time constant and improves motor performance at high speeds. The increased power consumption makes this type of operation inefficient. The bilevel and chopper drives offer best high-speed motor performance without sacrificing the drive power. These circuits are complex and expensive.

14. Mechanical resonance, which is accompanied by loss of torque at some operating speed, occurs with the VRSM but not with HSM or PM motors. Mechanical or electronic damping may be used in cases where resonance occurs.

15. Several driver configurations are illustrated in Fig. 3–40. In all cases the phase current rise is limited at high operating speeds by the time constant L/R. This results in a lower torque because the phase current is unable to reach the rated value at the end of the excitation period. The external load resistor helps in such cases but places a heavier burden on the power supply. Bilevel and chopper drives, which are expensive, provide excellent high-speed performance without sacrificing the drive power.

16. PM stepping motors generally have large stepping angles, as it is difficult to manufacture one with a large number of poles. Large stepping angles work best in applications where the load must be moved a long distance in the shortest time and in as few steps as possible. The torque per unit volume of the PM motor is poor.

The HSM, on the other hand, produces the highest torque per unit volume. It also has the best resolution, due to its small stepping angle (as low as 1.8 or 0.9° in half-stepping mode). The permanent magnet used by the HSM produces a detent torque which provides some holding torque when the power is OFF. The permanent magnet also increases the rotor inertia and therefore the mechanical time constant, thus limiting the response time and acceleration of the HSM. The VRSM, on the other hand, does not use a permanent magnet, and therefore has a better response and acceleration times, but it has no detent torque, which may result in the load displacing the rotor's required position when the power is OFF. The stepping angle of VRSM is larger than that of HSM (typical values are 7.5 and 15°), and the torque-producing capability per unit volume of the VRSM is moderate. The VRSM has a resonance problem, requiring the use of damping. If a mechanical damper is used, the damper inertia may degrade the response and acceleration times of the VRSM. Electronic damping, which is more complex and more expensive, has no effect on normal motor performance.

QUESTIONS

3–1. Describe the operation of the dc generator.

3–2. Compare the ripple content for a generator using slip rings and two-, four-, and six-segment commutators. Under what operating conditions would the signal-to-noise ratio of the generator be poor?

3–3. Describe all factors that have an effect on dc generator output voltage.

3–4. Describe the operation of the dc PM motor. What is the difference between the PM and field-wound types of dc motor?

3–5. Name four types of field-wound dc motors and explain the difference between them.

3–6. Explain how Lenz's law may be applied in determining the polarity of induced voltage in a generator and the direction of armature rotation in a dc motor.

3–7. Explain the function of the commutator used in a dc motor and illustrate by a diagram the operation of a six-segment commutator. Show the direction of conductor currents at an instant when the brushes are making contact with two segments of your choice.

3–8. Explain the motor's speed–torque characteristics. Explain the difference in operation of two motors with load applied if the speed–torque characteristics of one motor have a steeper slope.

3–9. Compare and contrast the two dc motor models.

3–10. Explain the function of the stepper motor controller.

3–11. Describe the mechanical construction and explain the operation of permanent-magnet, one- and three-stack variable reluctance, and hybrid stepper motors.

3–12. Explain the effect on rotor position, torque production, and power consumption of one phase and more than one phase at a time, and half-stepping excitation methods.

3–13. Explain the speed–torque characteristics of the stepping motor, including the start–stop and slew ranges.

3–14. Explain the ramping method used to operate the stepper motor in the slew range.

3–15. Explain the main reason for pull-out torque reduction at high speeds.

3–16. Discuss the relationship between the torque angle, the dynamic torque developed by the stepper motor, and the load torque.

3–17. Explain the causes and cures for the resonance problem in stepping motors. Which stepper motor type is most susceptible to resonance, and why?

3–18. Compare the unifilar and bifilar phase-winding constructions.

3–19. Compare and contrast the unipolar and the bipolar drivers, including the bilevel and chopper types, from the standpoint of operation and complexity.

3–20. Compare and contrast permanent-magnet, variable-reluctance, and hybrid stepper motors from the standpoint of construction, operation, performance characteristics, and practical applications.

PROBLEMS

3–1. The armature of a dc generator rotates between two poles of a magnet at 3600 rpm. The flux produced by each pole is 235.7×10^4 lines. The armature contains N turns wound over two parallel paths and the commutator has two segments. If the 1-kΩ load resistor dissipates 40 W, calculate the value of N and the frequency of induced voltage.

3–2. The armature of a dc generator uses four coils wound at equal angular intervals and an eight-segment commutator (two segments per coil). If the peak generator voltage is 100 V, calculate (by integration) the peak value of the ripple voltage across the load.

3–3. Given a dc PM motor whose speed–torque characteristics are as shown in Fig. P3–3, its inertia $J_m = 0.004$ oz.-in.-s^2, $R_a = 1.5$ Ω, load viscous friction coefficient $B_L = 14$ oz-in./1000 rpm, and load inertia $J_L = 0.014$ oz-in.-s^2. Neglect Coulomb friction.
  (a) Calculate the motor constants k_t, k_i, k_b, B_m, k_m, and τ_m.
  (b) Determine the motor transfer function using the values of applicable constants.
  (c) Calculate the value of V_a for the no-load steady-state speed of 1200 rpm.
  (d) What must be the value of V_a in part (c) if the load is included?
  (e) Assuming that the load is applied and the input is $V_a = 6u(t)$ volts, and $\omega_m(0) = (0)$ calculate the armature current at $t = 0$, and the steady-state armature current.
  (f) Suppose that the given load B_L, J_L is replaced by the pulley and weight shown in Fig. 3–16a. Assume that the combined inertia of the armature and the pulley is 0.087 oz-in.-s^2, the pulley diameter is 2 in., the weight is $\frac{1}{2}$ lb, and $\omega_m(0) = 0$. If the weight is to be raised at 10 ft/s, calculate the initial and steady-state armature currents and the initial acceleration of the load (in ft/s^2).
  (g) Calculate in part (f) above (1) the distance the load advances in 0.5s, after the application of a step input to the motor, (2) the velocity of the load at $t = 0.5$s, and (3) the acceleration of the load at $t = 0.5$s.

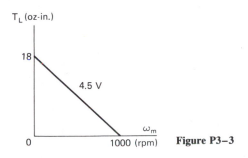

Figure P3–3

3–4. The phase-winding inductance of the stepper motor is 10 mH and the resistance is 10 Ω. The rated current is established in the phase winding from a 20-V power supply. It is required that the motor be driven at 1000 steps/s by pulses with a 50% duty cycle, and the phase current must be 1.9A at the end of the excitation time period. Calculate the value of the resistor that must be added in series with the phase winding and the value of the power supply that is capable of maintaining the same rated current as the 20-V supply. Compare the output powers of the two power supplies when the rated current is being drawn.

3–5. Design a sequencer to provide the sequential excitation of the phase windings (one at a time) of the five-phase VRSM. Follow the design procedure established in this chapter.

3–6. The stator of a two-phase bipolar HSM has 40 teeth, and the rotor has 50 teeth. The frequency of the excitation is 400 Hz. Calculate **(a)** the step angle; **(b)** the stepping rate R_s in steps/rev; **(c)** the motor speed in rpm; **(d)** the time required to advance the rotor 9°.

3–7. Suppose that the motor in Problem 3–6 drives the rack-and-pinion load shown in Fig. 3–16b (the pinion is coupled to the motor shaft). The rack has 8 teeth/in. and the pinion has 100 teeth. The weight is fixed to the rack. If the load is to be moved a distance of 5 ft, calculate **(a)** the number of steps made by the motor; **(b)** the time required to move the load. Assume the same stepping rate as in Problem 3–6.

3–8. Suppose that in Problem 3–7, instead of using the two-phase HSM, a three-phase VRSM with $N_r = N_s = 8$ teeth is used.

(a) Recalculate parts (a) and (b) in Problem 3–7 assuming the same stepping rate.

(b) Which of the two load-positioning systems is better from a practical standpoint? Why?

3–9. Consider the rack-and-pinion load described in Problem 3–7 with 8 teeth/in. on the rack and 100 teeth on the pinion. The driving motor is the two-phase HSM as described in Problem 3–6. The combined inertia of the rotor and the pinion is 0.2 oz-in.-s^2. Assume that the speed–torque characteristics of Fig. 3–33 apply to the given motor. The rack moves on rollers, resulting in a constant (independent of speed) friction torque of 10 oz-in. Initially, the motor establishes its operation at 400 steps/s, and shortly thereafter the constant-acceleration ramp shifts the operating point to 900 steps/s in the slew range. The acceleration is done in 210 ms. Calculate the value of the weight W (in lb) which is fixed to the rack, assuming that the rack weighs 20% of W.

LAB EXPERIMENT 3–1
SPEED–TORQUE CHARACTERISTICS OF A DC PM MOTOR

Objective.   To determine experimentally the speed–torque characteristics of a permanent-magnet dc motor.

Equipment

Frequency counter

Digital voltmeter

Dc power supply; its voltage and current rating depend on the motor used

Impedance bridge or accurate digital ohmmeter

Materials

2 dc PM motors

Optical speed-sensing system, consisting of the disk with 120 holes (a disk with different number of holes may be used)

OID

If necessary, an op amp buffer or a Schmidt trigger

1-Ω, 1%, 2-W resistor

Procedure

1. Connect components as shown in Fig. LE3–1a. If the OID does not have a Schmidt trigger on the chip, a ST must be inserted in series with the OID output to provide pulse shaping. (*Note*: Unless otherwise required, all voltages must be measured on the dc digital voltmeter and the pulse frequency from OID on the frequency counter. Refer to Chapter 2 for converting the OID frequency to motor speed in rpm.)

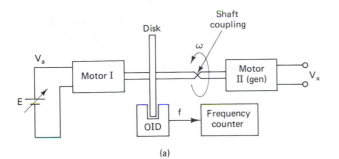

(a)

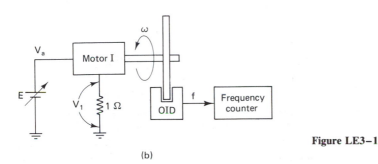

(b)

Figure LE3–1

2. Adjust V_a in Figure LE3–1a for the motor speed of 1000 rpm. Measure and record V_x.

3. Connect the motor test circuit as shown in Fig. LE3–1b. Adjust E for a motor speed of 250 rpm. Measure and record V_a, V_1, and ω_m.

4). Repeat step 3 for motor speeds of 500 to 2000 rpm in steps of 250 rpm.

Test data. The format for the test data and the calculated results, as well as the report format, must comply with and meet the requirements of your lab instructor.

Evaluation. As the ultimate objective of any lab experiment is to verify the theory-based predictions through measurements, significant differences between theory and measurements (generally, differences over 10%) must be resolved through additional experimentation, investigation, or retest.

1. Using the data from step 2 of the Procedure, calculate the motor torque constant k_i. Refer to the chapter text for applicable information.

2. Plot ω_m versus V_a using the test data from steps 3 and 4 of the Procedure. The slope of the resulting straight line is the motor speed constant k_m, which under no-load conditions is equal to $1/k_b$. Obtain the value of the slope by the least-squares method.

3. How does the product of k_b obtained in step 2 and 1000 rpm compare with the measured value of V_x from step 2 of the Procedure? Explain.

4. Write the equation based on the Kirchhoff voltage law for the armature circuit of motor I in Fig. LE3–1b. Note that in this equation, R_a is unknown, but the

values of V_a, V_1, and ω_m have been measured in steps 3 and 4 of the Procedure. Using the eight measured sets of V_a, V_1, and ω_m in the equation derived above, calculate the corresponding eight values of R_a. Record the eight R_a values together with the measured V_a, V_1 and ω_m values. Also record the average of the eight calculated R_a values.

5. Calculate and record the values of k_t and B_m. Consult the text for the definitions of these two constants, if necessary.

6. Construct the motor speed–torque characteristics as shown in Fig. 3–14. The graph must include four curves, corresponding to 500-, 1000-, 1500-, and 2000-rpm motor speeds.

LAB EXPERIMENT 3–2
TRANSFER FUNCTION OF A DC PM MOTOR

Objective. To determine experimentally the transfer function of the permanent-magnet dc motor. The values of the motor speed constant k_m and the motor time constant are required to specify the motor transfer function completely. K_m was determined in Lab Experiment 3–1, and the motor time constant is measured in this experiment. Two approaches to the measurement of the time constant are presented here: one based on the measurement of the armature inertia, and the other technique based on the use of the microprocessor.

Equipment

Dc power supply for driving the motor
Low-voltage power supply
Accurate clock for time measurement
Frequency counter

Materials

Dc PM motor
Optical disk (n holes)
OID
Op amp buffer
8085 microprocessor
EPROM firmware
Piano wire
Weights (10 to 100 g)
Clamp and supporting board
Protractor
Motor's armature

Procedure. Two procedures are presented here for the measurement of motor's mechanical time constant. One procedure is based on the measurement of armature inertia. To use this procedure, an armature that is identical to the one inside the motor must be available. The configurations shown in Fig. LE3–2.1 apply to the inertia measurement. The block diagram shown in Fig. LE3–2.2 applies to the second procedure, which is based on use of the 8085 microprocessor. The use of one of these procedures is sufficient to determine the mechanical time constant of the motor.

A. Inertia Measurement Procedure

1. The spring constant k is measured first. A 2-ft length of piano wire is used as a spring. Bend one end of this wire, approximately 3 in., at a right angle, and clamp the other end to the supporting board as shown in Fig. LE3–2.1a. Attach a known weight W at a distance R from the axis of the spring to cause the angular deflection θ, making sure that the clamped end does not rotate. R should be 2 to 3 in, and θ should not exceed 30°. Measure and record R and θ (a protractor may be used to measure θ).

2. Suspend the motor armature by the spring wire as shown in Fig. LE3–2.1b, using the spring from step 1. Align the pointer with point C, then rotate the armature until the pointer is aligned with point B and release the armature. Upon release, the potential energy due to torsion stored by the spring wire develops a simple harmonic motion which is characterized by maximum potential energy at points A and B and maximum kinetic energy midway between A and B. The time required for the pointer to describe the path $A–B–A$ or $B–A–B$ is the period of the oscillatory motion. As the frictional effects are almost negligible, the oscillatory motion will continue for some time. The release step from point B must be repeated if wobbling develops during the motion. The oscillatory motion must be smooth. A magnet below the armature axis might help in maintaining the armature along the vertical axis during the motion. Using an accurate clock or a stopwatch, measure and record at least 10 period times. The average of these will be taken later.

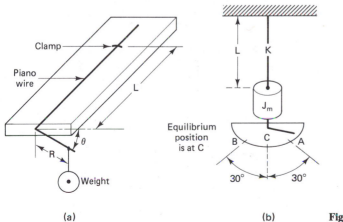

(a)

(b)   **Figure LE3–2.1**

B. Time Constant Measurement by Microprocessor

1. Construct the circuit as shown in Fig. LE3–2.2. The EPROM chip stores the response time control program shown in Figure LE3–2.3. The program written for the 8085-based microprocessor may be keyed-in manually. Storing the program on the ROM chip eliminates the need for manual entry, and therefore saves some time.

2. Close the switch and adjust E for the motor speed of 1000 rpm as measured on the frequency counter. For an optical disk with 120 holes, the counter will read 2000 Hz when the motor speed is 1000 rpm. Open the switch and allow the motor to come to rest.

3. Set the microprocessor to the starting address of the program. Close the switch and push the START key on the microprocessor simultaneously. The measurement takes less than 1 s. Read and record the contents of register pairs BC and DE.

Test data. The format for the test data and the calculated results, as well as the report format must comply with and meet the requirements of your lab instructor.

Evaluation. As the ultimate objective of any lab experiment is to verify the theory-based predictions through measurements, significant differences between theory and measurements (generally, differences over 10%) must be resolved through additional experimentation, investigation, or retest.

1. Calculate the spring constant $k = (W \cos \theta)(R)/\theta$ oz-in./rad using the values W (oz), R (in.), and θ (rad) from step 1 of Procedure A.

2. Assuming negligible viscous friction effects in the motion produced in step 2, the motion may be represented by a differential equation of the form $J_m \ddot{\theta} + k\theta = 0$, whose solution consists of a sine term and a cosine term. The argument of the sine and cosine is in terms of the period T, the spring constant k, and the armature inertia J_m. Using the values of k (in oz-in./rad) and the average value of the period T from step 1 of this Evaluation and step 2 of Procedure A, respectively, calculate the value of J_m in oz-in-s^2.

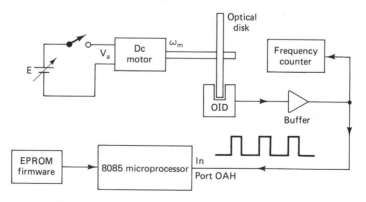

Figure LE3–2.2

```
LOCATION   OBJECT CODE  LINE  SOURCE LINE
                          1   "8085"
                          2           ORG   500H
0500       00             3   START   NOP
0501       3F01           4           MVI   A,01H
0503       D308           5           OUT   08H
0505       3117FF         6           LXI   SP,17FFH
0508       010000         7           LXI   B,0000H
050B       110000         8           LXI   D,0000H
050E       210000         9           LXI   H,0000H
0511       00            10           NOP
0512       00            11           NOP
0513       DB0A          12   WAIT:   IN    0AH
0515       E601          13           ANI   01H
0517       CA6D13        14           JZ    WAIT + 6800H
051A       23            15   ONE:    INX   H
051B       DB0A          16           JN    0AH
051D       E601          17           ANI   01H
051F       C26D1A        18           JNZ   ONE + 6800H
0522       23            19   ZERO:   INX   H
0523       DB0A          20           IN    0AH
0525       E601          21           ANI   01H
0527       CA6D22        22           JZ    ZERO + 6800H
052A       7C            23           MOV   A,H
052B       B7            24           ORA   A
052C       C26D36        25           JNZ   ROLL + 6800H
052F       3A1400        26   ENTER:  LDA   1400H
0532       BD            27           CMP   L
0533       D26D5A        28           JNC   INCR + 6800H
0536       CD6D4C        29   ROLL:   CALL  DELAY + 6800H
0539       03            30   POLL:   INX   B
053A       DB0A          31           IN    0AH
053C       E601          32           ANI   01H
053E       C26D39        33           JNZ   POLL + 6800H
0541       03            34   CHECK:  INX   B
0542       DB0A          35           IN    0AH
0544       E601          36           ANI   01H
0546       CA6D41        37           JZ    CHECK + 6800H
0549       C36D1A        38           JMP   ONE + 6800H
054C       210200        39   DELAY:  LXI   H,0200H
054F       2D            40   LOOP:   DCR   L
0550       00            41           NOP
0551       C26D4F        42           JNZ   LOOP + 6800H
0554       25            43           DCR   H
0555       C26D4F        44           JNZ   LOOP + 6800H
0558       1C            45           INR   E
0559       C9            46           RET
055A       211401        47   INCR:   LXI   H,1401H
055D       77            48           MOV   M,A
055E       23            49           INX   H
055F       70            50           MOV   M,B
0560       23            51           INX   H
0561       71            52           MOV   M,C
0562       23            53           INX   H
0563       73            54           MOV   M,F
0564       FF            55           RST   7
```

Figure LE3–2.3

3. Using the value of B_m from Lab Experiment 3–1, and J_m from step 2 of this Evaluation, calculate the value of the mechanical time constant τ_m.

4. Express the motor transfer function and its differential equation model using the measured motor parameter values. The K_m value is available from Lab Experiment 3–1.

5. Using the register contents values from step 3 of Procedure B, calculate the mechanical time constant T_m.

6. Wobbling of the armature was not permitted during the motion in step 2 of Procedure A. Why? Explain.

7. Construct the flowchart for the program in Fig. LE3–2.3 and explain the approach used in the program to determine the motor time constant.

8. Verify the register coefficient values in the equation of step 5.

9. Using the initial displacement, and assuming zero initial velocity, solve the equation in step 2 and evaluate the two constants. Obtain the expressions for the angular displacement and velocity as a function of time. Using the solutions, calculate the maximum velocity of the tip of the pointer and the time (first time) at which it occurs. Express the velocity in in./s.

10. In the first half-period of motion in step 2 of Procedure A, calculate the maximum potential energy, maximum kinetic energy, and the times at which these energies occur. Are the two energies equal? Does the difference between them remain constant as the motion continues? Explain.

LAB EXPERIMENT 3–3
STEPPER MOTOR MOTION CONTROL

Objective. To measure several characteristics of the stepper motor and to investigate the performance of a typical stepper motor motion control system. The speed and the position control are investigated when the system is controlled by the sequencer and when it is under the control of the microprocessor. The user is also required to write software that meets specific motion requirements.

Equipment

8085 microprocessor
Digital voltmeter
Dual-trace oscilloscope
Pulse generator
Frequency counter

Materials

As shown in the test system
Optical disk
Components required by the OID chip

Test system. The practical stepper motor motion control system shown in Fig. 3–44.

Procedure

1. Construct the test system shown in Fig. 3–44. Apply the optical disk (120 holes is suggested, but other disk configurations may also be used) to the motor shaft and position the OID so that the optical disk moves freely inside the grove of the OID, as shown in the systems of Figs. 8–11 and 8–12. Omit this step if a complete system is available. The stepper motor used in this experiment must be capable of a no-load stepping rate of 1000 steps/s/min.

2. With switch S_1 not contacting gate 7 or port A, apply an external $+5$ V dc to the input of Q_1. Adjust the power supply E so that $I_{\phi 1} = I_{rated}$. $I_{\phi 1}$ may be deduced from the voltage across the 100-Ω external resistor in series with the ϕ_1 winding. This condition (i.e., value of E and 100-Ω resistor in series with the ϕ_1 winding) will be referred to as E_{100}.

3. Repeat step 2 for the 0-Ω, 20-Ω ohm 10-W, and 50-Ω 10-W resistors in series with ϕ_1 winding and identify the corresponding conditions as E_0, E_{20}, and E_{50}.

4. Disconnect the stepper motor and measure the resistance and inductance of each of the four phases. Record all values.

5. With S_1 not contacting gate 7 or port A, apply 50% duty cycle, 5-V_{peak} (0 to $+5$ V dc), 100-Hz pulses to the input of Q_1, and establish an E_0 condition. Monitor the input pulse train and $I\hat{\phi}_1$ current on the dual-trace oscilloscope. Measure and record (a) the $I_{\phi 1}$ waveform, include all critical values; (b) the time required for $I_{\phi 1}$ to reach 63.2% of its rated value, if possible; (c) the value of $I_{\phi 1}$ at the instant Q_5 turns off.

6. Repeat step 5 for each of the following conditions: E_0, E_{20}, and E_{50}.

7. Repeat steps 5 and 6 for each of the following pulse frequencies: 200 to 1000 Hz in steps of 100 Hz.

8. Set switches S_1 to S_4 to connect the four outputs of the sequencer to the input of the power driver. Close the program inhibit switch, apply a HIGH ($+5$ V dc) to the Z input, establish the E_0 condition, and apply a 50% duty cycle, 5-V_{peak} (0 to $+5$ V dc) 120-Hz square wave to the clock input of the sequencer. Measure and record the motor speed in rpm (measure the frequency of the OID pulses on the frequency counter and then convert the measured frequency to rpm).

9. Repeat step 8 for each of the following clock frequencies: 180 to 900 Hz in steps of 60 Hz.

10. Repeat steps 8 and 9 for each of the following conditions: E_{20}, E_{50}, and E_{100}.

11. Clear the 8-bit counter, open the PROGRAM INHIBIT switch, manually load $(10000111)_2$ into the counter, establish the E_{100} condition, apply a 50% duty cycle, 5-V_{peak} (0 to $+5$ V dc), 120-Hz square wave to the clock input of the sequencer, and apply a LOW (0 V dc) to the Z input. Measure and record the angular displacement of the stepper motor shaft within $\pm 3°$ (the OID pulses may be counted, for example, as is done by the position control system in Fig. 8–12).

12. Repeat step 11, except determine and load the counter with the binary value that will cause the motor to advance $90 \pm 3°$. Measure and record the actual angle advanced by the motor, and record the binary value.

13. Set switches S_1 to S_4 to connect port A of the microprocessor to the input of the power driver. Establish condition E_{100} and load the speed control program of Fig. 3–45 into the microprocessor. Determine and load the memory locations reserved for the delay, the value of the delay corresponding to a motor speed of 100 rpm. Execute the program. Measure and record the actual motor speed.

14. Repeat step 13 for motor speeds of 200 and 500 rpm.

15. Maintain the system conditions as specified in step 13 and load the position control program of Fig. 3–45. Load the appropriate memory locations with (a) the delay corresponding to motor speed of 200 rpm; (b) the number of pulses necessary to advance the motor shaft $90 \pm 3°$. Execute the program. Measure and record the actual angular displacement of the motor shaft.

16. Write a program that causes the motor speed to ramp linearly from 500 steps/s to 1000 steps /s in 200 ms. Show the flowchart and load the program into the micropro-cessor RAM. Establish the E_{100} condition on the power supply and execute the program. Measure and record the motor speed response waveform (the optical disk, OID, and FVC combination used by the system in Fig. 8–11 is one way of accomplishing this). The linearity (maximum difference between the actual response waveform and the ideal ramp) must not exceed 50 steps/s.

Test data.   The format for the test data and the calculated results, as well as the report format, must comply with and meet the requirements of your lab instructor.

Evaluation.   As the ultimate objective of any lab experiment is to verify the theory-based predictions through measurements, significant differences between theory and measurements (generally, differences over 10%) must be resolved through additional experimentation, investigation, or retest.

1. Using the test data (step 4 of the Procedure), calculate the value of the phase-winding time constant and compare this value against measurements in step 5b of the Procedure. Explain your conclusions.

2. Using the test data, plot on suitable graph paper $I_{\phi1}(t_w)/I_r$ versus time for each of the conditions E_0, E_{20}, E_{50}, and E_{100}, where t_w is the time where Q_5 turns off and I_r is the rated phase current. Include on the same graph the four $I_{\phi1}(t_w)/I_r$ versus time curves based on theory. (Use $f = 100$ Hz.)

3. Repeat step 2 for each of the frequencies 500 Hz and 1000 Hz.

4. Using the test data, plot on suitable graph paper $I_{\phi1}(t_w)/I_r$ versus frequency for each of the following conditions: E_0, E_{20}, E_{50}, and E_{100}. Include on the same graph the four $I_{\phi1}(t_w)/I_r$ versus frequency curves as based on theory.

5. Using the test data, plot on suitable graph paper the motor speed ω_m versus frequency curves for each of the conditions E_0, E_{20}, E_{50}, and E_{100}.

The responses to the following questions and/or conclusions must be supported whenever possible by the applicable theory.

6. Are all four curves in step 5 straight lines over the test frequency range? If not, explain.

7. Do all the curves in step 5 have the same slope over the linear range? If not, explain. If so, what is the value of the slope, and which stepper motor parameters are implicit in the slope?

8. Using the graphical results of this section, determine the following stepper motor parameters: θ_s, R_s, P_r, P_s, N_r, and N_s.

9. Discuss the results of the position control with the use of the sequencer in steps 11 and 12 of the Procedure. Describe the method used in measuring position. Explain the determination of the binary value in step 12 for 90° displacement.

10. Discuss the results of the speed and position control by the microprocessor in steps 13 to 15. Explain the determination of the time delay for 100-, 200-, and 500-rpm motor speeds in steps 13, and 14. How did you determine the values in step 15? Explain.

11. Show the flowchart and hard copy of the program in assembly language that meets the requirements of step 16 of the Procedure. Discuss the approach and the problems encountered in the design of this software. How was the linearity measured? What value of linearity was achieved?

REPRESENTING CONTROL SYSTEMS

chapter 4 _____

In Chapters 2 and 3 we presented the characteristics and operating limitations of some of the most commonly used components. These components, referred to as the *control components*, are the basic building blocks of all control systems. A control system engineer designs the system under consideration according to its specifications and the desired performance criteria. The key to successful design and performance of a system is proper and accurate evaluation before fabrication. There are two different techniques that can be used to evaluate the system: graphical and analytical. The choice between the two techniques depends primarily on the performance specifications and the time and cost constraints. Sometimes it is necessary to use both graphical and analytical techniques. The root locus method, Bode plots, and Nyquist plot are some of the most commonly used graphical techniques; the Routh–Hurwitz criterion and the Laplace transform are the most commonly used analytical techniques. However, before any one of these techniques is used, the system has to be represented in a ''standard form'' referred to as the transfer function. Generally the transfer function is obtained by using a mathematical tool called Laplace transform.

In this chapter we represent specific building blocks in terms of their transfer functions. In the latter part of the chapter we present block diagram and signal flow graph methods which are necessary to represent a control system. Finally, block diagram and signal flow graph simplification methods are discussed so that the desired overall transfer function of the system can easily be obtained.

4–2 LAPLACE TRANSFORM

In this section we review the Laplace transform and its properties. Our major interest is in the use of Laplace transforms and transform operations. Hence the derivation of Laplace transform operations is not presented in this chapter. However, for convenience, Laplace transforms and Laplace transform operations are illustrated in Table 4–1.

TABLE 4–1 LAPLACE TRANSFORMS AND TRANSFORM OPERATIONS

a. Laplace Transforms

Function of time, $f(t)$	Laplace transform of $f(t) = F(s)$
1. Impulse function, $A\delta(t)$, where $A =$ amplitude	A
2. Step function, $Au(t)$, where $A =$ amplitude	$\dfrac{A}{s}$
3. Ramp function Kt, where $K =$ slope	$\dfrac{K}{s^2}$
4. General case, Kt^n, where $K =$ constant	$\dfrac{Kn!}{n+1}$
5. Decaying exponential function, $Ae^{-\alpha t}$, where $A =$ amplitude at $t = 0$	$\dfrac{A}{s+\alpha}$
6. Increasing exponential function, $Ae^{\alpha t}$, where $A =$ constant	$\dfrac{A}{s-\alpha}$
7. Sinusoidal function, $A \sin \omega t$, where $A =$ amplitude	$\dfrac{A\omega}{s^2 + \omega^2}$
8. Cosine function, $A \cos \omega t$, where $A =$ amplitude	$\dfrac{As}{s^2 + \omega^2}$
9. Damped sinusoidal function, $e^{-\alpha t}(A \sin \omega t)$, where $A =$ amplitude, $\alpha =$ damping constant	$\dfrac{A\omega}{(s+\alpha)^2 + \omega^2}$
10. Damped cosine function, $e^{-\alpha t}(A \cos \omega t)$, where $A =$ amplitude, $\alpha =$ damping constant	$\dfrac{A(s+\alpha)}{(s+\alpha)^2 + \omega^2}$
11. Gust function, $Kte^{-\alpha t}$, where $K =$ slope, $\alpha =$ damping constant	$\dfrac{K}{(s+\alpha)^2}$
12. General case, $Kt^n e^{-\alpha t}$	$\dfrac{Kn!}{(s+\alpha)^{n+1}}$

b. Laplace Transform Operations

Function of time, $f(t)$	Laplace transform of $f(t) = F(s)$
1. First derivative of $f(t)$, $\dfrac{df(t)}{dt}$	$sF(s) - f(0+)$ where $f(0+) =$ value of $f(t)$ at time $t = 0+$

TABLE 4–1 LAPLACE TRANSFORMS AND TRANSFORM OPERATIONS (*Continued*)

Function of time, $f(t)$	Laplace transform of $f(t) = F(s)$
2. Second derivative of $f(t)$, $$\frac{d^2f(t)}{dt^2}$$	$$s^2F(s) - s\frac{df(0+)}{dt} - f(0+)$$ where $f(0+)$ = value of $f(t)$ at time $t = 0+$
3. First integral of $f(t)$, $$\int_0^t f(t)\,dt$$	$$\frac{F(s)}{s} + \frac{\int_0^t f(0+)\,dt}{s}$$
4. Second integral of $f(t)$, $$\int_0^t\int_0^t f(t)\,dt$$	$$\frac{F(s)}{s^2} + \frac{\int_0^t f(0+)\,dt}{s^2} + \frac{\int_0^t\int_0^t f(0+)\,dt}{s^2}$$
5. Delayed function $$f(t-a)u(t-a)$$	$$e^{-as}F(s)$$
6. Multiplication by $e^{-\alpha t}$, $$e^{-\alpha t}f(t)$$	$$F(s+\alpha)$$
7. Multiplication by t, $$tf(t)$$	$$\frac{-dF(s)}{ds}$$
8. $f(at)$	$$\frac{1}{a}F\left(\frac{s}{a}\right)$$
9. Initial value theorem, $$\lim_{t\to 0} f(t)$$	$$\lim_{s\to\infty} sF(s)^*$$
10. Final value theorem, $$\lim_{t\to\infty} f(t)$$	$$\lim_{s\to 0} sF(s)^*$$

c. Laplace Transform of Some Electrical Quantities

Quantity	Formula
1. Dc voltage, v	$\dfrac{v}{s}$
2. Ac voltage, $v(t)$	$V(s)$
3. DC current, I	$\dfrac{I}{s}$
4. Ac current, $i(t)$	$I(s)$
5. Uncharged capacitor, C	$\dfrac{1}{Cs}$
6. Unfluxed inductor, L	Ls

* If the initial or final value theorem yields an indeterminate form, use L'Hôpital's rule; that is, take the ratio of the derivatives with respect to s.

4–3 TRANSFER FUNCTION

The *transfer function* for a given network or a system is defined as the ratio of Laplace transform of an output to the Laplace transform of an input with all initial conditions assumed to be zero. Remember that input and output are user defined; hence the transfer function offers a lot of flexibility for system design and analysis. Obviously, the transfer function of a given system when transformed into the time domain (using the inverse Laplace transform) gives the magnitude and phase relationship between the input and output of that system. The transfer function of a given network or a system is represented by a block diagram, as shown in Fig. 4–1. There are basically two types of transfer functions:

1. Open-loop transfer function
2. Closed-loop transfer function

The transfer function of a system without feedback is referred to as an *open-loop transfer function*; the transfer function of a system with feedback is referred to as a *closed-loop transfer function*. Note that here the word "system" is used in a broader sense, which may mean a simple network or even a group of networks. In equation form the transfer function is expressed as

$$\text{transfer function} = \frac{\mathscr{L}[\text{output}]}{\mathscr{L}[\text{input}]}\bigg|_{\text{all ICs zero}} \tag{4–1}$$

In Fig. 4–1,

$$\text{transfer function} = \frac{C(s)}{R(s)} = G(s)$$

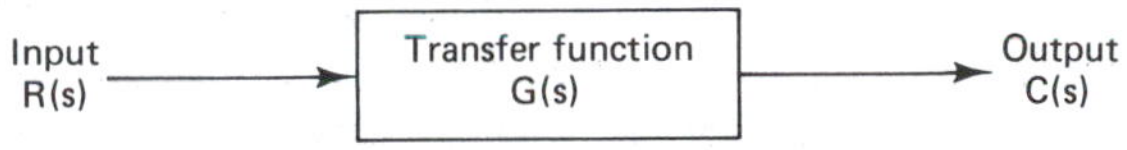

Input R(s) → [Transfer function G(s)] → Output C(s)

Figure 4–1. Representing a transfer function.

4–4 BLOCK DIAGRAMS

A block diagram is a graphical method of representing a system. Block diagrams simplify the representation and analysis of control systems. Often, it is difficult, if not impossible, to draw a schematic diagram of a control system each time it is analyzed or evaluated. In addition, not everyone is able to draw or understand the schematic diagrams of systems, especially since there are so many different types of control systems: electrical, mechanical, hydraulic, and so on. In a block diagram each block receives a signal, referred to as an input signal, from some part of the system or from an external source. The input signal is then acted upon by a block to produce an output signal, which in turn is an input signal for another part of the system; and so on.

The input and output signals may be physical variables such as a temperature, an air pressure, liquid flow rate, position, velocity, voltage, or current. The various blocks in a given system are connected by signal paths, called *directed line segments*. As shown in Fig. 4–2, the directed line segments representing signal paths may be physical media, such as electrical wires and mechanical pipes or linkages.

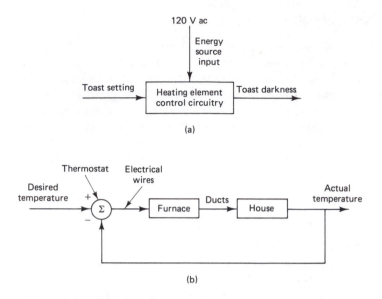

Figure 4–2. Block diagrams: (a) bread toaster; (b) home heating system.

4–5 BLOCK DIAGRAM CLASSIFICATION

Block diagrams can be classified as either open-loop or closed-loop (Fig. 4–3). This classification is based on the way the block diagrams are drawn. Sometimes even a closed-loop control system may be represented by a single block, which then may look like an open-loop block diagram. In fact, the way the block diagram is drawn is dictated by the purpose of the block diagram as well as the complexity of the control system.

In general, when a system is being designed, a detailed block diagram that includes feedbacks as well as important component transfer functions is necessary. On the other hand, if the system is to be analyzed, its desired overall transfer function represented by a single block may be sufficient. Thus the classification of block diagrams as open-loop and closed-loop is somewhat superficial. However, since the control systems are open-loop and closed-loop, it is common practice to identify the block diagrams as either open-loop or closed-loop, as shown in Fig. 4–3.

Next, let us define some important terms that will be used with block diagrams throughout. Let

$$R(s) = \text{input signal in the } s \text{ domain} = \mathcal{L}[r(t)]$$

$$C(s) = \text{output signal in the } s \text{ domain} = \mathcal{L}[c(t)]$$

$$G(s) = \text{transfer function in the forward path}$$

$$H(s) = \text{transfer function in the feedback path}$$

$$T(s) = \text{closed-loop transfer function}$$

$$E(s) = \text{error signal in the } s \text{ domain}$$

$$G(s)H(s) = \text{loop transfer function}$$

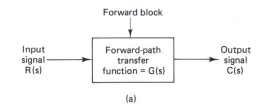

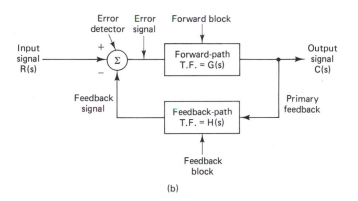

Figure 4–3. Block diagrams: (a) open-loop; (b) closed-loop.

4–6 TYPES OF BLOCKS

Block diagrams of given systems may contain one or more blocks, depending on the composition and complexity of the system. Normally, the transfer function is determined for each major part of the system. Each transfer function is then enclosed inside a block which represents that part. Also, a block is defined based on the type of transfer function it contains. For example, a block whose transfer function is a constant is called a proportional block. Following is a list of some of the most commonly used blocks.

1. Proportional block
2. Integration block
3. Proportional plus integration block
4. Derivative block
5. Proportional plus derivative block
6. Simple phase-lag block
7. Simple phase-lead block
8. Quadratic block
9. Combination block

Since this book emphasizes applications, we shall consider some specific electrical networks that exhibit characteristics of the blocks mentioned above.

Proportional Block

The characteristic of a *proportional block* is that its output signal is directly proportional to its input signal. Therefore, the transfer function of a proportional block is a constant (Fig. 4–4). Note that any system whose transfer function is a constant may be represented by a proportional block.

Figure 4–4. Proportional block, $C(s)/R(s) = K$.

EXAMPLE 4–1

Determine the transfer function for the amplifier shown in Fig. E4–1a and then draw the block diagram to represent it. Assume that the internal gain of the op amp remains constant over the desired operating frequency range.

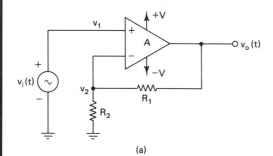

(a)

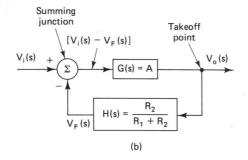

(b)

Figure E4–1. (a) Amplifier of Example 4–1; (b) closed-loop block diagram representation of the network in part (a).

SOLUTION   To obtain the transfer function, two approaches are possible:

   Method 1.   Write as many equations as possible for the given network and then transform them into the s domain.

   Method 2.   Transform the given network into the s domain and write the equations in the transform domain.

In this example we use the first method. Using the basic op amp theory, we know that

$$v_o(t) = \frac{A(R_1 + R_2)}{R_1 + R_2 + AR_2} v_i(t)$$

Taking the Laplace transform and rearranging, we get

$$\frac{V_o(s)}{V_i(s)} = \frac{A(R_1 + R_2)}{R_1 + R_2 + AR_2} \tag{4-2}$$

For a given op amp, A is constant and R_1 and R_2 are fixed resistors. Hence the transfer function $V_o(s)/V_i(s)$ is a constant and can be represented by a block diagram of Fig. 4–4. Let the op amp be a 741C, $R_1 = 10 \text{ k}\Omega$ and $R_2 = 1 \text{ k}\Omega$. Hence

$$\frac{V_o(s)}{V_i(s)} = \frac{2 \times 10^5 (11 \text{ k}\Omega)}{1 \text{ k}\Omega + 11 \text{ k}\Omega + (2 \times 10^5)(1 \text{ k}\Omega)} = 11$$

The network in Fig. E4–1a is commonly referred to as a closed-loop noninverting amplifier. It uses a feedback; hence its block diagram representation may be closed-loop, as shown in Fig. E4–1b.

Note that the block diagram of Fig. E4–1b yields the same transfer function as given in Eq. (4–2), as we will see in Section 4–7. Thus the overall transfer function does not indicate whether the system uses feedback or not; hence it is necessary to identify the transfer function as open-loop or closed-loop.

Integration Block

An integration block diagram is shown in Fig. 4–5. The characteristic of the *integration block* is that its output is equal to the integration of its input. In the s domain, integration is indicated by a pole at the origin in the transfer function. The integration block is most commonly used to compensate a control system, especially if its steady-state error has to be reduced to zero.

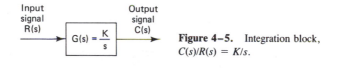

Figure 4–5. Integration block, $C(s)/R(s) = K/s$.

EXAMPLE 4–2

Determine the transfer function $V_o(s)/V_i(s)$ for the network shown in Fig. E4–2. Draw the block diagram to represent this network. Assume that A remains constant over the desired frequency range.

SOLUTION In Example 4–1, to obtain the network transfer function we first wrote the equations in the time domain and then transformed them into the s

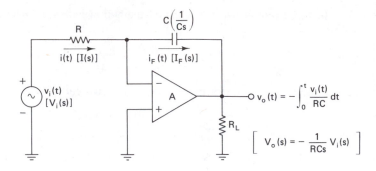

Figure E4–2. Integrator. Quantities in parentheses are Laplace transform equivalents.

domain. In this example we use a different approach. First we transform the given network into the s domain, and then to obtain the desired transfer function write the equations in the transform domain. Referring to Fig. E4–2, we have

$$I(s) \simeq I_f(s)$$

or

$$\frac{V_i(s)}{R} = -\frac{V_o(s)}{1/Cs}$$

That is,

$$\frac{V_o(s)}{V_i(s)} = -\frac{1}{RCs}$$

$$= -\frac{K}{s} \tag{4-3}$$

where $K = 1/RC$. Note that the negative sign in Eq. (4–3) indicates that the input is applied to the inverting input. The block diagram for the network is shown in Figure 4–5.

Proportional plus Integration Block

As its name indicates, the *proportional plus integration block* transfer function is the combination of proportional and integration block transfer functions, as shown in Fig. 4–6. Obviously, the proportional plus integration block transfer function contains one finite zero and a pole at the origin of the s plane.

Often in compensating applications, a proportional plus integration block is preferred over an integration block, because it offers more flexibility to improve the perfor-

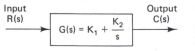

Figure 4–6. Proportional plus integration block. K_1 and K_2 are constants.

Representing Control Systems   Chap. 4

mance of a system being compensated. A system is said to be *compensated* if its response may be altered with modifications in its overall transfer function.

EXAMPLE 4–3

Determine the transfer function for the network shown in Fig. E4–3a. Also draw the block diagram to represent it. Assume that A_1, A_2, and A_3 remain constant over the desired frequency range.

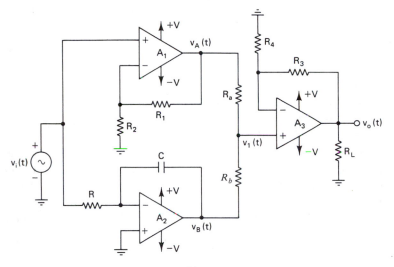

·(a)

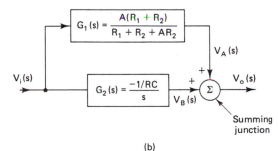

(b)

Figure E4–3. (a) Network of Example 4–3, $R_a = R_b$ and $R_3 = R_4$; (b) block diagram for the network of part (a).

SOLUTION   Close examination of the network in Fig. E4–3a reveals that op amp A_1 is configured as a noninverting amplifier, op amp A_2 is configured as an integration amplifier, and A_3 is configured as a summing amplifier. From the results of Examples 4–1 and 4–2, the transfer functions of noninverting and integration amplifiers are

$$\frac{V_A(s)}{V_i(s)} = \frac{A(R_1 + R_2)}{R_1 + R_2 + AR_2} \qquad (4-4a)$$

and

$$\frac{V_B(s)}{V_i(s)} = \frac{-1/RC}{s} \qquad (4-4b)$$

This means that we need to determine the transfer function of the summing amplifier formed by op amp A_3 and R_a, R_b, R_3, and R_4. Hence

$$V_o(s) = (2)\left[\frac{V_A(s)}{2} + \frac{V_B(s)}{2}\right]$$

or

$$V_o(s) = V_A(s) + V_B(s) \qquad (4-4c)$$

as expected. Equation (4–4c) indicates that the output $V_o(s)$ is the sum of two input signals $V_A(s)$ and $V_B(s)$ and may be represented by a summing junction. The summing junction is denoted by a small circle, as indicated in Fig. E4–3b. Note that the input signals are accompanied by their appropriate signs. In fact, summing junction is analogous to the node in an electrical circuit. Specifically, for a given summing junction the sum of all incoming signals is equal to the sum of all outgoing signals.

Substituting the values of $V_A(s)$ and $V_B(s)$ in Eq. (4–4c), we get

$$V_o(s) = \frac{A(R_1 + R_2)}{R_1 + R_2 + AR_2} V_i(s) + \frac{-1/RC}{s} V_i(s)$$

Therefore, the transfer function

$$\frac{V_o(s)}{V_i(s)} = K_1 + \frac{K_2}{s} \qquad (4-5)$$

where

$$K_1 = \frac{A(R_1 + R_2)}{R_1 + R_2 + AR_2}$$

$$K_2 = -\frac{1}{RC}$$

The network can be represented by a single block, as shown in Fig. 4–6.

Derivative Block

The characteristic of the *derivative block* is that its transfer function contains a zero which is at origin of the $s = \alpha + j\omega$ plane. (See the block diagram of Fig. 4–7.)

Because of its inherent unstable characteristic a derivative control is very seldom used in compensating a system. Specifically, the magnitude of the derivative transfer

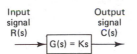

Figure 4–7. Derivative block, $G(s) = Ks$.

function increases with an increase in frequency. However, a derivative plus proportional control is commonly used in compensating control systems because it contains a pole and a zero, which provide a more stable frequency response.

EXAMPLE 4–4

Derive the transfer function for the network shown in Fig. E4–4, and draw the block diagram to represent it.

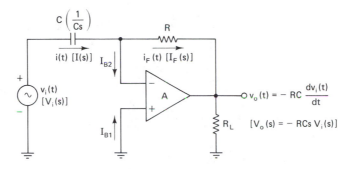

Figure E4–4. Ideal differentiator. Quantities in parentheses are Laplace transform equivalents.

SOLUTION Since the input bias currents I_{B1} and I_{B2} are negligibly small, we have

$$I(s) \simeq I_F(s)$$

$$\frac{V_i(s)}{1/Cs} = \frac{-V_o(s)}{R}$$

or

$$\frac{V_o(s)}{V_i(s)} = Ks \qquad (4-6)$$

where $K = -RC$. The negative sign indicates that the input and output signals are out of phase by 180°. The block diagram of an ideal differentiator is shown in Fig. 4–7.

Proportional plus Derivative Block

The *proportional plus derivative block* transfer function contains a finite zero as shown in Fig. 4–8. It is commonly used when a cancellation compensation technique is used to compensate a system.

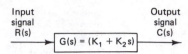

Figure 4–8. Proportional plus derivative block. K_1 and K_2 are constants.

EXAMPLE 4–5

Derive the transfer function for the network shown in Fig. E4–5a. Also, draw the block diagram to represent it.

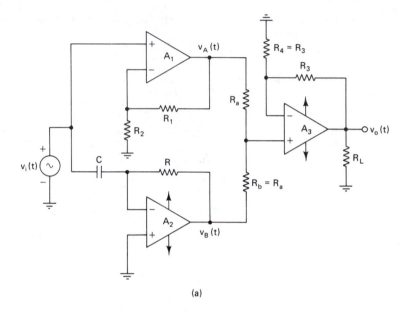

(a)

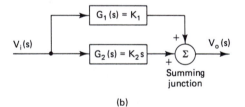

(b)

Figure E4–5. (a) Network for Example 4–5; (b) block diagram for the network of part (a).

SOLUTION Close examination of the network in Fig. E4–5a reveals that op amp A_1 is configured as a fixed-gain amplifier, op amp A_2 as a differentiator, and op amp A_3 as a summing amplifier. The transfer functions of these amplifiers are determined in Examples 4–1, 4–2, and 4–3, respectively. Therefore, for a fixed-gain amplifier A_1,

$$V_A(s) = K_1 V_i(s)$$

where

$$K_1 = \frac{A(R_1 + R_2)}{R_1 + R_2 + AR_2}$$

For the differentiator A_2,

$$V_B(s) = K_2 s V_i(s)$$

where $K_2 = -RC$. For the summing amplifier A_3,

$$V_o(s) = V_A(s) + V_B(s)$$

or

$$V_o(s) = K_1 V_i(s) + K_2 s V_i(s)$$

Therefore, the overall transfer function is

$$\frac{V_o(s)}{V_i(s)} = K_1 + K_2 s$$

where

$$K_1 = \frac{A(R_1 + R_2)}{R_1 + R_2 + AR_2}$$

$$K_2 = -RC$$

The block diagram for the network in Fig. E4–5a is shown in Fig. E4–5b.

Simple Phase-Lag Block

The transfer function of a simple *phase-lag block* has a real and simple pole. Also, there always exists a finite phase difference between the input and output signals, which is a function of the frequency. In fact, output signal always lags in phase to the input signal by a certain amount—hence the name *phase-lag network*. In a phase-lag network the phase difference between the output signal and the input signal is denoted by a negative angle. In other words, a phase-lag network may be used to add a negative angle to the total phase shift of an uncompensated system, which may make it stable. In addition, when the phase-lag network is used, it affects the overall gain of the original uncompensated system.

EXAMPLE 4–6

Determine the transfer function for the network shown in Fig. E4–6a. Then draw the block diagram to represent it.

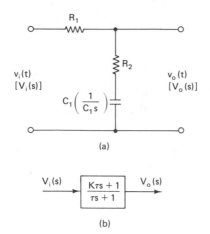

(a)

$$V_i(s) \longrightarrow \boxed{\dfrac{K\tau s + 1}{\tau s + 1}} \longrightarrow V_o(s)$$

(b)

Figure E4–6. (a) Phase-lag network with its Laplace transform representation; (b) block diagram for a phase-lag network where $K = R_2/(R_1 + R_2)$ and $\tau = (R_1 + R_2)C_1$.

SOLUTION   Using the voltage-divider rule, we have

$$V_o(s) = \frac{R_2 + 1/sC_1}{R_1 + R_2 + 1/sC_1} V_i(s)$$

Hence the transfer function is

$$\frac{V_o(s)}{V_i(s)} = \frac{K\tau s + 1}{\tau s + 1} \qquad (4\text{--}7)$$

where

$$K = \frac{R_2}{R_1 + R_2} = \text{gain constant}$$

$$\tau = (R_1 + R_2)C_1 = \text{lag-time constant}$$

Note that the transfer function of a lag network has one pole and one zero, similar to the proportional plus integration network transfer function of Fig. 4–6. Specifically, the transfer function of the lag network has a pole at $s = -1/\tau$ and a zero at $s = -1/K\tau$. However, if $K \leq 0.1$, the transfer function of the lag network in Eq. (4–7) can be approximated as having only a pole, and the block is then said to be a *simple lag*. The block diagram for the phase-lag network is shown in Fig. E4–6b.

Generally, when the phase-lag network is used to compensate a control system it reduces the overshoot and makes the uncompensated system more stable.

Simple Phase-Lead Block

The transfer function of a *simple phase-lead block* has a zero that is real and simple. The phase-lead block, on the other hand, has a finite zero and a finite pole. The characteristic of a phase-lead network is that the phase shift between the output and the input signals is positive. However, the amount of phase shift depends on the frequency of operation and is $\leq 90°$. Since output signal leads input signal in phase, the phase difference is denoted by a positive angle. A phase-lead network is used to reduce the phase lag of an uncompensated system.

The phase-lead network is often formed by using passive components. It can also be formed by using active components such as op amps.

EXAMPLE 4–7

Determine the transfer function for the passive phase-lead network of Fig. E4–7a. Then draw a block diagram to represent it.

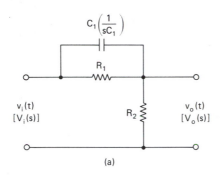

(a)

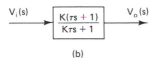

(b)

Figure E4–7. (a) Passive phase-lead network with its Laplace transform representation; (b) block diagram for a phase-lead network where $K = R_2/(R_1 + R_2)$ and $\tau = R_1 C_1 s$.

SOLUTION Using the voltage-divider rule, we have

$$V_o(s) = \frac{R_2 V_i(s)}{\dfrac{R_1(1/C_1 s)}{R_1 + 1/C_1 s} + R_2}$$

or

$$\frac{V_o(s)}{V_i(s)} = \frac{K[\tau s + 1]}{K \tau s + 1} \qquad (4–8)$$

where

$$K = \frac{R_2}{R_1 + R_2}$$

$$\tau = R_1 C_1 s$$

According to Eq. (4–8), the zero is at $s = -1/\tau$ and the pole is at $s = -1/K\tau$. However, if $K \leq 0.1$, the transfer function may be approximated as having only a zero and the phase-lead block is referred to as a *simple phase-lead*. Figure E4–7b shows the block diagram of the phase-lead network.

Generally, when a phase-lead network is used for compensating a control system, it improves the rise time to a step input and reduces the overshoot of the uncompensated system.

Quadratic Block

Ideally, the transfer function of a *quadratic block* has either two zeros or two poles. Since a practical system transfer function does not have more zeros than poles, we shall consider a quadratic block with two poles only. Remember that these two poles may be

1. Real and simple
2. Real and repeated (multiple order)
3. Complex conjugate
4. Pure imaginary

Obviously, the output response of a quadratic block depends on the nature of its poles and the type of input signal used.

EXAMPLE 4–8

Determine the transfer function of the network shown in Fig. E4–8a; draw the block diagram to represent it.

SOLUTION In Fig. E4–8a there are three op amps; op amp A_1 is used as a summing amplifier, and op amps A_2 and A_3 are configured as integrators. Therefore, to determine the overall transfer function we will find the transfer function of each of these op amp stages. First let us consider op amp A_1, which is configured as a summing amplifier. The transfer function of the summing amplifier can be obtained by using the superposition theorem. The output $V_{o2}(s)$ due to all inputs is

$$V_{o2}(s) = K_1 K_2 V_{o1}(s) - K_3 V_o(s) - K_4 V_i(s) \qquad (4-9a)$$

where

$$K_1 = \frac{R_4}{R_4 + R_5}$$

$$K_2 = 1 + \frac{R_2}{R_1 R_3/(R_1 + R_3)}$$

$$K_3 = \frac{R_2}{R_3}$$

$$K_4 = \frac{R_2}{R_1}$$

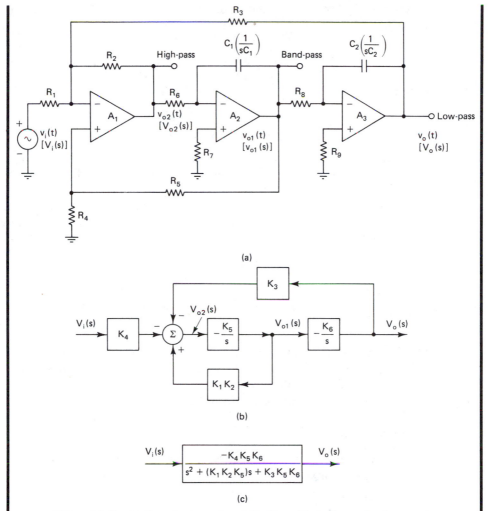

Figure E4-8. (a) Second-order state-variable filter. Block diagram for the second-order state-variable filter of part (a): (b) closed-loop; (c) open-loop.

Next, we will determine the transfer function of the integrators formed by op amps A_2 and A_3. Therefore, from Example 4–2 the transfer functions of these integrators are as follows. For the first integrator, which uses op amp A_2,

$$V_{o2}(s) = -\frac{s}{K_5} V_{o1}(s) \qquad (4\text{–}9b)$$

where $K_5 = 1/R_6C_1$. Similarly, for the second integrator, which uses op amp A_3,

$$V_{o1}(s) = \frac{-s}{K_6} V_o(s) \qquad (4\text{–}9c)$$

where $K_6 = 1/R_8C_2$. Substituting the values of $V_{o2}(s)$ and $V_{o1}(s)$ in Eq. (4–9a), we get

$$\frac{-s}{K_5}\frac{-s}{K_6} V_o(s) = K_1K_2\left[-\frac{s}{K_6} V_o(s)\right] - K_3V_o(s) - K_4V_i(s)$$

Hence

$$\frac{V_o(s)}{V_i(s)} = -\frac{K_4K_5K_6}{s^2 + (K_1K_2K_5)s + K_3K_5K_6} \tag{4–10}$$

In Eq. (4–10) the negative sign indicates the phase inversion of the input signal. As expected, the transfer function in Eq. (4–10) contains a denominator polynomial, a quadratic. Obviously, the roots of this quadratic (poles of the transfer function) depend on the values of constants K_1 through K_6, and may be real and simple, repeated, or complex conjugates. Usually Eq. (4–10) is expressed in a standard form as follows.

$$\frac{V_o(s)}{V_i(s)} = \frac{K\omega_n}{s^2 + 2\zeta\omega_n s + \omega_n^2} \tag{4–11}$$

where ω_n = natural angular frequency of oscillation
ζ = damping ratio
K = gain constant

The block diagram for the circuit of Fig. E4–8a can be obtained using Eqs. (4–9a), (4–9b), and (4–9c). Equation (4–9a) can be represented by a summing junction and the associated proportionality blocks, whereas Eqs. (4–9b) and (4–9c) can be represented by integrator blocks as shown in Fig. E4–8b. The block diagram of Fig. E4–8b can also be represented by a single block as shown in Fig. E4–8c.

Combination Block

A *combination block* is defined as a block that contains two or more of the basic building blocks, such as proportional, integration, derivative, simple phase lag, simple phase lead, and quadratic. For example, the block diagram of Fig. 4–9 consists of a

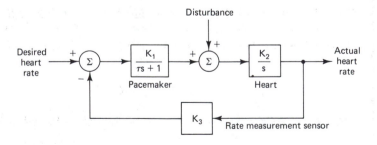

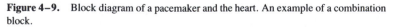

Figure 4–9. Block diagram of a pacemaker and the heart. An example of a combination block.

Representing Control Systems   Chap. 4

simple phase-lag block, an integration block, and a proportional block. The closed-loop transfer function of the block diagram obviously will be the combination of these blocks.

4-7 BLOCK DIAGRAM SIMPLIFICATION METHODS

In this section we examine various methods of simplifying given block diagrams so that a desired transfer function can easily be obtained.

A typical block diagram consists of one or more summing junction(s), directed line segment(s), forward and feedback blocks, and takeoff point(s). The *summing junction* is used to add the signals algebraically, directed line segments are used to represent the direction of the signals, *forward and feedback blocks* are used to represent the transfer function of the important parts of the system, and the *takeoff point* indicates the physical point in the system from where the desired signal is tapped off.

Block diagram simplification methods may involve the manipulation or rearrangement of summing junctions, forward and/or feedback blocks, and takeoff points such that the overall transfer function can easily be obtained. However, the block diagram simplification methods do not alter the nature of the given system or its transfer function.

Cascaded or Series Summing Junctions

Figure 4–10a shows two summing junctions that are connected in *series*. The output signal of the first summing junction is

$$D(s) = R(s) + A(s) - B(s) \qquad (4-12a)$$

whereas that of the second summing junction is

$$C(s) = D(s) - E(s) \qquad (4-12b)$$

Substituting the value of $D(s)$ from Eq. (4–12a) into Eq. (4–12b), we get

$$C(s) = R(s) + A(s) - B(s) - E(s) \qquad (4-12c)$$

Close examination of Eq. (4–12c) reveals that $C(s)$ is the output signal of the summing junction and $+R(s)$, $+A(s)$, $-B(s)$, and $-E(s)$ are the input signals into the summing junction. In other words, Eq. (4–12c) can be represented by a single summing

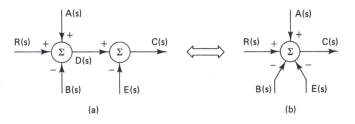

(a)   (b)

Figure 4–10. (a) Summing junctions in series; (b) its equivalent.

junction as shown in Fig. 4–10b. Note that the bidirectional arrow in Fig. 4–10 indicates that the two summing junctions in series can be represented by a single summing junction, or vice versa. The concept of replacing two summing junctions by a single summing junction can be extended to three or more summing junctions as well.

Cascaded Blocks

Figure 4–11 shows two blocks that are connected in series or *cascaded*. Recall that the output signal of a given block is equal to the input signal times the transfer function of the block. Applying this concept to the blocks in Fig. 4–11, we get

$$A(s) = R(s)G_1(s)$$

$$C(s) = A(s)G_2(s)$$

or

$$C(s) = R(s) [G_1(s)G_2(s)] \qquad (4–13)$$

where $G_1(s)$ and $G_2(s)$ are the transfer functions of the blocks. Thus from Eq. (4–13) it is obvious that the two blocks in series can be replaced by a single block whose transfer function is equal to the product of the transfer functions of the two original blocks (Fig. 4–11b). The foregoing concept can also be applied to three or more blocks which are connected in series.

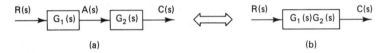

(a)   (b)

Figure 4–11.   (a) Two blocks in series; (b) its equivalent.

Parallel Blocks

Figure 4–12 shows two blocks connected in parallel. Again the output signals of each of the blocks and the summing junction can be expressed as follows:

$$A(s) = R(s)G_1(s)$$

$$B(s) = R(s)G_2(s)$$

$$C(s) = \pm A(s) \pm B(s)$$

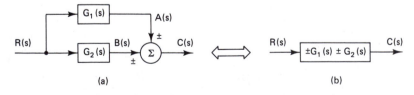

(a)   (b)

Figure 4–12.   (a) Two blocks in parallel; (b) its equivalent.

where $\pm$ signs indicate that signals $A(s)$ and $B(s)$ may be positive or negative.

$$C(s) = \pm R(s)G_1(s) \pm R(s)G_2(s)$$

or

$$C(s) = R(s)\,[\pm G_1(s) \pm G_2(s)] \tag{4-14}$$

Equation (4–14) indicates that the two blocks in parallel whose output signals are fed into a summing junction can be represented by a single block whose transfer function is equal to the algebraic sum of the transfer functions of the two original blocks (Fig. 4–12b). Again the concept can be extended to more than two blocks in parallel.

Moving a Takeoff Point

Moving a takeoff point to the right of a block.　In Fig. 4–13 a takeoff point P is to be moved to the right of a block $G_1(s)$. In this figure only a portion of the block diagram is shown, just to illustrate the concept. Before the takeoff-point is moved,

$$A(s) = R(s)G_3(s)$$

$$B(s) = R(s)G_1(s)$$

$$C(s) = B(s)G_2(s)$$

The output signals $A(s)$, $B(s)$, and $C(s)$ should not change in value when the takeoff point P is moved to the right of $G_1(s)$. Now, let us assume that the takeoff point P is moved to the right of block $G_1(s)$. The only block that is affected by this change is $G_3(s)$. Therefore,

$$A(s) = R(s)G_1(s)X(s)$$

where $X(s)$ is the transfer function of the block whose output is $A(s)$. However, $A(s)$ is to remain the same, hence $X(s)$ must be equal to $G_3(s)/G_1(s)$ (Fig. 4–13b). Thus any time a takeoff point is moved to the right of a block, the transfer function of the block into which the takeoff signal feeds must be *divided* by the transfer function of the block to the right of which the takeoff point is moved.

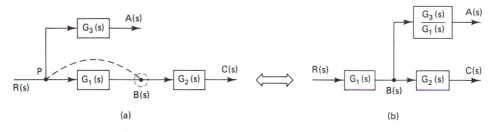

Figure 4–13.　Moving a takeoff point to the right of a block.

Moving a takeoff point to the left of a block.　Figure 4–14 shows the block diagrams before and after the takeoff point is moved to the left of a block. By

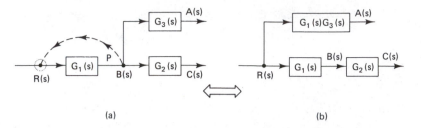

Figure 4-14. Moving a takeoff point to the left of a block.

using the same procedure as that used for moving a takeoff point to the right of a block, the reader should be able to verify that any time a takeoff point is moved to the left of a block, the transfer function of the block into which the takeoff signal feeds must be *multiplied* by the transfer function of the block to the left of which the takeoff point is moved (Fig. 4-14b).

Moving a Summing Junction

Moving a summing junction to the right of a block. Figure 4-15 shows the partial block diagram of a given system in which the summing junction is to be moved to the right of a block whose transfer function is $G(s)$. Before the summing junction is moved to the right of $G(s)$,

$$C(s) = R(s)G(s) - F(s)G(s)$$

or

$$C(s) = R(s)G(s) - A(s)G(s)H(s) \qquad (4-15a)$$

After the summing junction is moved to the right of $G(s)$,

$$C(s) = R(s)G(s) - A(s)X(s) \qquad (4-15b)$$

where $X(s)$ is the transfer function of the block whose output is $F(s)$. By comparing Eqs. (4-15a) and (4-15b) we conclude that $X(s)$ must be equal to $G(s)H(s)$ (Fig. 4-15b).

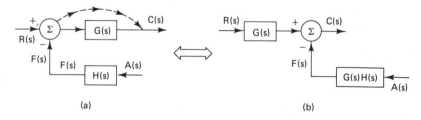

Figure 4-15. Moving a summing junction to the right of a block.

Moving a summing junction to the left of a block. Suppose that we need to move a summing junction to the left of a block with transfer function $G(s)$, as shown in Fig. 4-16a. Then we must *divide* the transfer function $H(s)$ of the block

Representing Control Systems Chap. 4

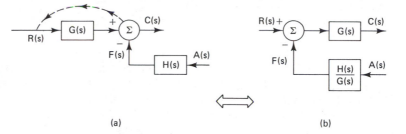

(a)   (b)

Figure 4–16. Moving a summing junction to the left of a block.

whose output signal is $F(s)$ by $G(s)$. The resulting block diagram is shown in Fig. 4–16b. The reader should be able to prove the result by using the same procedure as that used for moving a summing junction to the right of a block by writing equations for the signal $C(s)$ in Fig. 4–16.

Transfer Function of a Standard Closed-Loop Block Diagram

Any closed-loop control system can be represented by the block diagram of Fig. 4–17; therefore, it has become a standard for all closed-loop systems. Note that in Fig. 4–17 $G(s)$ is the transfer function in the forward path between $R(s)$ and $C(s)$; $H(s)$ is the transfer function in the feedback path between $C(s)$ and $F(s)$. Referring to Fig. 4–17, we have

$$E(s) = R(s) - F(s)$$

$$E(s) = R(s) - C(s)H(s)$$

However,

$$C(s) = E(s)G(s)$$

Therefore,

$$C(s) = [R(s) - C(s)H(s)]G(s)$$

or

$$C(s)[1 + G(s)H(s)] = R(s)G(s)$$

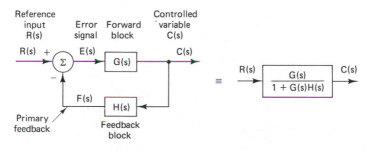

Figure 4–17. Block diagram of a closed-loop control system.

Hence the closed-loop transfer function

$$T(s) = \frac{C(s)}{R(s)} = \frac{G(s)}{1 + G(s)H(s)} \qquad (4-16a)$$

Equation (4–16a) indicates that the transfer function of a closed-loop system of Fig. 4–17 is equal to the forward transfer function $G(s)$ divided by 1 plus the product of forward transfer function $G(s)$ and the feedback transfer function $H(s)$.

If the feedback transfer function $H(s) = 1$, the system in Fig. 4–17 is called a *closed-loop system with unity feedback*, the transfer function of which is

$$T(s) = \frac{C(s)}{R(s)} = \frac{G(s)}{1 + G(s)} \qquad (4-16b)$$

If the reference input signal $R(s)$ is negative, the closed-loop transfer function is

$$T(s) = \frac{C(s)}{R(s)} = - \frac{G(s)}{1 + G(s)H(s)} \qquad (4-17)$$

On the other hand, if the feedback signal $F(s)$ is positive, the closed-loop transfer function $T(s)$ is

$$T(s) = \frac{C(s)}{R(s)} = \frac{G(s)}{1 - G(s)H(s)} \qquad (4-18)$$

EXAMPLE 4–9

Determine the closed-loop transfer function for the block diagram of Fig. E4–9a.

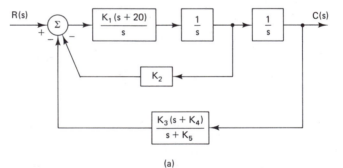

(a)

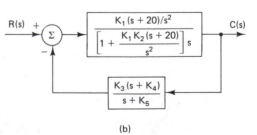

(b)

Figure E4–9. Closed-loop system for passenger-carrying mass-transit system.

SOLUTION The following steps will be used to determine the closed-loop transfer function for the block diagram of Fig. E4−9a.

1. Represent the summing junction as two summing junctions in series.
2. Combine two blocks with transfer functions $K_1(s + 20)/s$ and $1/s$ into a single block.
3. Replace the inner closed loop by a single block and combine it with series block having transfer function $1/s$ (Fig. E4−9b).
4. Finally, replace the closed-loop block diagram of Fig. E4−9b by a single block. Thus the transfer function is

$$
T(s) = \frac{C(s)}{R(s)}
$$

$$
= \frac{K_1(s + 20)(s + K_5)}{s[s^2 + K_1K_2(s + 20)][s + K_5] + K_1K_3(s + 20)(s + K_5)}
$$

EXAMPLE 4−10

For the block diagram shown in Fig. E4−10a, determine the transfer function $C(s)/R(s)$.

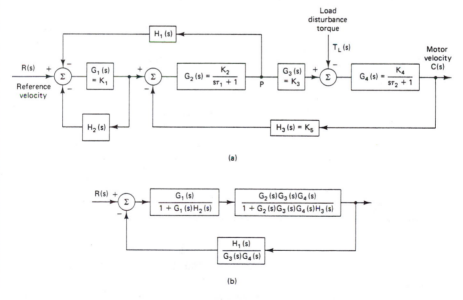

Figure E4−10. Dc motor speed control system.

SOLUTION The following steps are employed to determine the closed-loop transfer function for the block diagram of Fig. E4−10a.

1. Assume that $T_L(s) = 0$ since the desired transfer function is $C(s)/R(s)$.
2. Represent the input summing junction by two summing junctions in series.
3. Move the takeoff point P two blocks to the right.
4. Combine the three blocks in series by a single block and replace the inside closed loop by a single block (Fig. E4–10b).
5. Replace the closed-loop block diagram of Fig. E4–10b by a single block. The reader should be able to do this simplification.

4–8 SIGNAL FLOW GRAPH

Another graphical technique that is used to represent control systems is a *signal flow graph*. The signal flow graph was originally introduced by S. J. Mason as an input/output representation of linear systems. The signal flow graph essentially is a one-line diagram, the basic elements of which are nodes and branches. The *nodes* are used to represent the system variables and are denoted by small circles. Based on the relationship between the variables, the nodes are connected by line segments called *branches*. The branches include associated branch gains between the nodes and also indicate the direction of signal flow between the nodes. The direction of signal flow is indicated by an arrow and the signal is transmitted through the branch only in the direction of the arrow. In short, a signal flow graph is a graphical representation of the relationship between the variables which are interrelated by a set of linear algebraic equations.

An equivalent signal flow graph can be drawn from the given block diagram, or vice versa. Sometimes the transfer function determination using the block diagram simplification methods is very time consuming and difficult, especially if the given system has several summing junctions, takeoff points, and feedback loops. However, the transfer function determination using the signal flow graph methods is possible for any given system regardless of how complex its block diagram is.

Another important difference between the signal flow graph and the block diagram is that the rules to determine the transfer function from the signal flow graph are more stringent because of the rigid mathematical relationships between the variables.

Before we study how to determine the transfer function from a given signal flow graph, let us examine the terms that are frequently used with signal flow graphs.

Definitions of Some Important Terms

Node. A node is a small circle that is used to represent a variable. Obviously, the number of nodes in a signal flow graph is equal to the number of variables in the system. However, a node may be repeated as many times as desired provided that the branches connecting the same nodes have the branch gain of unity.

Branch. A branch is a directed line segment that connects nodes.

Path. A path between nodes is one or more branches (line segments) that are traversed in the same direction (direction of arrows).

Forward path. A forward path is a path that starts at one node and ends at the other. The peculiarity of the forward path is that the signal travels through it from left to right and no node is transversed more than once.

Path gain. The path gain is equal to the product of all the branch gains of the branches that constitute the path.

Forward-path gain. Forward-path gain is the gain of the forward path.

Loop. A loop is a path that starts and ends on the same node. However, in following the loop, no node other than the start node is encountered more than once.

Loop gain. Loop gain is the path gain of a loop.

Signal Flow Graph Construction

The following procedure is used to draw a signal flow graph for a given system.

1. Identify the number of variables in the system in addition to the input and output variables. Remember that the output of each summing junction is also a variable.
2. Denote these variables from input to output by small circles with appropriate space left between the nodes for branches and corresponding branch gains.
3. Connect the nodes with the necessary branches.
4. Label the branch gains connecting the nodes.

Example 4–11 illustrates the use of these steps in drawing a signal flow graph from a given block diagram.

EXAMPLE 4–11

Draw the signal flow graph for the block diagram of Fig. E4–11a.

SOLUTION

1. A close examination of the block diagram of Fig. E4–11a reveals that there are five variables: $R(s)$, $E(s)$, $A(s)$, $C(s)$, and $B(s)$.
2. These variables are represented by the nodes $R(s)$, $E(s)$, $A(s)$, $C(s)$, and $B(s)$ as shown in Fig. E4–11b.
3. Connect the nodes with appropriate branches, making sure that the proper signal directions are maintained.
4. Finally, indicate the branch gains over its corresponding branches. Make sure that the signs of the signals are preserved. For example, the output signal of the proportional block K_3 is negative; hence the branch gain of the path between $B(s)$ and $E(s)$ is $-K_3$. Thus the final signal flow graph is as shown in Fig. E4–11b.

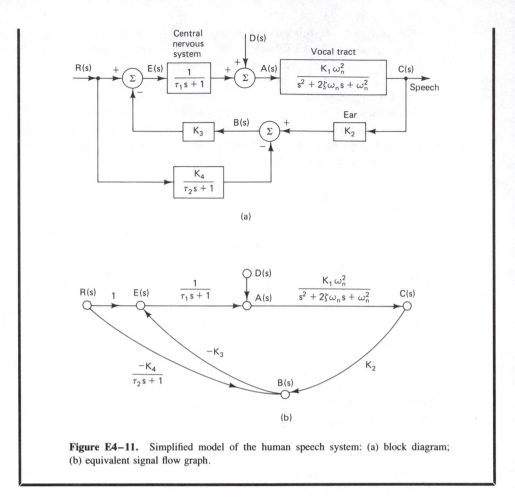

Figure E4–11. Simplified model of the human speech system: (a) block diagram; (b) equivalent signal flow graph.

Transfer Function Determination of a Signal Flow Graph

In Section 4–7 we studied how to determine the transfer function of a given block diagram through the use of block diagram simplification methods. In this section we use *Mason's rule* (*gain formula*) to determine the transfer function of a signal flow graph. Mason's transfer function gain formula is as follows:

$$T(s) = \frac{C(s)}{R(s)} = \sum_{K=1}^{n} \frac{M_K \Delta_K}{\Delta} \qquad (4-19)$$

where $T(s)$ = transfer function (or gain) between output node $C(s)$ and input node $R(s)$
$\quad C(s)$ = output node variable
$\quad R(s)$ = input node variable
$\quad n$ = total number of forward paths
$\quad M_K$ = gain of the Kth forward path
$\quad \Delta = 1 -$ (sum of all individual loop gains) + (sum of gain products of all

possible combinations of two nontouching loops) − (sum of gain products of all possible combinations of three nontouching loops) + · · ·

$\Delta_K = \Delta$ for that part of the signal flow graph that is not touching the Kth forward path

Although Eq. (4−19) seems complicated, it is straightforward to use, as we will see in the following examples.

EXAMPLE 4−12

Draw an equivalent signal flow graph for the block diagram of Fig. E4−10a and determine the closed-loop transfer function $C(s)/R(s)$, using Mason's transfer function formula.

SOLUTION Let $T_L(s) = 0$ because we need to determine transfer function $C(s)/R(s)$. There are six variables, including input and output variables. Therefore, there will be six nodes in all. These are connected by directed line segments with appropriate path gains, as shown in Fig. E4−12a.

Note that to verify the accuracy of a signal flow graph, we can always write the equation(s) at each node in terms of its node variable and the incoming signals. These equations can then be compared with the corresponding equations which are obtained from the original block diagram. For example, referring to

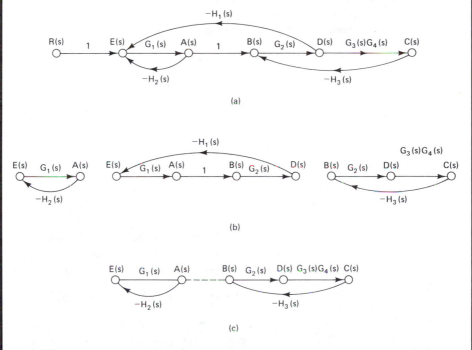

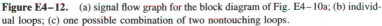

Figure E4−12. (a) signal flow graph for the block diagram of Fig. E4−10a; (b) individual loops; (c) one possible combination of two nontouching loops.

the block diagram of Fig. E4–10a, the signal $E(s)$ at the output of input summing junction is

$$E(s) = R(s) - A(s)H_2(s) - D(s)H_1(s)$$

Now, referring to the signal flow graph of Fig. E4–12a, the equation for the variable $E(s)$ is

$$E(s) = R(s) + [-H_1(s)D(s)] + [-H_2(s)A(s)] = R(s) - D(s)H_1(s) - A(s)H_2(s)$$

The preceding two equations are exactly the same as expected. The nicest feature of the signal flow graph is that even if a variable is repeated more than once, it does not affect the overall transfer function.

Now let us apply Mason's rule to Fig. E4–12a to determine the overall transfer function $C(s)/R(s)$. Note that according to Mason's rule, we have

$$\frac{C(s)}{R(s)} = \sum_{K=1}^{n} \frac{M_K \Delta_K}{\Delta}$$

where K = number of forward paths = 1. Since there is only one forward path, Eq. (4–19) becomes

$$\frac{C(s)}{R(s)} = \frac{M_1 \Delta_1}{\Delta}$$

where M_1 = gain of the forward path
$\quad = G_1(s)G_2(s)G_3(s)G_4(s)$
There are three individual loops, as shown in Fig. E4–12b:

$$E(s) \longrightarrow A(s) \longrightarrow E(s)$$

$$E(s) \longrightarrow A(s) \longrightarrow B(s) \longrightarrow D(s) \longrightarrow E(s)$$

$$B(s) \longrightarrow D(s) \longrightarrow C(s) \longrightarrow B(s)$$

The individual loop gains of these loops are

$$- G_1(s)H_2(s)$$

$$- G_1(s)G_2(s)H_1(s)$$

$$- G_2(s)G_3(s)G_4(s)H_3(s)$$

Remember that an individual loop is identified by its unique loop gain. In other words, a loop that is repeated more than once is identified by a repeated loop gain. The sum of individual loop gains equals

$$- G_1(s)H_2(s) - G_1(s)G_2(s)H_1(s) - G_2(s)G_3(s)G_4(s)H_3(s)$$

Is there any possible combination of two nontouching loops? Yes, there is one (Fig. E4–12c). The gain product of the two nontouching loops equals

$$G_1(s)H_2(s)G_2(s)G_3(s)G_4(s)H_3(s)$$

However, there is no possible combination of three nontouching loops; hence

$\Delta = 1 -$ (sum of all individual loop gains) + (sum of gain products of all possible combination of two nontouching loops) + 0

$= 1 - [-G_1(s)H_2(s) - G_1(s)G_2(s)H_1(s) - G_2(s)G_3(s)G_4(s)H_3(s)]$
$$+ [G_1(s)H_2(s)G_2(s)G_3(s)G_4(s)H_3(s)] + 0$$

Since there is only one forward path, we need to determine only Δ_1. However, Δ_1 is that part of Δ which is not touching the forward path $R(s) \rightarrow E(s) \rightarrow A(s) \rightarrow B(s) \rightarrow D(s) \rightarrow C(s)$. Since all the loops and their possible combinations touch the forward path, $\Delta_1 = 1$. Thus substituting the values of M_1, Δ_1, and Δ, we get the transfer function

$$\frac{C(s)}{R(s)} = \frac{[G_1(s)G_2(s)G_3(s)G_4(s)][1]}{[1 + G_1(s)H_2(s) + G_1(s)G_2(s)H_1(s) + G_2(s)G_3(s)G_4(s)H_3(s)]}$$
$$+ [G_1(s)G_2(s)G_3(s)G_4(s)H_3(s)]$$

The reader should verify that this transfer function is the same as that was obtained in Example 4–10.

EXAMPLE 4–13

Using Mason's rule, determine the closed-loop transfer function $C(s)/R(s)$ for the signal flow graph of Fig. E4–11b.

SOLUTION   Let $D(s) = 0$ since the required transfer function is $C(s)/R(s)$. There are two forward paths between $R(s)$ and $C(s)$: $R(s)E(s)C(s)$ and $R(s)B(s)E(s)C(s)$. Therefore, the gains M_1 and M_2 of these forward paths are (Fig. E4–13a):

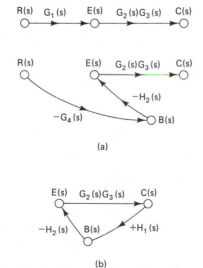

(a)

(b)

Figure E4–13.   Signal flow graph for Example 4–13: (a) two forward paths between input $R(s)$ and output $C(s)$; (b) individual loop $E(s) \rightarrow C(s) \rightarrow B(s) \rightarrow E(s)$.

$$M_1 = G_1(s)G_2(s)G_3(s)$$

$$M_2 = [-G_4(s)][-H_2(s)]G_2(s)G_3(s)$$

$$= G_2(s)G_3(s)G_4(s)H_2(s)$$

where

$$G_1(s) = 1$$

$$G_2(s) = \frac{1}{\tau_1 s + 1}$$

$$G_3(s) = \frac{K_1 \omega_n^2}{s^2 + 2\zeta\omega_n s + \omega_n^2}$$

$$G_4(s) = \frac{K_4}{\tau_2 s + 1}$$

$$H_1(s) = K_2$$

$$H_2(s) = K_3$$

Close examination of Fig. E4–11b reveals that there is only one loop: $E(s)$ $C(s)B(s)E(s)$. Hence the loop gain is

$$G_2(s)G_3(s)[+H_1(s)][-H_2(s)]$$

Since there is only one loop, there is no possible combination of two non-touching loops. Hence

$$\Delta = 1 - \text{(sum of all individual loop gains)}$$

or

$$\Delta = 1 + G_2(s)G_3(s)H_1(s)H_2(s)$$

There are two forward paths; therefore, we need to determine Δ_1 and Δ_2. Since the loop $E(s)C(s)B(s)E(s)$ touches both the forward paths,

$$\Delta_1 = \Delta_2 = 1$$

Thus substituting the values of M_1, M_2, Δ_1, Δ_2, and Δ in Eq. (4–19), we get

$$\frac{C(s)}{R(s)} = \frac{M_1\Delta_1}{\Delta} + \frac{M_2\Delta_2}{\Delta}$$

$$= \frac{G_1(s)G_2(s)G_3(s)}{1 + G_2(s)G_3(s)H_1(s)H_2(s)} + \frac{G_2(s)G_3(s)G_4(s)H_2(s)}{1 + G_2(s)G_3(s)H_1(s)H_2(s)}$$

or

$$\frac{C(s)}{R(s)} = \frac{G_1(s)G_2(s)G_3(s) + G_2(s)G_3(s)G_4(s)H_2(s)}{1 + G_2(s)G_3(s)H_1(s)H_2(s)}$$

Note that it would be difficult to obtain the closed-loop transfer function $C(s)/R(s)$ for the block diagram of Fig. E4–11a by using the block diagram simplification methods. However, the signal flow graph method using Mason's rule makes it easier to determine the transfer function of such a block diagram.

4–9 BLOCK DIAGRAM VERSUS SIGNAL FLOW GRAPH

Generally, block diagram representation of a given system is easy to understand, and also at first glance it gives more information about the system than does the corresponding signal flow graph. Hence, traditionally, a block diagram is often used to represent control systems. However, it is easier to draw an equivalent signal flow graph from a given block diagram than to draw an equivalent block diagram from a given signal flow graph. The signal flow graph is strictly a one-line diagram, whereas the block diagram is a combination of transfer function blocks and directed line segments. Normally, the transfer function determination of a simple block diagram is easy and straightforward. However, if the block diagram is relatively complex and uses multiple feedbacks, its transfer function determination using block diagram simplification methods becomes more complex and very time consuming. In a situation like this, it is relatively easy and less time consuming to draw an equivalent signal flow graph and use Mason's rule to determine its transfer function. As a general rule, to determine the transfer function of a given block diagram, first try block diagram simplification methods; if simplification is tedious, use the signal flow graph method.

SUMMARY

1. The performance evaluation of a given system may be done using graphical and or analytical techniques. However, before these techniques are employed, the system must be represented in a standard equation form referred to as a transfer function. The transfer function is obtained by using a mathematical tool called the Laplace transform. Specifically, for a given system the transfer function is the ratio of the Laplace transform of the output to the Laplace transform of the input with all initial conditions assumed to be zero.

2. When a system is to be designed or analyzed, it may be broken down into several important parts and each of these parts may then be represented by an appropriate transfer function. A block diagram is a graphical method composed of blocks, summing junctions, takeoff points; and directed line segments that are used to represent a given system. Each of these blocks contains a transfer function which represents a specific part of the system. Thus block diagrams not only simplify the representation but also aid the analysis of control systems. The block diagrams may be drawn as open-loop or closed-loop. Also, a block is labeled based on the type of transfer function it contains.

3. Block diagram simplification methods are used so that a desired transfer function can easily be obtained from a given block diagram. Basically, block diagram simplification methods involve one or more of the following operations: moving a takeoff point, moving a summing junction, and combining series or parallel blocks.

4. Another useful graphical method that is often used to represent a control system is the signal flow graph. The signal flow graph is a one-line diagram composed of nodes and branches. Specifically, the nodes represent the system variables, and the branches are the directed line segments that connect the nodes. An equivalent signal flow graph can be drawn from a given block diagram, or vice versa. Mason's gain formula is used to determine a desired transfer function from a given signal flow graph.

5. Generally, block diagram representation of a given system is easy to understand and at first glance gives more information about the system than does the corresponding signal flow graph. If the block diagram is relatively complex and uses multiple feedbacks, the transfer function determination using Mason's gain formula is easier than the block diagram simplification methods.

QUESTIONS

4–1. Why are all initial conditions assumed to be zero in determining the transfer function of a given network?

4–2. What is the significance of a transfer function?

4–3. What is the difference between closed-loop and open-loop transfer functions?

4–4. What criterion is used to label different blocks? Explain.

4–5. What is the difference between the phase-lead and simple lead blocks?

4–6. Explain the difference between the phase-lead and phase-lag networks.

4–7. What is the significance of block diagram simplification methods?

4–8. What is a signal flow graph? How is it different from a block diagram?

4–9. What is Mason's gain formula? Explain its significance.

4–10. Compare and contrast the usefulness of block diagrams and signal flow graphs.

PROBLEMS

4–1. Determine the transfer function for the practical integrator of Fig. P4–1 and represent it by a block diagram. What is the difference between the networks of Figs. P4–1 and E4–2? Explain the effect of R_2 on the transfer function.

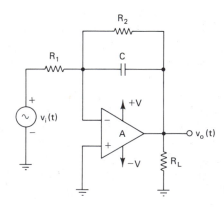

Figure P4–1. Practical integrator of Problem 4–1.

4-2. Calculate the transfer function for the network of Fig. P4–2. Also, draw a block diagram to represent it. Assume that A_1 and A_2 remain constant over the desired frequency range.

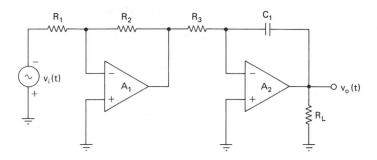

Figure P4-2. Network of Problem 4–2.

4-3. Derive the transfer function for the network of Fig. P4–3 and draw the closed-loop block diagram to represent it. Explain the difference between the transfer function of this network and that obtained in Example 4–4.

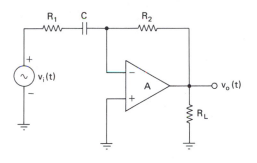

Figure P4-3. Network of Problem 4–3.

4-4. Calculate the transfer function for the low-pass filter network shown in Fig. P4–4. Also draw a closed-loop block diagram for the network. Assume that the gain A_1 remains constant over the desired operating frequency range. What is the difference between the transfer function of the simple lead, simple lag, and low-pass filter networks? Explain.

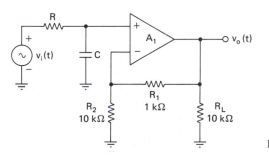

Figure P4-4. Low-pass filter.

4-5. Calculate the transfer function for the Sallen–Key filter shown in Fig. P4–5 and represent it with a closed-loop block diagram. Assume that the gain of an op amp is constant over

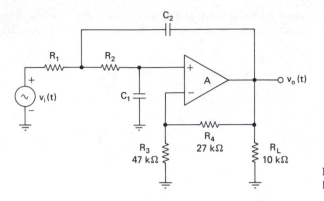

Figure P4–5. Sallen–Key filter of Problem 4–5.

the desired operating frequency range and that the input base currents are approximately zero.

4–6. Calculate the transfer function for the multiple feedback filter network shown in Fig. P4–6. Assume that the gain A is constant over the desired operating frequency range and that the input base currents are negligibly small. Draw a closed-loop block diagram to represent the filter.

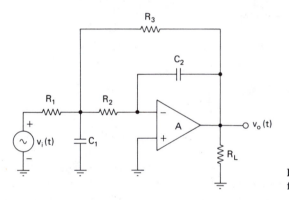

Figure P4–6. Second-order multiple feedback filter.

4–7. Simplify the block diagram shown in Fig. P4–7 to determine the transfer function $C(s)/R(s)$. Represent the final transfer function by a single block.

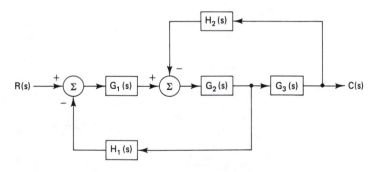

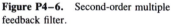

Figure P4–7. Block diagram for Problem 4–7.

Representing Control Systems   Chap. 4

4–8. Repear Problem 4–7 for the block diagram of Fig. P4–8.

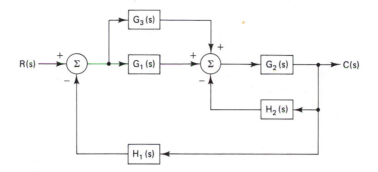

Figure P4–8. Block diagram for Problem 4–8.

4–9. Repeat Problem 4–7 for the block diagram of Figure P4–9.

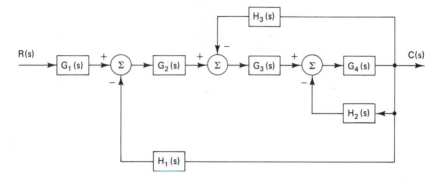

Figure P4–9. Block diagram for Problem 4–9.

4–10. Simplify the block diagram shown in Fig. P4–10 to determine the equation for the controlled variable $C(s)$ as a function of both the reference input $R(s)$ and the disturbance $D(s)$.

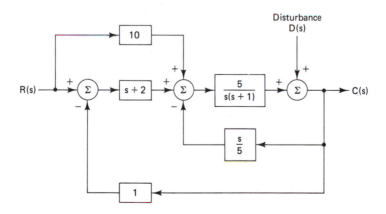

Figure P4–10. Block diagram for Problem 4–10.

4–11. Repeat Problem 4–10 for the block diagram of Fig. P4–11.

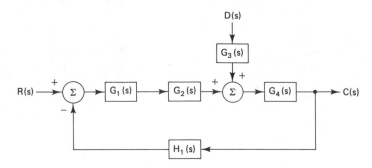

Figure P4–11. Block diagram for Problem 4–11.

4–12. Draw an equivalent signal flow graph for the block diagram of Fig. P4–7. Find the transfer function $C(s)/R(s)$ using Mason's gain formula. Is the transfer function the same as that obtained in Problem 4–7? Explain.

4–13. Draw an equivalent signal flow graph for the block diagram of Fig. P4–8. Find the transfer function $C(s)/R(s)$ using Mason's gain formula. Is the transfer function the same as that obtained in Problem 4–8? Explain.

4–14. For the block diagram shown in Fig. P4–9 determine the transfer function $C(s)/R(s)$ using Mason's gain formula. Is the transfer function the same as that obtained in Problem 4–9? Explain.

FEEDBACK SYSTEM CHARACTERISTICS AND ANALYSIS

chapter 5 _____

5-1 INTRODUCTION

History shows that negative feedback was used by the ancient civilizations in a variety of practical systems. The early Greeks, for instance, realized more than 2000 years ago that a simple float can be used to sense the liquid level and thus provide means for operating a system without human supervision. They applied this concept to the oil lamp, where the float was used to control the opening of the valve, thus regulating the flow of oil into the lamp from a hidden reservoir and maintaining a constant oil level within the lamp, allowing the lamp to burn indefinitely. This and other examples bear evidence to the fact that the concept of feedback is not a product of recent technological development. Its origin is to be found in man's early attempts in creating a mechanical system that operates in an unsupervised manner.

The potential benefits that may be derived from the use of feedback in the control environment sparked the imagination of scientists through the centuries. Probably the first significant application of feedback to the automatic control of an industrial process is attributed to James Watt for his invention, the flyball governor, in the latter part of the eighteenth century. The flyball governor, a mechanical device, maintains the speed of a steam engine at a desired value by regulating the flow of steam to the engine, increasing the flow as the engine slows down, and vice versa. The flyball governor was never fully analyzed by its inventor; the full mathematical treatment, first by Maxwell and then by Wishnegradsy, followed almost a century later.

The absence of a unified theory involving feedback and its effects on the performance of a closed-loop system added to the uncertainty and confusion that surrounded the design of a system with feedback for many centuries. Feedback was generally regarded as a mysterious concept, misunderstood and misused in most design applications, which were almost always based on trial and error. This situation existed throughout recorded history until the present century. Some of the early contributions to the theory of feedback originated at Bell Laboratories, where Nyquist, Bode, and Black investigated the properties of feedback amplifiers which were used in communication equipment. Their findings were applied to the general control systems by Hall and Harris.

The demands of World War II created many new applications for feedback control systems. These include aircraft control, radar tracking, gun positioning, and others. The traditional approach to system design based on trial and error could no longer be applied to the critical military systems, as these systems had to have predictable performance. This added momentum and intensity to research and the application of feedback control theory.

In the early years following World War II, the industrial community exploded with activity; books were published, articles written, and systems designed, all promoting the control system with feedback. This was due in part to the declassification by the government of the theory and of some systems used during the wartime years. The other reason, and probably a more significant one, is the realization by the industry that feedback control systems is an emerging technology with a promising future. It is not surprising that control system engineering emerged at that time as an independent discipline and a profession.

Modern control is based on the use of feedback. We owe a debt of gratitude to Bode, Nyquist, Black, and others, including Evans, who in the late 1940s explained the use of the s plane in the design and analysis of control systems. These pioneers explained the properties of feedback and laid a firm foundation for the sound design of control systems in the years to come.

The proliferation and the availability of computers in modern times has made it both possible and feasible to include the computer in the closed loop of some of the more elaborate control systems. As the computer is capable of making quick computations and logical decisions, it is an ideal element for the multivariable processing environment. In the commercial auto-pilot system, for instance, once the destination coordinates together with other information are programmed into the airborne computer, it automatically guides the flight of the aircraft. Actually, the computer is part of complex feedback, accepting the current aircraft coordinate information from the appropriate sensors, comparing this information against the desired heading and then in the case of error, outputting the control signal to the rudder or other controllers to control the flight course of the aircraft.

Computer-based feedback control system plays an important role in a modern industrial plant, controlling simultaneously many parameters that are critical to a particular process. Robotics, an emerging technology, is based on the use of negative feedback in the mechanical robot configurations. Its limited industrial application includes, for example, an assembly plant in Japan that uses fully automatic robot arms. Manually operated or automatic robot arms are often used in a hazardous environment or to perform an operation that may present dangers to a human operator. Robotics is perhaps in its infancy today, but it is almost sure to make an important impact on both the industrial and consumer applications of tomorrow.

Feedback control systems are not limited to electrical and mechanical configurations. Feedback is used by our bodies to control various biological functions. Our body temperature, for example, is regulated to within a fraction of a degree. Our bodies also maintain the chemical balance of various elements, such as zinc, which are essential for our survival and regulate their concentration levels, which range from a fraction of 1% to a tolerance of several parts per million. Just as incredible is the coordination of body motions in the execution of various daily tasks which we take for granted. They are part of a complex process that involves the interaction of the visual feedback with the brain and the muscles.

Feedback may be used to model human behavior. The student–teacher relationship, for example, may be represented by a closed-loop system (which uses negative feedback) whose input by the teacher is the desired knowledge to be gained by the student and whose output is the knowledge actually gained by the student. The feedback mechanism may be obtained through examinations which offer a quantitative measure of the actual knowledge gained and which may be used to compare the actual knowledge with the desired knowledge.

In 1970, Paul Samuelson of MIT received the Nobel prize for his contribution to economic science. His model, which is based on feedback, provides an analytical approach to solving complex dynamic problems in economy. As can be seen, the concept of feedback has a broad spectrum of applications in many disciplines. With some thought, readers can probably provide their own examples based on feedback. In this chapter we explore the fundamental properties of negative feedback, analyze feedback-based systems, and show the effects of negative feedback on system performance.

5–2 OPEN-LOOP AND CLOSED-LOOP SYSTEMS

Any open-loop control system may be represented by the general block diagram shown in Fig. 5–1. It includes the controller, whose output affects the process response. The controlled variable C is the process variable whose value is to be maintained within some predetermined limits. In responding to the reference input R, the controller acts on the process forcing C to attain a desired value. Accordingly, R may be regarded as the desired value of C.

A practical example of an open-loop system is shown in Fig. 5–2. In this system the process being controlled is the oven temperature T_0, and the power amplifier–heater combination serves the function of the controller. In responding to the input V_i, the controller applies the input heat flow q_i to the oven chamber. Heat also flows out of the oven due to the imperfect insulation (finite thermal resistance) of its walls. The heat that flows out of the chamber is the heat loss and represents the thermal load on the system. It increases with a decrease in the thermal resistance of the wall material

Figure 5–1. General open-loop system.

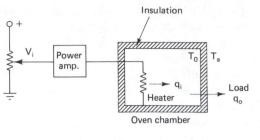

(a)

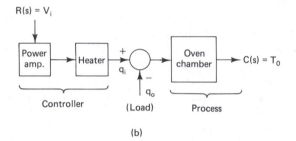

(b)

Figure 5–2. Open-loop temperature control system: (a) system diagram; (b) block diagram.

or with the decrease in the outside temperature T_a, and vice versa. For a selected input V_i, the difference in heat flow $q_i - q_o$ is the heat that accumulates inside the chamber, causing its temperature to rise. When the system reaches its steady state, T_0 attains a constant value resulting in $q_i = q_o$ (i.e., the heat loss is replaced exactly by the input heat flow). As long as this equilibrium condition is maintained, T_0 will remain at a constant value indefinitely.

In a practical operating environment, the power amplifier output may drift over a period of time and the temperature T_a outside the chamber may also change. The former will cause q_i to change and the latter, q_o to change. This will disturb the thermal equilibrium, causing a change in $q_i - q_o$ and, consequently, a change in T_0. If it is required that T_0 be held constant under such conditions, the operator must be present and supervise the system by visually monitoring T_0 and making adjustments in V_i whenever T_0 changes.

This brings to light the following properties, which apply not only to the system discussed in the example, but to any open-loop system:

1. An open-loop system is sensitive to changes in load.
2. An open-loop system is sensitive to variations in its own parameters and external disturbances.

The answer to these problems is a closed-loop system. The general closed-loop system configuration is shown in Fig. 5–3. In addition to the process and the controller, a closed-loop system includes two new components: the feedback and the summing junction. The feedback component, whose gain or transfer function is K_f, produces the

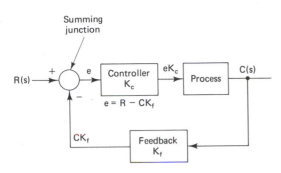

Figure 5–3. General closed-loop system.

signal CK_f, which is applied together with the input R to the summing junction. As negative feedback is used, the instantaneous polarity of CK_f must always be opposite to that of the input R. Actually, the signal CK_f gives us a measure of the actual value of the output, being related to the output through K_f and R, once again, may be regarded as the desired value of the output C.

The summing junction (a summing op amp circuit) produces the signal e, which is the difference between the input and the feedback signals, and applies it to the controller. The controller, in turn, acts on the process with the signal eK_c. The process responds to the input eK_c and changes C in the direction to reduce the value of e. This, in turn, reduces the input to the process, resulting in a smaller change in C. This chain of events continues until a time is reached when C approximately equals R (the difference between C and R is the steady-state error). At this point all changes in the process cease.

The operation of a closed-loop system is affected considerably less by changes in load or changes in controller gain than by the operation of the same system without feedback under same disturbance conditions. This is evident from the closed-loop-system block diagram of Fig. 5–3. Should the value of C increase above its steady-state value due to a disturbance, the value of the actuating signal e also increases in the negative direction because $CK_f > R$. The controller, then, applies a negative signal eK_c to the process, thus restoring C to its original value. If the disturbance lowers C below its equilibrium value, the controller applies, this time, a positive signal eK_c to the process and again restores C to its equilibrium value. The closed-loop system is thus able to regulate itself in the presence of disturbance or in the presence of variations in its own characteristics. In this respect, a closed-loop system has a distinct advantage over an open-loop system. The important characteristics of a closed-loop system are treated in later sections.

The open-loop temperature control system of Fig. 5–2 is readily converted to a closed-loop system through the addition of a summing amplifier and feedback, as shown in Fig. 5–4. The output $K_t T_0$ of the temperature sensor is amplified by K_2 and becomes the negative feedback voltage V_f. V_f is the analog signal representing the actual oven temperature. It is applied together with the input V_i, which represents the desired oven temperature, to the summing amplifier, whose gain is K_1. As shown in the block diagram, the summing and power amplifiers are combined into a single amplifier with gain $K_1 K_p$; together with the heater element, they perform the function of controller. The summing junction, which also appears in the block diagram, is a summing amplifier with unity

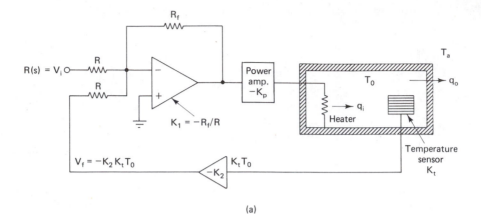

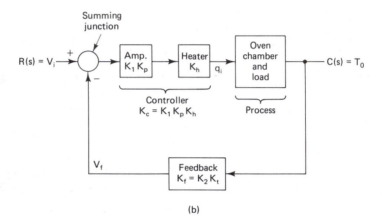

Figure 5–4. Closed-loop temperature control system: (a) system diagram; (b) block diagram.

gain. The operation of this system follows closely the operation described for the general closed-loop system of Fig. 5–3.

Any system, regardless of its complexity, may be reduced through the use of the block manipulation technique, to the basic form shown in Fig. 5–5. The closed-loop transfer function may be readily derived from this diagram as follows:

$$e = R - CH$$

Then

$$C = eG = C(R - CH)$$

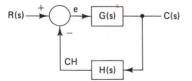

Figure 5–5. Negative feedback system: basic configuration.

Feedback System Characteristics and Analysis   Chap. 5

Solving for the transfer function gives us

$$T(s) = \frac{C(s)}{R(s)} = \frac{G(s)}{1 + GH(s)} \qquad (5-1)$$

where $G(s)H(s)$ is the loop gain and $T(s)$ is the closed-loop gain. If the loop gain is large in relation to unity [i.e., $GH(s) > 1$], Eq. (5-1) reduces to

$$T(s) \approx \frac{1}{H(s)}$$

indicating that the closed-loop gain depends only on the value of feedback, being almost independent of system parameters which are included in the forward-path transfer function.

5-3 SYSTEM ACCURACY

When an input is suddenly applied to a closed-loop system such as that shown in Fig. 5-5, the system response consists of the transient part which occurs first, followed by the steady state. Recalling that the input $R(s)$ may be regarded, from the practical standpoint, as the "desired output," the system error may be defined as the difference between the desired output $R(s)$ and the actual output $C(s)$. Stated mathematically,

$$E(s) = R(s) - C(s) \qquad (H = 1) \qquad (5-2)$$

This difference is dimensionally homogeneous, that is, the units of R and C are the same only in the case of the unity feedback system ($H = 1$). Equation (5-2) loses its meaning when $H \neq 1$ because the units of R and C are different. This may be illustrated by the following simple example. Suppose that the block diagram of Fig. 5-5 represents a speed control system. Under steady-state conditions, $R = 10$ V, $C = 950$ rpm, and $H = 10$ V/1000 rpm. Clearly, the difference $R - C$ has no practical meaning. The input R may, on the other hand, be converted to the units of C through division by H and then the difference may be taken between R/H and C. Alternatively, the output of the summing junction $R - CH$ may be divided by H to obtain the same result. Thus

$$\frac{R}{H} - C = \frac{R - CH}{H} = 50 \text{ rpm}$$

The error of the system in this example is 50 rpm.

Equation (5-2) may thus be modified to include nonunity feedback systems as follows:

$$E(s) = \frac{R - CH}{H} = \frac{R}{H} - C \qquad (H \neq 1) \qquad (5-3)$$

Using Eq. (5-1) in Eq. (5-3) yields

$$E(s) = \frac{R}{H} - \frac{GR}{1 + GH}$$

$$= \frac{R}{H(1 + GH)} \qquad (5-4)$$

For the unity feedback system $H = 1$ and expression (5–4) for the system error reduces to the following simpler form:

$$E(s) = \frac{R(s)}{1 + G(s)} \tag{5-5}$$

The open-loop gain G and the input R are functions of s and H is generally constant, especially for simple feedback configurations.

As the system output C undergoes considerable fluctuations during the transient part of the response, the system error also varies with time. Consequently, it is impractical to specify the system error during the transient part of system's response. It is generally agreed that system error is evaluated only when the system attains its steady state; hence it is called the *steady-state error*. The steady-state error may be defined mathematically as

$$e_{ss} = \lim_{t \to \infty} L^{-1}[E(s)] = \lim_{t \to \infty} e(t) \tag{5-6}$$

The use of the final value theorem eliminates the inverse Laplace transform operation in Eq. (5–6), thus simplifying the evaluation of e_{ss}. Substituting Eq. (5–4) in Eq. (5–6) and applying the final value theorem to the resulting equation, we obtain

$$e_{ss} = \lim_{s \to 0} sE(s) = \lim_{s \to 0} \frac{sR(s)}{H(1 + HG(s))} \tag{5-7}$$

An alternative approach to evaluating e_{ss} makes use of the system's time-domain response $c(t)$. Assuming that $c(t)$ is available, it is a relatively simple task to find e_{ss} as follows:

$$* \quad e_{ss} = \lim_{t \to \infty} \left(\frac{r(t)}{H} - c(t) \right) \tag{5-8}$$

The steady-state error as expressed by Eq. (5–7) depends on the input $R(s)$ and also on the type of system, which is characterized by its open-loop gain $G(s)$. The standard inputs generally include the step, ramp, and parabolic waveforms.

Systems, on the other hand, may be arranged or categorized according to *type*. System type may be determined from its $G(s)$, which is expressed in the following form:

$$G(s) = \frac{K(s - z_1)(s - z_2) \cdots (s - z_m)}{s^j(s - p_1)(s - p_2) \cdots (s - p_n)} \tag{5-9}$$

as the order of the pole at the origin. The system is thus assigned a type value equal to the value of j in Eq. (5–9), where the values of z represent the zeros of G and the values of p represent the poles of G.

From the practical standpoint, the value of j represents the number of integrations that occur in the system's forward path. A *type 0 system* has no integrations in its forward path, and a *type-1 system* has a single integration in the forward path. The speed control system described in Chapter 8, for example, is a type 0 system, whereas the position control system discussed in the same chapter is a type 1 system.

As stated earlier, the steady-state error depends on the form of the input and on the system type. The steady-state error takes a more explicit form, when the input and the system type are specified. We shall evaluate next the steady-state error for system types 0, 1 and 2 using the step, ramp and parabolic input waveforms.

Type-0 System (j = 0)

Step input. Substituting the step input

$$R(s) = \frac{E}{s}$$

in Eq. (5–7) and assuming that H is constant, the following result is obtained:

$$e_{ss} = \frac{E}{H(1 + K_0)} \tag{5–10}$$

where K_0 is the error constant for the type 0 system and a step input. It is defined by

$$K_0 = \lim_{s \to 0} HG(s) \tag{5–11}$$

Ramp input. Substituting the ramp input

$$R(s) = \frac{E}{s^2}$$

in Eq. (5–7), we obtain

$$e_{ss} = \lim_{s \to 0} \frac{E}{H[s + sHG(s)]} = \frac{E}{0} = \infty$$

The denominator in the equation above approaches 0 in the limit as s approaches 0. Clearly, the first term s in the denominator must approach 0 and the second term consists of H, which is assumed to be independent of s and $G(s)$, which from Eq. (5–9) must approach a constant value as s approaches 0 because $j = 0$. Consequently, the second term also, being multiplied by s, must approach 0 as s approaches 0. Thus, the type 0 system cannot follow the ramp input.

Parabolic input. Substituting the parabolic input

$$R(s) = \frac{E}{s^3}$$

In Eq. (5–7) and once again assuming H to be constant yields

$$e_{ss} = \lim_{s \to 0} \frac{E}{H[s^2 + s^2HG(s)]} = \infty$$

Again, as in the case of the ramp input, the type 0 system is unable to follow the parabolic input.

Type 1 System ($j = 1$)

Step input. The step input

$$R(s) = \frac{E}{s}$$

is substituted in Eq. (5–7) as follows:

$$e_{ss} = \lim_{s \to 0} \frac{E}{H[1 + HG(s)]} = \frac{E}{\infty} = 0$$

The denominator in the equation above becomes infinite due to $G(s)$, which approaches infinity for $j = 1$ in Eq. (5–9) as s approaches 0. The type 1 system is thus able to follow a step input without error.

Ramp input. The ramp input

$$R(s) = \frac{E}{s^2}$$

is substituted in Eq. (5–7) with the following result:

$$e_{ss} = \lim_{s \to 0} \frac{E}{H[s + sHG(s)]} = \frac{E}{HK_1} \tag{5–12}$$

where K_1 is the error constant for the type 1 system and the ramp input. It is defined as

$$K_1 = \lim_{s \to 0} sHG(s) \tag{5–13}$$

The value of K_1 is finite and nonzero because the product $sG(s)$ approaches a finite, nonzero value for $j = 1$ in Eq. (5–9) as s approaches 0. The type 1 system is thus able to follow a ramp input, but with an error.

Parabolic input. When the parabolic input

$$R(s) = \frac{E}{s^3}$$

is used in Eq. (5–7), the resulting steady-state error

$$e_{ss} = \lim_{s \to 0} \frac{E}{H[s^2 + s^2 HG(s)]} = \infty$$

increases without limit as both terms in the denominator approach 0. The first term naturally approaches 0 and the second term approaches 0 after the $G(s)$ in Eq. (5–9) is multiplied by s^2 and the limit taken of the resulting product, allowing s to approach 0.

As in the case of the type 0 system, the type 1 system is unable to follow the parabolic input with a finite error.

Type 2 System (j = 2)

Step and ramp inputs. It is left as an exercise for the reader to show, by following the procedure established in the preceding sections, that the type 2 system has a 0 steady-state error for both the step and ramp inputs. The type 2 system, which has two integrations in its forward path, is able to follow the step and ramp inputs without error.

Parabolic input. Upon substituting the parabolic input

$$R(s) = \frac{E}{s^3}$$

in Eq. (5–7) and applying the limit as follows,

$$e_{ss} = \lim_{s \to 0} \frac{E}{H[s^2 + s^2 HG(s)]} = \frac{E}{HK_2} \qquad (5-14)$$

the steady-state error approaches in the limit a finite nonzero value, where K_2 is the error constant of the type 2 system and the parabolic input, defined as

$$K_2 = \lim_{s \to 0} s^2 HG(s) \qquad (5-15)$$

Once again the first term in the denominator of Eq. (5–14) immediately approaches 0 as $s \to 0$, and the second term approaches a constant value after the $G(s)$ in Eq. (5–9) is multiplied by s^2 and the limit taken of the resulting product, treating H as a constant. The type 2 system is thus able to follow a parabolic input with a constant, finite error.

The results of the preceding section are summarized in Table 5–1.

TABLE 5–1 STEADY-STATE ERROR VERSUS SYSTEM TYPE AND INPUT

Type	Step, $Eu(t)$	Ramp, $Etu(t)$	Parabolic, $\frac{1}{2}Et^2u(t)$
0	$\dfrac{E}{H(1 + K_0)}$	∞	∞
1	0	$\dfrac{E}{HK_1}$	∞
2	0	0	$\dfrac{E}{HK_2}$

Input

$^*K_0 = \lim\limits_{s \to 0} HG(s),\ K_1 = \lim\limits_{s \to 0} sHG(s),\ K_2 = \lim\limits_{s \to 0} s^2 HG(s).$

The results show that the system of higher type which has more integrations in the forward path is more accurate for lower-order inputs. The following general representation of the input

$$r(t) = Et^n u(t) \qquad n = 0, 1, 2, \ldots \tag{5-16}$$

reduces to the step for $n = 0$, the ramp for $n = 1$, and parabolic input for $n = 2$.

It may be concluded from Table 5–1 and Eq. (5–16) that

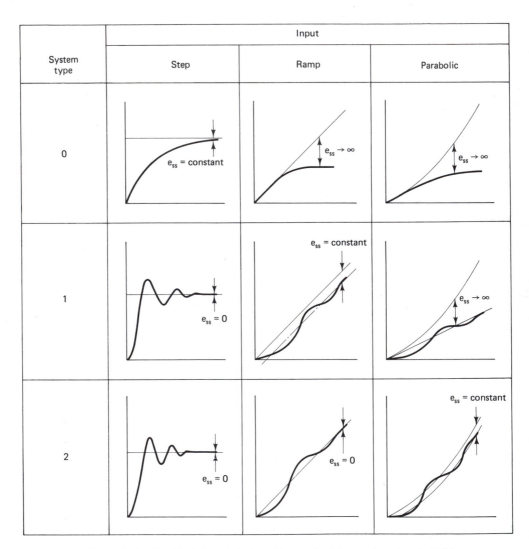

Figure 5–6. Graphical characterization of the steady-state error for selected conditions of input and system type.

$$
e_{ss} = \begin{cases} 0 & n < j \\ \text{const} & n = j \\ \infty & n > j \end{cases} \qquad (5\text{--}17)
$$

where j represents the system type and n identifies the input in Eq. (5–16).

As the number of integrations in the system's forward path (the value of j) exceeds the value of the exponent n in Eq. (5–16), the steady-state error vanishes and the output of the system becomes equal to its input under steady-state conditions. This occurs when a step is applied to the type 1 system or when either a step or a ramp is applied to the type 2 system.

As expressed by Eqs. (5–10), (5–12), and (5–14), the steady-state error assumes a finite nonzero value. This occurs when the system type designator j is equal to the input exponent n. In each case the steady-state error expression contains the loop gain GH in its denominator. This makes it possible to reduce the steady-state error by increasing the value of the loop gain. The loop gain may be adjusted by adding an amplifier stage to the system's forward path. The ability to control the value of the system's steady-state error may be attributed to the use of feedback. Other important features of feedback systems are considered in later sections.

When the input is such that its exponent n exceeds the system type designator j, the steady-state error of the system increases without limit, that is, the difference between the input and the output of the system increases without limit with increasing time. For the system to attain an infinite steady-state error, the input to the system must be applied for a very long time. Accordingly, an infinite steady-state error has more meaning mathematically, and much less so from a practical standpoint. For example, when a ramp is applied to a type 0 system, the steady-state error at the end of the ramp is not infinite, although it has been increasing up to that point, and would have continued to increase if the ramp did not terminate. An input that leads to an infinite steady-state error is generally not used in a practical case, not because it leads to an infinite steady-state error but, more than likely, because it does not serve a useful application function or a useful system test function. A graphical characterization of the steady-state error for selected conditions of input and system type is shown in Fig. 5–6.

EXAMPLE 5–1

Consider the feedback configuration which represents the position control system shown in Fig. E5–1. G_1 and G_2 are defined as

$$G_1 = K \text{ (V/rad)}$$

$$G_2 = \frac{20}{s(0.1s + 1)} \qquad \text{rad/V}$$

Calculate (a) The steady-state error for each of the following inputs: $\theta_{i1} = 60u(t)$ degrees, $\theta_{i2} = 400t^2u(t)$ rad/s^2, and $K = 100$ V/rad; (b) the value of K such that the steady-state error is $10°$ if the input is $\theta_i = 400tu(t)$ rad/s.

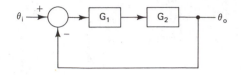

Figure E5-1.

SOLUTION (a) The given system is a type 1 system. From Table 5-1, the steady-state error is 0 for the first input θ_{i1}, which is a step input, and ∞ for the second input, which is a parabolic input.

(b) From Eq. (5-13),

$$K_1 = \lim_{s \to 0} \frac{20K}{0.1s + 1} = 20K$$

Substituting this result in Eq. (5-12) gives us

$$10° = 0.175 \text{ rad} = \frac{400}{20K}$$

Solving for K yields

$$K = 114.3$$

5-4 SYSTEM SENSITIVITY

In Section 5-3 it was shown that a feedback system can be made more accurate through adjustment in its loop gain. This is one example where a system performance characteristic is improved through the use of negative feedback. Another important characteristic of any control system, which must also be considered, is the sensitivity of the system to the variation in its parameters. Suppose that the components included in $G(s)$ or $H(s)$ in Fig. 5-5 change their values in the course of system's operation. Such changes in component values may be attributed to power supply fluctuations, component aging, or simply to the normal drift in component characteristic. Regardless of the reason for such changes, we must be aware that they can have a significant effect on system performance.

The sensitivity of the closed-loop system to the variation in its forward-path parameters is represented by the sensitivity factor S_G, which is defined as follows:

$$S_G = \frac{\partial T/T}{\partial G/G} \tag{5-18}$$

where the numerator represents the percent change in the closed-loop transfer function as defined by Eq. (5-1) and the denominator represents the percent change in the forward-path transfer function. For a constant input, $\partial T/T$ reduces to a percent change in the output $\partial C/C$ after T is replaced by C/R. The partial derivative symbol ∂, which represents a differential change, implies that changes in Eq. (5-18) must be small (10% or less). The partial derivative also means that T is a function of more than one independent variable. Equation (5-18) may be rewritten as

$$S_G = \frac{\partial T}{\partial G} \frac{G}{T} \tag{5-19}$$

Using Eq. (5–1) and performing the required differentiation, we have

$$\frac{\partial T}{\partial G}\bigg|_{H=\text{const}} = \frac{1}{(1 + GH)^2}$$

and substituting the result in Eq. (5–19) yields

$$S_G = \frac{1}{(1 + GH)^2} \frac{G}{G/(1 + GH)} = \frac{1}{1 + GH(s)} \tag{5-20}$$

The result above reflects the sensitivity of the closed-loop system to variation in the forward-path parameters. The sensitivity of the system to variation in the feedback-path parameters is expressed by

$$S_H = \frac{\partial T/T}{\partial H/H} = \frac{\partial T}{\partial H} \frac{H}{T} \tag{5-21}$$

as the ratio of the percent change in the closed-loop transfer function to the percent change in the feedback-path transfer function. Differentiating Eq. (5–1) with respect to H while holding G constant, we have

$$\frac{\partial T}{\partial H}\bigg|_{G=\text{const}} = \frac{-GH}{(1 + GH)^2}$$

and substituting this result in Eq. (5–21) gives us

$$S_H = \left|\frac{-GH}{(1 + GH)^2}\right| \left|\frac{H}{G/(1 + GH)}\right| = \frac{-GH(s)}{1 + GH(s)} \tag{5-22}$$

$$S_H \approx -1 \qquad \text{if } GH \gg 1$$

The sensitivity of the open-loop system to variations in G may be obtained by letting $H = 0$ in Fig. 5–5 and Eq. (5–1). The resulting output,

$$C = GR$$

may be differentiated to yield

$$\partial C = R \, \partial G$$

and

$$\frac{\partial C}{C} = \frac{\partial G}{G}$$

These results may be combined to express the sensitivity factor for the open-loop system as follows:

$$S_G = \frac{\partial C/C}{\partial G/G} = 1 \qquad H = 0 \tag{5-23}$$

The open-loop system is thus 100% sensitive to variations in the value of G, which are reflected to the output on the one-to-one basis. All components selected for an open-loop system must therefore be very stable to ensure a minimum change in the output C due to the system parameter variations. An operational amplifier, for example, may be included in G. As all op amps do, it must exhibit an offset voltage at its output. The offset voltage, however, is of little concern, as it may be nulled out at the beginning of the operation. The drift in the offset voltage that occurs during the operation, on the other hand, may cause serious problems. This is especially true in applications where small dc signals are involved, and the drifting offset voltage in such cases may constitute a significant part of the output. In general, to reduce the sensitivity of an open-loop system to the variation in its parameters, high-quality, and therefore more expensive, components must be used. Higher cost is the price that must be paid for a less sensitive system.

The remarks above also apply to a closed-loop system and its sensitivity to the variations in feedback parameters. This is supported by Eq. (5–22) for large values of loop gain. Generally, the loop gain is larger than unity, which results in S_H being close to unity. Any fluctuations in the values of the feedback components are thus reflected to the output on almost a one-to-one basis. As in the case of an open-loop system, high-quality, stable components must be used in the feedback path to reduce the sensitivity of the system to the feedback parameter variations.

The role that the feedback plays in affecting the sensitivity of the system to parameter variations is best expressed by Eq. (5–20). The parameter variations referred to are in forward path of the closed-loop system. The form of Eq. (5–20) suggests that the loop gain GH may be used to reduce the sensitivity value. For example, a 20% change in the amplifier gain, which is part of G, results in a 20% change in G; consequently, if the steady-state value of loop gain is 9, the change in the output is only 2%. Even for a moderate value of loop gain, the system is relatively insensitive to a sizable change in its forward-path parameter. The sensitivity is further improved for a larger value of loop gain. The use of feedback is thus responsible for reducing the sensitivity of the system to the variation in its forward-path parameters. The components used in the forward path may therefore be less stable and consequently less expensive. The following example illustrates the use of Eq. (5–20).

EXAMPLE 5–2

In the speed control system shown in Fig. E5–2, the amplifier gain K_a decreases 10%. Calculate the value of the amplifier gain before the change occurs such that the change in the motor speed is less than 0.1%.

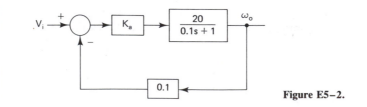

Figure E5–2.

SOLUTION   Under the steady-state conditions, $G = 20K_a$ and $H = 0.1$: hence the steady-state value of loop gain GH is $2K_a$. Substituting this in Eq. (5–20), we obtain

$$S_G = \frac{\partial T/T}{\partial G/G} = \frac{\partial(\omega_o/V_i)/(\omega_o/V_i)}{\partial(20K_a)/20K_a}$$

$$= \frac{\partial\omega_o/\omega_o}{\partial K_a/K_a} = \frac{-0.001}{-0.1} = 0.01 = \frac{1}{1 + 2K_a}$$

Solving for K_a yields

$$K_a = 49.5$$

This is the minimum value of K_a for which the loop gain is 99. This value of loop gain attenuates any changes in the forward path by 100 to 1, before they appear in the output. The relative insensitivity of the system to changes in its forward path is made possible by the use of the feedback. The amplifier, together with its power supply, need not be very stable for satisfactory system operation. This translates directly to reduced system costs.

The situation is very different if the value of feedback is reduced by 10%, from 0.1 to 0.09. Using 49.5 for K_a and Eq. (5–22), the resulting change in the output speed is $+(99/100)(.1)$ or almost a 10% increase in the output speed for a 10% decrease in the value of feedback. As noted previously, the system is very sensitive to parameter variations in its feedback path. The components in the feedback path must therefore be very stable and generally of high quality.

The relationships that were derived for system sensitivity are based on small changes in G. These relationships are less accurate for large changes in G. The use of differential quantities ∂T and ∂G, which must be small by definition, is the reason for this. One can argue that a closed-loop system will immediately compensate any initial changes in G, and will not allow these changes to become large. This is true for changes in G that occur under drift conditions; however, should G inadvertently undergo a sudden large change, the system is then forced to deal with a large change in G. With this in mind, we wish to derive a system sensitivity expression that applies for large changes in G.

In general, large changes may be described by the use of the increment Δ. An incremental change in the function of a single independent variable $y = f(x)$ may be expressed by

$$\Delta y = f(x + \Delta x) - f(x)$$

and an incremental change in the function of two independent variables $z = g(x, y)$ may be expressed by

$$\Delta z = g(x + \Delta x, y + \Delta y) - g(x, y)$$

There is no limit on how large Δx or Δy can be. Applying this to the closed-loop system, we obtain

$$S_G = \frac{\Delta T/T}{\Delta G/G} = \frac{\Delta T}{\Delta G}\frac{G}{T} \tag{5-24}$$

From Eq. (5–1),

$$\Delta T = T(G + \Delta G) - T(G)$$

$$= \frac{G + \Delta G}{1 + H(G + \Delta G)} - \frac{G}{1 + GH}$$

$$= \frac{(1 + GH)(G + \Delta G) - G(1 + HG + H\,\Delta G)}{(1 + GH)(1 + GH + H\,\Delta G)}$$

$$= \frac{\Delta G}{(1 + GH)(1 + GH + H\,\Delta G)}$$

Dividing the above result by T gives us

$$\frac{\Delta T}{T} = \left|\frac{\Delta G}{(1 + GH)(1 + GH + H\,\Delta G)}\right|\left|\frac{1 + GH}{G}\right|$$

$$= \frac{\Delta G/G}{1 + GH + H\,\Delta G}$$

Hence

$$S_G = \frac{\Delta T/T}{\Delta G/G} = \frac{1}{1 + GH + H\,\Delta G} = \frac{1}{1 + GH[1 + (\Delta G/G)]} \tag{5-25}$$

If $\Delta G/G \ll 1$, Eq. (5–25) reduces to

$$S_G = \frac{1}{1 + GH}$$

This result clearly shows that when the change in G is small in relation to unity, Eq. (5–25) reduces to Eq. (5–20), which was derived for differential changes in G.

EXAMPLE 5–3

(a) Recalculate the value of K_a in Example 5–2 on the assumption that the 10% change in K_a is a large change.

(b) The initial value of K_a in the configuration of Example 5–2 is 20. Suppose that K_a is suddenly decreased by a large amount to some lower value K_{af}. Calculate K_{af} if maximum permissible change in the output speed is 5%.

SOLUTION (a) Using the values from Example 5–2 in Eq. (5–25) gives us

$$0.01 = \frac{1}{1 + 2K_a(1 - 0.1)}$$

Solving for K_a yields

$$K_a = 55$$

(b) Using $x = \Delta G/G = \Delta K_a/20$ in Eq. (5–25), we have

$$\frac{-0.05}{x} = \frac{1}{1 + 40(1 + x)}$$

Solving for x gives us

$$x = -0.68 = \frac{K_{af} - 20}{20}$$

Solving for K_{af} yields

$$K_{af} = 6.3$$

5–5 EFFECT OF DISTURBANCES

In the course of its operation, any control system may experience the effect of disturbances. Depending on the system, the disturbances may take the form of noise or other undesirable electrical signals. The disturbances may also be mechanical, acting on the system as external torques. A gust of wind impacting a servo-driven dish antenna is an example of a disturbance where the servo system is subjected to a sudden external disturbance torque. Most disturbances are unpredictable in time of occurrence, duration, or the exact point of their application. This type of randomness makes it difficult to model the disturbances for the purposes of analysis. In the feedback system diagram shown in Fig. 5–7, the disturbance is assumed to act on the process G_2. In this model the effect of the disturbance D is combined with the output of the controller G_1 in the symbolic or virtual summing junction. Thus D can influence the signal that drives the process and consequently, the process itself.

The effect of the disturbance on the open-loop system is considered first. For an open-loop system, $H = 0$, and the output may be expressed as

$$\begin{aligned} C &= (RG_1 + D)G_2 = (G_1G_2)R + (G_2)D \qquad H = 0 \\ &= C_r + C_d \end{aligned} \tag{5–26}$$

where $C_r = (G_1G_2)R$, that part of the output due to the input R, and $C_d = (G_2)D$, that part of the output due to the disturbance D. The ratio C_r/C_d, whose counterpart in a communication system is the signal-to-noise ratio (S/N), may be expressed as

$$DR = \frac{C_r}{C_d} = G_1 \frac{R}{D} \qquad H = 0 \tag{5–27}$$

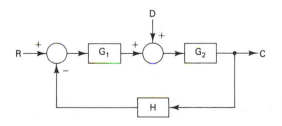

Figure 5–7. Feedback system with disturbance input.

This relationship suggests that the effect of the disturbance on the output may be reduced by increasing the input R or the controller gain G_1, or both.

For a closed-loop system where $H \neq 0$, the expression for the output may be obtained by applying Eq. (5–1) and the superposition principle on the system in Fig. 5–7, or simply by going around the loop as follows:

$$[(R - CH)G_1 + D]G_2 = C$$

and then solving for C:

$$C = \left| \frac{G_1 G_2}{1 + G_1 G_2 H} \right| R + \left| \frac{G_2}{1 + G_1 G_2 H} \right| D \qquad H \neq 0 \qquad (5–28)$$

where C_r, the first term in Eq. (5–28), represents that part of the output that is due to the input R, and C_d, the second term, represents that part of the output that is due to the disturbance D. The disturbance rejection ratio

$$\text{DR} = \frac{C_r}{C_d} = G_1 \frac{R}{D} \qquad H \neq 0 \qquad (5–29)$$

is identical to that expressed by Eq. (5–27) for the open-loop system. Thus the use of feedback offers no apparent disturbance rejection advantage over the open-loop configuration. In both cases DR is independent of the process gain G_2 because it amplifies the input and the disturbance signals by the same amount. It is indeed surprising, and contrary to intuitive expectations, that closed-loop operation offers no improvement in disturbance rejection over the open-loop system. However, the closed-loop structure allows for special conditions which can be used effectively to improve its disturbance rejection ratio.

It is apparent from Eq. (5–28) that an increase in G_1 reduces the value of the second term (which is due to disturbance) and has almost no effect on the first term. The disturbance rejection ratio is thus improved if G_1 or R are increased. The improvement in the disturbance rejection, however, is the same for the open-loop system [Eq. (5–27)] and the closed-loop system [Eq. (5–29)].

It is possible to achieve an improvement in the disturbance rejection ratio of the closed-loop system over that of the open-loop system if certain conditions are satisfied. Suppose that we insist, for example, that the disturbance rejection is compared for the two systems when the output of the closed-loop system due to the input R is equal to or is a multiple of the output of an open-loop system. Since the gain of the closed-loop system is reduced due to the use of negative feedback, G_1 and R of the closed-loop system must be increased to achieve comparable outputs.

Suppose that G_2 is the same for both the open- and closed-loop systems. If G_1 and R are used for the open-loop system, $k_1 R$ and $k_2 G_1$ are used for the closed-loop system, where k_1 and k_2 are positive constants greater than 1. Equating the outputs of the closed- and open-loop systems due to R, we have

$$C_r = \frac{K_1 K_2 G_1 G_2 T}{1 + K_2 G_1 G_2 H} = K_3 G_1 G_2 R \qquad (5–30)$$

The disturbance rejection ratio is formed next, using the second term in Eq. (5–28), which represents the output of the closed-loop system due to the disturbance:

$$DR = \frac{C_r}{C_d} = G_1 \frac{R}{D} [K_3(1 + K_2G_1G_2H)]$$

$$(5-31)$$

$$= K_1K_2G_1 \frac{R}{D}$$

If the outputs of the closed- and open-loop systems are made equal, $K_3 = 1$ and the following equation must be satisfied:

$$K_1K_2 = 1 + K_2G_1G_2H$$

This equation shows that K_1 and K_2 are not independent. If K_1 is chosen, for example, K_2 is determined from

$$K_2 = \frac{1}{K_1 - G_1G_2H}$$

$$(5-31a)$$

and K_1 must satisfy

$$K_1 > G_1G_2H$$

if K_2 is to remain positive. According to Eq. (5–31), an improvement in disturbance rejection of a closed-loop system over that of an open-loop system is realized when

$$K_1K_2 > 1$$

These conditions are easily satisfied, as illustrated by the next example.

EXAMPLE 5–4

In Fig. 5–7, $G_1 = 2$, $H = 0.2$, $R = 1$, $D = 0.5$, and $G_2 = 10/(s + 1)$. Both R and G_1 are adjusted so that the outputs of the system shown and the output of the same system with $H = 0$ are made equal. R is increased 4.5 times for the closed-loop system. Calculate (a) the disturbance rejection for the closed-loop system; (b) the disturbance rejection improvement over the open-loop system.

SOLUTION   (a) From Eq. (5–31a),

$$K_2 = \frac{1}{4.5 - 2(10)(0.2)} = 2$$

Using the given values and $K_2 = 2$ in Eq. (5–31), we obtain

$$DR = 2(4.5)(2)(2) = 36$$

(b) The improvement in the disturbance rejection is

$$K_1K_2 = 9$$

5–6 EFFECT OF FEEDBACK ON SYSTEM TRANSIENT RESPONSE

Another advantage derived from the use of feedback is an improvement in a system's transient response. The step response of a closed-loop system, for example, is faster than the response of the same system without feedback. This is supported by the analysis

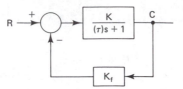

Figure 5-8. First-order feedback system.

in Section 6–5. The results of the analysis are used here. The reader should consult Section 6–5 for details.

Figure 5–8 shows a first-order feedback system where the controller and the process are combined into a single block with the dc gain K and the time constant τ, which is associated with the time delay. The response of this system to the step input $r(t) = Eu(t)$ may be expressed as

$$c(t) = K_s E(1 - e^{-t/\tau s})$$

where

$$K_s = \frac{K}{1 + KK_f} \qquad \text{closed-loop system dc gain}$$

$$\tau_s = \frac{\tau}{1 + KK_f} \qquad \text{closed-loop system time constant}$$

The ratio $\tau/\tau_s = 1 + KK_f$ clearly shows that the closed-loop time constant τ_s is smaller than the open-loop time constant. The smallness of τ_s relative to τ depends on the loop gain KK_f, whose value may be adjusted, for example, by adjusting the amplifier gain in the system's forward path. If the loop gain is adjusted to 19, the closed-loop system is 20 times faster than the same system configured as an open loop.

5-7 PLL CONTROL SYSTEM

The conventional analog feedback control system depends on voltage comparison for its operation. The temperature control system in Fig. 5–4, for example, uses the summing operational amplifier, which combines the input voltage representing the desired output temperature with the negative feedback voltage, which reflects the current temperature. The actuating or difference signal derived at the output of the summing junction is then used to control the process.

Another approach, which differs considerably from that described above, is based on frequency or phase comparison. As shown in Fig. 5–9, the closed-loop speed control system generates at the output of the optical sensor a pulse train whose frequency is proportional to the speed of the motor.

The feedback frequency f_o and the pulses with reference frequency f_i are applied together to the phase comparator. The phase comparator then generates an output that is proportional to the frequency difference of the two signals. This signal is converted to a dc voltage which drives the motor. Any difference in frequencies of the two signals forces the motor to adjust its speed until the frequencies are the same. Consequently, the motor runs in synchronism with the input pulse train.

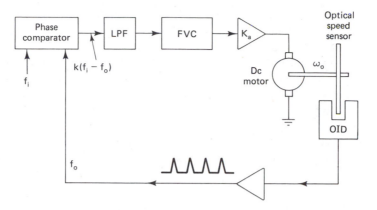

Figure 5–9. PLL control system.

The control system described here bears a strong resemblance to the phase-locked loop (PLL) circuit. The main difference lies in the device that is being controlled. In the case of the PLL circuit, the frequency of the voltage-controlled oscillator (VCO) is slaved to the frequency of the input signal, and in the control system shown, the frequency of the optical speed sensor is slaved to the frequency of the input signal; hence the motor speed is synchronized to the frequency of the input signal. In both instances the comparator plays an important role in developing the error signal, which is proportional to the difference in the two frequencies, in the same way that the summing circuit does in developing an output that is proportional to the difference between the input voltage and the feedback voltage.

In comparing the two types of control system, the PLL version does not appear to have any important advantages over the analog type; it merely offers another alternative to a control system design. The PLL control system does, however, have one advantage in a digital environment. In an application where the motor must operate in synchronism with the pulses that originate in a microprocessor or in a computer system, the PLL system has an advantage because it can accept input pulses directly without additional interfacing. In a similar application, an analog system would require additional interface hardware.

5–8 STABILITY

Any ideal open-loop system is unconditionally stable, and hence one need not be concerned about its stability. The situation is totally different in the case of a feedback system. There is always a possibility that it may be unstable even if negative feedback had been used. A possible instability is a distinct disadvantage which is associated with the use of feedback. As noted in Section 5–1, the concept of feedback and its application to a practical design was poorly understood for many centuries, and consequently, trial and error was the commonly used design approach. This situation persisted up to some 40 years ago, when the demands of World War II stimulated a great deal of research in this area. The results of this research made it possible, for the first time in the history of controls, to design a stable system mathematically or to test a given system and accurately predict its stability. So significant were these results that to the present day, they are still used, unsurpassed in their original form.

Generally, in the digital world, where microprocessors and computers are readily available, the question of system stability becomes academic. In today's industrial environment of computer control, software may be used to maintain the desired level of system stability. Many analog systems are also in use, of course, and that justifies the attention given here to the stability of such systems.

There are several methods that may be used to design the desired level of stability into the system or to test a given system for stability. These include:

1. The time-domain response
2. The root locus method, which is restricted to the s plane
3. The Bode, Nyquist, and Nichols methods, which are performed in the frequency domain
4. The Routh–Hurwitz stability criterion, which is used to determine only whether or not the system is stable

It is not the intent of this book to present a detailed quantitative coverage of the methods noted above, as they generally involve system compensation or an implementation of a required level of system performance in addition to the treatment of system stability. Instead, the reader will be exposed to the semiquantitative, symptomatic treatment of stability. The Routh-Hurwith method is presented here in greater detail because it is simple to understand and apply.

In an attempt to characterize system instability in the time domain, consider the system transfer function of the form

$$\frac{C(s)}{R(s)} = \frac{N(s)}{(s - s_1)(s - s_2) \cdots (s - s_n)} \tag{5-32}$$

and the step input $R(s) = E/s$. The response is obtained in the usual manner by following the procedure of Chapter 6. The first step usually involves partial-fraction expansion as follows:

$$C(s) = \frac{EN(s)}{s(s - s_1) \cdots (s - s_n)} = \frac{C_0}{s} + \frac{C_1}{s - s_1} + \cdots + \frac{C_n}{s - s_n} \tag{5-33}$$

where the zeros of the transfer function are included in $N(s)$ and the poles are s_1, s_2, $\ldots$, s_n. The inverse Laplace transform is taken next, instead of the usual next step when the constants are evaluated,

$$c(t) = C_0 + C_1 e^{s_1 t} + C_2 e^{s_2 t} + \cdots + C_n e^{s_n t} \tag{5-34}$$

The poles of the closed-loop transfer function appear as exponents in the resulting time-domain response. They must necessarily dictate the form of the response. The zeros of the closed-loop transfer function, on the other hand, together with the poles, determine the amplitudes C_0, C_1, $\ldots$, C_n of the individual response terms.

The poles may be real or complex. Pole s_1 in Fig. 5–10 has a negative real part

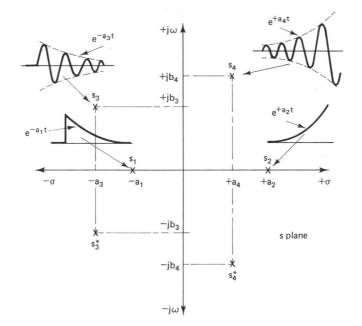

Figure 5–10. The time-domain responses corresponding to the given poles illustrate a stable and unstable system. To assure a stable response, the poles of the closed transfer function must be on the LHS of the s plane.

and its time-domain response is, as shown, a decaying exponential. Pole s_2, on the other hand, is real positive and its associated time-domain response is an increasing exponential. Poles s_3 and s_4 are both complex conjugates. The real part of s_3 is negative while that of s_4 is positive. Although both time-domain responses are sinusoidal, the response due to s_3 is decaying and that due to s_4 is increasing with time.

One way of characterizing a stable system is to insist that its time-domain response to a step input be bounded at all times. This, of course, means that each term of the response must be bounded. To satisfy this requirement, the poles of the closed-loop transfer function must have "negative real parts." This will ensure a decaying response because the real part of the pole appears in the exponent.

The criterion for a closed-loop system stability based on the time-domain response may be expressed as follows: *A closed loop system is stable if all its poles are on the left-hand side (LHS) of the s plane. In fact, the farther the poles are from the $j\omega$ axis, the more stable the system is.* The poles s_2 and s_4 are on the LHS of the s plane; hence their time-domain responses are unbounded. Any system that contains poles such as these must necessarily be unstable.

The poles of

$$\frac{C(s)}{R(s)} = \frac{G(s)}{1 + GH(s)}$$

must be the roots of

$$GH + 1 = 0 \qquad \text{or} \qquad GH = -1 \qquad\qquad (5-35)$$

Thus all closed-loop system poles must satisfy this equation, which is also known as the *characteristic equation*. The independent variable in Eq. (5–35) is s, which is a complex variable; hence the quantity GH may be represented by its magnitude and phase as

$$|GH(s)| = 1 \qquad\qquad (5-36)$$

and

$$\underline{/GH(s)} = 180° \pm k(360°) \qquad k = 1, 2, 3 \ldots \qquad\qquad (5-37)$$

Equations (5–36) and (5–37) form the basis of the *root locus method*, invented by Evans in late 1940s. Evans proposed that $GH(s)$ may be expressed as

$$GH(s) = \frac{K[N(s)]}{D(s)}$$

Then, according to Eq. (5–35), the characteristic equation becomes

$$D(s) + (K)[N(s)] = 0 \qquad\qquad (5-38)$$

where K represents the open-loop gain, whose value is to be varied from 0 to ∞. For a given value of K, Eq. (5–38) is an nth-order polynomial in s which must have n roots or n values of s which satisfy Eqs. (5–36) to (5–38). Substituting another value of K in Eq. (5–38) results in another set of n roots. For each value of K, the corresponding set of roots is plotted as points in the s plane, resulting in loci of roots (or poles of the closed-loop transfer function)—hence the root locus method.

The root locus method offers the user several distinct advantages. First, it shows graphically the position of the poles in the s plane for a specific value of system gain K. The desired level of stability can thus be designed into the system. Second, for a simple second-order system (or even a higher-order system which is approximated by the second-order system) the user may choose graphically the value of gain to satisfy some system performance requirement, such as the percent overshoot or the settling time. The root locus is thus a graphical tool that may be used to

1. Compensate an unstable system.
2. Design a desired level of stability into the system.
3. Design into the system a desired performance level.

For these reasons the root locus method has been preferred for decades by many control system design engineers. There are several disadvantages associated with the use of the root locus method, however. One of these is the task of sketching the root locus. This can be a tedious and involved task, especially if the system being sketched is complex. To assist the user in this respect, there are guidelines that simplify the sketching procedure considerably. Another possible problem with this method is related to the availability of the system transfer function. To sketch the root locus, the system transfer function must be known. The transfer function may not be readily available for a large and complex system. The present-day technology relieves the engineer of sketching the root locus by offering a broad range of software which can be used effectively in optimizing system performance.

To illustrate the root locus properties, consider a transfer function of the form

$$GH(s) = \frac{K}{s(s + 2)(s + 4)} \qquad (5-39)$$

where K represents the system gain parameter whose value is to be varied. Readers are again reminded that they are not to be concerned with the details of sketching the root locus but instead, with the meaning of a given root locus and its interpretation or its properties. Accordingly, the given transfer function may be rearranged in the characteristic equation form as follows:

$$s(s + 2)(s + 4) + K = 0$$

As shown in Fig. 5-11a, there are three loci. Each locus begins at the pole of the given $GH(s)$, that is, $s = 0, -2, -4$, where the value of K, according to the characteristic equation above, is 0. For each value of K there must be three roots that satisfy the cubic characteristic equation. As the value of K is increased, two loci progress until they meet at $s = -0.85$, while the third locus moves along the negative real axis away from the origin. At $K = 3.1$, all three roots are real and negative; two of these are equal to -0.85, while the third root (not shown) is to the left of $s = -4$. As K is increased above 3.1, two roots become complex and the third root remains real and negative. Accordingly, the two loci break away at $s = -0.85$ into the complex s plane, where the corresponding roots are complex conjugates with negative real parts, while the third locus moves along the negative real axis toward $-\infty$.

At $K = 20$, for instance, two roots are described by $s = -0.4 \pm j1.9$ and the third root by $s = -5.2$. As K continues to increase, two loci approach the $j\omega$ axis. At $K = 48$, two roots are exactly on the $j\omega$ axis at $s = \pm j2.8$, whereas the third root is at $s = -6$. For values of K above 48 the two roots move into the RHS of the s plane, where the real parts of the roots are positive. It may therefore be concluded that for $K < 48$ the system is stable, at $K = 48$ the system is marginally stable, and for values of $K > 48$ the system becomes unstable.

As mentioned earlier, the root locus graph may be used effectively to achieve a desired system performance. For the sake of an illustration, let us assume that $K = 20$. At this value of K the two complex conjugate roots are $-0.4 \pm j1.9$ and the third root is -5.2. At this point we shall approximate the given third-order system by a second-order system simply by ignoring the real root. The justification for this is based on the analysis in Chapter 6. It is shown there that the reciprocal of the real part of the root (or the reciprocal of the root itself if it is real) determines the time constant, which dictates the rate of decay of the transient response term. Accordingly, in the present case, one time constant is $1/0.4 = 2.5$, while the other time constant is $1/5.2 = 0.19$ s. Clearly, the time constant associated with the complex conjugate roots is more than 10 times larger than the time constant of the real root. Consequently, the larger time constant, which is due to the complex conjugate roots, dominates the transient response. As the third root recedes away from the origin along the negative real axis, its effect on the overall transient response becomes less significant. It may generally be

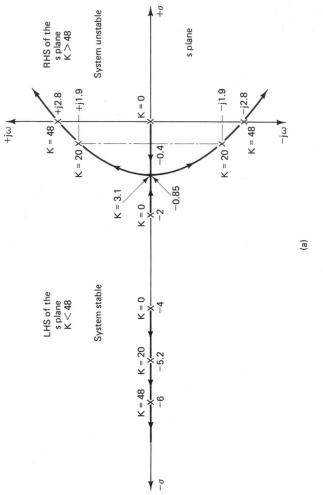

Figure 5–11. $GH(s) = K/[s(s + 2)(s + 4)]$ represented on: (a) root locus diagram in the s plane; (b) Bode diagram in the frequency domain; (c) Nyquist diagram in the complex GH plane. All three representations show that the system is stable for $K \leq 48$ and unstable for $k > 48$. The relative stability is expressed on the Bode diagram by the phase margin M_o or the gain margin M_G.

256

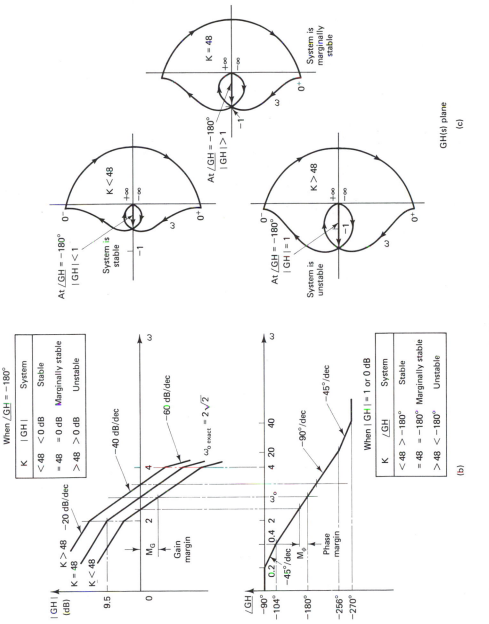

Figure 5-11. (*continued*)

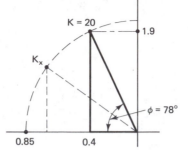

Figure 5-12. Use of the root locus method for system performance design.

ignored if its magnitude is at least 10 times greater than the magnitude of the real part of the complex conjugate roots.

If the third-order system considered in this example is approximated by a second-order system, the root locus may be used effectively to design a specific level of system performance. Suppose that the gain is 20 and the two complex conjugate roots that correspond to this gain are $-0.4 \pm j1.9$, as shown on the root locus plot in Fig. 5-11a. If you will look ahead to Fig. 6-14, you will see that we can construct a triangle in the s plane which relates the general second-order system parameters: ζ and ω_n. The details of the triangle are shown in Fig. 5-12, where the angle ϕ may easily be calculated as 78°. The angle ϕ, as shown in Fig. 6-14, is related to ζ through

$$\phi = \cos^{-1} \zeta$$

Hence ζ is calculated as 0.2 from the equation above using 78°. The percent overshot (POT), which characterizes the step response of the underdamped second-order system, is only a function of ζ [we show this in Eq. (6-29)]. For the value of $\zeta = 0.2$, the value of POT is calculated from Eq. (6-29) as 53%. This value of overshoot is perhaps too excessive, as it falls outside the system's specification limits. The designer is then faced with the task of choosing the new value of the gain K to reduce the overshoot. This becomes a simple matter, particularly if the root locus plotted to scale is available. The designer simply chooses another triangle whose angle ϕ meets the required overshoot specification, and obtains from the plot the corresponding value of K. This is shown by the dashed lines in Fig. 5-12, where the new value of gain is K_x.

In addition to system performance adjustments, the root locus plot may be used to indicate, and adjust if necessary, the relative system stability. As the poles recede from the origin (or the $j\omega$ axis), the system becomes more stable. As the value of K dictates the position of the roots in the s plane, it must also give us a measure of system stability. In Fig. 5-12, for example, the system is more stable for the gain K_x than it is for $K = 20$.

Bode Diagram

As we have seen in the preceding section, the root locus method is completely restricted to the s plane, where the independent variable is s, the Laplace variable. This method requires a priori knowledge of the transfer function that forms the basis of the analysis. An alternative approach which is due to Bode is based on the frequency-domain response.

To perform the frequency response of a system experimentally, a sinusoidal signal at a selected frequency is applied to the system and the output is measured. The output is also sinusoidal, differing from the input only in amplitude and in phase. This procedure is repeated for many frequencies, and at each frequency, the magnitude of the output and its phase are measured and represented graphically. The graphical representation consists of two plots: one is for the magnitude and the other is for the phase; both are plotted as a function of frequency. The two plots are generally referred to as the *Bode plot* or *Bode diagram*.

There are several advantages to Bode's method. First, system stability may be determined from the experimentally derived Bode diagram without knowledge of the system transfer function. Bode diagram may also be determined, of course, from the system transfer function, if it is available. Second, the system transfer function may often be derived from the experimental data. In addition, the system time-domain performance may be inferred from the frequency response plots. The performance of the system in the time domain may be predicted on the basis of the frequency response data. (We show this in Section 6–9.)

Bode's criteria for stability may be expressed in one of two equivalent ways:

1. The closed-loop system is stable if $\underline{/GH} > -180°$ when $|GH| = 0$ dB (unity gain).
2. The closed-loop system is stable if $|GH| < 0$ dB when $\underline{/GH} = -180°$.

The system is thus stable if either of the foregoing conditions is met. The system is unstable if its gain exceeds unity (> 0 dB) when the phase is $-180°$, or conversely, it is unstable if the phase is more negative than $-180°$ when the gain is 0 dB. The system is marginally stable if the gain is 0 dB when the phase is $-180°$.

The relative stability of the feedback system may be measured directly on the Bode diagram and expressed in terms of the gain margin M_G or in terms of the phase

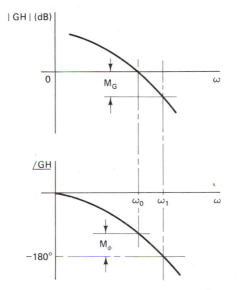

Figure 5–13. Stability on a Bode diagram.

margin M_ϕ. As shown in Fig. 5–13, the gain margin is measured on the magnitude response curve at the frequency for which the phase is $-180°$. Expressed mathematically,

$$M_G = -20 \log |GH(j\omega_1)| \qquad (5-40)$$

The phase margin is measured as the difference between $-180°$ and the phase at the frequency for which the gain is unity, or

$$M_\phi = |-180 - \underline{/GH(j\omega_0)}| \qquad (5-41)$$

The transfer function of Eq. (5–39) for which the root locus is shown in Fig. 5–11a is also represented on the Bode diagram in Fig. 5–11b. As expected, the system stability information obtained from the root locus is exactly the same as that obtained from the Bode diagram. The magnitude response curve for $K = 48$ crosses the 0-dB gain at the frequency for which the phase is $-180°$, showing that the system is marginally stable. The exact value of the crossover frequency ω_0 is calculated as 2.8 (the asymptotic response, which is less accurate, predicts this frequency to be 3.46), which matches the frequency at which the root locus crosses the $j\omega$ axis. The magnitude response curve for $K > 48$ crosses the 0-dB gain at the frequency for which the phase is more negative than $-180°$, or conversely, at $-180°$ phase the gain is > 0 dB. Thus both the root locus and the Bode diagram predict the system to be unstable for $K > 48$. Finally, the magnitude response curve for $K < 48$ crosses the 0-dB gain at the frequency for which the phase is more positive than $-180°$, thus clearly showing, as did root locus, that the system is stable. The stability of the system is also confirmed by the less-than-unity gain at the frequency for which the phase is $-180°$. The gain and phase margins are also shown for the response that is associated with $K < 48$.

EXAMPLE 5–5

Using the Bode plot in Fig. 5–11b (asymptotic response), calculate the value of K for the gain margin of 2 dB. Also calculate the associated phase margin.

SOLUTION The frequency at which the phase is $-180°$ is determined from the following equation based on the linear frequency decrease at $90°$ per decade between $\omega = 0.4$ and $\omega = 20$:

$$-104 - 90 \log \omega_o/0.4 = -180$$

Solving for ω_o yields

$$\omega_o = 2.8$$

The gain at $\omega = 2$ is determined from the linear gain fall-off at the rate of -40 dB/decade between the frequencies $\omega = 2$ and $\omega = 2.8$ as follows:

$$20 \log |GH(j2)| - 40 \log \frac{2.8}{2} = -2$$

The right-hand side in the equation above must be the given gain margin. Solving for the gain at $\omega = 2$ gives us

$$20 \log |GH(j2)| - 6 = -2$$

$$|GH(j2)| = 10^{-2} = 1.58$$

The gain from Eq. (5–39) for $\omega \leq 2$ may be expressed by the asymptotic approximation as

$$|GH(j\omega)| = \frac{K/8}{\omega}$$

Then

$$|GH(j2)| = \frac{K}{16} = 1.58$$

$$K = 25.3$$

Since

$$|GH(j2|_{dB} = +4$$

the crossover frequency is determined from

$$4 - 40 \log \frac{x}{2} = 0$$

Solving for the crossover frequency x, we have

$$x = 2.5 \text{ rad/s}$$

Then from the asymptotic phase response

$$-104 - 90 \log \frac{2.5}{0.4} = -175.6°$$

and

$$M_\phi = 180 - 175.6 = 4.4°$$

Nyquist Diagram

As has been shown, Bode's diagram is comprised of the magnitude and phase of GH, plotted individually as a function of frequency. Nyquist, on the other hand, combines the magnitude and the phase of GH in a single graph. The coordinate system that is used for the Nyquist contour is shown in Fig. 5–14a. This is a complex plane which is generally referred to as the GH plane. In this coordinate representation, GH may be expressed in rectangular form as

$$GH(j\omega) = x + jy$$

or in polar form as

$$GH(j\omega) = |GH(j\omega)| \angle \phi$$

or in exponential form as

$$GH(j\omega) = |GH(j\omega)| e^{j\phi}$$

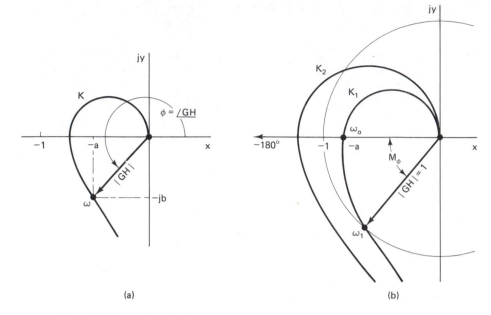

Figure 5–14. (a) Nyquist diagram coordinate system; (b) relative stability on Nyquist diagram.

The task of generating the Nyquist contour is considerably more involved than plotting the Bode diagram or the root locus.

Nyquist's contour is based (as in root locus and the Bode plot) on the characteristic equation

$$F(s) = 1 + GH(s) = 0$$

where $GH(s)$ generally takes the form

$$GH(s) = \frac{KN(s)}{D(s)}$$

The stability of the system is ensured if the roots of $F(s)$ [or the zeros of $F(s)$], which must necessarily also be the poles of the closed-loop transfer function

$$T(s) = \frac{G(s)}{1 + GH(s)}$$

are all in the left-hand side of the s plane (i.e., the roots must have negative real parts).

To determine the possible existence of zeros of $F(s)$ in the RHS of the s plane, the entire RHS must be scanned by means of an appropriately chosen contour. The contour that is normally used for this purpose begins at the origin and extends the entire length of the $j\omega$ axis to $+j\infty$; it continues CW along a semicircular arc of infinite radius to $-j\infty$ and back to the origin along the negative $j\omega$ axis.

This contour is then mapped into the complex GH plane by means of the given $F(s)$, thus generating a corresponding contour in the GH plane. A sample point is shown in Fig. 5–14a, which is represented by a vector of length $|GH|$ and angle ϕ. The magnitude and the phase of this vector are evaluated at the selected frequency ω, which is one of the points on the contour in the s plane. One contour is obtained in the GH plane for each value of K, where K is the system gain constant associated with $GH(s)$. In Fig. 5–14 two contours are shown, one for each value of K.

Nyquist's stability criteria may be expressed by

$$N_Z = N_C + N_P \qquad (5\text{--}42)$$

where N_Z = number of zeros of $F(s)$ that are in the RHS of the s plane

N_C = number of times that the contour in the GH plane encircles the point $-1 + j0$; N_C is positive for CW encirclements and negative for CCW encirclements

N_P = number of poles of $F(s)$ that are in the RHS of the s plane

Then according to Nyquist:

1. If $N_P = 0$, which is generally the case, the system is stable if $N_Z = 0$. In this case the contour in the GH plane *must not* encircle the point $-1 + j0$.
2. If $N_P \neq 0$, the system is stable if $N_Z = 0$ and from Eq. (5–42) $N_C = -N_P$; that is, the number of CCW encirclements of the point $-1 + j0$ in the GH plane must equal the number of poles of $F(s)$ that are on the RHS of the s plane.

In Fig. 5–14b, the partial contour associated with gain K_1 is not likely to encircle the -1 point and the contour for the gain K_2 is very likely to encircle the -1 point once it is completed. The former represents a system that is probably stable and the latter a system that is probably unstable.

The relative stability of the system may also be expressed on the Nyquist diagram just as it was shown on the Bode diagram. As shown in Fig. 5–14b, the phase margin M_ϕ is the difference in phase between the -1 vector and the $|GH(j\omega_1)| = 1$ vector, whose phase must be $> -180°$ for a stable system. The gain margin is shown here as the length of the vector $-a$, where $|a| < 1$. Expressed mathematically,

$$M_G = 20 \log \left|\frac{1}{a}\right| = -20 \log |a|$$

It would be interesting to compare the results obtained on the Bode and root locus diagrams for the system of Eq. (5–39) with those of the Nyquist diagram. The previous conclusion was that the system is stable for $K < 48$, marginally stable for $K = 48$, and unstable for $K > 48$. These conclusions are supported by the Nyquist contours shown in Fig. 5–11c. In the case of $K > 48$, the contour encircles the -1 point once in the CW direction, indicating that the system is unstable and that its closed-loop transfer function has a single pole on the RHS of the s plane since $N_Z = 1$.

Routh–Hurwitz Method

We have considered thus far the root locus method, which is attributed to Evans, as well as the Bode and Nyquist methods. As shown, Bode concentrates his analysis in the frequency domain, the Nyquist contour is generated in the complex GH plane, and the root locus method is done in the s plane. These methods have several aspects in common. First, they can be used to analyze system stability. Second, they may be used to compensate a given system to achieve a desired level of system performance. Finally, relative stability may be incorporated into a given system by these methods in terms of the gain margin or the phase margin. In addition, the methods were all introduced in a relatively short period of time in the 1940s.

The Routh–Hurwitz method, on the other hand, dates back to the latter part of the nineteenth century, when Routh and Hurwitz reported independently a method for investigating the stability of a linear system. Their approach to system stability analysis differs considerably from the methods described above. First, their method depends on the availability of the characteristic equation for the given system. Second, this method cannot predict the relative stability of the system. It is only able to answer the question regarding a system's absolute stability: "yes," the system is stable, or "no," the system is not stable. The Routh–Hurwitz criterion is sufficient to establish the stability of a linear system.

The Routh–Hurwitz criterion is based on the characteristic equation of a given system as follows:

$$F(s) = 1 + GH(s) = 0$$

which may also be expressed as an nth-order polynomial as

$$a_1 s^n + a_2 s^{n-1} + a_3 s^{n-2} + a_4 s^{n-3} + \cdots + a_n s + a_{n+1} = 0 \qquad (5\text{--}43)$$

For the system to be stable, the roots of the polynomial above must all be on the left-hand side of the s plane (i.e., their real parts must be negative). The Routh–Hurwitz criterion is sufficient to determine the quantity of roots which are on the RHS of the s plane, thus establishing whether or not the system is stable.

According to Routh–Hurwitz, the constant coefficients of the characteristic equation (5–43) are formed into the following array:

$$
\begin{array}{c|ccccc}
s^n & a_1 & a_3 & a_5 & a_7 & \cdots \\
s^{n-1} & a_2 & a_4 & a_6 & a_8 & \cdots \\
s^{n-2} & b_1 & b_2 & b_3 & & \cdots \\
s^{n-3} & c_1 & c_2 & c_3 & & \cdots \\
& \cdot & \cdot & & & \\
& \cdot & \cdot & & & \\
& \cdot & \cdot & & & \\
s^1 & d_1 & d_2 & & & \cdots \\
s^0 & e_1 & e_2 & & & \cdots
\end{array}
\qquad (5\text{--}44)
$$

where

$$b_1 = \frac{a_2 a_3 - a_1 a_4}{a_2} = -\frac{1}{a_2} \begin{vmatrix} a_1 & a_3 \\ a_2 & a_4 \end{vmatrix}$$

$$b_2 = \frac{a_2 a_5 - a_1 a_6}{a_2} = -\frac{1}{a_2} \begin{vmatrix} a_1 & a_5 \\ a_2 & a_6 \end{vmatrix}$$

$$b_3 = \frac{a_2 a_7 - a_1 a_8}{a_2} = -\frac{1}{a_2} \begin{vmatrix} a_1 & a_7 \\ a_2 & a_8 \end{vmatrix} \qquad (5\text{--}45)$$

$$c_1 = \frac{b_1 a_4 - b_2 a_2}{b_1} = -\frac{1}{b_1} \begin{vmatrix} a_2 & a_4 \\ b_1 & b_2 \end{vmatrix}$$

$$c_2 = \frac{b_1 a_6 - b_3 a_2}{b_1} = -\frac{1}{b_1} \begin{vmatrix} a_2 & a_6 \\ b_1 & b_3 \end{vmatrix}$$

This process is continued until all constants of the array, including the s^0 row, are evaluated. The Routh–Hurwitz stability criterion states: *The system is unconditionally stable if there are no changes in sign in the first column of the array*, Eq. (5–44). The system is, on the other hand, unstable if there are changes in sign in the first column of the array. Each change in sign corresponds to the root of the characteristic equation (5–43), which is on the RHS of the s plane.

The conclusion regarding system stability may readily be made as long as it is possible to form the array. Problems arise when one of the entries in the first column is a 0. This means that a division by 0 will occur in calculating the following entry. For this reason, there are special cases that must be treated individually. These are summarized below.

Case 1. One of the entries in the first column is a 0, and the remaining entries in the row that contains a 0 in the first column are not 0. This problem is resolved by first replacing the 0 by ϵ, then proceeding with the formation of the rest of the array, and finally applying the limit as $\epsilon \to 0$ to those terms that contain ϵ. At this point the conclusion regarding system stability can readily be made by searching for the changes in sign in the first column.

Case 2. The Routh–Hurwitz array contains an entire row of zeros. This situation generally arises when the poles of the closed-loop transfer function are symmetrically arranged in the s plane. Pure imaginaries and complex conjugates are examples of such poles. This case is resolved by forming the "auxiliary equation" $A(s)$, which is based on the row immediately above the row containing all zeros. Since $A(s)$ is one of the polynomials (when characteristic equation is expressed as the product of polynomials), it can be used as the divisor polynomial of the characteristic equation, without a remainder. The polynomial that is the result of the division may be solved for the roots immediately (if it is a lower-order polynomial such as a quadratic) or it may be used as a basis for another Routh–Hurwitz array. If the second array again contains a row of zeros (which does not occur very frequently), case 2 may be repeated.

The examples that follow provide an adequate illustration of the Routh–Hurwitz method and its special cases. The first example is based on Eq. (5–39), which has been done by the root locus, Bode, and Nyquist methods, and by this time the results should be very familiar to the reader. It remains to show that the same results may also be obtained by the Routh–Hurwitz method.

EXAMPLE 5–6

Determine the range of K for which the system of Eq. (5–39) is stable.

SOLUTION The characteristic equation

$$s(s + 2)(s + 4) + K = 0$$

or

$$s^3 + 6s^2 + 8s + K = 0$$

which is obtained from Eq. (5–39), is used to construct the Routh–Hurwitz array as follows:

$$
\begin{array}{c|cc}
s^3 & 1 & 8 \\
s^2 & 6 & K \\
s^1 & \dfrac{48 - K}{6} & 0 \\
s^0 & K & 0
\end{array}
$$

Clearly, K must be less than 48 and it must also be greater than 0, or

$$0 < K < 48$$

The endpoints must, of course, be tested to see if the system is stable when $K = 0$ or when $K = 48$. Setting K to 0, the characteristic equation above has three roots: 0, −2, and −4. These are the poles of $GH(s)$, which are shown by crosses in the root locus plot of Fig. 5–11a; the loci begin at these points. The system is, then, also stable for $K = 0$.

Next letting $K = 48$ results in all zeros in the third row of the array above. In applying case 2, we form the auxiliary equation on the basis of the row above the row that contains all zeros (s^2 row) as follows:

$$A(s) = 6s^2 + 48 = 6(s^2 + 8)$$

The significance of the auxiliary equation is that it must be one of the factors of the characteristic equation. This may be expressed as follows:

$$6(s^2 + 8)[X(s)] = s^3 + 6s^2 + 8s + 48 = 0$$

where $X(s)$ is the unknown polynomial which may be determined by dividing longhand the auxiliary equation into the characteristic equation as demonstrated below:

$$
\begin{array}{r}
s + 6 \\
s^2 + 8 \overline{)\, s^3 + 6s^2 + 8s + 48} \\
\underline{s^3 + 8s } \\
6s^2 + 48 \\
\underline{6s^2 + 48}
\end{array}
$$

As expected, there is no remainder, and the characteristic equation may be expressed in the following form:

$$s^3 + 6s^2 + 8s + 48 = 6(s^2 + 8)(s + 6) = 0$$

where the roots are -6 and $\pm j2\sqrt{2} = \pm j2.8$. The results thus coincide with those obtained previously. The root locus graph, for instance, shows these roots for $K = 48$. Inasfar as the Routh–Hurwitz analysis is concerned, we have tested the end points and found that the system is also stable at these endpoints (i.e., $K = 0$ and $K = 48$). Consequently, according to the Routh–Hurwitz method, the system is stable when the value of K is in the range $0 \leqslant K \leqslant 48$.

EXAMPLE 5–7

Determine the stability of the system whose characteristic equation is

$$s^6 + 2s^5 + 3s^4 + 4s^3 + 3s^2 + 2s + 1 = 0$$

by the Routh–Hurwitz method.

SOLUTION The Routh–Hurwitz array is formed on the basis of the characteristic equation as follows:

$$
\begin{array}{c|cccc}
s^6 & 1 & 3 & 3 & 1 \\
s^5 & 2 & 4 & 2 \\
s^4 & 1 & 2 & 1 \\
s^3 & 0 & 0 & 0
\end{array}
$$

The array cannot be completed because the fourth row contains all zeros. Case II applies here and the auxiliary equation is formed on the basis of the s^4 row as follows:

$$s^4 + 2s^2 + 1 = 0$$

Dividing the auxiliary equation above into the characteristic equation, the following result is obtained:

$$
\require{enclose}
\begin{array}{r}
s^2 + 2s + 1 \\
s^4 + 2s^2 + 1 \enclose{longdiv}{s^6 + 2s^5 + 3s^4 + 4s^3 + 3s^2 + 2s + 1} \\
\underline{s^6 \qquad + 2s^4 \qquad + s^2} \\
2s^5 + s^4 + 4s^3 + 2s^2 + 2s + 1 \\
\underline{2s^5 \qquad + 4s^3 \qquad + 2s} \\
s^4 \qquad + 2s^2 \qquad + 1
\end{array}
$$

and the characteristic equation may be expressed as

$$s^6 + 2s^5 + 3s^4 + 4s^3 + 3s^2 + 2s + 1 = (s^4 + 2s^2 + 1)(s^2 + 2s + 1)$$
$$= (s^2 + 1)^2(s + 1)^2$$

Hence there is a double root at $s = +j$, a double root at $s = -j$, and a double root at $s = -1$. The system is therefore stable because there are no roots on the RHS of the s plane.

EXAMPLE 5–8

Using the Routh–Hurwitz method, evaluate the stability of the system whose characteristic equation is $s^5 + 2s^4 + 2s^3 + 4s^2 + 5s + 4 = 0$.

SOLUTION   Based on the given characteristic equation, the Routh–Hurwitz array is constructed in the usual way:

$$
\begin{array}{c|ccc}
s^5 & 1 & 2 & 5 \\
s^4 & 2 & 4 & 4 \\
s^3 & \epsilon & 3 & 0 \\
s^2 & \dfrac{4\epsilon - 6}{\epsilon} & 4 & 0 \\
s^1 & A & 0 & 0 \\
s^0 & 4 & 0 & 0
\end{array}
$$

and since the third row entry in the first column is a zero, case 1 applies, and the zero is replaced temporarily by a positive infinitesimal quantity ϵ. The array is then completed in terms of ϵ. Taking the limit of those terms that contain ϵ yields

$$
\lim_{\epsilon \to 0} \frac{4\epsilon - 6}{\epsilon} = -\infty \quad \text{and} \quad \lim_{\epsilon \to 0} [A] = \lim_{\epsilon \to 0} \frac{-\epsilon^2 + 3\epsilon - 4.5}{\epsilon - 1.5} = +3
$$

In the limit as $\epsilon \to 0$, the first column, fourth row entry becomes very large and negative. The magnitudes in the first column may be ignored; however, the signs must be considered, as the Routh–Hurwitz criterion is based on the sign change in the first column. Consequently, the negative entry in the fourth row of the first column is responsible for two sign changes and hence two poles on the RHS of the s plane, thus making the system unstable.

EXAMPLE 5–9

The characteristic equation for some system is described by $s^5 + 2s^4 + 7s^3 + 12s^2 + 7s + 4 = 0$. Determine the system stability through use of the Routh–Hurwitz criterion

SOLUTION   The Routh–Hurwitz array is constructed from the characteristic equation

$$
\begin{array}{c|ccc}
s^5 & 1 & 7 & 7 \\
s^4 & 2 & 12 & 4 \\
s^3 & 1 & 5 & \\
s^2 & 2 & 4 & \\
s^1 & 3 & & \\
s^0 & 4 & &
\end{array}
$$

all remaining entries in the array are pre-sumed to be 0

There are no sign changes in the first column; hence the system has no poles on the RHS of the s plane and is therefore presumed to be stable.

SUMMARY

1. Any open-loop system may be represented by the controller and the process. The controller responds to the reference input R and drives the process to attain the desired value of the process variable C, also called the controlled variable.

2. As shown in Fig. 5–3, any closed-loop system consists of the controller and the process blocks (also the open-loop system components), and in addition it also has negative feedback. The negative feedback block is a transducer that converts the present value of C to another form of energy which is compatible with the input R. The output of the summing junction that compares the input R with the feedback signal drives the controller, which in turn affects process behavior in the direction to reduce the difference between the input and the feedback.

3. The use of negative feedback as shown in Fig. 5–3 provides the basis for automatic control. A feedback system is, then, a self-regulating system as it is able to maintain its output C at a relatively constant value under varying load conditions or system parameter fluctuations without the intervention of the operator. A home heating system is a good example of this type of feedback system. The thermostat setting (the input) dictates the value of the temperature inside the room (the process variable). The system responds automatically to a change in load, such as the opening and closing of the door (which initially lowers the room temperature slightly), and restores the room temperature to the thermostat setting.

4. The disadvantages of the open-loop system include:

    a. Sensitivity to external disturbances
    b. Sensitivity to variations in load
    c. Sensitivity to changes in system parameters

The advantages of the open-loop system include:

    a. No feedback, therefore fewer components and less expensive.
    b. Unconditionally stable.

5. The advantages of the closed-loop system include:

    a. Relative insensitivity to changes in load
    b. Relative insensitivity to changes in system parameters
    c. High accuracy is possible.
    d. Relative insensitivity to disturbances
    e. Faster transient response
    f. Automatic tracking capability

The disadvantages of the closed-loop system include:

    a. The addition of feedback makes the system more complex. It has more components and it is more difficult to analyze.

b. The additional feedback components add to the system cost. The additional cost may be insignificant if the feedback component is a simple and inexpensive chip. On the other hand, the added cost may be considerable if the feedback component is a microprocessor or computer.

c. The use of feedback presents a real possibility that the system may be unstable.

The advantages of a closed-loop system generally far outweigh its disadvantages. Feedback systems are therefore more popular and more widely used than open-loop systems.

6. As expressed in Eq. (5–2), the accuracy of the system is measured as the difference between the input and the output. The accuracy is actually measured under steady-state conditions, when all transients have subsided. The final form for the steady-state error is expressed by Eq. (5–7).

7. The steady-state error depends on (a) type of system and (b) type of input. The commonly used inputs include step, ramp, and parabolic waveforms. The steady-state error results are summarized in Table 5–1. In cases where the steady-state error is finite and nonzero, the error is inversely related to the loop gain. Hence the accuracy of the system is improved as the loop gain is increased.

8. System sensitivity, which is expressed by Eq. (5–20), may be used to predict the effect on the output due to small changes in the values of the forward-path system components. Equation (5–25) applies for large changes (as well as small changes) in forward-path component values. Generally, it may be concluded that as the loop gain GH is increased, the system becomes less sensitive to variations in the forward-path component values. Equation (5–22) shows that the system is extremely sensitive to variation in the feedback component values. Changes in the feedback component values are translated to the output on almost a one-to-one basis. Consequently, the feedback components must be very stable. On the other hand, the system's forward-path components need not be as stable as long as the loop gain is sufficiently high.

9. A closed-loop system may reject external disturbances better than the open-loop system only if specific conditions are satisfied; otherwise, there is no significant improvement in disturbance rejection. Equation (5–31) states the disturbance rejection ratio, and the conditions that follow must be satisfied to realize an improvement by a closed-loop system over an open-loop system.

10. Increasing the loop gain GH reduces the value of the closed-loop time constant, thus making the closed-loop system faster than its open-loop counterpart.

11. The PLL control system uses phase comparison between the feedback and input signals as opposed to the conventional voltage comparison in the summing amplifier. The PLL control system is better adapted to digital applications than is its analog counterpart. For example, in applications where the motor must operate in synchronism with the control pulses from a microprocessor or computer, the PLL system has an advantage because it can accept the control pulses directly without any additional interfacing. An analog control system would require additional interface hardware in similar applications.

12. System stability may be evaluated in the time domain. As shown in Fig. 5–10, the system is stable if its time-domain response decreases in amplitude with time. This criterion is met if the poles of the closed-loop system transfer function are located on the LHS of the s plane. Any poles on the RHS of the s plane contribute to system's instability.

13. The root locus method maps the zeros of $1 + GH(s)$, which are also the poles of the closed-loop transfer function in the s plane as the value system gain K is varied from 0 to ∞. System stability, as well as its relative stability, may be inferred from this graph.

14. The Bode diagram consists of the magnitude and the phase of $GH(j\omega)$ individually plotted against frequency for a selected value of the gain K. A system's stability and its relative stability may be inferred from this diagram. Bode's stability criteria are expressed in Fig. 5–13, and the relative stability by Eqs. (5–40) and (5–41), which express the gain and phase margins, respectively.

15. Nyquist's contour is based on the characteristic equation $F(s) = 1 + GH(s) = 0$. It is generated by mapping the entire RHS of the s plane into the complex GH plane by means of the given $F(s)$. System stability may be determined from such a contour in accordance with the Nyquist's stability criterion as expressed by Eq. (5–42).

16. Unlike other methods, the Routh–Hurwitz method can predict only the absolute stability of the system; it is unable to predict the relative stability. It also depends totally on the availability of the system's transfer function. The Routh–Hurwitz method is based on the array, which is based on the characteristic equation. According to Routh–Hurwitz, the system is absolutely stable if there are no sign changes in the first column of the array. Special cases arise when the array contains a zero in the first column or an entire row of zeros; these problems are resolved by the procedures described in cases 1 and 2, respectively.

QUESTIONS

5–1. Discuss the advantages and the disadvantages of the open-loop system and the closed-loop system.

5–2. Draw a block diagram of a simple closed-loop system and describe the function of each block, the input, and the output.

5–3. State the closed-loop system transfer function and explain the significance of each term.

5–4. Which factors affect the accuracy of the system?

5–5. Explain how system type value is determined mathematically. Explain the practical significance of system type value.

5–6. Define the error constants K_0, K_1, and K_2 and identify for each constant the applicable input and the system type.

5–7. Identify the input and system type for which the steady-state error is 0. Consider all combinations of step, ramp, and parabolic inputs and system types 0, 1, 2, and 3.

5–8. Repeat Question 5–7 for an infinite steady-state error.

5–9. Define system sensitivity. Explain the difference between S_G and S_H.

5–10. Why must the feedback components of a closed-loop system be very stable and those of the forward path much less stable? Explain.

5–11. Under what conditions does the S_G defined for large changes in system component values become almost the same as the S_G, which is defined for small changes in component values?

5–12. Which parameters of the closed-loop system must be varied (increased or decreased) to reduce the effects of external disturbance? Explain.

5–13. State the condition that must be satisfied to improve the disturbance rejection ratio of the closed-loop system over that of an open-loop system.

5–14. Explain by means of equations the effect that loop gain has on the closed-loop system speed of response.

5–15. Explain the operation of the PLL control system. In what type of application does this system have an advantage over its analog counterpart? Explain.

5–16. State the stability criterion that is based on the location of the closed-loop transfer function poles in the s plane. Sketch the underdamped time-domain responses for a stable and an unstable second-order system and relate these responses to the location of the poles in the s plane.

5–17. Explain the root locus diagram: its construction and its application to stability analysis.

5–18. How can the root locus diagram be used to obtain a desired performance level in the time domain?

5–19. State the mathematical basis for the Bode diagrams and explain their construction.

5–20. State Bode's two equivalent forms of stability criteria.

5–21. State the gain and phase margin equations and explain how these can be determined from Bode's diagram.

5–22. Explain the mathematical basis and the construction of Nyquist diagram.

5–23. State Nyquist's stability criterion.

5–24. Explain the mathematical basis of the Routh–Hurwitz array and its construction.

5–25. State the Routh–Hurwitz stability criterion. Explain the special problems that are included in the two special cases and manner in which these problems are resolved.

PROBLEMS

5–1. Consider the closed-loop system shown in Fig. 5–3. The process is described by $200/(s + 10)$, $K_f = 0.1$, and $r(t) = 10u(t)$.
   (a) Determine the system type.
   (b) Calculate the value of K_c such that the steady-state error is 2.
   (c) Calculate the value of K_c such that the steady-state value of the output is 80.
   (d) Calculate the closed-loop-system time constant and compare it to the open-loop-system time constant for $K_c = 4.5$.

5–2. Suppose that the process is defined by $200/(s + 10)$ for the feedback system in Fig. 5–3, $K_f = 0.1$, and the input is $Eu(t)$. If the value of K_c decreased from 10 to 8, calculate the percent change in the output through the use of the appropriate sensitivity factor.

5–3. In the system of Fig. 5–3 the process is $200/(s + 10)$, $K_f = 0.1$, and the input is $Eu(t)$. After the steady-state conditions are reached, K_c is reduced by 20% and the corresponding change in the output is from 100 to 98. Calculate the values of E and K_c.

5–4. In the feedback control system shown in Fig. P5–4, the inertia of the motor armature is 0.001 oz-in.-s^2 and both potentiometers have an active angle of 343°. Assume that the summer op amp is ideal and there is no load.

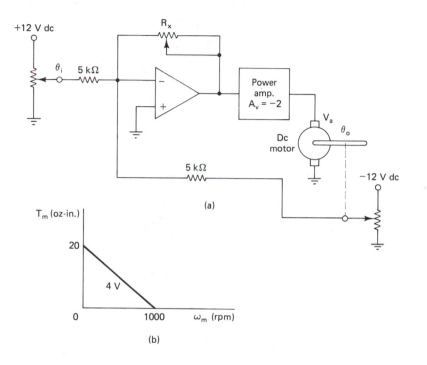

(a)

(b)

Figure P5–4. (a) Feedback control system; (b) motor speed–torque characteristics.

(a) If the given system is to be represented by the block diagram of Fig. 5–3, determine the transfer functions for the controller, the process, and the feedback.

(b) Determine the transfer function for the given closed-loop system. Simplify it as much as possible and express it in the form where the coefficient of the highest power of s in the denominator is unity.

(c) Determine the type and order of the given system.

5–5. Suppose that the system in Problem 5–4 has reached the steady state for some input θ_i. The wiper arm of the input pot is then suddenly increased by 30°. Determine the value of the output after the system is allowed to reach the steady state again.

5–6. Consider the feedback system of Problem 5–4.

(a) The wiper arm of the input potentiometer is advanced at a steady rate of 6.25° for each 10 ms. It is required that the output follow this input within 30 minutes of arc. Calculate the value of R_x in kilohms to satisfy this requirement.

(b) Suppose that the wiper arm of the input potentiometer begins its upward motion from the bottom of the potentiometer, where $\theta_i = 0°$, at a constant rate of 6° for each 10 ms. Calculate the exact value of the motor shaft displacement at the instant when the input reaches 20°. Assume that the system reaches its steady state in three system time constants and that $R_x = 5$ kΩ.

5–7. Consider the system configuration shown in Fig. P5–7. It may be operated as a feedback system with switch S in position 1 or as an open-loop system with the switch in position 2. V_i represents the reference input, and the input D represents the effect of the disturbance.

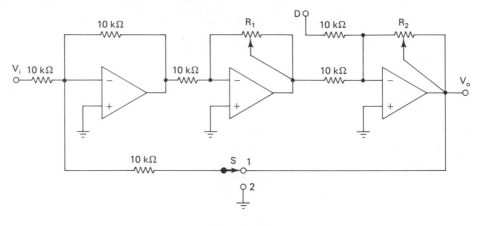

Figure P5–7.

(a) Construct the simple block diagram for the given system, assuming that the switch is in position 1, and determine the transfer functions for each block. What is the value of H?

(b) If $R_1 = R_2 = 20$ kΩ, $V_i = -0.6$ V dc, and $D = +0.4$ V dc, calculate the value of V_o when the switch is in position 1; repeat the calculation for the switch in position 2.

(c) Calculate the disturbance rejection ratio DR for each position of the switch in part (b). Does the feedback system reject the disturbance better than the open-loop system? Explain.

(d) It is required that 5% of the output V_o be due to the disturbance and the remaining 95% be due to the input V_i, assuming that V_i and D values are those given in part (b). To meet this requirement, V_i is changed to -3 V dc and R_1 is not changed. Calculate the value of R_1 to satisfy the requirement. What should be the value of R_2? Justify your choice of its value. The switch is in position 1.

(e) If $R_1 = 100$ kΩ, $V_i = +5$ V dc, and $V_o = +4.95$ V dc, calculate the value of R_2 using the steady-state error theory. Assume that $D = 0$ and the switch is in position 1.

(f) If $R_1 = 100$ kΩ and its value is changed to 80 kΩ, calculate the value of R_2 using the appropriate sensitivity factor so that the change in the output is -0.5%. Assume that $D = 0$ and the switch is in position 1.

5–8. (a) Assess the stability of each of the feedback systems represented by the root locus diagrams shown in Fig. P5–8. Indicate whether the system is unconditionally stable, always unstable, or conditionally stable. If conditionally stable, determine the gain range over which the system is stable. Justify your answers.

(b) If the poles for the unity feedback system shown in Fig. P5–8g are at 0, -2, and $-4 \pm j5$, determine the closed-loop system transfer function.

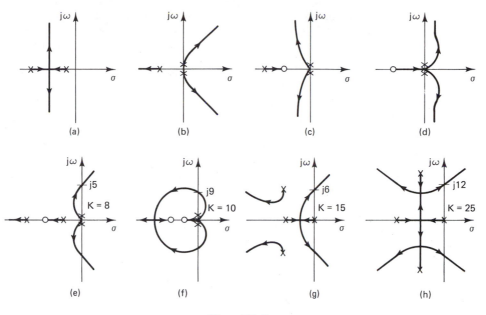

Figure P5–8.

5–9. In the Bode diagram shown in Fig. P5–9, the three gain responses have the same phase response as shown.

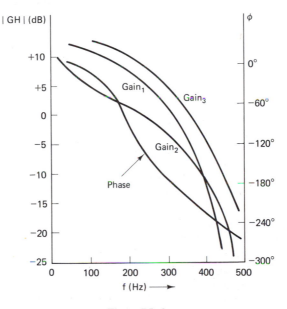

Figure P5–9.

(a) Which of the gain response curves represent an unstable system? Explain.

(b) Determine the gain margin (as a ratio, not in decibels) and phase margin for each of the responses that represent a stable system.

5–10. **(a)** Assess the stability of the closed-loop system for each of the Nyquist diagrams shown in Fig. P5–10. Indicate whether the system is unconditionally stable, always unstable, or conditionally stable. If conditionally stable, under what conditions does the system become unstable?

(b) For each of the given systems that are stable, estimate from the given diagrams the gain margin and the phase margin.

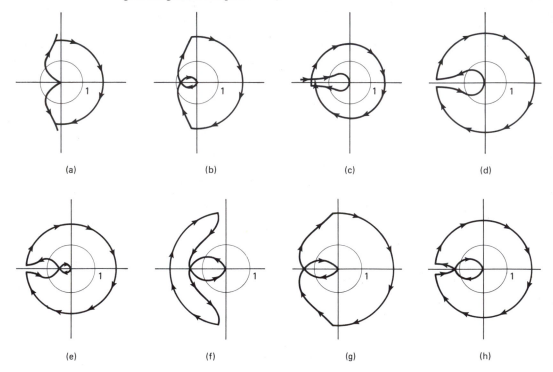

(a)   (b)   (c)   (d)

(e)   (f)   (g)   (h)

Figure P5–10. GH has one pole on the RHS of the s plane in part (f) and no poles on the RHS of the s plane in the remaining parts.

5–11. Apply the Routh–Hurwitz criterion to each of the following characteristic equations:
(a) $s^3 + 3s^2 + 3s + 1 = 0$
(b) $2s^3 + s^2 + 2s + 1 = 0$
(c) $s^4 + s^3 + 3s^2 + 2s + 1 = 0$
(d) $s^5 + s^4 + 3s^3 + 2s^2 + 4s + 1 = 0$
(e) $s^6 + s^5 + 3s^4 + 2s^3 + 3s^2 + 2s + 1 = 0$
In each of the problems above, how many roots are on the RHS of the s plane, the LHS of the s plane, and on the $j\omega$ axis?

5–12. Apply the Routh–Hurwitz criterion to each of the problems below, and determine the range of K for which the given system is stable (test the endpoints of the range):
(a) $s^4 + s^3 + 12s^2 + 2s + K = 0$

(b) $GH(s) = \dfrac{K}{s(s^2 + s + 20)}$

5–13. **(a)** Determine the relationship between the parameters K_1, K_2, K_f, τ_1, and τ_2, to assure that the response of the system shown in Fig. P5–13 is stable.

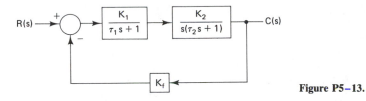

Figure P5–13.

(b) Determine the expression for the frequency ω in terms of system parameters if the steady-state response of the given system is a constant-amplitude sinusoid.

(c) If $K_f = 1$, $K_1 = 5$, and $\tau_1 = \tau_2 = 0.1$ s, determine the maximum value of K_2 for which the system is stable.

(d) If $\tau_2 = 0.01$ s, determine the value of τ_1 for which the frequency of the constant-amplitude oscillations is 100 Hz.

LAB EXPERIMENT 5–1
OPEN-LOOP SYSTEM

Objective.   To investigate the characteristics of an open-loop system. Sensitivity, accuracy, disturbance rejection, and transient response are considered in this experiment.

Equipment

Dual-trace oscilloscope
Pulse generator
Digital voltmeter

Materials.   As shown in the test circuit.

Procedure

1. Construct the test circuit shown in Fig. LE5–1 and null out the offset voltages of the operational amplifiers.

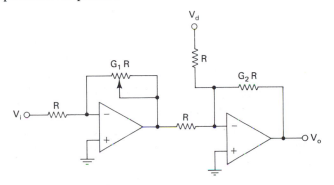

Figure LE5–1.   Test circuit, $R = 2.0$ kΩ.

2. Adjust the gains $G_1 = G_2 = 1$, and apply $V_i = 0.500$ V dc. Measure V_o on the digital voltmeter (a) without the disturbance input (i.e., $V_d = 0$ V); (b) with the disturbance input $V_d = 0.250$ V dc. Record the measured values.

3. Repeat step 2 for each of the following G_2 values while maintaining $G_1 = 1$: 2 to 20 in steps of 2.

4. Repeat step 2 for each of the following G_1 values while maintaining $G_2 = 1$: 2 to 20 in steps of 2.

5. Adjust the gains $G_1 = G_2 = 1$, and apply a 50% duty cycle, 1 V_{peak} (0 to 1.00 V), 1 kHz to the input V_i. Display V_i and V_o on a dual-trace oscilloscope. Measure and record the times corresponding to the following points on the response waveform: (a) 10%; (b) 50%; (c) 90%.

6. Repeat step 5 for each G_2 gain specified in step 3 with $G_1 = 1$.

7. Repeat step 5 for each G_1 gain specified in step 4 with $G_2 = 1$.

Test data. The format for the test data and the calculated results, as well as the report format, must comply with and meet the requirements of your lab instructor.

Evaluation. As the ultimate objective of any lab experiment is to verify the theory-based predictions through measurements, significant differences between theory and measurements (generally, differences over 10%) must be resolved through additional experimentation, investigation, or retest.

1. On suitable graph paper, plot the disturbance rejection DR as a function of G_2 with $G_1 = 1$ using the data obtained in steps 2 and 3 of the Procedure. Include on the same graph the DR curves based on theory (consult the text). Significant differences between measurements and theory (in excess of 10%) must be resolved through additional investigation, or retest. Show all theory-related work.

2. Repeat step 1 for the data taken in steps 2 and 4 of the Procedure.

3. On suitable graph paper, plot the percent change in the output $\Delta V_o/V_o$ as a function of $\Delta G/G$, which is plotted along the abscissa from 0 to 20, using the data from steps 2 to 4 of the Procedure for $V_d = 0$. Include on the same graph the $\Delta V_o/V_o$ curve based on the theory, and show all theory-related work. (Note that $G = G_1 G_2$.)

4. On suitable graph paper, plot the t_{50} time which corresponds to the 50% point on the response curve as a function of $G = G_1 G_2$ using the data from steps 2 to 4 of the Procedure. Scale the abscissa from 1 to 20. Include on the same graph the t_r curve (the time between the 10 and 90% points on the response curve) as a function of G. Include on the graph the rise time t_r curve versus G based on theory. Provide documentation on all theory-related work.

The responses to the following questions and/or conclusions must be supported whenever possible by the applicable theory.

5. From Evaluation step 1 deduce the effect of G_2 on DR.

6. From Evaluation step 2 deduce the effect of G_1 on DR.

7. What does the slope of the curve in step 2 represent? What is the value of the slope? Is the value of the slope consistent with theory?

8. What does the slope of the curve in step 3 represent? What is the value of the slope? Is the value of the slope consistent with theoretical predictions? What system characteristic does the slope represent?

9. From step 4 deduce the effect of system gain G_1G_2 on t_{50} and t_r. Is this consistent with the theory-based predictions?

10. What does the slope of t_r curve in step 4 depend on?

11. How does the amplifier (in the test circuit) gain–bandwidth product affect the system rise time?

12. Would your answer to question 11 be the same if a simple common-emitter transistor amplifier were used instead of the operational amplifier?

LAB EXPERIMENT 5–2
CLOSED-LOOP SYSTEM

Objective. To evaluate the performance of a closed-loop system. The performance characteristics to be considered include disturbance rejection, sensitivity to variation in forward-path gain, accuracy, closed-loop-system disturbance rejection improvement, and the extent to which loop gain affects these performance characteristics. Closed-loop system performance will also be compared with that of the open-loop experiment (Lab Experiment 5–1).

Equipment

Dual-trace oscilloscope
Digital voltmeter

Materials. As shown in the test circuit.

Procedure

1. Construct the test circuit shown in Fig. LE5–2 and null out the offset voltages of the operational amplifiers, $R = 2.0$ kΩ.

2. Adjust the gains $G_1 = 1$ and $G_2 = 1$, and apply $V_i = -4.00$ V dc; measure and record as follows: (a) V_o with S_1 open and $V_d = 0$ V dc; (b) V_o with S_1 closed

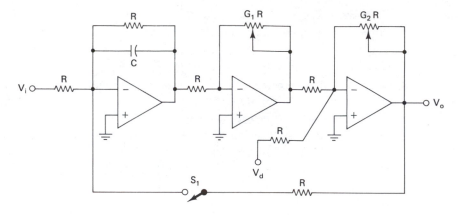

Figure LE5–2. Test circuit.

and $V_d = 0$ V dc; (c) V_o with S_1 closed and $V_d = 2.00$ V dc. With S_1 closed there is a possibility of an oscillatory response. Choose a value of C shown in the circuit through experimentation to just suppress the oscillations (0.1 μF or more).

3. Set the op amp gains to $G_1 = 1$, $G_2 = 1$. (a) With S_1 closed and $V_d = 0$ V dc, adjust V_i until V_o is equal to the value measured in step 2a. Monitor V_i and V_o on a digital voltmeter. Measure and record V_i. (b) With V_i just measured still applied to the test circuit and S_1 closed, adjust $V_d = 2.00$ V dc. Measure and record V_o.

4. With $G_2 = 1$, repeat step 3 for each of the following values of G_1: 4 to 20 in steps of 2.

5. With G_1 and G_2 interchanged (i.e., G_2 assumes G_1 values, and vice versa), repeat steps 2 to 4.

6. Close the feedback switch S_1, set $V_d = 0$ V dc, and adjust the gain G_2 to unity. Apply $V_i = -5.00$ V dc, measure on the digital voltmeter, and record the value of V_o for each of the following values of G_1: 2 to 20 in steps of 2.

Test data. The format for the test data and the calculated results, as well as the report format, must comply with and meet the requirements of your lab instructor.

Evaluation. As the ultimate objective of any lab experiment is to verify the theory-based predictions through measurements, significant differences between theory and measurements (generally, differences over 10%) must be resolved through additional experimentation, investigation, or retest.

1. Plot on suitable graph paper the output V_{o1} (due to V_i and V_d) and V_{o2} (due only to V_d) with S_1 closed as a function of G_1. Include on the same graph the V_{o1} and V_{o2} curves based on theory. Use the test data from steps 2 to 4 of the Procedure.

2. Repeat step 1 using data from step 5 of the Procedure, except plot V_{o1} and V_{o2} versus G_2.

3. Using the test data from steps 2 to 4 of the Procedure, calculate and plot the closed-loop disturbance rejection ratio DR_c as a function of G_1. Include on the graph the DR_c curve based on theory.

4. Repeat step 3 using the data from step 5 of the Procedure, except plot DR_c versus G_2.

5. Using the test data from step 6 of the Procedure, calculate and plot on suitable graph paper the steady-state error e_{ss} (expressed as a percent of V_i) as a function of loop gain $G = G_1 G_2$. Include on the graph the e_{ss} curve based on theory.

6. Repeat step 5 except use the test data from steps 2 to 4 of the Procedure and plot e_{ss} as a function of G_1.

7. Repeat step 5 except use the test data from step 5 of the Procedure and plot e_{ss} as a function of G_2.

8. Using the test data from step 6 of the Procedure, plot V_o versus loop gain $G_1 G_2$. Include on the graph the V_o curve based on theory.

9. Using the test data from step 6 of the Procedure, calculate and plot on the graph of step 8 the closed-loop sensitivity factor S_G. Include on the graph the S_G curve based on theory.

10. Using the test data from steps 2 to 4 of the Procedure, calculate and plot the disturbance rejection improvement (closed loop over open loop) $X = DR_c/DR_o$, where DR_o is the open-loop disturbance rejection. Recall from theory that the disturbance rejection improvement for the closed-loop system over that of the open-loop system may be achieved when $V_{o(\text{closed loop})} = V_{o(\text{open loop})}$, where $V_{o(\text{open loop})}$ is measured under reference gain condition (in this experiment $G_1 = G_2 = 1$ has been used as the reference gain), $V_{o(\text{closed loop})}$ must satisfy the equation above for any loop gain, V_i is changed to $K_1 V_i$, and G_1 is changed to $K_2 G_1$. Include on the graph the X curve based on theory.

The responses to the following questions and/or conclusions must be supported whenever possible by the applicable theory.

11. For each of the following, use the graphical results (steps 1 to 10) and draw conclusions regarding the closed-loop system performance and the effect on that performance due to the parameter variation along the abscissa. (a) V_{o1}, V_{o2}, and DR_c from steps 1 and 3; (b) V_{o1}, V_{o2} and DR_c from steps 2 and 4; (c) the disturbance rejection improvement X (step 10); (d) the e_{ss} from steps 5 to 7 (How do these responses differ? Should they? Explain.); (e) the S_G response from step 9. (f) On the basis of the shape of the curve obtained in step 8, how can one conclude whether one system, as compared to another system, is more or less accurate? is more or less sensitive to variation in the forward-path gain?

12. (a) Derive the equation $V_o = f(V_i, V_d)$. (b) Consider the closed-loop disturbance rejection improvement, where G_1 and G_2 are held constant and V_i changes to KV_i. Determine K in terms of loop gain. (c) If V_i in step 11b is held constant and G_1 changes to KG_1, determine K in terms of loop gain. What condition must loop gain satisfy if K is to remain positive? (d) Derive the normalized steady-state

error e_{ss}/V_i in terms of loop gain. (e) Do not derive, but state S_G in terms of system parameters.

13. Explain the effect of (a) percent change in loop gain; (b) the initial value of loop gain on e_{ss} and S_G. Include in your discussion low values of loop gains (of the order of unity) and large values of loop gains (much greater than unity). Do the test data support your conclusions? Does either S_G or e_{ss} functionally depend on the percent change in loop gain? If so, under what conditions does this dependence become negligible? Explain.

14. Suppose in step 12b that V_i changes from -0.5 V dc to -2.0 V dc. Calculate the improvement in the disturbance rejection $X = DR_c/DR_o$, and the value of loop gain $G_1 G_2$.

15. Why is the test circuit used in this experiment capable of an oscillatory response? [*Hint*: Consider the transfer function $GH(S)$ and its poles.] Consult the text on the Routh–Hurwitz and root locus stability test methods. In this respect, what function does the capacitor C perform in the test circuit? Explain.

16. Compare and contrast the performance of an open-loop system in Lab Experiment 5–1 with that of the closed-loop system in this experiment.

TIME-DOMAIN ANALYSIS

chapter 6 _____

6-1 INTRODUCTION

The term *response* is generally used to describe the output of a system or a circuit for a given input. Responses fall into several categories. A sinusoidal input results in a response that is also sinusoidal. Bode plots characterize this type of a response, which is commonly referred to as the *frequency-domain response*. Frequency, the independent variable, is scaled along the abscissa in Bode plots. Bode plots thus offer information about the system in the frequency domain.

When a step or a ramp is applied to the system, its response can be broken down into two parts: the transient and steady-state portions. The transient part of the response is generally characterized by an exponential term that decreases or increases with time, and the steady-state portion is that part of the response that remains in most cases constant or independent of time. There are exceptions to this rule, however. One that quickly comes to mind is the ramp response of the second-order system; its steady-state response increases with time in ramplike fashion; however, the difference between the input and the output remains constant in the steady state. In any case the independent variable now is time; it is the variable that is scaled along the abscissa in the time-domain response graph.

The reader may with some justification question the practical benefits to be derived from time-domain analysis of the system for a very specific input such as the step or the ramp. In a practical application the input to the analog control system may be complex and not easily defined by a step or a ramp; however, a segment of the complex

283

waveform taken over a short time interval may be approximated by some combination of the step, ramp, and parabolic waveforms. Since we are dealing with linear systems, any combination of the standard inputs must result in a combination of the corresponding responses at the output of the system. This, of course, is the statement of superposition. It is then possible to predict system performance for an irregular input if its responses to the standard inputs, such step, ramp, and parabolic waveforms, are known. From this standpoint the system time-domain analysis are well justified. The results of such analysis enables the user to predict a given system performance, to infer system parameter values from the measured response, and to compare the performance of similar systems.

Control systems and circuits are generally classified according to the order of the differential equations that represent these systems. The speed control system presented in Chapter 8 is a first-order system, and the position control system is shown in the same chapter to be a second-order system. There are numerous other practical systems that fall into the first-order or second-order system category. There are also higher-order systems that result from a more complex system configuration. Depending on the location of their poles in the s plane, these systems can often be approximated by a first-order or second-order system. This is important, as the time-domain analysis of systems higher than second order become rather involved; therefore, this chapter covers the analysis of first-order and second-order system only.

As indicated earlier, the system time-domain response consists of the steady state and the transient components. Both components possess important information pertaining to system characteristics; for example, the steady-state error of the system is determined as the difference between the input and the steady-state component, and the system speed is inferred from the transient component. Other important information about the system can also be extracted from the time-domain response.

In addition to system "order," control systems are also classified according to system "type." This classification indicates the number of integrations contained in the forward-path transfer function of the system and is very useful in predicting the system steady-state error for a given input.

There exists a relationship between the time-domain response, the frequency-domain response, and the location of the system transfer function poles in the s plane. As the values of the poles are invariably related to the system parameter values, their adjustment results in a "predictable" system performance.

6–2 TEST SIGNALS

Consider an irregular and aperiodic analog waveform shown in Fig. 6–1. Although it would be difficult to generate this type of waveform repeatedly, sections of the waveform may be represented as shown by simpler and much-easier-to-generate waveforms, such as the impulse, step, ramp, and parabolic. Hence at t_1 the waveform may be represented by a ramp, at t_2 by a step, at t_3 by an impulse, and at t_4 by a parabola. It is therefore possible to represent any complex waveform by these four simple waveforms.

By applying one of the standard test signals—step, ramp, impulse, or parabolic—it is possible to predict from the response the behavior of the system under actual operating conditions. Such important system performance characteristics as the transient

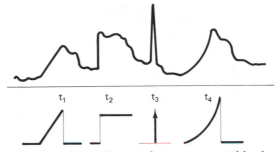

Figure 6–1. Segments of an analog waveform are represented by the standard test signals: ramp, step, impulse, and a parabola.

response and accuracy may be deduced from the response. It is for these reasons that the test signals are important because they provide a standard and a repeatable means of testing any control system.

Included in Table 6–1 are the step, ramp, and parabolic test functions. They fit the standard mathematical form

$$f(t) = At^n u(t) \tag{6-1}$$

whose Laplace transform is

$$\mathcal{L}[f(t)] = \frac{An!}{s^{n+1}} \tag{6-2}$$

Accordingly, Eq. (6–1) represents a step for $n = 0$, a ramp for $n = 1$, and a parabola for $n = 2$.

Other test signals are included in Table 6–2—the step at $t = a$, the impulse at $t = 0$ and at $t = a$. As shown, the impulse is derived from a pulse whose height approaches infinity and whose width approaches zero but whose area is unity. Since the Laplace transform of a unit impulse is 1, it is often used as a test waveform for determining experimentally the system transfer function.

TABLE 6–1 TEST SIGNALS

Test signal	$f(t) = At^n u(t)$	$\mathcal{L}[f(t)] = An!/s^{n+1}$	Waveform
$n = 0$: Step	$Au(t)$	A/s	
$n = 1$: Ramp	$Atu(t)$	A/s^2 A = ramp slope	
$n = 2$: Parabolic	$\frac{1}{2}At^2 u(t)$	A/s^2 A = rate of slope change	

TABLE 6–2 OTHER TEST SIGNALS

Test signal	f(t)	$\mathcal{L}[f(t)]$	Waveform
Step at t = a a > 0	$Au(t - a)$	s/Ae^{-as}	
Impulse at t = 0	$\delta(t)$	1	
Impulse at t = a a > 0	$\delta(t - a)$	e^{-as}	

As is evident from Eq. (6–1), the step and ramp waveforms differ in the power n by 1. The ramp and parabolic waveforms also differ in n by 1. Consequently, all waveforms in Table 6–1 may be practically derived from a squarewave through successive integration (an op amp integrator may be used). A ramp waveform, for example, may be generated by integrating a square wave, and a parabolic waveform is the result of integrating the ramp. The impulse function may be obtained by differentiating a pulse, although other methods, such as suddenly interrupting the current through an inductor, may also be used.

6–3 POLES AND ZEROS

A given system can be represented or modelled in a variety of ways. The transfer function approach was discussed in Chapter 4. By this method the ratio of system's output to input is expressed as a ratio of polynomials. One useful feature of the transfer function is that it is independent of the input thus making it possible to predict the system's output for any input. Transfer function is commonly used to represent a system or its component by a block diagram. In its most general form, a transfer function $G(s)$ may be expressed by

$$G(s) = \frac{N(s)}{D(s)} = \frac{K(s - z_1)(s - z_2) \cdots (s - z_m)}{(s - p_1)(s - p_2) \cdots (s - p_n)} \qquad (6-3)$$

By definition, the *zeros* of the transfer function are those values of s that make the numerator $N(s)$ equal to zero; the *poles* of the transfer function are those values of s that make $D(s)$ equal to zero. Thus $z_1, z_2, \ldots, z_m$ are the zeros of $G(s)$ and the values $p_1, p_2, \ldots, p_n$ are the poles of $G(s)$. Zeros may be finite or infinite. Poles may be real or complex. If complex, they occur in conjugate pairs. The complex s plane is generally used to represent poles and zeros; the symbol $\times$ designates a pole and $\bigcirc$ designates a zero. It is thus possible to represent a system in the s plane by its poles and zeros. System stability considerations require that all complex poles have a

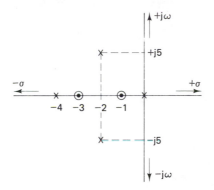

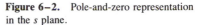

Figure 6–2. Pole-and-zero representation in the s plane.

negative real part and all real poles be negative. Thus all poles of a stable system must be located on the left-hand side of the s plane. The system shown in Fig. 6–2 has two zeros: one at $s = -1$ and the other at $s = -3$. Its poles are at the origin ($s = 0$), at $s = -4$, and a complex conjugate pair at $s = -2 \pm j5$. The system's transfer function corresponding to these poles and zeros may be expressed as

$$G(s) = \frac{K(s + 1)(s + 3)}{s(s + 4)(s^2 + 4s + 29)}$$

6–4 FIRST-ORDER-SYSTEM MODELS

There are two first-order-system models that may be used for the purposes of analysis: the differential equation model, and the transfer function model. The most general form of the *differential equation model*, which fits all first-order systems, may be expressed as

$$\tau \dot{y} + y(t) = Kx(t) \tag{6–4}$$

where τ is the time constant of the system under consideration and K represents the dc value of system gain.

The *transfer function model* is derived from Eq. (6–4) by first taking the Laplace transform of both sides of the equation, and then solving for the ratio of the output $Y(s)$ to the input $X(s)$. The initial condition, $y(0)$ in the present case, must be equated to zero, or otherwise it is not possible to solve algebraically for the desired ratio. Having done that, the reader should be able to verify that the final result is

$$\frac{Y(s)}{X(s)} = \frac{K}{\tau s + 1} \tag{6–5}$$

The two mathematical models above are expressed in terms of general variables: the output y, the input x, and the system parameters τ and K. When an actual application is considered, these variables and parameters assume a form that applies to the system under consideration. For example, in the case of a simple low-pass filter, one of several first-order systems shown in Fig. 6–3a, the input x is represented by V_i, the output by V_o, $K = 1$, and the electrical time constant $\tau = RC$. In the case of a dc motor shown

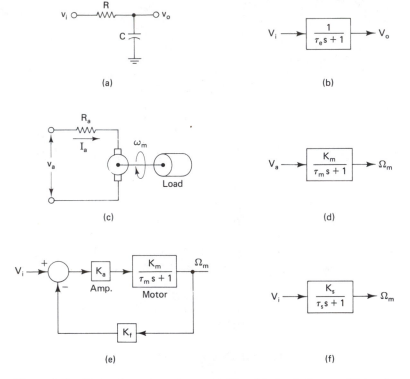

Figure 6–3. First order systems. Low pass filter: (a) circuit diagram, (b) transfer function. DC motor: (c) circuit diagram, (d) transfer function. Speed control system: (e) system diagram, (f) transfer function.

in Fig. 6–3c, the input x is the armature voltage V_a, the output y the motor speed ω_m, K the motor speed constant K_m, and τ the mechanical motor time constant τ_m, which depends on the inertia and viscous friction. Another first-order system is a closed-loop speed control system which uses negative feedback as shown in Fig. 6–3e. Here $x = V_i$; $y = \omega_m$; $K = K_s$, the closed loop system speed constant; and $\tau = \tau_s$, the closed-loop system time constant.

The two mathematical models of the first-order system may be used in the analysis of any first-order system. The reader must bear in mind that if the transfer function model is used, all system initial conditions must be zero (i.e., the system must be at rest), whereas the differential equation model permits the use of initial conditions and in that sense is more general.

6–5 FIRST-ORDER-SYSTEM STEP RESPONSE

In this section we wish to derive the response of any first-order system to a step input. For the sake of generality, initial conditions are admitted in the analysis, and therefore the differential equation model Eq. (6–4) will be used. Suppose that the input x_0 has been applied to the system for a long time and the system produces an output y_0 in the steady state. Then at $t = 0$ a step input of the form $Eu(t)$ is applied. The combined input

$$x(t) = x_0 + Eu(t)$$

is substituted next in Eq. (6–4), and the Laplace transform is taken of both sides of the resulting equation. Solving for the output $Y(s)$, and also expressing it by means of the partial-fraction expansion, the following result is obtained:

$$Y(s) = \frac{(K/\tau)(x_0 + E)}{s(1 + 1/\tau)} = \frac{A}{s} + \frac{B}{s + 1/\tau}$$

The reader should be aware that the reason for the partial-fraction expansion is to break up a more complex function, such as $Y(s)$, into simpler form and thus simplify the process of taking the inverse Laplace transform. The constants A and B in the partial-fraction expansion above are evaluated next using the Heaviside method, with the following result:

$$A = K(x_0 + E) \qquad \text{and} \qquad B = -K(x_0 + E)$$

Substituting A and B in the expression for $Y(s)$ above and taking the inverse Laplace transform,

$$y(t) = (x_0 + E)K - [(x_0 + E)K - y(0)]e^{-t/\tau} \tag{6–6}$$

Equation (6–6) may be expressed in a more compact form in terms of the initial and steady-state values of y as follows:

$$y(t) = y_f - (y_f - y_i)e^{-t/\tau} \tag{6–7}$$

where

$$y_f = y_{ss} = \lim_{t\to\infty} y(t) = (x_0 + E)K \qquad \text{steady-state (final) value}$$
$$\text{of the system's output } y \tag{6–8}$$

$$y_i = y(0) \qquad\qquad\qquad\qquad \text{initial value of the output}$$

If the system attains its steady state value at the time when the step is applied, then the initial value of the output may be expressed as $y(0) = x_0 K$. Under such conditions Eq. (6–6) reduces to

$$y(t) = y(0) + KE(1 - e^{-t/\tau}) \tag{6–9}$$

In Eq. (6–9) the transient term is dominated by the exponential term $e^{-t/\tau}$, which decreases with time. In three time constants $t = 3\tau$ the value of the exponential term is approximately 0.05 and the corresponding value of the response is

$$y(3\tau) = y(0) + 0.95KE$$

differing from the limiting value of the response as expressed by Eq. (6–8) by 5% of KE [assuming that $y(0) = x_0 K$]. Although it is assumed in many applications that the system assumes its steady state at this point in time ($t = 3\tau$), it is not an absolute rule. If, for example, the steady state for a particular system is defined as the point in time where the actual response differs by 1% from the limiting value of

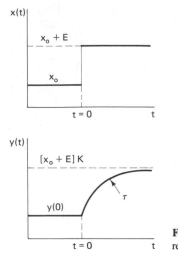

Figure 6–4. First-order system step response.

the response as defined by Eq. (6–8), the system will require 4.6 time constants to reach its steady state. The graphical illustration of the first-order system's step response, which is consistent with Eqs. (6–6), (6–7), and (6–9), is shown in Fig. 6–4.

All that has been said thus far regarding the step response applies to any first-order system, including closed-loop control systems with negative feedback. The use of negative feedback is particularly important in control system applications, as it offers numerous system performance advantages. As discussed in Chapter 5, some of these advantages include better accuracy, less sensitivity to variations in the forward-path parameters, and under special conditions, less sensitivity to random disturbances. Also, wider bandwidth and reduced gain is to be expected from a feedback system which is a consequence of the constant gain–bandwidth product limitation of a system under consideration.

At this time, and quite appropriately, we wish to consider the effects (if any) due to feedback on the transient response of a control system. Consider a simplified block diagram of a typical first-order control system where the output is represented by the controlled variable $C(s)$, the reference input by $R(s)$, the forward-path transfer function as shown, and feedback by K_f. Any of the three first-order control systems (speed, temperature, and liquid-level control) discussed in Chapter 8 may be described by this type of simplified block diagram. Using Eq. (5–10), where $G(s) = K_1/(\tau_1 s + 1)$ and $H(s) = K_f$, the transfer function for the system in Fig. 6–5 may be expressed in the following simplified form:

$$\frac{C(s)}{R(s)} = \frac{K_s}{\tau_s s + 1} \tag{6–10}$$

where

$$K_s = \frac{K_1}{1 + K_1 K_f} \qquad \text{dc gain of the closed-loop system}$$

$$\tau_s = \frac{\tau_1}{1 + K_1 K_f} \qquad \text{closed-loop system time constant}$$

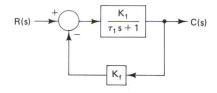

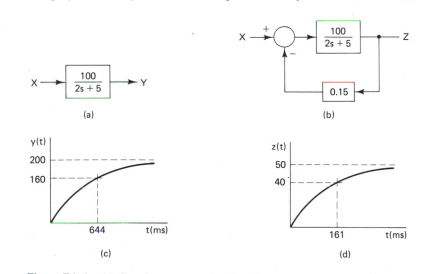

Figure 6–5. Typical first-order control system with negative feedback.

K_1 and τ_1 are the open-loop gain and time constant, respectively, and $K_1 K_f$ is the system loop gain. When the following ratios are considered

$$\frac{K_1}{K_s} = \frac{\tau_1}{\tau_s} = 1 + K_1 K_f \approx K_1 K_f \qquad (6-11)$$

and coupled with the realization that the loop gain $K_1 K_f$ is generally much larger than unity, $K_s < K_1$ and $\tau_s < \tau_1$. As expected, the closed-loop dc gain is less than the open-loop gain, and the closed-loop system time constant is less than the corresponding open-loop system time constant. This means that the step response of the closed-loop system is faster than if the same system was configured as an open-loop system, and the improvement in speed depends directly on the value in loop gain; the response time decreases with an increase in loop gain. Similar observations were made in Chapter 5, where both the closed-loop-system accuracy and its sensitivity to variations in the forward-path parameters were improved due to an increase in loop gain. These are some of the most important reasons for using negative feedback in a control system. Some of these observations are illustrated in the next example.

EXAMPLE 6–1

Given the open-loop system in Fig. E6–1a, and the same system configured as a closed-loop system in Fig. E6–1b. The input to each system is $x(t) = 10u(t)$.

Figure E6–1. (a) Open-loop system; (b) closed-loop system; (c) step response of system in part (a); (d) step response of system in part (b).

(a) Calculate the time required for the open-loop system to attain 80% of the steady-state value of the output y.

(b) Repeat part (a) for a closed-loop system, and compare the response times of the two systems.

(c) Calculate the new value of the K_f such that the closed-loop system reaches 80% of its steady-state value in 100 ms; also, recalculate the input so that the steady-state value of the output remains unchanged.

SOLUTION (a) After dividing by 5, the open-loop transfer function becomes $20/(0.4s + 1)$, from which $\tau = 0.4$, $K = 20$. Substituting these values, $y(t)/y_{ss} = 0.8$ and $y(0) = 0$ in Eq. (6–9):

$$0.8 = 1 - e^{-t/0.4}$$

Solving for t yields

$$t = 644 \text{ ms}$$

(b) Substituting the given values $\tau_1 = 0.4$, $K_1 = 20$, and $K_f = 0.15$ in Eq. (6–10), we obtain

$$K_s = \frac{20}{1 + 20(0.15)} = 5$$

$$\tau_s = \frac{0.4}{1 + 20(0.15)} = 0.1 \text{ s}$$

Substituting these values, $z(t)/z_{ss} = 0.8$, and $z(0) = 0$ in Eq. (6–9) gives us

$$0.8 = 1 - e^{-t/0.1}$$

Solving for t yields

$$t = 161 \text{ ms}$$

The steady-state value of the closed-loop response is

$$z_{ss} = K_s E = 5(10) = 50$$

From the results of parts (a) and (b), the closed-loop system is four times as fast in reaching the 80% point on the response curve.

(c) It is required now that the closed-loop system attain the 80% point on the response curve in 100 ms instead of 161 as calculated in part (b). To achieve this, the time constant must have a different value. Using the information from part (b), we get

$$0.8 = 1 - e^{-0.1/\tau_s}$$

The value of the time constant from the equation above is calculated as 62 ms. From Eq. (6–10),

$$0.062 = \frac{0.4}{1 + 20K_f}$$

Solving for K_f, we obtain

$$K_f = 0.27$$

To maintain $z_{ss} = 50$ as calculated in part (b), the new value of E is determined from $z_{ss} = EK_s$ using the K_s expression from Eq. (6–10) and all given values of K_f calculated above; hence

$$50 = \frac{20E}{1 + 20(0.27)}$$

Solving for E yields

$$E = 16$$

The larger value of feedback raises the loop gain from 3 to 5.4, causing a reduction in the time constant from 100 ms to 62 ms. The input must be increased from 10 to 16 to ensure that the steady-state value of the output remains at 50 in the presence of greater negative feedback, which has increased in the negative direction from -7.5 to -13.5.

6–6 SECOND-ORDER-SYSTEM MODELS

As in the case of first-order systems, there are a number of systems that belong to a general category known as *second-order systems*. Although their physical structure may vary considerably, all second-order systems share the same mathematical model and the same characteristics. It is important that such system commonality be understood and recognized when a particular system is being considered.

The differential equation that may be used to represent any second-order system takes the following general form:

$$\tau \ddot{y} + \dot{y} + Ky(t) = Kx(t) \tag{6–12}$$

where τ and K depend on the parameters of the system under consideration, x is the input, and y is the output. Several second-order control systems are illustrated in Fig. 6–6. Included in the illustration are a simple *RLC* electrical circuit, a mechanical rotational and a translational system, and a position control servomechanism. Although these systems are obviously very different in structure, each of these systems is described by the same transfer function as shown in Fig. 6–6, where the form of τ and K [same τ and K as used in Eq. (6–12)] are indicated for each of the systems illustrated.

The transfer function used in the illustration is derived from Eq. (6–12) by taking the Laplace transform of both sides of the equation with all initial conditions set to zero [i.e., $\dot{y}(0) = 0$ and $y(0) = 0$], and then solving for the ratio of output to input. The final form of the transfer function may be expressed as follows:

$$\frac{Y(s)}{X(s)} = \frac{K}{\tau s^2 + s + K} \tag{6–13}$$

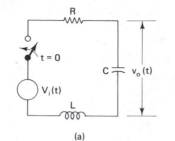

(a)

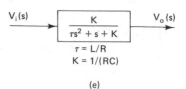

(e)

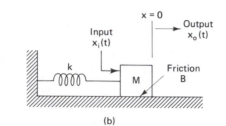

(b)

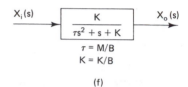

(f)

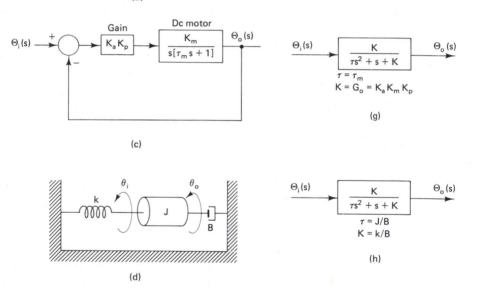

(c)

$\Theta_i(s)$ → $\dfrac{K}{\tau s^2 + s + K}$ → $\Theta_o(s)$

$\tau = \tau_m$
$K = G_o = K_a K_m K_p$

(g)

(d)

$\Theta_i(s)$ → $\dfrac{K}{\tau s^2 + s + K}$ → $\Theta_o(s)$

$\tau = J/B$
$K = k/B$

(h)

Figure 6–6. Typical second-order systems: (a) *RLC* circuit; (b) mechanical translational; (c) position control servo; (d) mechanical rotational; (e) to (h) their corresponding transfer functions.

Although the transfer function form above is general, the universal form of the second-order system transfer function which is used in the literature more often is expressed by the following transfer function:

$$\frac{C(s)}{R(s)} = \frac{\omega_n^2}{s^2 + 2\zeta\omega_n s + \omega_n^2} = \frac{Y(s)}{X(s)} \qquad (6-14)$$

where $C(s)$ [also $Y(s)$] represents the system output or the controlled variable in the case of a control system, and $R(s)$ [also $X(s)$] is the input, a reference input in a control system, ω_n is the natural resonant frequency, and ζ is the damping ratio. The damping ratio is a factor that represents a measure of losses within the system. The energy losses take the form of friction in a mechanical system, and in an electrical system the resistive element produces losses.

In comparing Eqs. (6–13) and (6–14), the parameter relationship can readily be established as follows:

$$\omega_n = \left(\frac{K}{\tau}\right)^{1/2}$$

(6–15)

and

$$\zeta = \tfrac{1}{2}(K\tau)^{-1/2}$$

Hence, in the case of the *RLC* circuit in Fig. 6–6, $\omega_n = (LC)^{-1/2}$ and $\zeta = R/R_c$, where the critical resistance being a function of L and C is commonly defined as $R_c = 2(L/C)^{1/2}$. The reader may wish to show that $(L/C)^{1/2}$ has, indeed, units of ohms. Similar relationships may also be established for the remaining systems in Fig. 6–6 and for any other second-order systems.

Once the system parameter values are specified, the constants K, τ, ω_n, and ζ assume specific values in the differential equation (6–12) or in the transfer functions (6–13) and (6–14), and the identity of the system being considered is lost. To illustrate this, consider the electrical *RLC* circuit in Fig. 6–7a, where $R = 50\ \Omega$, $L = 0.5$ H, and $C = 200\ \mu$F, and the mechanical rotational system in Fig. 6–7d, where $k = 10$ oz-in./rad, $J = 0.001$ oz-in.-s^2, and $B = 0.1$ oz-in.-s. For both systems $K = 100\ \text{s}^{-1}$, $\tau = 10$ ms, $\omega_n = 100$ rad/s, and $\zeta = 0.5$. Both systems may therefore be represented by the differential equation $\ddot{y} + 100\dot{y} + 10{,}000y = 10{,}000x(t)$, or the transfer function $Y(s)/X(s) = 10{,}000/(s^2 + 100s + 10{,}000)$, and it is clearly impossible to tell which system is being represented by these equations. This is, of course, the point of the example, mainly that all second-order systems have the same mathematical form. Therefore, the result of the analysis done on a general mathematical form such as Eq. (6–12) or Eqs. (6–13) and (6–14) applies to all second-order systems.

6–7 SECOND-ORDER-SYSTEM STEP RESPONSE

Considered in this section is the response of a second-order system to a step input. As noted earlier, the result of the analysis done on a general second-order system applies to any second-order system. With this in mind, the transfer function in Eq. (6–14) is used as the mathematical model for the analysis. As part of the analysis procedure, the step input $r(t) = Eu(t)$ whose Laplace transform is $R(s) = E/s$ is substituted in Eq. (6–14). Solving next for the output $C(s)$, the following results:

$$C(s) = \frac{E\omega_n^2}{s(s^2 + 2\zeta\omega_n s + \omega_n^2)}$$

(6–16)

TABLE 6–3 CHARACTER OF THE SECOND ORDER SYSTEM STEP RESPONSE

Damping ratio	Roots of $s^2 + 2\zeta\omega_n s + \omega_n^2 = 0^*$	Response
$\zeta < 1$	$s_{1,2} = -\zeta\omega_n \pm j\omega_n\beta$	Underdamped
$\zeta = 1$	$s_{1,2} = -\omega_n$	Critically damped
$\zeta > 1$	$s_{1,2} = -\zeta\omega_n \pm \omega_n\alpha$	Overdamped

$^*\beta = (1 - \zeta^2)^{1/2}; \alpha = (\zeta^2 - 1)^{1/2}.$

The inverse Laplace transform of Eq. (6–16) represents the time-domain response that we seek. To simplify the process of taking the inverse Laplace transform, Eq. (6–16) is expanded in partial fractions. The constants in the partial-fraction expansion are evaluated using any of the several techniques that are available (the Heaviside method is probably the simplest), and finally the inverse Laplace transform is taken. To expand Eq. (6–16) in partial fractions, the poles (roots of the denominator) must be determined. There are three poles because the denominator is a cubic equation. One pole is at the origin ($s = 0$), and the other two are determined from

$$s^2 + 2\zeta\omega_n s + \omega_n^2 = 0 \tag{6–17}$$

as

$$s_1 = -\zeta\omega_n + \omega_n(\zeta^2 - 1)^{1/2}$$
$$s_2 = -\zeta\omega_n - \omega_n(\zeta^2 - 1)^{1/2}$$

Corresponding to the value of $\zeta < 1, = 1, > 1$, the two roots are complex conjugates, real equal, and real unequal, respectively. The shape of the time-domain response, which depends on the character of the transfer function poles, is therefore determined by the value of the damping ζ. The three cases of damping which are summarized in Table 6–3 are commonly referred to as underdamped ($\zeta < 1$), critically damped ($\zeta = 1$), and overdamped ($\zeta > 1$). Once again, the damping, which determines the shape of the response, depends on the energy losses within the system. In the analysis that follows we shall consider the step and ramp responses of the second-order system for the three cases of damping.

Underdamped System Step Response ($\zeta < 1$)

The partial-fraction expansion for Eq. (6–16) is as follows:

$$C(s) = \frac{E\omega_n^2}{s[s^2 + (2\zeta\omega_n)s + \omega_n^2]} = \frac{A}{s} + \frac{B}{s - s_1} + \frac{C}{s - s_2} \tag{6–18}$$

where the roots of the quadratic s_1 and s_2 are defined in Table 6–3, the constants A, B, and C are evaluated by the Heaviside method as follows:

Time-Domain Analysis   Chap. 6

$$A = \frac{E\omega_n^2}{s^2 + (2\zeta\omega_n)s + \omega_n^2}\bigg|_{s=0} = E$$

$$B = \frac{E\omega_n^2}{s(s - s_2)}\bigg|_{s=s_1} = \frac{E\omega_n^2}{s_1(s_1 - s_2)} = \frac{-E}{2\beta(\beta + j\zeta)}$$

Since the roots s_1 and s_2 are complex conjugates, C is also a conjugate of B, hence

$$C = B^* = \frac{-E}{2\beta(\beta - j\zeta)}$$

Substituting the results obtained for A, B, C in Eq. (6–18) yields

$$C(s) = \frac{E}{s} - \frac{E}{2\beta}\left[\frac{(\beta + j\zeta)^{-1}}{s - s_1} + \frac{(\beta - j\zeta)^{-1}}{s - s_2}\right]$$

Taking the inverse Laplace transform

$$c(t) = E - \frac{E}{2\beta}\left[\frac{(e^{-\zeta\omega_n t})(e^{j\omega_n\beta t})}{\beta + j\zeta} + \frac{(e^{-\zeta\omega_n t})(e^{-j\omega_n\beta t})}{\beta - j\zeta}\right]$$

$$= E - \frac{Ee^{-\zeta\omega_n t}}{2\beta}\left[\frac{e^{j\omega_n\beta t}}{\beta + j\zeta} + \frac{e^{-j\omega_n\beta t}}{\beta - j\zeta}\right]$$

and combining the two terms inside the brackets under a common denominator gives us

$$c(t) = E - \frac{Ee^{-\zeta\omega_n t}}{2\beta}\left[\frac{(\beta - j\zeta)e^{j\omega_n\beta t} + (\beta + j\zeta)e^{-j\omega_n\beta t}}{(\beta + j\zeta)(\beta - j\zeta)}\right]$$

We next replace the denominator inside the brackets by unity sine $\beta^2 + \zeta^2 = 1$ from the definition of ζ. Combining the real and imaginary terms inside the brackets, we have

$$c(t) = E - \frac{Ee^{-\zeta\omega_n t}}{2\beta}[(e^{j\omega_n\beta t} + e^{-j\omega_n\beta t}) - j\zeta(e^{j\omega_n\beta t} - e^{-j\omega_n\beta t})]$$

and applying Euler's identity $(e^{\pm j\theta} = \cos\theta \pm j\sin\theta)$ to the terms inside the brackets:

$$c(t) = E - \frac{Ee^{-\zeta\omega_n t}}{\beta}[\zeta\sin(\omega_n\beta t) + \beta\cos(\omega_n\beta t)]$$

We obtain the result above where the sine and the cosine terms may be combined because their arguments are the same, as follows. Since

$$\sin(A + B) = \sin A \cos B + \cos A \sin B$$

let $A = \omega_n\beta t$ and $B = \phi$, where the angle ϕ is defined by the right triangle shown here, which is based on $\zeta^2 + \beta^2 = 1$. The adjacent leg in the triangle must be equal in length to ζ to assure that the coefficient of the sine (which is equal to $\cos\phi$) in $c(t)$ is ζ. The final form of $c(t)$ may then be expressed as follows:

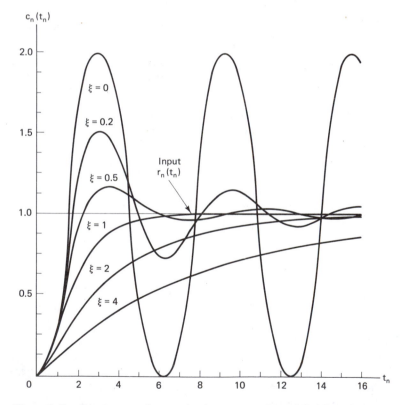

$$c(t) = E\left[1 - \frac{e^{-\zeta\omega_n t}}{\beta} \sin(\omega_n \beta t + \phi)\right] \qquad (6-19)$$

where $\phi = \cos^{-1} \zeta$.

The plot of $c(t)$ to scale is not possible without the knowledge of parameter values. This includes the values of ζ and ω_n and the magnitude of the input E. Through the process of normalization it is possible, however, to reduce the number of parameters without altering the wave shape. This process is sometimes called amplitude and time scaling. Amplitude scaling is applied to Eq. (6–19) by dividing both sides of the equation by E. Time scaling is defined by the relationship between the scaled time t_n (a dimensionless quantity) and real time as follows:

$$t_n = \omega_n t \qquad (6-20)$$

Figure 6–7. Step response of a second-order system; undamped ($\zeta = 0$), underdamped ($\zeta < 1$), critically damped ($\zeta = 1$), overdamped ($\zeta > 1$).

Time-Domain Analysis   Chap. 6

Applying this type of normalization to Eq. (6–19), the following result is obtained:

$$c_n(t_n) = \frac{c(t_n/\omega_n)}{E} = 1 - \frac{e^{-\zeta t_n}}{\beta} \sin(\beta t_n + \phi) \qquad (6\text{–}21)$$

The resulting normalized equation is a function of t_n with ζ as the only parameter. A family of curves is thus possible. This is shown in Fig. 6–7, where c_n is plotted to scale for three values of $\zeta < 1$: $\zeta = 0$, 0.2, and 0.5.

The curve corresponding to $\zeta = 0$ is a cosine function superimposed on the average value of 1. This becomes clear when one realizes that for $\zeta = 0$ in Eq. (6–21), $\beta = 1$, the exponential term is unity, $\phi = 90°$, and $\sin(t_n + \pi/2)$ reduces to $\cos t_n$. The final result is

$$c_n(t_n) = 1 - \cos t_n \qquad \zeta = 0 \qquad (6\text{–}22)$$

Other responses for $\zeta < 1$ in Fig. 6–7 appear to oscillate about the input and decay with time. This is a typical response of an underdamped system to a step input. The first peak of the response is the largest in amplitude. It is commonly used to describe the extent to which the response exceeds the input or the percent overshoot (POT).

At any of the response peaks, the slope must be zero. By applying the min-max theory from calculus we can determine the time at which the peak occurs and the maximum or the minimum value of the function c_n at the peak.

We begin by differentiating c_n in Eq. (6–21), equating the result to zero and finally solving for t_n, which corresponds to the maximum or minimum peak of c_n as follows:

$$\frac{dc_n}{dt_n} = \frac{e^{-\zeta t_n}}{\beta} \sin(\beta t_n + \phi) - e^{-\zeta t_n} \cos(\beta t_n + \phi)$$

$$= \frac{e^{-\zeta t_n}}{\beta} [\zeta \sin(\beta t_n + \phi) - \beta \cos(\beta t_n + \phi)]$$

Expanding gives us

$$\zeta \sin(\beta t_n + \phi) = \zeta^2 \sin \beta t_n + \zeta\beta \cos \beta t_n$$

$$\beta \cos(\beta t_n + \phi) = -\beta^2 \sin \beta t_n + \zeta\beta \cos \beta t_n$$

The difference above, which corresponds to the contents within the brackets, reduces to $\sin \beta t_n$. The derivative, then, may be expressed in its simplest form as follows:

$$\frac{dc_n}{dt_n} = \frac{e^{-\zeta t_n}}{\beta} \sin \beta t_n \qquad (6\text{–}23)$$

When the derivative is equated to zero, the resulting equation is satisfied only if

$$\sin \beta t_n = 0$$

This is possible if

$$\beta T_{nk} = k\pi \qquad k = 1, 2, 3, \ldots$$

Hence

$$T_{nk} = \frac{k\pi}{\beta} \tag{6-24}$$

The index k identifies the particular peak. The first peak occurs at $k = 1$, the second at $k = 2$, and so on. The value of the function at any of its peaks is obtained by substituting Eq. (6–24) in Eq. (6–21) for t_n, with the following result:

$$C_{nk} = 1 - \frac{e^{-k\pi\zeta/\beta}}{\beta} \sin (k\pi + \phi) \tag{6-25}$$

Equation (6–25) may be simplified as a result of the expansion $\sin (k\pi + \phi) = \sin k\pi \cos \phi + \cos k\pi \sin \phi = \beta \cos k\pi = \beta(-1)^k$, where the term $(-1)^k$ is equivalent to $\cos k\pi$, which is equal to -1 for odd multiples of π and $+1$ for even multiples of π. Hence Eq. (6–25), which represents the normalized value of the kth response peak, may be expressed in a more compact form as follows:

$$C_{nk} = 1 - (-1)^k e^{-k\pi\zeta/\beta} = 1 + (-1)^{k+1} e^{-k\pi\zeta/\beta} \tag{6-26}$$

In the normalized form, the first peak time and amplitude are determined from Eqs. (6–24) and (6–26) as follows (for $k = 1$):

$$T_{n1} = \frac{\pi}{\beta} \tag{6-27}$$

$$C_{n1} = 1 + e^{-\pi\zeta/\beta}$$

The actual (denormalized) time and amplitude are obtained using Eqs. (6–20) and (6–21) as

$$T_1 = \frac{T_{n1}}{\omega_n} = \frac{\pi}{\omega_n\beta} \tag{6-28}$$

and

$$C_1 = EC_{n1} = E(1 + e^{-\pi\zeta/\beta})$$

The first peak, being the largest, is used to determine the percent overshoot POT, which gives us a measure of the amount by which the response exceeds the input. Hence

$$POT = \frac{C_1 - E}{E} \times 100 = 100e^{-\pi\zeta/\beta} \tag{6-29}$$

POT depends only on the value of ζ, being largest in the undamped case (i.e., POT = 100% for $\zeta = 0$), and attaining its lowest value of zero as ζ approaches unity.

In contrast to POT, T_1, which depends on the values of ω_n and ζ, assumes its lowest value at $\zeta = 0$ and increases with ζ. Figure 6–7 shows several underdamped response curves plotted to scale in the normalized form, and Fig. 6–8 shows the variation of POT and T_{n1} with ζ.

A practical application would typically require that POT be as low as possible and that system response be as fast as possible. Since T_1 may be used as a measure of

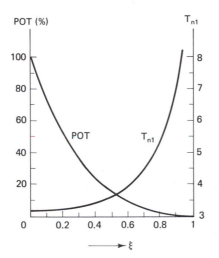

Figure 6–8. Variation of the POT and T_{n1} with damping.

the speed of response, it is clear from Fig. 6–8 that it is not possible to achieve a low POT and a low T_1 at the same time (i.e., it is not possible to achieve a fast system with low overshoot), and therefore a compromise must be made.

As stated above, T_1 may serve as a satisfactory measure of system's speed. The speed of response may also be conveniently measured by means of the rise time T_r. T_r is the time at which the response is equal to the input for the first time. From Eq. (6–19) it is evident that the response is equal to the input only when the transient term vanishes. The transient term vanishes when the argument of the sine term is equated to a multiple of π. Hence $C(t_m) = E$ when $\omega_n \beta t_m + \phi = m\pi$, where m is an integer. Solving for t_m yields

$$t_m = \frac{m\pi - \phi}{\omega_n \beta} \qquad m = 1, 2, 3, \ldots \qquad (6\text{–}30)$$

t_m thus represents all time points where the response is equal to the input. The first occurrence of this ($m = 1$) is, by definition, the rise time T_r (i.e., $t_1 = T_r$). It follows then from Eq. (6–30) that

$$T_r = \frac{\pi - \phi}{\omega_n \beta} \qquad (6\text{–}31)$$

In practical applications it is important to define the settling time T_s, which marks the point at which the system's response reaches the steady state and the transient no longer has any effect (or at least a negligible effect). The transient part of the response, which is represented by the exponential term in Eq. (6–19), decays with the time constant

$$\tau = \frac{1}{\zeta \omega_n} \qquad (6\text{–}32)$$

From a purely mathematical viewpoint, it takes infinite time for the transient term to vanish, which offers limited practical information in defining the steady state. In adopting

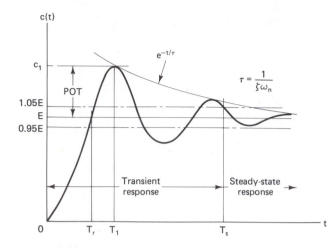

Figure 6–9. Step response characteristics of an underdamped second-order system (unscaled format is used).

a more realistic approach, consider a band that is centered on the input and whose width is a predetermined fraction of the input. Then the time at which the response enters this band for the first time and is thereafter contained by the band may be used to represent the settling time, the time at which the steady state begins. The narrower the band is, the larger will be the settling time. The size of the band generally depends on a particular systems requirements.

In this book we use a band whose width is 10% of the input (i.e., 5% above the input and 5% below the input) as shown in Fig. 6–9. The time that it takes the response to enter and be bounded by the band of width $0.1E$ may be estimated from Eq. (6–19). It is assumed in this approximate solution that the coefficient of the exponential term is approximately unity and the resulting equation is equated to $(1 \pm 0.05)E$ at $t = T_s$ [i.e., $E(1 \pm e^{-T_s/\tau}) = (1 \pm 0.05)E$], from which it may be concluded that $e^{-T_s/\tau} = 0.05$ and the settling time

$$T_s = 3\tau = \frac{3}{\zeta\omega_n} \tag{6–33}$$

The step response characteristics of a second-order system, which include the POT, T_r, T_1, and T_s, are illustrated in Fig. 6–9.

EXAMPLE 6–2

Given a second-order system transfer function $Y(s)/X(s) = 2500/(s^2 + 20s + 2500)$ and the step input $x(t) = 40u(t)$, calculate (a) POT; (b) the values of the first two response peaks and their corresponding times; (c) T_r and the time at which the response is equal to the input for the second time; (d) T_s.

SOLUTION (a) Comparing the given transfer function with the standard form in Eq. (6–14), it is evident that $\omega_n = (2500)^{1/2} = 50$, and $2\zeta\omega_n = 20$ or $\zeta = 20/[2(50)] = 0.2$. From this we can conclude that the response is underdamped.

The quantity $\beta = (1 - 0.2^2)^{1/2} = 0.98$ and $\pi\zeta/\beta = \pi(0.2)/0.98 = 0.64$. Substituting 0.64 in Eq. (6–29) yields

$$\text{POT} = 100e^{-0.64} = 52.7\%$$

(b) Substituting the given input $E = 40$ and 0.64 for $\pi\zeta/\beta$ calculated in part (a) in Eq. (6–28), we have

$$y_1 = 40(1 + e^{-0.64}) = 61.1 \qquad \text{and} \qquad T_1 = \frac{\pi}{50(0.98)} = 64.1 \text{ ms}$$

The values of y_2 and T_2 are obtained from Eqs. (6–26) and (6–24), respectively, after denormalization [i.e., $y_2 = Y_{n2}E$ and $T_2 = T_{n2}/\omega_n$ (the general output variable c is replaced by y in this problem)]. Hence letting $k = 2$, we have

$$T_2 = \frac{2\pi}{\omega_n\beta} = \frac{2\pi}{(50)(0.98)} = 2T_1 = 128.2 \text{ ms}$$

$$y_2 = 40[1 + (-1)^3 e^{-(2)(0.64)}] = 40(1 - e^{-1.28}) = 28.9$$

(c) From Eq. (6–31),

$$T_r = \frac{\pi - 1.37}{(50)(0.98)} = 36.2 \text{ ms} \qquad \text{where } \cos^{-1} 0.2 = 1.37 \text{ rad}$$

From Eq. (6–30), the time at which the response is equal to the input for the second time; that is, $m = 2$ is

$$t_2 = \frac{2\pi - 1.37}{50(0.98)} = 100.3 \text{ ms}$$

(d) From Eq. (6–33), the settling time is

$$T_s = \frac{3}{0.2(50)} = 300 \text{ ms}$$

Critically Damped System Step Response ($\zeta = 1$)

Substituting $\zeta = 1$ in Eq. (6–16), and expanding the resulting equation in partial fractions where the roots of the partial-fraction expansion are obtained from Eq. (6–17), we have

$$C(s) = \frac{E\omega_n^2}{s(s^2 + 2\omega_n s + \omega_n^2)} = \frac{E\omega_n^2}{s(s + \omega_n)^2} = \frac{A}{s} + \frac{B}{(s + \omega_n)^2} + \frac{C}{s + \omega_n}$$

The constants A, B, and C are evaluated by the Heaviside method as follows:

$$A = \left.\frac{E\omega_n^2}{(s + \omega_n)^2}\right|_{s=0} = E \qquad B = \left.\frac{E\omega_n^2}{0!(s)}\right|_{s=-\omega_n}$$

$$= -E\omega_n \qquad C = \frac{1}{1!}\left.\left(\frac{d}{ds}\right)\frac{E\omega_n^2}{s}\right|_{s=-\omega_n} = \left.\frac{-E\omega_n^2}{s^2}\right|_{s=-\omega_n} = -E$$

Substituting these results for A, B, C in $C(s)$ and taking the inverse Laplace transform, the following response is obtained:

$$c(t) = E - Ee^{-\omega_n t}(\omega_n t + 1) \tag{6-34}$$

or in the normalized form as

$$c_n(t_n) = \frac{c(t_n/\omega_n)}{E} = 1 - e^{-t_n}(t_n + 1) \tag{6-35}$$

Equation (6-35), which represents the response of the critically damped system to the step input, is plotted to scale in Fig. 6-7. It is exponentially rising toward its steady-state value of E, the first term in Eq. (6-34). Its rise time is lower than that of any other response for $\zeta > 1$. It is, therefore, the fastest response without the overshoot. The rise time between the 10% and 90% points may readily be evaluated with the following results:

$$c_n(0.53) = 0.1$$

and

$$c_n(3.88) = 0.9$$

Hence the difference between the two times is the rise time

$$T_{nr} = 3.35 \tag{6-36}$$

or

$$T_r = \frac{3.35}{\omega_n} = 3.35\tau \tag{6-37}$$

which is equal to 3.35 time constants.

The second term in Eq. (6-34) or (6-35) is the transient term, which in time decays to zero with the time constant

$$\tau = \frac{1}{\omega_n} \tag{6-38}$$

In 3 time constants the response reaches 95% of its final value.

Overdamped System Step Response ($\zeta > 1$)

In the case of the overdamped system, the roots of the characteristic equation are real and nonrepeated, as shown in Table 6-3. Using these roots and the step input in Eq. (6-14), we obtain the following result:

$$C(s) = \frac{E\omega_n^2}{s(s - s_1)(s - s_2)} = \frac{A}{s} + \frac{B}{s - s_1} + \frac{C}{s - s_2}$$

where

$$s_1 = -\zeta\omega_n + \omega_n\alpha \qquad \alpha = (\zeta^2 - 1)^{1/2}$$

$$s_2 = -\zeta\omega_n - \omega_n\alpha$$

from Table 6-3.

The constants A, B, and C in the partial-fraction expansion are again evaluated by the Heaviside method as follows:

$$A = \frac{E\omega_n^2}{(s - s_1)(s - s_2)}\bigg|_{s=0} = \frac{E\omega_n^2}{(s_1)(s_2)} = E$$

$$B = \frac{E\omega_n^2}{s(s - s_2)}\bigg|_{s=s_1} = \frac{E\omega_n^2}{s_1(s_1 - s_2)} = \frac{-E}{2\alpha(\zeta - \alpha)}$$

$$C = \frac{E\omega_n^2}{s(s - s_1)}\bigg|_{s=s_2} = \frac{E\omega_n^2}{s_2(s_2 - s_1)} = \frac{E}{2\alpha(\zeta + \alpha)}$$

These results are substituted back into $C(s)$, the inverse Laplace transform taken, and after some algebraic simplification such as combining the two exponential terms under common denominator, the following results:

$$c(t) = E(1 - C_1 e^{-t/\tau_1} + C_2 e^{-t/\tau_2}) \tag{6–39}$$

where

$$C_1 = \frac{\zeta + \alpha}{2\alpha} \qquad \tau_1 = \frac{1}{\omega_n(\zeta - \alpha)}$$

$$C_2 = \frac{\zeta - \alpha}{2\alpha} \qquad \tau_2 = \frac{1}{\omega_n(\zeta + \alpha)} \tag{6–40}$$

It can also be shown that

$$C_1 = 1 + C_2 \qquad \text{and} \qquad \tau_1\tau_2 = \frac{1}{\omega_n^2}$$

C_1 and C_2 are the peak amplitudes of the two transient terms in Eq. (6–39), and τ_1 and τ_2 are the corresponding time constants. The ratios of the amplitudes and the corresponding time constant are equal and may be expressed as follows:

$$\frac{C_2}{C_1} = \frac{\tau_2}{\tau_1} = \frac{\zeta - \alpha}{\zeta + \alpha} = (\zeta - \alpha)^2 \tag{6–41}$$

The scaling procedure used before may also be applied to $c(t)$. Amplitude scaling is accomplished by dividing both sides of Eq. (6–39) by E, the amplitude scaling factor, thus reducing the equation to the dimensionless form. The dependence of the resulting equation on ω_n may be eliminated through time scaling, which relates the dimensionless, scaled time t_n to the real time t (measured in seconds) by

$$t_n = (\omega_n)t \tag{6–42}$$

Implementing the foregoing procedure, the resulting scaled equation may be expressed as follows:

$$c_n(t_n) = \frac{c(t_n/\omega_n)}{E} = 1 - C_1 e^{-t_n/\tau_{n1}} + C_2 e^{-t_n/\tau_{n2}} \tag{6–43}$$

where

$$\tau_{n1} = \frac{1}{\zeta - \alpha} \qquad \text{and} \qquad \tau_{n2} = \frac{1}{\zeta + \alpha} \qquad (6\text{–}44)$$

are the normalized time constants. The independent variable in Eq. (6–43) is the dimensionless time t_n and ζ is the parameter. A single response curve is defined by choosing a value of $\zeta > 1$ and using it in Eq. (6–43). A family of response curves can thus be generated with ζ as the parameter. One such curve is shown in Fig. 6–7 for $\zeta = 4$.

The first term in Eq. (6–39) or (6–43) is the steady-state term and the other two terms represent the transient response. The rise time is not as easily determined here as it was in the critically damped case. The reason for that is that we have two transient terms here, and the time constants which are associated with each of these terms depend on the value of ζ. As the value of ζ is increased so is the rise time of the response, which is measured between the 10 and 90% points. This may be understood at least in part from Eq. (6–41), where the value of the term $(\zeta - \alpha)^{1/2}$ decreased with increasing ζ. As a consequence of this or Eqs. (6–40) and (6–44), the time constant τ_1 increases and τ_2 decreases with increasing ζ. This emphasizes one of the very important properties of overdamped systems—that the overdamped system becomes slower in its response to the step input if its damping ratio is increased.

As indicated above, τ_1 increases and τ_2 decreases with increasing ζ. This suggests that there are instances where the last term in Eq. (6–43) or (6–39) may be negligible in relation to other terms. The relative importance of the last term may be evaluated by considering the approximate solution C_a, which consists of the first two terms in Eq. (6–43), and the exact solution C_e, which includes all the terms in Eq. (6–43) and then forming the ratio X defined as follows:

$$X(t_n) = \frac{1 - C_1 e^{-t_n/\tau_{n1}}}{C_2 e^{-t_n/\tau_{n2}}} = \frac{C_a}{C_e - C_a} \qquad (6\text{–}45)$$

Equation (6–45) may be rearranged and solved for

$$\frac{C_a}{C_e} = \frac{X}{X + 1} \qquad (6\text{–}46)$$

When Eq. (6–46) is plotted as a function of t_n for different values of ζ, a family of curves result, as shown in Fig. 6–10. It is clear from the graph that for values of ζ which are very close to unity, it takes a much longer time for the ratio C_a/C_e to attain values close to 100% than it does for larger values of ζ; for example, when $C_a/C_e = 90$, this means that the approximate solution is equal to 90% of the exact solution. The curve $\zeta = 1.01$ attains this value at $t_n = 3.4$, whereas the curve $\zeta = 3$ attains this value at $t_n = 0.35$. It is also clear that during the initial phase of the response, when t_n is close to zero, the approximate and exact solutions are very far apart regardless of the value of ζ. Also when t_n is sufficiently large, the approximate and the exact solutions are almost the same, even for low values of ζ; for example, for $\zeta > 1.05$ all approximate solutions are 98% or higher of their corresponding exact solutions. The curves in Fig. 6–10 may thus be used as a guide to decide whether the last term in Eq. (6–43) can be ignored when the overdamped response is evaluated at a given point in time.

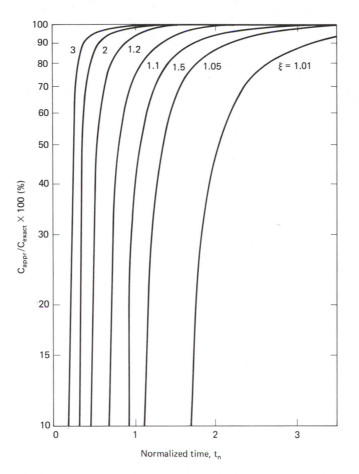

Figure 6–10. Variation of C_a/C_e with t_n and with the damping ratio ζ. C_a is the sum of the first two terms and C_e is the sum of all terms in Eq. (6–43).

EXAMPLE 6–3

Consider the position control system in Fig. 6–6c; $K_p = 2$ V/rad, $K_m = 20$ rad/s/V, $\tau_m = 0.1$s, and $\theta_i = 40u(t)$ degrees. Calculate (a) the value of K_a for a critically damped response and POT = 40%; (b) T_1 in part (a); (c) K_a for a critically damped response; (d) K_a if $\zeta = 1.5$; (e) $\theta_o(0.15)$ and $\theta_o(0.3)$ using $\zeta = 1.5$. Can the last transient term be ignored in these calculations? Explain.

SOLUTION   (a) From Eq. (6–29),

$$0.4 = e^{-\pi\zeta/\beta}$$

Solving for ζ, we have

$$\zeta = 0.28$$

From the block diagram in Fig. 6–6c,

$$G(s) = \frac{K_a K_m K_p}{s(\tau_m s + 1)} = \frac{40 K_a}{s(0.1s + 1)}$$

$$H(s) = 1$$

Then the closed-loop transfer function may be expressed as follows:

$$\frac{\theta_o(s)}{\theta_i(s)} = \frac{G(s)}{1 + GH(s)} = \frac{400 K_a}{s^2 + 10s + 400 K_a}$$

Comparing the transfer function above to that of Eq. (6–14), the following may be concluded:

$$\omega_n^2 = 400 K_a$$

and

$$2\zeta\omega_n = 10$$

Using the value of ζ calculated above, we obtain

$$2(0.28)(20)(K_a)^{1/2} = 10$$

$$K_a = 0.8$$

(b) From Eq. (6–28),

$$T_1 = \frac{\pi}{(320)^{1/2}(1 - 0.28^2)^{1/2}} = 0.18 \text{ s}$$

(c) From part (a),

$$2(1)\omega_n = 10$$

Substituting for ω_n from part (a), we have

$$2(400 K_a)^{1/2} = 10$$

Therefore,

$$K_a = \frac{1}{16}$$

(d) From part (a),

$$2(1.5)\omega_n = 10$$

Substituting from part (a) the value of ω_n, we have

$$4(400 K_a)^{1/2} = 10$$

Solving for K_a yields

$$K_a = \frac{1}{64}$$

(e)

$$\alpha = (1.5^2 - 1)^{1/2} = 1.118 \qquad \text{and} \qquad \omega_n = \frac{10}{3}$$

Substituting these values in Eqs. (6–40) and (6–44) gives us

$$C_2 = 0.17 \qquad \tau_{n1} = 2.62$$

$$C_1 = 1 + C_2 = 1.17 \qquad \tau_{n2} = \frac{1}{\tau_{n1}} = 0.38$$

Substituting these values in Eq. (6–43), we have

$$\theta_n(t_n) = 1 - 1.17e^{-t_n/2.62} + 0.17e^{-t_n/0.38}$$

From Eq. (6–40), the normalized time corresponding to $t = 0.15$s is

$$t_n = \frac{10}{3}(0.15) = 0.5$$

Hence

$$\theta_n(0.5) = 1 - 0.967 + 0.046 = 0.079 = C_e$$

$$\theta_o(0.15) = E\theta_n(0.5) = 0.079(40) = 3.16°$$

and

$$\frac{C_a}{C_e} = \frac{0.033}{0.079} = 0.42$$

The normalized time corresponding to $t = 0.3$ is

$$t_n = \frac{10}{3}(0.3) = 1$$

Hence

$$\theta_n(1) = 1 - 0.799 + 0.012 = 0.213 = C_e$$

The actual angular displacement at 0.3 s may be computed as

$$\theta_o(0.3) = E\theta_n(1) = 0.213(40) = 8.52°$$

and

$$\frac{C_a}{C_e} = \frac{0.201}{0.213} = 0.94$$

The sum of the first two terms in $\theta_n(0.5)$ is comparable in value to the third term; hence the third term should be included in the calculations; furthermore, the ratio C_a/C_e indicates that the approximate solution is 42% of the exact solution. On the other hand, the sum of the first two terms in $\theta_n(1)$ is considerably larger than the third term and the ratio C_a/C_e indicates that the approximate solution is 94% of the exact solution. If 6% error can be tolerated, the third term may be excluded from the calculations. The conclusions above are supported by Fig. 6–10.

In this section we present an analysis of the second-order system's response to the ramp input of the form

$$r(t) = Etu(t)$$

whose Laplace transform is

$$R(s) = \frac{E}{s^2}$$

Three cases of damping are considered in the analysis. The characteristic equation and its roots as expressed by Eq. (6-17) or by Table 6-3 also apply to the analysis here. The underdamped response is considered first in the next section.

Underdamped-System Ramp Response ($\zeta < 1$)

Substituting the ramp input in Eq. (6-14) gives us

$$C(s) = \frac{E\omega_n^2}{s^2[s^2 + (2\zeta\omega_n)s + \omega_n^2]} = \frac{A}{s^2} + \frac{B}{s} + \frac{C}{s - s_1} + \frac{D}{s - s_2}$$

The constants of the partial-fraction expansion are evaluated by the Heaviside method as follows:

$$A = \left(\frac{1}{0!}\right) \frac{E\omega_n^2}{s^2 + (2\zeta\omega_n)s + \omega_n^2}\bigg|_{s=0} = E$$

$$B = \left(\frac{1}{1!}\right) \frac{d}{ds} \frac{E\omega_n^2}{s^2 + (2\zeta\omega_n)s + \omega_n^2}\bigg|_{s=0} = \frac{-2E\omega_n^2(s + \zeta\omega_n)}{[s^2 + (2\zeta\omega_n)s + \omega_n^2]^2} = -\frac{2\zeta E}{\omega_n}$$

$$C = \frac{E\omega_n^2}{s^2(s - s_2)}\bigg|_{s=s_1} = \frac{E\omega_n^2}{s_1^2(s_1 - s_2)} = \frac{E}{2\omega_n\beta[2\zeta\beta + j(\zeta^2 - \beta^2)]}$$

$$D = C^* = \frac{E}{2\omega_n\beta[2\zeta\beta - j(\zeta^2 - \beta^2)]}$$

Substituting the results for A, B, C, and D in and taking the inverse Laplace transform

$$c(t) = Et - \frac{2\zeta E}{\omega_n} + \frac{E}{2\omega_n\beta}\left[\frac{e^{s_1t}}{2\zeta\beta + j(\zeta^2 - \beta^2)} + \frac{e^{s_2t}}{2\zeta\beta - j(\zeta^2 - \beta^2)}\right]$$

and replacing s_1 and s_2 by their corresponding expressions from Table 6-3, the following result is obtained after the two terms inside the brackets are combined under a common denominator:

$$c(t) = Et - \frac{2\zeta E}{\omega_n} + \frac{Ee^{-\zeta\omega_n t}}{2\omega_n\beta}\left[\frac{(2\zeta\beta - j(\zeta^2 - \beta^2))e^{j\omega_n\beta t} + (2\zeta\beta + j(\zeta^2 - \beta^2))e^{-j\omega_n\beta t}}{(2\zeta\beta)^2 + (\zeta^2 - \beta^2)^2}\right]$$

The denominator inside the brackets reduces to $(\zeta^2 + \beta^2)^2$, which is equal to unity. The real and imaginary terms inside the brackets are combined next:

$$c(t) = Et - \frac{2\zeta E}{\omega_n} + \frac{Ee^{-\zeta\omega_n t}}{2\omega_n\beta} [2\zeta\beta(e^{j\omega_n\beta t} + e^{-j\omega_n\beta t}) - j(\zeta^2 - \beta^2)(e^{j\omega_n\beta t} - e^{-j\omega_n\beta t})]$$

Euler's identity,

$$e^{\pm j\theta} = \cos\theta \pm j\sin\theta$$

is used to combine the exponential terms inside the brackets as follows:

$$c(t) = Et - \frac{2\zeta E}{\omega_n} + \frac{Ee^{-\zeta\omega_n t}}{\omega_n\beta} [2\zeta\beta\cos\omega_n\beta t + (\zeta^2 - \beta^2)\sin\omega_n\beta t]$$

The two terms inside the brackets may be replaced by $\sin(\omega_n\beta t + 2\phi)$. This may be verified as follows:

$$\sin(\omega_n\beta t + 2\phi) = \sin\omega_n\beta t \cos 2\phi + \cos\omega_n\beta t \sin 2\phi$$
$$= \sin\omega_n\beta t (\cos^2\phi - \sin^2\phi) + \cos\omega_n\beta t (2\sin\phi\cos\phi)$$
$$= (\zeta^2 - \beta^2)\sin\omega_n\beta t + 2\zeta\beta\cos\omega_n\beta t$$

The final form of the solution may be expressed as

$$c(t) = Et - \frac{2\zeta E}{\omega_n} + \frac{Ee^{-\zeta\omega_n t}}{\beta\omega_n}\sin(\omega_n\beta t + 2\phi) \qquad (6\text{--}47)$$

Equation (6–47) is magnitude scaled by E/ω_n and time scaled by $t_n = \omega_n t$ as follows:

$$c_n(t_n) = \frac{\omega_n}{E}c\left(\frac{t_n}{\omega_n}\right) = t_n - 2\zeta + \frac{e^{-\zeta t_n}}{\beta}\sin(\beta t_n + 2\phi) \qquad (6\text{--}48)$$

where $\beta = (1 - \zeta^2)^{-1/2}$ and $\phi = \cos^{-1}\zeta$. The last term in Eq. (6–47) or (6–48) is the transient response term, which decays to zero with increasing time. Its rate of decay is dictated by the time constant $1/\zeta\omega_n$. The first term, Et, is the input ramp. Under the steady-state conditions the last term becomes negligible in relation to the sum of the first two terms (this is true when the time is sufficiently large, 3 time constants or more). The difference, then, between the input and the output is equal to a constant, the second term in Eq. (6–47) or (6–48). This may be expressed mathematically as

$$\lim_{t\to\infty}[Et - c(t)] = \frac{2\zeta E}{\omega_n}$$

In the normalized form, this limit is 2ζ, the second term in Eq. (6–48). This difference between the input and the output under the steady-state conditions is commonly referred to as the steady-state error,

$$e_{ss} = \frac{2\zeta E}{\omega_n} \qquad (6\text{--}49a)$$

or in the normalized form as

$$e_{nss} = \frac{\omega_n}{E}e_{ss} = 2\zeta \qquad (6\text{--}49b)$$

If in a given configuration there are no losses, the damping ratio ζ must be zero. Although this situation is unlikely to occur in a practical system, the ramp response of such a system may still be considered from the mathematical standpoint. Equating ζ to zero in Eq. (6–48) results in $\beta = 1$ and $\phi = \pi/2$ and

$$c_n(t_n) = t_n + \sin(t_n + \pi) = t_n - \sin t_n \tag{6–50}$$

This is plotted to scale in Fig. 6–11 as a sinusoidal waveform oscillating about the ramp input (which may be expressed as t_n upon normalization) and never decreasing in amplitude. The steady-state error is zero due to zero damping; hence the sinusoidal waveform oscillates about the input ramp, which is effectively the average value of the waveform.

Figure 6–11 shows another underdamped ramp response curve for $\zeta = 0.2$. This damped sinusoid oscillates about its average value, represented by the dashed line, and for sufficiently large values of time converges to the dashed line. The dashed line is parallel to and 0.4 unit below the ramp input. According to Eq. (6–49), this is precisely

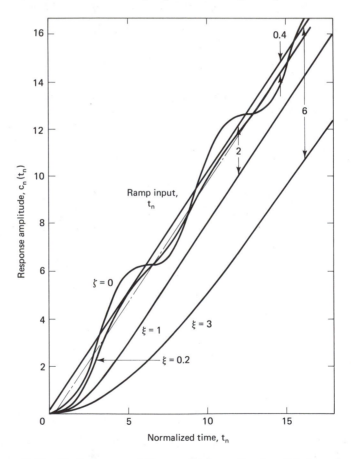

Figure 6–11. Ramp response of the second-order system; $\zeta = 0$ is the undamped response, $\zeta < 1$ is the underdamped response, $\zeta = 1$ is the critically damped response, and $\zeta > 1$ is the overdamped response.

the steady-state error for the response curve $\zeta = 0.2$. The underdamped response curves for larger values of ζ oscillate about the average, which is farther away from the input ramp, and the amplitude of the oscillations is smaller.

Critically-Damped-System Ramp Response ($\zeta = 1$)

For $\zeta = 1$, the roots of the characteristic equation are both equal to $-\omega_n$. These roots and the ramp input are substituted in Eq. (6–14), which is expanded in partial fractions as follows:

$$c(s) = \frac{E\omega_n^2}{s^2(s + \omega_n)^2} = \frac{A}{s^2} + \frac{B}{s} + \frac{C}{(s + \omega_n)^2} + \frac{D}{s + \omega_n}$$

The constants in the partial-fraction expansion are then evaluated as

$$A = \left(\frac{1}{0!}\right) \frac{E\omega_n^2}{(s + \omega_n)^2}\bigg|_{s=0} = E$$

$$B = \left(\frac{1}{1!}\right) \frac{d}{ds} \frac{E\omega_n^2}{(s + \omega_n)^2}\bigg|_{s=0} = \frac{-2E\omega_n^2}{(s + \omega_n)^3}\bigg|_{s=0} = \frac{-2E}{\omega_n}$$

$$C = \left(\frac{1}{0!}\right) \frac{E\omega_n^2}{s^2}\bigg|_{s=-\omega_n} = E$$

$$D = \left(\frac{1}{1!}\right) \frac{d}{ds} \frac{E\omega_n^2}{s^2}\bigg|_{s=-\omega_n} = \frac{-2E\omega_n^2}{s^3}\bigg|_{s=-\omega_n} = \frac{2E}{\omega_n}$$

These results are substituted into $C(s)$ and the inverse Laplace transform taken. The critically damped time-domain response may then be expressed in the simplified form

$$c(t) = Et - \frac{2E}{\omega_n} + \frac{Ee^{-\omega_n t}}{\omega_n}(\omega_n t + 2) \tag{6–51}$$

and the scaled version of this response as follows:

$$c_n(t_n) = \frac{\omega_n}{E} c\left(\frac{t_n}{\omega_n}\right) = t_n - 2 + e^{-t_n}(t_n + 2) \tag{6–52}$$

This equation is plotted to scale as the $\zeta = 1$ curve in Fig. 6–11. The last term in Eq. (6–51) or (6–52) is the transient term, which decays to zero with the time constant $\frac{1}{\omega_n}$ or 1 when it is expressed in the normalized form. The first term is the ramp input and the second term is the steady-state error as expressed by Eq. (6–49).

Overdamped-System Ramp Response ($\zeta > 1$)

In the case where the damping ratio is greater than unity, the roots of the characteristic equation as expressed in Table 6–3 are both real and unequal. These roots and the ramp input are substituted in Eq. (6–14), which is expanded in partial fractions with the following result:

$$C(s) = \frac{E\omega_n^2}{s^2(s - s_1)(s - s_2)}$$

$$= \frac{A}{s^2} + \frac{B}{s} + \frac{C}{s - s_1} + \frac{D}{s - s_2}$$

The constants in the partial-fraction expansion are evaluated, as before, by the Heaviside method as

$$A = \left(\frac{1}{0!}\right) \frac{E\omega_n^2}{s^2 + 2\zeta\omega_n s + \omega_n^2}\bigg|_{s=0} = E$$

$$B = \left(\frac{1}{1!}\right) \frac{d}{ds} \frac{E\omega_n^2}{s^2 + 2\zeta\omega_n s + \omega_n^2}\bigg|_{s=0} = \frac{-2E\omega_n^2(s + \zeta\omega_n)}{(s^2 + 2\zeta\omega_n s + \omega_n^2)^2}\bigg|_{s=0} = \frac{-2\zeta E}{\omega_n}$$

$$C = \frac{E\omega_n^2}{s^2(s - s_2)}\bigg|_{s=s_1} = \frac{E\omega_n^2}{s_1^2(s_1 - s_2)} = \frac{E}{2\omega_n\alpha(\zeta - \alpha)^2}$$

$$D = \frac{E\omega_n^2}{s^2(s - s_1)}\bigg|_{s=s_2} = \frac{E\omega_n^2}{s_2^2(s_2 - s_1)} = \frac{-E}{2\omega_n\alpha(\zeta + \alpha)^2}$$

These results are substituted back into $C(s)$ and the inverse Laplace transform taken, with the following result:

$$c(t) = Et - \frac{2\zeta E}{\omega_n} + \frac{E(\zeta + \alpha)^2}{2\omega_n\alpha}e^{-\omega_n(\zeta - \alpha)t} - \frac{E(\zeta - \alpha)^2}{2\omega_n\alpha}e^{-\omega_n(\zeta + \alpha)t} \qquad (6-53)$$

The constants C_1, C_2, τ_1, and τ_2, which are defined by Eq. (6-40), may be used here as follows:

$$c(t) = Et - \frac{2\zeta E}{\omega_n} + \frac{2\alpha E C_1^2}{\omega_n}e^{-t/\tau_1} - \frac{2\alpha E C_2^2}{\omega_n}e^{-t/\tau_2} \qquad (6-54)$$

The normalized form of this equation,

$$c_n(t_n) = \frac{\omega_n}{E}c\left(\frac{t_n}{\omega_n}\right) = t_n - 2\zeta + 2\alpha C_1^2 e^{-t_n/\tau_{n1}} - 2\alpha C_2^2 e^{-t_n/\tau_{n2}} \qquad (6-55)$$

where the normalized time constants τ_{n1} and τ_{n2} are defined by Eq. (6-44) and the response is plotted to scale in Fig. 6-11 for $\zeta = 3$. The initial curvature of the response resembling a parabola, is the transient part of the response. It is due to the last two terms in Eqs. (6-53) or (6-54). These terms decay in time and at some point along the response are much smaller than the sum of the first two terms. The last transient term becomes negligible first because its time constant is smaller than that of the second term. The response may then be represented by the first three terms. Eventually, however, the third term may be neglected and the response at that time is represented by the first two terms, which are associated with the steady state. At this time the response may be represented by the straight line (first two terms) parallel to the input ramp. The steady-state error being the difference between the input and the response, is equal to 2ζ on the normalized scale as predicted by Eq. (6-49). The response curve for $\zeta = 3$ in Fig. 6-11 differs from the input by not quite 6 units at $t_n = 15$. The steady state is obviously attained by this response for times greater than 15. On the other hand, the response curve for $\zeta = 1$ attains its steady state at t_n approximately equal to 4. It may be concluded that the transient time interval increases for larger values of ζ. The critically damped response attains its steady state in the shortest time without oscillations.

EXAMPLE 6–4

The second-order system transfer function is given by

$$\frac{C(s)}{R(s)} = \frac{10,000}{s^2 + 600s + 10,000}$$

and the input is

$$r(t) = 25tu(t)$$

Calculate the time where the difference between the given input and the response is 4 on the normalized scale, and the value of the response at this time.

SOLUTION Comparing the given transfer function with that of Eq. (6–14), the values of ω_n and ζ are obtained as

$$\omega_n = 100 \text{ s}^{-1}$$

$$\zeta = 3$$

From Fig. 6–11, the difference between the input ramp and the curve $\zeta = 3$ is 4 at $t_n = 6.1$ and $c_n(6.1) = 2.13$. These values must be denormalized to obtain the actual values of time and the response; hence

$$t = \frac{t_n}{\omega_n} = \frac{6.1}{100} = 61 \text{ ms}$$

and

$$c(0.061) = \frac{E}{\omega_n} c_n(6.1) = \frac{25}{100}(2.13) = 0.53$$

where the value of E, the slope of the input ramp, is given as 25 s^{-1}. When the values of $\zeta = 3$, $\omega_n = 100$, and $t = 0.061$ s are substituted into Eq. (6–53), the exact value of $c(0.061)$ is calculated as 0.57.

6–9 THE RELATIONSHIP BETWEEN THE POLES AND ZEROS IN THE S-PLANE AND THE CORRESPONDING RESPONSES IN THE TIME AND FREQUENCY DOMAINS

The location of the given system's poles and zeros on the pole–zero diagram dictates the behavior of the system not only in the time domain but also in the frequency domain. There exists, then, this not altogether unexpected relationship between the s plane, the time domain, and the frequency domain. As a consequence of this, it is possible to predict the behavior of a system in the time domain on the basis of measurements performed in the frequency domain, and vice versa.

Single-Pole System

We consider first the single-pole system characterized by the following transfer function:

$$\frac{C(s)}{R(s)} = \frac{1}{\tau s + 1} \tag{6-56}$$

This system has no finite zeros and a pole at $-1/\tau$ as shown in Fig. 6–12a, where the pole is designated as $-1/\tau_1$. The time-domain response of this system to a step input may readily be obtained by the method of Section 6–5 as

$$c(t) = E - Ee^{-t/\tau}$$

Figure 6–12b shows the total response $c(t)$ and Fig. 6–12c shows the transient part of the response, the second term in the equation above.

The frequency-domain response, the Bode plot, for the single-pole system is shown in Fig. 6–12d and e. The reader is referred to Chapter 7, where Bode plots are explained.

Suppose that τ is reduced from its original value τ_1 to τ_2. The corresponding changes in the frequency-domain and time-domain responses are shown by the dashed lines. The value of the pole $-1/\tau_2$ is now more negative; hence, effectively, the pole has shifted away from the $j\omega$ axis to its new position at $-1/\tau_2$. The transient response corresponding to the smaller time constant τ_2, as expected, now takes less time, allowing

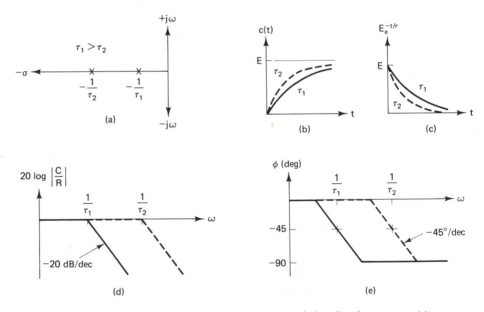

Figure 6–12. Single-pole system: (a) s plane; (b) total time-domain response; (c) transient response; (d) and (e) frequency-domain response. Solid response lines are associated with τ_1 and the dashed lines with τ_2.

the steady state to be attained sooner. The Bode plots are also affected. The 3-dB point and $-45°$ phase are shifted to a higher frequency.

The time-domain and frequency-domain responses are clearly related. It is thus possible to determine τ by measuring the break frequency and to predict the step response in the time domain. The s plane, which may seem trivial in this case, does offer a representation of the system in terms of its poles and zeros. In the sections that follow, the s plane will play a more important role.

Single-Pole, Single-Zero System

The system characterized by a single pole at $s = -1/\tau_2$ and a single zero at $s = -1/\tau_1$ may be described mathematically by the following transfer function:

$$\frac{C(s)}{R(s)} = \frac{(\tau_1)s + 1}{(\tau_2)s + 1} \tag{6-57}$$

The time-domain and frequency-domain responses differ depending on whether τ_1 is smaller or greater than τ_2. We consider first the case where τ_1 is greater than τ_2. The reciprocal of τ_1 is then less than that of τ_2 (as an absolute value). The pole and the zero are then on the negative real axis as shown in Fig. 6-13a, with the zero being closer to the origin or $j\omega$ axis.

The time-domain response to the step input $r(t) = Eu(t)$ may be obtained by following the procedure of Section 6-5 as

$$c(t) = E - E\left(1 - \frac{\tau_1}{\tau_2}\right)e^{-t/\tau_2} \tag{6-58}$$

From Eq. (6-58), the initial value of the response is

$$c(0) = E\frac{\tau_1}{\tau_2}$$

and the final value of the response results when t approaches infinity; hence

$$c_f = E$$

As $\tau_1 > \tau_2$, it follows that

$$c(0) > c_f$$

The response then instantly jumps to the initial value $c(0)$ and decays exponentially to the final value c_f with the time constant τ_2.

The frequency-domain response (the Bode plots) may be obtained by the methods of Chapter 7. The results are shown graphically in Fig. 6-13e and g. For convenience, the equations that represent these Bode graphs are repeated here. The magnitude (exact equation)

$$\left|\frac{C(j\omega)}{R(j\omega)}\right| = \left[\frac{(\omega\tau_1)^2 + 1}{(\omega\tau_2)^2 + 1}\right]^{1/2} \tag{6-59}$$

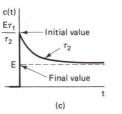
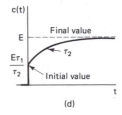

(a) (b)

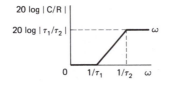
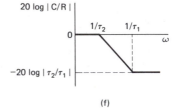

(c) (d)

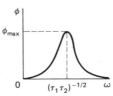
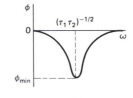

(e) (f)

(g) (h)

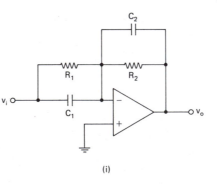

(i)

Figure 6–13. Single-pole, single-zero system: (a) s-plane diagram for $\tau_1 > \tau_2$ (zero is closer to $j\omega$ axis); (b) s-plane diagram for $\tau_1 < \tau_2$ (pole is closer to $j\omega$ axis); (c) time-domain response for $\tau_1 > \tau_2$; (d) time-domain response for $\tau_1 < \tau_2$; (e) and (g) Bode plots for $\tau_1 > \tau_2$; (f) and (h) Bode plots for $\tau_1 < \tau_2$; (i) single-pole, single-zero system example.

The magnitude graph in Fig. 6–13e does not represent Eq. (6–59); instead, it is the approximate asymptotic representation, with the lower break frequency being $1/\tau_1$ and the upper break frequency $1/\tau_2$. The phase

$$\phi(\omega) = \tan^{-1} \omega\tau_1 - \tan^{-1} \omega\tau_2 = \tan^{-1} \frac{\omega(\tau_1 - \tau_2)}{1 + \omega^2\tau_1\tau_2} \qquad (6-60)$$

The magnitude response is that of a high-pass filter with the lower break frequency determined by the numerator of Eq. (6–57) and the upper break frequency determined by the denominator for the case where $\tau_1 > \tau_2$.

The phase, on the other hand, does not have the simple linear response that is normally associated with the asymptotic representation. Its value is zero at both the low and high frequencies. It is always positive because $\tau_1 > \tau_2$; therefore, it must reach a peak at some finite, nonzero frequency. By differentiating Eq. (6–60) with respect to ω, equating the result to zero, and solving for ω_{max}, we obtain the frequency

$$\omega_{max} = \left(\frac{1}{\tau_1\tau_2}\right)^{1/2} \qquad (6-61)$$

which is the geometric mean (the square root of the product) of the two break frequencies, at which the peak phase occurs. The value of the peak phase is obtained by substituting ω_{max} in Eq. (6–60), with the following result:

$$\phi_{max} = \tan^{-1} \frac{\tau_1}{\tau_2} - \tan^{-1} \frac{\tau_2}{\tau_1} = \tan^{-1} \frac{\tau_1 - \tau_2}{2(\tau_1\tau_2)^{1/2}} \qquad (6-62)$$

In the case where $\tau_1 < \tau_2$, Eqs. (6–58) to (6–61) still apply; however, the time-domain and frequency-domain responses are completely different. As shown in Fig. 6–13b, the pole instead of the zero is now closer to the origin in the s plane. This makes all the difference.

The initial value of the time-domain response is now lower than the final value and the response exponentially increases toward the final value E with the time constant τ_2 (Fig. 6–13d).

The frequency-domain response is now also different. The lower break frequency is at $1/\tau_2$ and the upper break frequency is at $1/\tau_1$ because $\tau_1 < \tau_2$. The resulting magnitude versus frequency response (Fig. 6–13f) is that of a low-pass filter. The phase response (Fig. 6–13h) is now negative for all frequencies, as predicted by Eq. (6–60), reaching the negative peak at the geometric mean of the two break point frequencies. Equation (6–62), which was used to evaluate the maximum positive phase may also be used to evaluate the maximum negative phase.

An example of a simple single-pole, single-zero system is shown in Fig. 6–13i. The transfer function $V_o(s)/V_i(s)$ by standard techniques is equal to the ratio $-Z_2/Z_1$, where Z_2 is the parallel combination of R_2 and the capacitive reactance due to C_2 and Z_1 is the parallel combination of R_1 and the capacitive reactance due to C_1. R_2/R_1 represents the low-frequency gain. If $R_2 = R_1 = R$, it can be shown that the transfer function for this circuit (except for the negative sign, which is due to the amplifier inversion) may be represented by Eq. (6–57), where $\tau_1 = RC_1$ and $\tau_2 = RC_2$. By adjusting C_1 or C_2, the circuit can be made to behave as described in the preceding discussion.

The following important conclusions may be reached in regard to the characteristics of the single pole, single zero system:

The case $\tau_1 > \tau_2$

1. The s-plane diagram shows the zero closer to the origin in relation to the pole.

2. The Bode diagram shows that the phase is always positive, reaching the positive peak at the geometric mean of the two break frequencies, which are in turn determined by the pole and the zero. The system is therefore a phase-lead system (the output leads in phase the input), with maximum phase realized only at one frequency.

3. The Bode diagram shows also that the magnitude response is that of the high-pass filter, where the lower break frequency is determined by the zero and the upper break frequency is determined by the pole. Adjustment in τ_1 or τ_2, or both, causes the pole and the zero to shift along the negative real axis of the s plane as well as the break point frequencies on the Bode diagram. Adjustment of the pole or the zero closer to the origin lowers the corresponding break frequency, and vice versa.

4. When the asymptotic representation is used to obtain the magnitude versus frequency response, as has been done in Fig. 6–13e, then in the frequency range between $1/\tau_1$ and $1/\tau_2$ the denominator of the transfer function Eq. (6–57) reduces to unity and the numerator may be approximated by $\tau_1 s$, the entire transfer function reduces to $\tau_1 s$; consequently, $C(s) = (\tau_1 s)R(s)$. Multiplication by s in the s plane is equivalent to "differentiation" in the time domain. Thus the system behaves as a high-pass filter in the frequency domain and also acts as a differentiator in the time domain for the input frequency components which are between the two break frequencies. Below the lower break frequency the gain is unity (the low-frequency gain can be higher as demonstrated by the system in Fig. 6–13i), and above the upper break frequency the high-frequency gain is τ_1/τ_2.

5. The spikelike appearance of the time-domain response is characteristic of the differentiated output. The presence of high frequencies is implicit in the sharp leading edge and the spike portion of the response. It also follows that the absence of the flat top in the response represents the loss of low frequencies. This is supported by the fact that the system acts as a high-pass filter, amplifying the upper-frequency spectrum and attenuating the lower end.

6. Reducing the value of τ_2 causes the pole to shift farther away from the origin. The pole shifts toward the origin if τ_2 is increased. The zero similarly shifts away from the origin if the value of τ_1 is reduced and toward the origin if τ_1 is increased.

7. Shifting the pole away from the origin increases the amplitude of the spike and reduces the transient decay time, while shifting the pole closer to the origin reduces the spike and increases the transient decay time.

8. The pole, not the zero, controls the transient decay. The zero affects the initial amplitude of the spike. A shift in the zero toward the origin increases the spike, and away from the origin, reduces the spike; the transient decay remains unaffected.

9. A simultaneous shift of the pole toward the origin and the zero away from the origin until they coincide results in $\tau_1 = \tau_2$. In this case the transfer function, Eq. (6–57) reduces to unity. The corresponding Bode plot is very simple—the gain is unity for all frequencies and the phase is zero for all frequencies. In the time domain the response is the same as the input.

The case $\tau_1 < \tau_2$

1. In the s plane, the pole is closer to the origin in relation to the zero.

2. The phase is always negative, reaching the maximum negative peak at the geometric mean of the two break frequencies, which in turn are determined by the pole and the zero. The system is therefore a phase-lag system (the output lags the input in phase) with the maximum negative phase realized only at one frequency.

3. The magnitude versus frequency response is that of the low-pass filter, where the lower break frequency is determined by the pole and the upper break frequency is determined by the zero. Adjustment in τ_1 or τ_2, or both, results in a shift in the pole and the zero along the negative real axis in the s plane and a corresponding shift in the break frequencies in the frequency domain. Reduction in the value of the pole or the zero lowers the corresponding break frequencies, and vice versa.

4. In the frequency range $1/\tau_2 < \omega < 1/\tau_1$ the asymptotic form of the transfer function (6–57) reduces to $1/\tau_2 s$ [according to the asymptotic model, the numerator $1 + \tau_1 s$ is equal to unity below the break point $1/\tau_1$ and the term $1/(\tau_2 s + 1)$ in Eq. (6–57) is equal to $1/\tau_2 s$ above the break point $1/\tau_2$]; hence $C(s) = R(s)/\tau_2 s$. Division by s in the s plane is equivalent to "integration" in the time domain. Thus a system that is a low-pass filter in the frequency domain is also an integrator in the time domain for the input frequency components which are between the two break frequencies.

5. The rounded initial part of the response in Fig. 6–13d and a somewhat flat top implicitly indicates the presence of low frequencies and the absence of high frequencies. This must be so because the low-pass filter amplifies or passes unattenuated the low frequencies and attenuates the high frequencies.

6. As was the case with the high-pass filter, the pole controls the transient part of the response.

7. Shifting the pole away from the origin by reducing the value of τ_2 or shifting the zero toward the origin by increasing the value of τ_1 increases the initial value of the time-domain response in the direction which is toward the final value, and vice versa.

8. Shifting the pole away from the origin in the s plane or the zero toward the origin until the zero becomes closer to the origin in relation to the pole transforms the integrator system to the differentiator system, which was discussed previously.

9. Adjusting the values of the pole and the zero until they overlap (i.e., $\tau_1 = \tau_2$) reduces the transfer function to a unity with the result that the time-domain response becomes equal to the step input. The corresponding Bode plot reduces to a flat magnitude versus frequency response without phase shift.

Second-Order System

Before attempting to relate the s plane diagram to the performance of the second-order system in the time and frequency domains, it is important to understand the dynamic behavior of the second-order-system poles in the s plane. The second-order-system transfer function as expressed in Eq. (6–14) will be repeated here for convenience; hence

$$G(s) = \frac{C(s)}{R(s)} = \frac{\omega_n^2}{s^2 + 2\zeta\omega_n s + \omega_n^2} = \frac{\omega_n^2}{(s - s_1)(s - s_2)}$$

where

$$s_1 = -\zeta\omega_n + \omega_n\sqrt{\zeta^2 - 1}$$
$$s_2 = -\zeta\omega_n - \omega_n\sqrt{\zeta^2 - 1}$$

(6–63)

are the poles of $G(s)$. For the value

$$\zeta = 0$$

then from Eq. (6–63), $\beta = j$ and

$$s_1 = j\omega_n = \omega_n \underline{/90°} \qquad \text{(vector notation is described later)}$$
$$s_2 = -j\omega_n = \omega_n \underline{/-90°}$$

(6–64)

The poles in this case are complex conjugates with the real part equal to zero. The pole s_1 lies along the $+j\omega$ axis in the s plane and s_2 along the $-j\omega$ axis as shown in Fig. 6–14. When the values of ζ fall between 0 and 1, that is,

$$0 < \zeta < 1$$

then from Eq. (6–63),

$$s_1 = -\zeta\omega_n + j\omega_n\beta$$
$$s_2 = -\zeta\omega_n - j\omega_n\beta$$

where

$$\beta = \sqrt{1 - \zeta^2}$$

The negative real part $-\zeta\omega_n$ with imaginary part $\pm -\omega_n\beta$ restricts the location of the poles to the left-hand side of the s plane (i.e., to the left of the $j\omega$ axis and off the real axis, as shown in Fig. 6–14). For values of ζ between 0 and 1, the poles will always occur in pairs, the pole pair always being symmetrical about the negative real axis as shown in Fig. 6–14.

It is often convenient to represent the complex conjugate poles as vectors (real poles may also be represented in this manner). The pole s_1, then, may be represented by the vector $\mathbf{0s}_1$, which extends from the origin to the coordinates of pole s_1. Since a vector is a directed line segment, it must be presented by a magnitude and the angle.

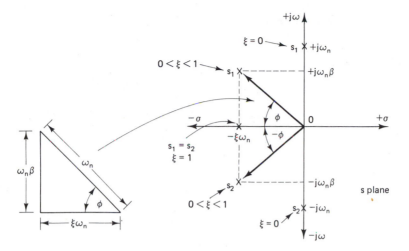

Figure 6–14. Second-order system pole configuration for values of ζ in the range $0 < \zeta \le 1$.

The magnitude of $\mathbf{0s}_1$ may be determined from the coordinates of the pole s_1 as

$$|\mathbf{0s}_1| = (\sigma^2 + \omega^2)^{1/2}$$
$$= [(\zeta\omega_n)^2 + (\omega_n\beta)^2]^{1/2} = \omega_n(\zeta^2 + \beta^2)^{1/2}$$
$$= \omega_n$$

and the angle of $\mathbf{0s}_1$, which is measured with respect to the negative real axis, is

$$\phi = \cos^{-1}\zeta$$

In Fig. 6–14, the right triangle is used to emphasize the relationship between the coordinates of the pole s_1 (the legs of the triangle) and the vector $\mathbf{0s}_1$, whose magnitude is represented by the hypotenuse of the triangle and whose angle is ϕ, as shown. Exactly the same triangle applies to the pole s_2 except that the angle becomes $-\phi$. In view of the above, the poles s_1 and s_2 may be represented by vectors as follows:

$$\mathbf{0s}_1 = \omega_n \underline{/\phi}$$
$$\mathbf{0s}_2 = \omega_n \underline{/-\phi}$$

(6–65)

When the system is critically damped, that is,

$$\zeta = 1$$

then from Eq. (6–63), $\beta = 0$ and

$$s_1 = s_2 = -\omega_n \qquad \text{or} \qquad \mathbf{0s}_1 = \mathbf{0s}_2 = \omega_n \underline{/0°}$$

(6–66)

The poles in this case are both real and equal and lie along the negative real axis as shown in Fig. 6–14.

When the system is overdamped, that is,

$$\zeta > 1$$

Then from Eq. (6–63),

$$\mathbf{0s}_1 = s_1 = -\zeta\omega_n + \omega_n\alpha = -\omega_n(\zeta - \alpha)$$
$$\mathbf{0s}_2 = s_2 = -\zeta\omega_n - \omega_n\alpha = -\omega_n(\zeta + \alpha)$$

$$(6-67)$$

where

$$\alpha = \sqrt{\zeta^2 - 1}$$

Both poles are real, unequal, and both are on the negative real axis. It is true that for all values of $\zeta > 1$,

$$\alpha = (\zeta^2 - 1)^{1/2} < \zeta$$

Hence

$$|\mathbf{0s}_1| < |\mathbf{0s}_2|$$

Consequently, the pole s_1 is closer to the origin than is s_2.

Figure 6–15 illustrates the dynamic behavior of the second-order system poles as ω_n is held constant and ζ is increased from 0 to ∞. Beginning at $\zeta = 0$, the system has no damping and the poles are on the $j\omega$ axis, in accordance with Eq. (6–64) and as shown in Fig. 6–15a.

As ζ is increased to any value below unity, the system becomes underdamped with complex conjugate poles restricted to the left-hand side of the s plane and off the real axis. Equation (6–65) expresses these poles as vectors in polar form, their length being equal to ω_n and their angles being a function of ζ. The length of each vector is fixed because ω_n is held at a constant value. The angle of $\mathbf{0s}_1$ decreases from its initial value of 90° at $\zeta = 0$ to 0° as ζ increases to 1 while the angle of $\mathbf{0s}_2$ varies at the same time from $-90°$ to 0°. Consequently, $\mathbf{0s}_1$ and $\mathbf{0s}_2$ rotate simulta-

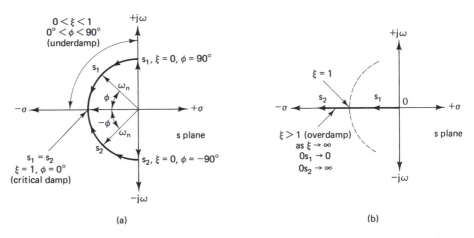

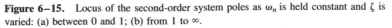

Figure 6–15. Locus of the second-order system poles as ω_n is held constant and ζ is varied: (a) between 0 and 1; (b) from 1 to ∞.

Time-Domain Analysis   Chap. 6

neously toward the negative real axis and in the process sweep out a semicircular arc which represents a locus of roots or pole pairs in the range $0 \leqslant \zeta < 1$. When ζ equals unity, the two vectors coincide along the negative real axis as predicted by Eq. (6–66) and the system becomes critically damped. Both cases are illustrated in Fig. 6–15a.

Equation (6–67) governs the location of the poles in the s plane as ζ exceeds unity for an overdamped system. Both poles are restricted to lie only on the negative real axis for $\zeta > 1$ as shown in Fig. 6-15b. Increasing ζ causes the two poles to recede in the opposite directions, s_1 toward the origin and s_2 away from the origin. As ζ increases without limit, magnitude $|0s_1|$ approaches 0, and $|0s_2|$ becomes infinite. The entire negative real axis thus becomes a locus of all possible pole pairs, which correspond to the values of ζ from 1 to infinity.

The contour in Fig. 6–15a was generated by maintaining ω_n at a constant value. Suppose that ω_n as well as ζ are allowed to vary over a given range of values. The result is a family of concentric circular contours whose radii are determined by the values of ω_n. As shown in Fig. 6–16, these contours are superimposed over a family of radial lines whose angles are determined by the values of ζ since $\phi = \cos^{-1} \zeta$. The circle of smallest radius corresponds to the lowest value of ω_n, while the radial line that is closest to the negative real axis is associated with the largest value of $\zeta < 1$. The negative real axis corresponds to the 0° line and the $j\omega$ axis to the 90° line.

This type of representation permits simultaneous changes in ω_n and ζ. A pole at A in Fig. 6–16, for example, whose coordinates are (ω_{n2}, ϕ_2), shifts to position B if ϕ_2 is changed to ϕ_3 and ω_{n2} to ω_{n4}.

As noted before, s may be expressed in rectangular coordinate form or in polar form as follows:

$$s = -\sigma \pm j\omega$$
$$= -\zeta\omega_n \pm j\omega_n\beta \qquad \text{where } \beta = \sqrt{1 - \zeta^2}$$
$$= \omega_n \underline{/\phi} \qquad \text{where } \phi = \cos^{-1} \zeta$$

Figure 6–16 illustrates the polar representation where s is expressed by a vector of length ω_n and angle ϕ which is determined by the value of ζ. Figure 6–17a, on the

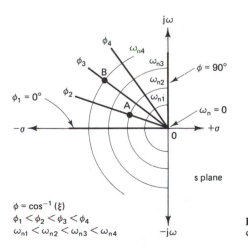

$\phi = \cos^{-1}(\xi)$
$\phi_1 < \phi_2 < \phi_3 < \phi_4$
$\omega_{n1} < \omega_{n2} < \omega_{n3} < \omega_{n4}$

Figure 6–16. Contours of constant ω_n and constant ζ in the s plane.

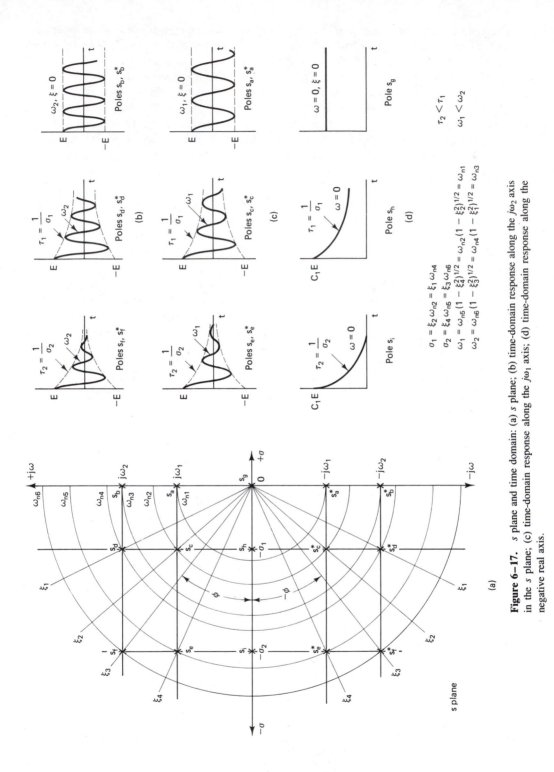

Figure 6–17. s plane and time domain: (a) s plane; (b) time-domain response along the $j\omega_2$ axis in the s plane; (c) time-domain response along the $j\omega_1$ axis; (d) time-domain response along the negative real axis.

$\sigma_1 = \xi_2\,\omega_{n2} = \xi_1\,\omega_{n4}$
$\sigma_2 = \xi_4\,\omega_{n5} = \xi_3\,\omega_{n6}$
$\omega_1 = \omega_{n5}(1 - \xi_4^2)^{1/2} = \omega_{n2}(1 - \xi_2^2)^{1/2} = \omega_{n1}$
$\omega_2 = \omega_{n6}(1 - \xi_3^2)^{1/2} = \omega_{n4}(1 - \xi_1^2)^{1/2} = \omega_{n3}$

$\tau_2 < \tau_1$
$\omega_1 < \omega_2$

other hand, shows the rectangular coordinate system (heavy lines) superimposed on the polar coordinate system, thus showing the relationship between the two forms. Both forms play an important role in establishing the relationship between pole location in the s plane and the response of the second-order system in the frequency and time domains.

The vertical lines are lines of constant real part (i.e., $\sigma = \zeta\omega_n$) and the horizontal lines are lines of constant frequency $\omega = \omega_n\beta$. Six complex conjugate pole pairs are included in the s plane of Fig. 6–17a: s_a, s_b, s_c, s_d, s_e, s_f and their respective conjugates. Also shown are three poles along the negative real axis: s_g, s_h, and s_i. Poles s_a and s_b are along the $j\omega$ axis; hence their real parts are zero while the imaginary parts of poles s_h and s_i are zero because they are on the real axis. Pole s_g is at the origin; therefore, its real and imaginary parts are both zero. The remaining poles have a nonzero real and imaginary parts. Poles s_c and s_d are on the same vertical coordinate line which intersects the real axis at $-\sigma_1$; hence they have the same real part, which is equal to $-\sigma_1$. The real part of poles s_e and s_f is $-\sigma_2$, for the same reason. Poles that lie on the same horizontal coordinate line must have the same imaginary part. This applies to poles s_c and s_e whose imaginary part is $j\omega_1$ and to poles s_d and s_f whose imaginary part is $j\omega_2$. The same applies to the conjugates of these poles.

Referring now to the polar coordinate representation of s whose position in the s plane is defined by the radial length ω_n and the angle that is determined by the value of ζ, the value of σ is equal to the product of ζ and ω_n. The constancy of σ, then, along any vertical line is assured as long as the product remains constant, despite the fact that both ζ and ω_n can vary individually. For example, the circle ω_{n2} and the radial line ζ_2 intersect at the pole s_c, which is on the vertical line $-\sigma_1$. The circle ω_{n4} and the radial line ζ_1 intersect at the pole s_d, which is also on the same vertical line. It follows, then, that $\sigma_1 = \zeta_2\omega_{n2} = \zeta_1\omega_{n4}$ because both of these poles have the same real part. In the same fashion, the constancy of $\omega = \omega_n(1 - \zeta^2)^{1/2}$ along any horizontal line is also maintained. This type of equality is summarized in Fig. 6–17a.

The relationship between the position of the second-order system poles in the s plane and the corresponding step response in the time domain may be established through the use of the s-plane diagram as shown in Fig. 6–17a and the results of the step response analysis which were developed in Section 6–7. Equations (6–19), (6–34), and (6–39) apply to the underdamped, critically damped, and the overdamped responses, respectively. The time-domain parameters which play an important role in establishing the desired relationship include:

$$\tau = \frac{1}{\zeta\omega_n}$$ time constant that dictates the rate of decay of the underdamped response in Eq. (6–19)

$$\omega = \omega_n(1 - \zeta^2)^{1/2} = \omega_n\beta$$ damped frequency of oscillation of the underdamped response in Eq. (6–19); depends on ζ and ω_n.

POT from Eq. (6–29) expresses the percent overshoot in the underdamped response. It depends only on the value of ζ.

$$\tau = \frac{1}{\omega_n(\zeta - \alpha)}$$ dominant time constant associated with the overdamped system step response in Eq. (6–39)

The parameters above clearly depend on the following parameters in the s plane:

$\zeta\omega_n$, the coordinate value along the negative real axis

$\omega_n\beta$, the coordinate value along the imaginary axis

ω_n, the radius of the circle

ζ, determines the angle of the radial line

Turning next to the time-domain response, the reader is referred to Fig. 6–17b, which shows three response curves along the constant $j\omega_2$ horizontal coordinate line. The curves represent the transient part of the response, the second term in Eq. (6–19). Clearly, this is the case of the underdamped response where the damped frequency of oscillation ω_2 is the same in all three cases, but the time constant that dictates the transient decay is different. The time constant τ_2 of pole f is less than the time constant τ_1 of pole d. The reason for this is due to the fact that the real part σ_2 of pole f is larger (as an absolute value) than the real part σ_1 of pole d; hence the transient part of the response which is associated with pole f decays to zero faster than that of pole d. Pole b is on the $j\omega$ axis, where the real part is zero because $\zeta = 0$ and consequently the time constant is infinite. This is the case of the undamped or purely sinusoidal type of a response. Thus the poles along the $j\omega_2$ line have the same imaginary part which determines the frequency of the transient term, but their real parts and hence their time constants are different; the poles closer to the $j\omega$ vertical axis have a larger time constant and a slower rate of decay of the transient term, as opposed to those poles which are farther away from the $j\omega$ axis.

Along the $j\omega_1$ line, the transient response associated with poles e, c, and a differs from the responses due to poles f, d, and b only in one respect: $\omega_1 < \omega_2$; hence the frequency of the transient response here is smaller.

Along the negative real axis $\omega = 0$; hence any poles on the real axis must have real values. The poles of the critically damped response are both real and equal because $\zeta = 1$. They may be represented by a pole such as i or h in Fig. 6–17. The value of both poles is equal to ω_n, and hence the time constant is the reciprocal of ω_n. The transient portion of the time-domain response, the second term in Eq. (6–34), is a decaying exponential which may be represented by one of the responses in Fig. 6–17d. Once again, as the poles recede from the origin, the time constant becomes smaller and the exponential decay faster.

For values of $\zeta > 1$, the time-domain step response is overdamped, both poles being real, unequal, and both located on the negative real axis. The transient portion of the time-domain response is described by the second and third terms in Eq. (6–39). The pole corresponding to the second transient response term is $-\sigma_1 = -\zeta\omega_n + \omega_n\alpha$ and $-\sigma_2 = -\zeta\omega_n - \omega_n\alpha$ is associated with the second transient term. As an absolute value, σ_1 is smaller than σ_2. The pole σ_1, then, is closer to the origin than is σ_2. The time constant, which is equal to the reciprocal of the real part of the pole (and in this case the value of the pole itself, because it is real), is therefore smaller, for the pole that is farther away from the origin and larger for the pole that is closer to the origin, as shown in Eq. (6–40). As a consequence of this, the exponential decay associated with σ_2, the pole farther from the origin, is faster than the exponential decay due to

σ_1. These responses are illustrated by poles h and i in Fig. 6–17d. As the value of ζ is increased toward infinity, one pole approaches negative infinity and the other approaches the origin, the time constants approaching zero and infinity, respectively, and the corresponding time-domain responses approaching a spike and a step, respectively.

The preceding discussion and theory is concerned mainly with the relationship between the s plane and the time-domain response of the second-order system. As noted before, from a practical standpoint it is important for the reader to understand the representation of the system in the s plane and its response in the frequency and time domains. We proceed, then, to relate the results above to the second-order-system response in the frequency domain. The reader is referred to Chapter 7 and to the developments established here.

We consider first the underdamped system, where

$$\zeta < 1$$

From Eq. (6–14),

$$\frac{C(s)}{R(s)} = \frac{\omega_n^2}{s^2 + 2\zeta\omega_n s + \omega_n^2} = \frac{\omega_n^2}{(s - s_1)(s - s_2)}$$

where the poles

$$s_1 = -\zeta\omega_n + j\omega_n\beta$$
$$s_2 = -\zeta\omega_n - j\omega_n\beta$$
$$\beta = (1 - \zeta^2)^{1/2}$$

Letting $s = j\omega$ and substituting s_1 and s_2 in the equation above, we obtain

$$\frac{C(j\omega)}{R(j\omega)} = G(j\omega) = \frac{1}{[\zeta + j(y - \beta)][\zeta + j(y + \beta)]} \tag{6–68}$$

$$= \frac{1}{(1 - y^2) + j(2\zeta y)}$$

The form of the result above is achieved by first dividing both the numerator and the denominator by ω_n^2 and then letting

$$y = \frac{\omega}{\omega_n}$$

where y represents the normalized frequency. The second part of Eq. (6–68) is obtained by multiplying the two bracketed terms and combining the real and imaginary parts. The complex form of Eq. (6–68) may be expressed in polar form (i.e., by magnitude and the angle) as follows:

$$G(j\omega) = M(y) \underline{/\phi}$$

where $M(y)$ is the magnitude of $G(j\omega)$:

$$M(y) = |G(j\omega)| = \{[\zeta^2 + (y - \beta)^2][\zeta^2 + (y + \beta)^2]\}^{-1/2}$$
$$= [(1 - y^2)^2 + 4\zeta^2 y^2]^{-1/2} \tag{6–69}$$

and ϕ is the phase of $G(j\omega)$:

$$\phi(y) = -\left(\tan^{-1}\frac{y-\beta}{\zeta} + \tan^{-1}\frac{y+\beta}{\zeta}\right)$$

$$= -\tan^{-1}\left|\frac{2\zeta y}{1-y^2}\right| \tag{6-70}$$

As $\omega \to 0$, then $y \to 0$ and

$$M(0) = 1 \quad \text{or} \quad M(0)_{\text{dB}} = 20\log(1) = 0 \text{ dB} \qquad y \ll 1 \tag{6-71}$$

on the other hand, as $y \to \infty$, the term y^4 in Eq. (6–69) dominates the magnitude; consequently, at high frequencies, M may be approximated by y^{-2}, or expressed in decibels,

$$M(y)_{\text{dB}} = 20\log|y^{-2}| = -40\log y \qquad y \gg 1 \tag{6-72}$$

This shows that when M_{dB} is plotted along the ordinate (linear scale) against y along the abscissa (logarithmic scale), then at high frequencies for each decade (10 times increase in frequency), M decreases by 40 dB in straight-line fashion. The slope of this straight line is -40 dB/decade. The straight line is the asymptote which M_{dB} approaches at high frequencies. The other asymptote, which was described above, is along the 0 dB line. The two asymptotes intersect at $y = 1$ because they both have the same value of 0 dB. The asymptotic (the approximate) system response is shown in Figure 6–18a by heavy lines.

On the other hand, the exact value of M at $y = 1$ from Eq. (6–69) is

$$M(1) = \frac{1}{2\zeta} \tag{6-73}$$

It is generally understood that an error (the difference between the asymptotic and the exact responses) of 3 dB is incurred for a single-pole system at the break frequency and a maximum of 6 dB for the double-pole system. Equation (6–73), on the other hand, suggests that as $\zeta \to 0$, M increases without limit. For values $0 < \zeta < 1$, M must reach a peak since it falls off above and below $y = 1$. If the peak occurs at $\omega = \omega_m$ or $y_m = \omega_m/\omega_n$, the slope of the line that is tangent to the response curve at $y = y_m$ is zero. Applying this to Eq. (6–69) yields

$$\frac{dM}{dy} = 0 = -\tfrac{1}{2}[(1-y^2) + 4\zeta^2 y^2]^{-3/2}[(-4y)(1-y^2) + 8\zeta^2 y]$$

The result above is equal to zero only if the numerator (the second bracketed term) is zero. Equating it to zero and solving for y, which is actually y_m, the following result is obtained:

$$y_m = \sqrt{1 - 2\zeta^2} \qquad \zeta \le \frac{1}{\sqrt{2}} \tag{6-74}$$

$$\omega_m = \omega_n\sqrt{1 - 2\zeta^2}$$

Equation (6–74) expresses the frequency at which the peak occurs. Practical reasons demand that the frequency be positive and real; hence ζ must be limited to values less

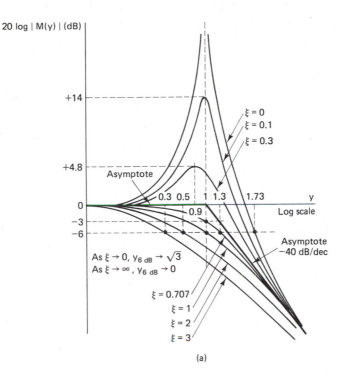

(a)

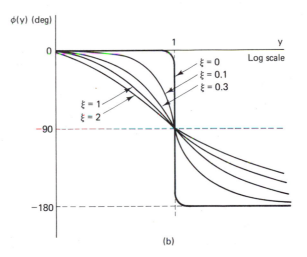

(b)

Figure 6–18. Second-order-system frequency-domain responses: (a) magnitude; (b) phase.

than 0.707. The value of y_m, the frequency at which the response peaks, clearly depends on the value of ζ, decreasing from 1 when $\zeta = 0$ to 0 when $\zeta = 0.707$. The response obviously does not peak for values of ζ above 0.707, as no real values of y_m exist for such values of ζ.

The expression for the peak value of the response is obtained by substituting Eq. (6–74) into Eq. (6–69):

$$M_{max} = M(y_m) = \frac{1}{2\zeta \sqrt{1 - \zeta^2}} = \frac{1}{2\zeta\beta} \qquad \zeta \leq \frac{1}{\sqrt{2}} \qquad (6\text{–}75)$$

The response is sketched for several values of ζ in Fig. 6–18a.

The phase of the underdamped response as defined by Eq. (6–70) approaches $0°$ as $y \to 0$ and $-180°$ as $y \to \infty$. At $y = 1$ it is equal to $-90°$. The phase is also a function of ζ, the phase response becoming more abrupt for lower values of ζ and more linear for larger values of ζ. Phase response is shown for several values ζ in Fig. 6–18b.

In the case of

$$\zeta = 1$$

the magnitude from Eq. (6–69) reduces to

$$M(y) = \frac{1}{1 + y^2} \qquad (6\text{–}76)$$

and the phase from Eq. (6–70) becomes

$$\phi(y) = -2 \tan^{-1} y \qquad (6\text{–}77)$$

The response corresponding to Eq. (6–76) is shown in Fig. 6–18a. At $y = 1$ ($\omega = \omega_n$), the response is down to -6 dB. This is the exact response (as opposed to the asymptotic response) of the critically damped system. The corresponding asymptotic response for the case of $\zeta = 1$ is shown by the heavy lines. The typical phase response for $\zeta = 1$ is shown in Fig. 6–18b in relation to other values of ζ.

For values of

$$\zeta > 1$$

the poles

$$s_1 = -\zeta\omega_n + \omega_n\alpha$$

and

$$\alpha = \sqrt{\zeta^2 - 1}$$

$$s_2 = -\zeta\omega_n - \omega_n\alpha$$

of Eq. (6–14) are real and unequal. Equation (6–14) may be expressed in the following form:

$$\frac{C(s)}{R(s)} = \frac{\omega_n^2}{(s - s_1)(s - s_2)}$$

$$= \frac{1}{[(s/\omega_1) + 1][(s/\omega_2) + 1]} \qquad (6\text{–}78)$$

where the break frequencies of the asymptotic response are

$$\omega_1 = \omega_n(\zeta - \alpha)$$
$$\omega_2 = \omega_n(\zeta + \alpha)$$

(6-79)

The frequency-domain representation is obtained in the usual way, by replacing s by $j\omega$ in Eq. (6-78). The magnitude may be expressed as

$$M(y) = \left| \frac{C(j\omega)}{R(j\omega)} \right|$$

$$= \{[(\zeta - \alpha)^2 + y^2][(\zeta + \alpha)^2 + y^2]\}^{-1/2}$$

(6-80)

$$= [(1 - y^2) + 4\zeta^2 y^2]^{-1/2} \qquad \text{where } y = \frac{\omega}{\omega_n}$$

The final form of the magnitude expression above is exactly the same as the final form of the magnitude in Eq. (6-69), which was derived for the underdamped case. The phase associated with the overdamped case may be expressed as follows:

$$\phi(y) = -\left(\tan^{-1} \frac{y}{\zeta - \alpha} + \tan^{-1} \frac{y}{\zeta + \alpha} \right)$$

(6-81)

$$= -\tan^{-1} \frac{2\zeta y}{1 - y^2}$$

The magnitude and phase responses may be represented in the frequency domain in one of two ways. By far the most common method is the asymptotic plot or graph, where the exact response is approximated by the straight-line asymptotes. As shown in Fig. 6-19a, the asymptotes intersect at the break frequencies, which are expressed in Eq. (6-79). The break frequencies are determined by each of the two real poles. Thus the gain is 0 dB up to the first break, decreasing then at the rate of -20 dB/decade up to the second break and at the rate of -40 dB/decade for all frequencies past the

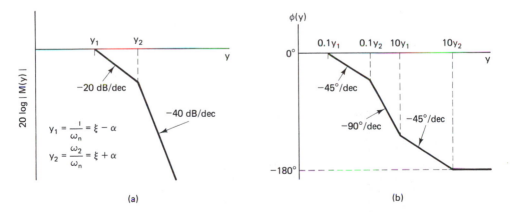

Figure 6-19. Second-order-system asymptotic responses: (a) magnitude; (b) phase ($\zeta > 1$).

second break. The asymptotic phase response is constructed in the usual way: Each pole spans two decades along the frequency axis (one decade above the break frequency and one decade below the break frequency) and 90° along the ordinate; the individual phase contributions are then summed as the frequency is increased, thus resulting in the asymptotic phase response as shown in Fig. 6–19b.

The exact magnitude and phase responses, which correspond to Eqs. (6–80) and (6–81), respectively, are shown in Fig. (6–18) for several values of ζ. Both the magnitude and the phase clearly depend on the value of ζ.

The form of the phase response, particularly in the neighborhood of $y = 0$ and $y = 1$, is best described by the slope $d\phi/dy$. The slope is obtained in the usual manner, by differentiating Eq. (6–70) with respect to y as follows:

$$\phi' = \frac{d\phi}{dy} = -\frac{d}{dy}\left[\tan^{-1}\left(\frac{2\zeta y}{1 - y^2}\right)\right]$$

$$= -\frac{(d/dy)\,[2\zeta y/(1 - y^2)]}{1 + [2\zeta y/(1 - y^2)]^2}$$

$$= -\frac{[(1 - y^2)\,(2\zeta y) + (2\zeta y)\,(2y)]/(1 - y^2)^2}{1 + [2\zeta y/(1 - y^2)]^2}$$

$$\phi'(y) = -\frac{2\zeta(1 + y^2)}{(1 - y^2)^2 + 4\,\zeta^2 y^2}$$

The slope at $y = 0$ and $y = 1$ may be expressed as

$$\phi'(0) = -2\zeta$$

and

$$\phi'(1) = \frac{-1}{\zeta}$$

These results indicate, as illustrated in Fig. 6–20, that the initial slope is directly related to the value of ζ, and the slope at $y = 1$ is inversely related to the value of ζ. It can also be shown that for large values of y,

$$\lim_{y \to \infty} \phi'(y) = 0$$

all phase response curves approach the $-180°$ asymptote, regardless of the value of ζ.

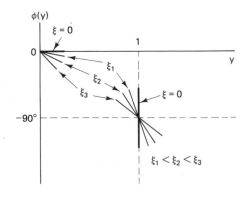

Figure 6–20. Character of the phase response in the neighborhood of $\zeta = 0$ and $\zeta = 1$.

Taking the critically damped response as a reference ($\zeta = 1$), its break frequency, the -6-dB point, occurs at $y = 1$. We can, then, predict quantitatively the variation of the -6-dB point with the value of ζ. The magnitude in Eq. (6–80) is -6 dB when the value of the discriminant is 4. Equating, then, the discriminant to 4,

$$(1 - y^2)^2 + 4\zeta^2 y^2 = 4$$

and solving the resulting equation

$$y^4 + 2(2\zeta^2 - 1)y^2 - 3 = 0$$

which is a quadratic in y^2, for y, the following result is obtained for the 6-dB point:

$$y_{6\,\text{dB}} = \sqrt{1 - 2\zeta^2 + \sqrt{(1 - 2\zeta^2)^2 + 3}} \qquad (6\text{–}82)$$

The value of $y_{6\,\text{dB}}$ decreases with increasing values of ζ; for example, as shown in Fig. 6–18a, the 6-dB point for the $\zeta = 2$ response curve occurs at $y = 0.5$. As $\zeta \to \infty$, $y_{6\,\text{dB}} \to 0$ [due to the dominance of the ζ^2 terms in Eq. (6–82) which cancel] and as $\zeta \to 0$, $y_{6\,\text{dB}} \to \sqrt{3}$. These observations are also supported by the asymptotic response results for large values of ζ. In Eq. (6–79), as $\zeta \to \infty$, the lower break frequency $\omega_1 \to 0$ and the upper break frequency $\omega_2 \to \infty$. The phase response increases in abruptness for lower values of ζ, as shown in Fig. 6–18b, becoming less abrupt as the value of ζ is increased.

Having established the necessary formalism in the time and frequency domains, we consider next the relationship between these domains and the location of the poles in the s plane. As shown in Fig. 6–21a, the shaded sector of the s plane represents all possible values of ζ between 0 and 0.707. Recalling the relationship $\phi = \cos^{-1} \zeta$, $\zeta = 0.707$ determines the 45° radial line, which bounds the sector on one end, and the $\zeta = 0$ line, which coincides with the $j\omega$ axis and limits the sector on the other end. In accordance with the preceding frequency-domain discussion, all poles of the second-order-system transfer function that fall within the shaded sector must have a resonant response M_{max} as defined in Eq. (6–75). The peak value of this response increases with decreasing values of ζ, approaching infinity as ζ approaches 0. The corresponding phase response becomes more abrupt (steplike) with decreasing values of ζ. Figure 6–21c and d show this type of a response for the complex conjugate poles s_1 and s_2. The resonant peak corresponding to s_1 is greater than that due to s_2 because ζ_1, which corresponds to s_1, is smaller than ζ_2.

The resonant peak in the frequency domain which occurs for values of $\zeta < 0.707$ has its dual in the time domain. As shown in Fig. 6–21b, the decaying oscillations of the underdamped response exhibit a maximum peak at T_{n1} for poles such as s_1 or s_2 which are in the shaded sector of the s plane. Other poles, such as s_3, also produce an oscillatory response. In fact, all poles that have a complex conjugate form with negative real part produce this type of decaying oscillatory response. However, since the first positive peak of the response is only a function of ζ, the pole with a lower value of ζ (or a larger angle ϕ) will produce a larger response peak C_{n1} than the pole with a larger value of ζ; hence s_1, which is on the line ζ_1, produces a larger peak than does s_2, and s_2 in turn produces a larger peak than does s_3.

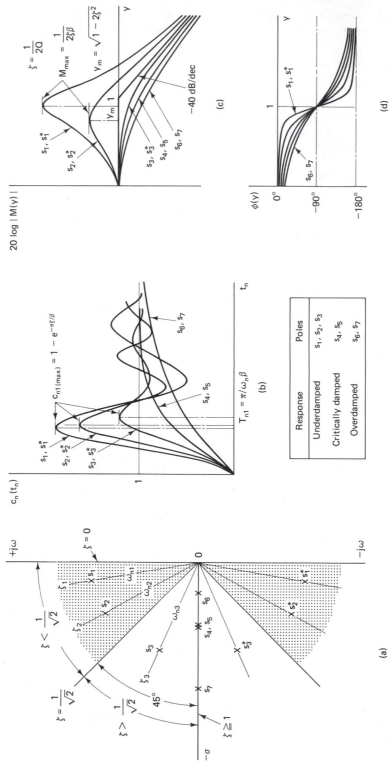

Figure 6–21. _s_ plane and the corresponding responses in the frequency and time domains: (a) _s_ plane (poles that are located in the dashed sector cause the resonant response in the frequency domain); (b) time-domain step responses due to the typical poles shown in the _s_ plane; (c) frequency-domain responses; (d) phase responses due to the same poles in the _s_ plane.

Thus an underdamped system exhibits resonance phenomenon in the time and frequency domains. In the time domain it occurs for all values of $\zeta < 1$, and in the frequency domain, for all values of $\zeta < 0.707$. The critically damped and overdamped responses that occur for $\zeta \geqslant 1$ assume an exponential form in the time domain and a logarithmically decaying form in the frequency domain, approaching the -40-dB/decade asymptote for sufficiently large frequencies above $y = 1$. The associated phase response assumes more of a linear transition with frequency as ζ increases. It may be noted that the -6dB point [defined by Eq. (6–82)] on the magnitude response curves in the frequency domain shifts toward the origin and the 90% point on the time-domain response curves shifts away from the origin as ζ increases above unity. This means that for increasing values of ζ above unity, the system is slower, taking longer to reach 90% of its steady-state value in the time domain, and in the frequency domain it exhibits a greater loss for larger values of ζ; that is, if one fixes attention on a particular value of frequency (or value of y), the magnitude response decreases below unity (or becomes more negative if expressed in decibels as ζ increases.

In the treatment of practical systems, we generally regard ζ as a parameter that is related to the energy losses within the given system. Energy losses or the power dissipated over a given time interval take the form of friction in mechanical systems. In electrical systems, on the other hand, they occur in resistive elements; that is, power dissipated by the resistor due to current flow radiates to the surroundings in the form of heat. This represents a loss of useful energy. Tuned circuits make use of Q, which is expressed as the ratio of energy stored (in the form of magnetic field due to current through the coil) to energy dissipated (in the resistive part of the coil) per cycle. Q is related to ζ by $\zeta = 1/(2Q)$ and thus may replace ζ in the coefficient of s in Eq. (6–14).

Resonance is a phenomenon that can occur in both electrical and mechanical systems. It is made possible by the presence of components that are capable of storing and expanding the stored energy. Spring is an example of such a mechanical component, and an inductor that is capable of storing energy in the form of a magnetic field by virtue of current flowing through it is an example of an electrical component. When the resonance condition occurs, several components within the system are in the process of exchanging energy, and the system response (displacement, for example, in a mechanical system or voltage across a tuned circuit in an electrical system) is sinusoidal in form, provided that the energy losses are sufficiently low. Addition of sufficient amount of damping or energy losses to the system can change the response from the sinusoidal to an exponential form. Both ζ and Q serve as indicators of such losses and may be used to predict the response type. Added damping, of course, increases the value of ζ, and hence the real part of the pole, thus causing the pole to shift in the direction away from the $j\omega$ axis in the s plane.

The preceding discussion offers a qualitative explanation and supports the type of time-domain responses shown in Figs. 6–7, 6–18, and 6–21 and frequency-do-main responses as shown in Figs. 6–18 and 6–21. The reader should bear in mind that a system whose energy losses are very low is characterized by complex conjugate poles, a time-domain oscillatory response which decays in time, and a resonant peak in the frequency domain.

EXAMPLE 6–5

The frequency-domain measurements on the second-order system reveal a 4.85-dB magnitude peak at 90.6 rad/s. Determine (a) the system transfer function, including all values; (b) the system poles in rectangular and polar form; (c) the value of the first peak and the time at which it occurs in the time domain and the period of the damped oscillations, assuming that the input is $10u(t)$; (d) the system settling time. The settling time for this particular system is defined as the time that it takes the time-domain response to be bounded for the first time by the region ± 0.03, which is centered on the step input $Eu(t)$. Do not assume the value of E.

SOLUTION

(a)
$$M_{max(dB)} = 4.85 = 20 \log M_{max}$$
$$M_{max} = 10^{4.85/20} = 1.748$$

From Eq. (6–75),

$$1.748 = \frac{1}{2\,\zeta\,\sqrt{1 - \zeta^2}}$$

Solving for ζ yields

$$\zeta = 0.3$$

From Eq. (6–74),

$$90.6 = \omega_n \sqrt{1 - 2(0.3)^2}$$

Hence

$$\omega_n = 100 \text{ rad/s}$$

Substituting these values in Eq. (6–14), the system transfer function becomes

$$\frac{C(s)}{R(s)} = \frac{10,000}{s^2 + 60s + 10,000}$$

(b)
$$\zeta\omega_n = 0.3(100) = 30$$
$$\omega_n\beta = 100(1 - 0.09)^{1/2} = 95.4$$
$$\phi = \cos^{-1} 0.3 = 72.5°$$

Using Eqs. (6–63) and (6–64), the poles may be represented in rectangular and polar forms as follows:

$$s_1 = -30 + j95.4 = 100 \underline{/75.4°}$$
$$s_2 = -30 - j95.4 = 100 \underline{/-75.4°}$$

(c)

From Eq. (6–29),
$$\frac{\pi\zeta}{\beta} = \frac{\pi(0.3)}{0.95} = 0.992$$
$$POT = 100e^{-(0.992)} = 37.1\%$$

Hence

$$\frac{C_{max} - 10}{10} = 0.371$$

$$C_{max} = 13.71 \qquad \text{response value at the first peak}$$

From Eq. (6–28),

$$T_1 = \frac{\pi}{100(0.95)} = 33.1 \text{ ms}$$

The period of damped oscillations

$$T_d = \frac{2\pi}{\omega_d} = \frac{2\pi}{\omega_n\beta} = 2T_1 = 27.42 \text{ ms}$$

(d) The term $e^{-\zeta\omega_n t}$ in Eq. (6–19) controls the rate at which the transient term decays. Since the steady-state value of C is E, it follows that when

$$|Ee^{-\zeta\omega_n t} \sin(\omega_n\beta t + \phi)| = 0.03E$$

the response is within the required limits. The equation above reduces to $e^{-\zeta\omega_n t} = 0.03$, and from this expression $\zeta\omega_n t = 3.5$ or $t = 3.5/\zeta\omega_n = 116.9$ ms. Thus the response requires 3.5 time constants to be within 3% of its steady-state value.

SUMMARY

1. Test signals are used to test a control system. Typical test signals include step, ramp and parabolic waveforms. These waveforms may be expressed by a general equation of the form $f(t) = At^n u(t)$, from which it is evident that the increasing powers of n beginning with $n = 0$ produce step, ramp, and parabolic waveforms. This suggests that all tests waveform may be generated from the step waveform through successive integration.

2. The system models are the differential equation and the transfer function. Any first-order system is described by a first-order differential equation, and a second-order system by the second-order differential equation.

3. The step response of a general first-order system is described by an exponentially rising waveform like that of a charging capacitor. The step response of a first-order feedback system is described also by a rising exponential; however, its dc gain and its time constant are affected by the system loop gain. The step response of a feedback first-order system is generally faster than that of its open-loop counterpart.

4. The step response of a second-order system depends on the value of system damping being oscillatory in character for damping ratios less than unity, and exponential when damping is in excess of unity. POT, T_{n1}, t_m, and T_r characterize the form of the underdamped response. Of the three responses, the underdamped response is the fastest. The overshoot and response time vary in opposite manner with the damping ratio.

5. The ramp response of a second-order system also exhibits oscillatory response damping ratios less than unity, and exponential response for damping in excess of

unity. In contrast to the step response which becomes equal to the input under the steady-state conditions, the ramp response differs from the input when steady-state conditions are reached. This difference, which depends on the value of the damping ratio, is related to the steady-state error in Chapter 5.

6. The location of system poles and zeros in the s plane dictates the character of the responses in the time and frequency domains. Thus there exists a unique relationship between the system response in the time domain and its response in the frequency domain. An underdamped system, for example, exhibits an overshoot to a step input in the time domain and a resonant peak in the frequency domain. The value and the time of the overshoot peak, as well as the value and frequency of the resonant peak, are related to the value of system damping. It is therefore possible to predict the time-domain response on the basis of the frequency-domain measurements, and vice versa.

QUESTIONS

6-1. Determine a circuit that may be used to generate a ramp waveform from a step input, and a parabolic waveform from a ramp input. Can the parabolic waveform be derived directly from the step input? Explain.

6-2. Determine a circuit that may be used to generate a ramp waveform from a parabolic input waveform and a step waveform from a ramp input.

6-3. Define a pole and a zero.

6-4. State the two models and explain the significance of the associated parameters for **(a)** a first-order system; **(b)** a second-order system.

6-5. Explain the effect of loop gain on the response time and the dc gain of a first-order feedback system.

6-6. Consider the step response of an underdamped second-order system. Which response parameters does the damping ratio affect? (Include all.) Which parameters does ω_n affect? Explain.

6-7. Why is it not possible to achieve a very fast underdamped response without overshoot? Explain.

6-8. What practical information regarding the performance of a second-order system may be extracted from the step and ramp input tests? Explain.

6-9. Compare and contrast the step and ramp responses of an underdamped second-order system on the basis of **(a)** response wave shape; **(b)** response mathematical form; **(c)** system performance characteristics.

6-10. Explain the effect due to the relative position of the pole and the zero in the s plane on the time- and frequency-domain responses of a first-order system. What would be the difference in the time- and frequency-domain performance of two systems characterized by their transfer functions $(s + 5)/(s + 10)$ and $(s + 10)/(s + 5)$?

6-11. Explain the effect due to the relative position of a pole in the s plane on the time- and frequency-domain responses of a second-order system **(a)** underdamped; **(b)** critically damped; **(c)** overdamped.

6-12. Suppose that a charged capacitor is suddenly connected across an ideal inductor (no resistance),

(a) Sketch the voltage across the parallel combination versus time.

(b) Calculate the value of system damping.

(c) Calculate the frequency of the waveform in terms of circuit parameters.

(d) How would the response differ if a real inductor were used?

(e) What additions to the circuit must be made to convert the waveform obtained with the real inductor to that obtained with the ideal inductor? Explain and include the concept of energy conservation into your explanation.

(f) What is the order of this simple system? Why?

PROBLEMS

6–1. The values of parameters for the system in Fig. 6–5 are $K_1 = 100$, $\tau_1 = 0.4$ s, $K_f = 0.5$, the input $r(t) = 5u(t)$, and $c(0) = 0$.

(a) Calculate the steady-state value of the response.

(b) Calculate the value of the response at $t = 5$ ms.

(c) Determine the expression for the system differential equation using the given values.

6–2. Given the system transfer function

$$\frac{C(s)}{R(s)} = \frac{100}{s^2 + Ks + 100}$$

the input is $r(t) = 10u(t)$. If the value of K is adjusted to 3, calculate **(a)** POT; **(b)** the values of the first three consecutive response peaks and the corresponding times at which they occur; **(c)** the value of the response at $t = 250$ ms; **(d)** system settling time; **(e)** system poles in rectangular form and in polar form or as vectors. **(f)** Sketch the response showing all important values and identify the steady state and the transient part of the response. **(g)** Determine the new value of K for a critically damped response.

6–3 The response of the system shown in Fig. P6–3a to the input $v_i(t) = -2u(t)$ is as shown in Fig. P6–3b. Calculate the values of R and C.

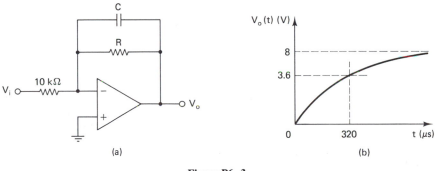

Figure P6–3.

6–4. If the input to the system in Problem 6–3 is $v_i(t) = 10tu(t)$, calculate the values of R and C such that the response is able to track the input and the steady-state delay is 100 μs. Sketch the response and show all important values.

6–5. The value of R is adjusted to 10 kΩ and C to 0.01 μF in the circuit of Problem 6–3 and a 0.05-μF capacitor is connected across the input 10-kΩ resistor. Sketch **(a)** the time-domain response to the input $v_i(t) = -5u(t)$; **(b)** the frequency-domain response, both the magnitude and the phase. In both cases include in the responses important numerical values which are necessary to characterize the responses.

6–6. Repeat Problem 6–5 for the 0.002-μF capacitor that is connected across the 10-kΩ input resistor.

6–7. Determine the value of C that is to be placed across the 10-kΩ input resistor in Problem 6–5 to achieve a maximum phase lead of 45°.

6–8. The response of the second-order system to the input $r(t) = 2u(t)$ is as shown in Fig. P6–8. Determine the system transfer function which includes parameter values such as ζ, ω_n, and so on.

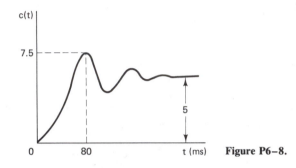

Figure P6–8.

6–9. Suppose that the response of the system whose block diagram is as shown in Fig. P6–9 is the same as the response of Problem 6–8; the same input is also assumed. Determine the values of A, B, and D.

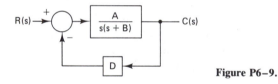

Figure P6–9.

6–10. Calculate the settling time for the system in Problem 6–9.

6–11. Given the second-order system response equation

$$c(t) = 5[1 - xe^{-yt} \sin (500t + 1)]$$

(a) Calculate the values of x and y.
(b) Sketch the input waveform $r(t)$; include values.
(c) Determine the system transfer function using the parameter values.

6–12. Suppose that the time-domain response given in Problem 6–11 is generated by the system of Problem 6–9. Assume that the input of Problem 6–9 applies. Determine the values of system parameters A, B, and D.

6–13. The response of the second-order closed-loop system of the type shown by the block diagram in Problem 6–9 to the ramp input is of the following form:

$$c(t) = 10t - 0.024 + \frac{12.5}{a} e^{-abt} \sin \left(\frac{4at}{5} + c \right)$$

(a) Determine the values of A, B, and D in the block diagram of Problem 6–9.
(b) Using all parameter values, determine the system transfer function.

(c) Sketch the response of this system to the input $r(t) = 10u(t)$. Show on the response curve all important values such as C_{max}, T_1, T_s, and so on.

6–14. The second-order system is defined by the transfer function of Problem 6–2. Suppose that K takes on the following values: 0, 1, 5, 10, 20, and 40. For each value of K show the position of the poles in the s plane, and sketch the corresponding time- and frequency-domain responses. Show all critical values on the response curves.

6–15. In the closed-loop system configuration shown in Fig. P6–15, $R_1 = R_4$, $R_2 = R_3$, and $R_7 = R_8$ and let $K = (R_5 + R_6)/R_7$, $\tau_1 = R_1C_1$, and $\tau_2 = R_2C_2$. Draw the system block diagram. (Assume ideal op amps)

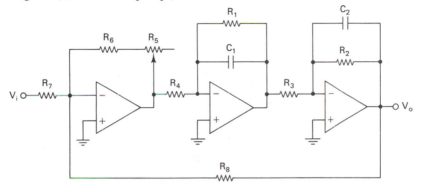

Figure P6–15.

(a) Show that the closed-loop system transfer function may be expressed as

$$\frac{V_o(s)}{V_i(s)} = \frac{K/\tau_1\tau_2}{s^2 + (1/\tau_e)s + (K + 1)/\tau_1\tau_2} \qquad \text{where } \tau_e = \frac{\tau_1\tau_2}{\tau_1 + \tau_2}$$

(b) Show that the system uses negative feedback. What is the value of the feedback?

(c) Show that

$$\zeta = \frac{1 + x}{2\sqrt{x(K + 1)}} \qquad \text{and} \qquad \omega_n = \sqrt{\frac{K + 1}{\tau_1\tau_2}} \qquad \text{where } x = \frac{\tau_1}{\tau_2}$$

Also show that as long as the gain K is above unity, it is not possible to achieve a critically damped or overdamped response for any combination of τ_1 and τ_2, and if K is held constant, the maximum value of ζ occurs at $x = 1$.

(d) If C_3 is connected across R_3, show that the resulting closed-loop system transfer function may be expressed as

$$\frac{V_o(s)}{V_i(s)} = \frac{(K/\tau_1\tau_2)(\tau_3 s + 1)}{s^2 + (1/\tau_e + \tau_3/\tau_1\tau_2)s + (K + 1)/\tau_1\tau_2} \qquad \text{where } \tau_3 = R_3C_3$$

Explain the effect of the added zero on the system step response. Is it possible now to achieve a critically damped or an overdamped response for $K > 1$?

(e) Using the values $\tau_1 = 2$ ms, $\tau_2 = 1$ ms, $K = 50$, and the step input $v_i(t) = 5u(t)$, sketch the time-domain response and the corresponding frequency-domain response showing all important values on the response curves. Repeat the above for $K = 35$. (Assume that C_3 is not included.)

(f) Including the value of C_3 such that $\tau_3 = 400$ μs and using the values of τ_1 and τ_2 and the input from part (e), sketch the time- and frequency-domain responses for three values of K: $K = 10$, 35, and 50.

LAB EXPERIMENT 6–1
SECOND-ORDER SYSTEM

Objective. To evaluate the performance of a second-order system. The system selected for this experiment is an active second-order Butterworth filter as shown in the test circuit. The step and the ramp response tests are done to investigate the accuracy and the transient behavior of an underdamped second-order system. The purpose of the frequency response tests, on the other hand, is to explore the resonant behavior of a second-order system. The relationship between the frequency- and time-domain responses is also considered.

Equipment

Dual-trace oscilloscope
Digital voltmeter ad/dc
Frequency counter
Pulse/ramp generator
Variable-frequency sinusoidal generator
Digital voltmeter

Materials. As shown in the test circuit.

Procedure

1. Null out the offset voltage of the operational amplifier and adjust $R_1 = R_2 = 50$ kΩ in the test circuit shown in Fig. LE6–1.

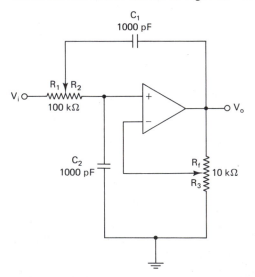

Figure LE6–1. Test circuit.

2. Adjust the value $R_f/R_3 = 2.0$, and apply a 50% duty cycle, 1-V_{peak} (0 to +1 V), 500-Hz square wave to the input V_i. Monitor V_i and V_o on a dual-trace oscilloscope.

Measure and record the following: (a) frequency f of the oscillatory response; (b) the value of the first peak V_{o1}; (c) the time T_1 of the first peak; (d) T_s based on 10% band.

3. Repeat step 2 for the following values of R_f/R_3: 1.95, 1.8, to 0.2 in steps of 0.2, 0.1, and 0.05.

4. Adjust the value $R_f/R_3 = 2.0$, and apply a 5-V_{peak} (0 to +5 V), 1-kHz ramp or a triangular waveform to the input V_i. Measure and record the steady-state error e_{ss}. To ensure the accuracy of this measurement, a difference amplifier should be used (construct one if not available). Remember that the dc gain of the test circuit is $A_v = 1 + R_f/R_3$; hence in measuring e_{ss} the output should be scaled by A_v and then compared to V_i. Some oscilloscopes may also be used in a differential mode.

5. Repeat step 4 for each of the R_f/R_3 values as specified in step 3.

6. Adjust the value $R_f/R_3 = 1.90$, and apply a 0.1-V_{rms}, 250-Hz sinusoidal waveform to V_i. Measure and record V_o on a digital ac voltmeter, and f on a frequency counter.

7. Repeat step 5 for frequencies of 250 Hz to 6 kHz in steps of 250 Hz, and from 6 to 42 kHz in steps of 3 kHz. Measure and record V_o and f at each test frequency.

8. Repeat steps 5 and 6 for the following values of R_f/R_3: 1.80, 1.60, 1.0, and 0.20. Measure and record all data.

Test data. The format for the test data and the calculated results, as well as the report format, must comply with and meet the requirements of your lab instructor.

Evaluation. As the ultimate objective of any lab experiment is to verify the theory-based predictions through measurements, significant differences between theory and measurements (generally, differences over 10%) must be resolved through additional experimentation, investigation, or retest.

1. Derive the expression for the given test circuit transfer function in the form $V_o(s)/V_i(s) = (K_1 K_2)/(s^2 + K_3 s + K_2)$ as suggested by Eq. (6–14), and from the result, extract the expressions for ω_n and ζ in terms of circuit parameters.

2. Using the test data of steps 2 and 3 of the Procedure, calculate the value of POT for each value of R_f/R_3 and record the calculated POT values in your data sheet along side the corresponding R_f/R_3 values. The dc gain of the test circuit is $A_v = 1 + R_f/R_3$; hence V_{o1} must be scaled by A_v before it can be compared to V_i in the POT calculation.

3. Plot on suitable graph paper the POT values obtained in step 2 as a function of R_f/R_3 (these POT values are based on measurements). Include on the graph the POT versus R_f/R_3 curve based on theory; use the expression for ζ from step 1 in POT calculations (this is the theory-based POT curve). Also plot on the graph the ζ versus R_f/R_3 curve by adding the second scale on the right side of the graph in the manner of Fig. 6–8 (where the right-hand scale is used for T_{n1}). Use the expression from step 1 for this purpose.

4. Plot on suitable graph paper T_1 as a function of R_f/R_3 using the measured values

of T_1. Include on the graph the T_1 versus R_f/R_3 curve based on theory (use the information from step 1).

5. Plot on suitable graph paper the damped frequency ω_d (based on measurements) as a function of R_f/R_3. Include on the graph the ω_d versus R_f/R_3 curve based on theory.

6. Plot on suitable graph paper the settling time T_s using the measured values as a function of R_f/R_3. Include on the graph the T_s versus R_f/R_3 curve based on theory.

7. Plot on semilog paper the normalized gain $|A_v(f)/A_{vo}|$ in decibels as a function of frequency f in hertz, where $A_v(f) = V_o(f)/V_i$ is the test circuit gain at a frequency f, and A_{vo} is the dc gain of the test circuit, using the data corresponding to $R_f/R_3 = 1.90$. Include on the same graph the frequency response curve for $R_f/R_3 = 1.90$ based on theory (see the text).

8. Repeat step 7 for each value of R_f/R_3 used in step 8 of the Procedure.

9. Plot on suitable graph paper the steady-state error e_{ss} using data from steps 4 and 5 of the Procedure as a function of R_f/R_3. Include on the graph the e_{ss} curve, which is based on theory.

10. Calculate POT, T_1, and T_s in the time domain using only the frequency response data from steps 6 to 8 of the Procedure for $R_f/R_3 = 1.8$. How do these values compare with the corresponding values obtained from the time-domain response?

11. Calculate the peak value $|V_o(f_m)/V_i|$ in decibels and the frequency f_m at which the peak occurs for $R_f/R_3 = 1.8$ in the frequency domain using only the time-domain response data from steps 2 and 3 of the Procedure. How do these values compare with the corresponding values obtained from the frequency-domain response tests?

12. Repeat steps 9 and 10 for $R_f/R_3 = 1.6$.

The responses to the following questions and/or conclusions must be supported whenever possible by the applicable theory.

13. The theory requires that the system damping ratio ζ vary in an inverse manner with the forward gain of the system. Do the results of step 1 support this requirement? How does ζ depend on R_f/R_3?

14. Draw conclusions on the basis of data regarding the effect of R_f/R_3 on (a) POT; (b) T_1; (c) T_s; (d) e_{ss}; (e) ω_d; (f) frequency response resonant peak M_m and the frequency of f_m of the resonant peak.

15. Would your conclusions in question 14 be the same if R_f/R_3 is replaced by (a) ζ; (b) forward gain of the test circuit? Explain.

16. Discuss the correlation between the frequency and time domains on the basis of results in questions 9 to 11, and the practical importance of such correlation.

17. What effect does the dc gain of the test circuit have on POT, T_1, T_s, ω_d, e_{ss}, M_m, and f_m?

18. For what value of R_f/R_3 does the system become critically damped? undamped?

19. How is the natural resonant frequency ω_n affected by variations in R_f/R_3?

20. Which system components have a dominant effect on POT, and which components have a small effect on POT?

21. Determine the effect on POT, M_m, T_1, f_m, and ω_n due to each of the following changes in the test circuit parameters: (a) R_1 and R_2 each double in value; (b) R_2 doubles and R_1 halves; (c) R_1C_1 and R_2C_2 each double; (d) C_1 and C_2 double; (e) C_2 doubles and C_1 halves. Use theory to answer these questions and express changes in POT, M_m, T_1, f_m, and ω_m as a percent of their initial values which are based on the test circuit parameter values as shown. Assume that $R_f/R_3 = 1.6$ and remains constant.

FREQUENCY-DOMAIN ANALYSIS

chapter 7

7–1 INTRODUCTION

In Chapter 6 we presented time-domain analysis of various systems. In this chapter we discuss frequency-domain analysis and its significance for control systems. Frequency-domain analysis is the set procedure that is used to determine the way the circuit behaves in response to the variation in its input frequency over a desired operating range.

Frequency-domain analysis is the procedure that is embodied in the following five steps.

1. For a given network identify the type of input to be used and the desired output.
2. Obtain the transfer function for the network in the *s* domain, that is,

$$T(s) = \frac{O(s)}{I(s)}$$

where $T(s)$ = desired transfer function
$O(s)$ = Laplace transform of an output
$I(s)$ = Laplace transform of an input

Note that all initial conditions are assumed zero in determining the transfer function $T(s)$.

3. Substitute $s = j\omega$ in the transfer function $T(s)$.

4. Vary the value of angular frequency ω over a desired operating range and determine the corresponding values of magnitude and phase angle of the transfer function $T(s)$.

5. Plot the magnitude versus phase angle plot in polar coordinates. Such a plot is referred to as a *polar plot*.

An alternative method is to plot the magnitude versus frequency and phase angle versus frequency as two independent plots. These plots are referred to as *Bode plots*.

The term *frequency response* represents the steady-state response of a system subjected to a sinusoidal input of fixed amplitude but variable frequency. For example, in an amplifier block diagram of Fig. 7–1a an input is $A \sin \omega t$, and the output is $B \sin (\omega t + \theta)$, where A and B are amplitudes of input and output signals, respectively, and θ is the phase angle difference between the input and output signals. The ratio of amplitudes B/A and the phase angle θ are both functions of angular frequency ω. The ratio B/A is defined as the *magnitude ratio* and will be denoted by $M(\omega)$. Also, the phase angle will be designated by $\theta(\omega)$.

Frequency-domain analysis of linear systems as such can be performed by using one or more of the following methods:

1. Polar plot–Nyquist diagram (Nyquist diagram is an extension of the polar plot)
2. Bode plots
3. Nichols chart

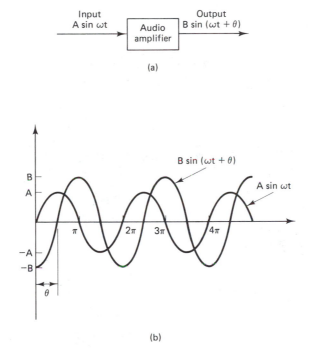

(a)

(b)

Figure 7–1. (a) Audio amplifier; (b) its input and output waveforms.

The Nyquist diagram and Bode plots require the open-loop transfer function of a control system to determine its frequency response. The Nichols chart, however, is constructed using the data off Bode plots and is used for the frequency response determination of closed-loop systems. The main advantage in using the Bode plots rather than the Nyquist diagram is in the relative ease of plotting. However, the steps required to obtain the data for the Nyquist plot and Bode plots are pretty much identical. For this reason we shall briefly examine the polar plot concepts, which will also help to identify the basic differences between the polar plot and the Bode plots.

7–2 POLAR PLOT

Let us consider the active first-order low-pass filter shown in Fig. 7–2a. To perform frequency-domain analysis, we will follow the five steps outlined earlier.

1. For the low-pass filter shown in Fig. 7–2a, the input is $A \sin \omega t$ and the output is $v_o(t)$.

2. Determine the transfer function $V_o(s)/V_i(s)$.

$$V_o(s) = \frac{V_i(s)A_F}{RCs + 1}$$

where $A_F = 2$ and $RC = 1$ ms. Therefore,

$$\frac{V_o(s)}{V_i(s)} = \frac{2000}{s + 1000} \tag{7–1}$$

3. Substituting $S = j\omega$ in Eq. (7–1), we get

$$\frac{V_o(j\omega)}{V_i(j\omega)} = \frac{2000}{j\omega + 1000}$$

4. Obtain the magnitudes and phase angles as the value of ω is varied from 0 to ∞ rad/s. Hence from Eq. (7–1) we have

$$\left| \frac{V_o(j\omega)}{V_i(j\omega)} \right| = \frac{2000}{\sqrt{\omega^2 + 10^6}}$$

or

$$M(\omega) = \frac{2000}{\sqrt{\omega^2 + 10^6}} \tag{7–2a}$$

and

$$\theta(\omega) = -\tan^{-1} \frac{\omega}{1000} \tag{7–2b}$$

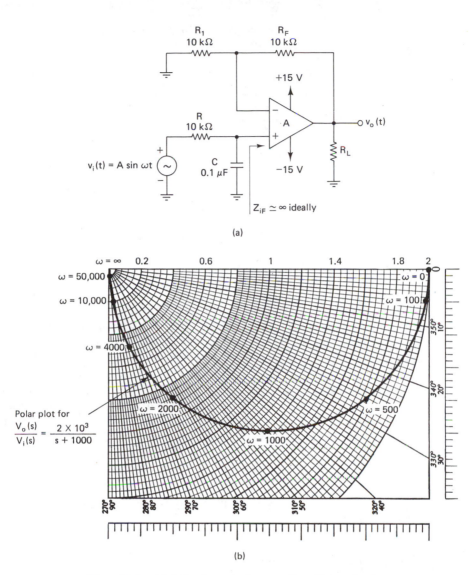

Figure 7–2. (a) Active first-order low-pass filter; (b) its polar plot.

Table 7–1 is constructed to include the magnitude ratio $M(\omega)$ and the phase angle $\theta(\omega)$ for different values of ω.

5. The polar plot is plotted using the data of Table 7–1 on the complex plane with ω as a parameter (variable). The resultant plot is shown in Fig. 7–2b.

The procedure above illustrates how to draw a polar plot which is the basis for the Nyquist diagram.

Angular frequency, ω (rad/s)	Magnitude, $M(\omega)$	Phase angle, $\theta(\omega)$ (deg)
0	2	0
100	1.99	−5.71
500	1.79	−26.57
1,000	1.41	−45
2,000	0.89	−63.43
4,000	0.49	−75.96
10,000	0.20	−84.29
50,000	0.04	−88.85
100,000	0.02	−89.43

7–3 BODE PLOTS

Another commonly used method to represent the frequency response data is the Bode plots or the rectangular-coordinate form. Bode plots constitute two plots, magnitude ratio $M(\omega)$ versus angular frequency ω and phase angle versus frequency ω. It is customary and convenient to plot magnitude ratio as $\log_{10} M(\omega)$ against ω and phase angle $\theta(\omega)$ against ω. The magnitude ratio is generally expressed in *decibels*. The decibels versus ω plot is referred to as a *magnitude plot*, in which the magnitude ratio in decibels is plotted on a vertical axis using a linear scale; and the angular frequency ω is plotted on a horizontal axis using a logarithmic scale. Thus the magnitude plot is made on semilog paper. Similarly, the phase angle $\theta(\omega)$ versus ω plot is referred to as a *phase angle plot* in which the phase angle $\theta(\omega)$ is plotted as a function of ω. The phase angle plot is also made on semilog paper with $\theta(\omega)$ in degrees on the vertical axis using a linear scale and ω on the horizontal axis using a logarithmic scale. Since both the magnitude and the phase angle plots require semilog paper, they can be made on the same sheet.

7–4 CONCEPTS OF BODE PLOTS

Let us reconsider the typical closed-loop control system represented by the block diagram of Fig. 7–3. The *closed-loop transfer function* is

$$T(s) = \frac{C(s)}{R(s)}$$

$$= \frac{G(s)}{1 + G(s)H(s)} \tag{7–3}$$

where $G(s)$ is the forward transfer function or the transfer function of the block(s) in the forward path, and $H(s)$ is the feedback transfer function or the transfer function of the block(s) in the feedback path. $G(s)$ is also referred to as the *open-loop transfer*

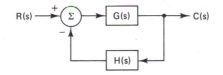

Figure 7–3. Block diagram of a typical closed-loop control system.

function; in addition, $G(s)H(s)$ is the *loop transfer function* and the *characteristic equation* is defined as

$$1 + G(s)(Hs) = 0 \qquad (7-4)$$

Almost all graphical techniques, including Bode plots and the Nyquist plot, require the loop transfer function $G(s)H(s)$. The next step is to substitute $s = j\omega$ in the transfer function $G(s)H(s)$. However, when s is replaced by $j\omega$, the transfer function becomes a complex variable, that is,

$$G(j\omega)H(j\omega) = A(\omega) + jB(\omega)$$

where $A(\omega)$ is the real part and $jB(\omega)$ is the imaginary part. For example, let

$$G(s)H(s) = \frac{2(s)}{s+1}$$

Replacing $s = j\omega$ yields

$$G(j\omega)H(j\omega) = \frac{2(j\omega)}{1 + j\omega}$$

To show that $G(j\omega)H(j\omega)$ is a complex variable, we multiply and divide $G(j\omega)H(j\omega)$ by $(1 - j\omega)$, the complex conjugate of $(1 + j\omega)$. Therefore,

$$G(j\omega)H(j\omega) = \frac{2(j\omega)}{(1 + j\omega)} \frac{(1 - j\omega)}{(1 - j\omega)}$$

$$= \frac{2(j\omega + \omega^2)}{1 + \omega^2}$$

$$= \frac{2\omega^2}{1 + \omega^2} + \frac{2\omega}{1 + \omega^2}$$

$$= A(\omega) + jB(\omega)$$

where

$$A(\omega) = \frac{2\omega^2}{1 + \omega^2} \qquad \text{and} \qquad B(\omega) \frac{2\omega}{1 + \omega^2}$$

Also, if we take the logarithm of $G(j\omega)H(j\omega)$, according to complex-variable theory the logarithm of a complex number is another complex number, that is,

$$\log_e G(j\omega)H(j\omega) = \log_e R(\omega) + j\theta(\omega)$$

where

$$R(\omega) = \text{real part}$$
$$= [A^2(\omega) + B^2(\omega)]^{1/2}$$

and

$$\theta(\omega) = \text{imaginary part}$$

$$= \tan^{-1} \frac{B(\omega)}{A(\omega)}$$

In other words, the $\log_e G(j\omega)H(j\omega)$ is composed of two separate functions of ω, the real part $\log_e R(\omega)$ and the imaginary part $\theta(\omega)$. Bode plots require to plot the real part $\log_e R(\omega)$ and imaginary part $\theta(\omega)$ versus the logarithm of ω. To illustrate the logarithm of complex numbers, let us consider the following example. Let

$$G(j\omega)H(j\omega) = re^{j\omega}$$

or

$$G(j\omega)H(j\omega) = r \cos \omega + jr \sin \omega$$

Taking the logarithm on both sides, we get

$$\log_e G(j\omega)H(j\omega) = \log_e \sqrt{(r \cos \omega)^2 + (r \sin \omega)^2} + j \tan^{-1} \frac{r \sin \omega}{r \cos \omega}$$

$$\log_e G(j\omega)H(j\omega) = \log_e r + j\omega$$

Thus the logarithm of complex variable $G(j\omega)H(j\omega)$ is composed of two separate functions of ω, the real part $\log_e r$ and the imaginary part $j\omega$.

Before we consider a general form of $G(j\omega)H(j\omega)$, it is important to remember the following axioms of logarithms:

$$\log_e ab = \log_e a + \log_e b$$

$$\log_e \frac{a}{b} = \log_e a - \log_e b$$

$$\log_e a^x = x \log_e a$$

$$\log_e x = \frac{\log_{10} x}{\log_{10} e}$$

The general form of $G(s)H(s)$ as a ratio of polynomials or in terms of poles and zeros is

$$G(s)H(s) = \frac{K(s + z_1)(s + z_2) \cdots (s + z_m)}{(s + p_1)(s + p_2) \cdots (s + p_n)} \qquad (7-5)$$

where $K = $ a constant and $z_1, z_2, \ldots, z_m$ are zeros and $p_1, p_2, \ldots, p_n$ are poles of $G(s)H(s)$ such that number of zeros $m \leq$ number of poles n. Replacing s by $j\omega$ in Eq. (7–5), we get

$$G(j\omega)H(j\omega) = \frac{K(j\omega + z_1)(j\omega + z_2) \cdots (j\omega + z_m)}{(j\omega + p_1)(j\omega + p_2) \cdots (j\omega + p_n)} \qquad (7-6)$$

Taking logarithms on both the sides yields

$$\log_e G(j\omega)H(j\omega) = \log_e \frac{K(j\omega + z_1)(j\omega + z_2) \cdots (j\omega + z_m)}{(j\omega + p_1)(j\omega + p_2) \cdots (j\omega + p_n)}$$

$$= \log_e K + (\log_e j\omega + z_1 + j\theta z_1)$$

$$+ (\log_e j\omega + z_2 + j\theta z_2) + \cdots + (\log_e j\omega + z_m + j\theta z_m)$$

$$- (\log_e j\omega + p_1 + j\theta p_1) - (\log_e j\omega + p_2 + j\theta p_2) - \cdots$$

$$- (\log_e j\omega + p_n + j\theta p_n)$$

Collecting magnitude and phase angle terms, we get

$$\log_e G(j\omega)H(j\omega) = \log_e K + \log_e j\omega + z_1 + \log_e j\omega + z_2$$

$$+ \cdots + \log_e j\omega + z_m - \log_e j\omega + p_1 - \log_e j\omega$$

$$+ p_2 - \cdots - \log_e j\omega + p_n + j(\theta z_1 + \theta z_2 + \cdots$$

$$+ \theta z_m - \theta p_1 - \theta p_2 - \cdots - \theta p_n)$$

$$= \log_e K + \log_e (z_1^2 + \omega^2)^{1/2} + \log_e (z_2^2 + \omega^2)^{1/2} + \cdots$$

$$+ \log_e (z_m^2 + \omega^2)^{1/2} - \log_e (p_1^2 + \omega^2)^{1/2} - \log_e (p_2^2 + \omega^2)^{1/2}$$

$$- \cdots - \log_e (p_n^2 + \omega^2)^{1/2} + j\left(\tan^{-1} \frac{\omega}{z_1} + \tan^{-1} \frac{\omega}{z_3} + \cdots\right.$$

$$\left. + \tan^{-1} \frac{\omega}{z_m} - \tan^{-1} \frac{\omega}{p_1} - \tan^{-1} \frac{\omega}{p_2} - \cdots - \tan^{-1} \frac{\omega}{p_n}\right) \qquad (7-7)$$

Equation (7–7) can now be used to obtain the two complete plots which represent the $\log_e G(j\omega(H(j\omega)$. These plots are

$$\log_e G(j\omega)H(j\omega) = \log K + \log_e (z_1^2 + \omega^2)^{1/2} + \log_e (z_2^2 + \omega^2)^{1/2}$$

$$+ \cdots + \log_e (z_m^2 + \omega^2)^{1/2} - \log_e (p_1^2 + \omega^2)^{1/2}$$

$$- \log_e (p_2^2 + \omega^2)^{1/2} - \cdots - \log_e (p_n^2 + \omega^2)^{1/2} \qquad (7-8)$$

and

$$\theta(\omega) = \tan^{-1} \frac{\omega}{z_1} + \tan^{-1} \frac{\omega}{z_2} + \cdots$$

$$+ \tan^{-1} \frac{\omega}{z_m} - \tan^{-1} \frac{\omega}{p_1} - \tan^{-1} \frac{\omega}{p_2}$$

$$- \cdots - \tan^{-1} \frac{\omega}{p_n} \qquad (7-9)$$

Normally, the amplitude relationship of Eq. (7–8) is expressed in decibels as follows:

$$20 \log G(j\omega)H(j\omega) = 20 \log k + 20 \log (z_1^2 + \omega^2)^{1/2}$$

$$+ 20 \log (z_2^2 + \omega^2)^{1/2} + \cdots + 20 \log (z_m^2 + \omega^2)^{1/2}$$

$$- 20 \log (p_1^2 + \omega^2)^{1/2} - 20 \log (p_2^2 + \omega^2) - \cdots$$

$$- 20 \log (p_n^2 + \omega^2)^{1/2} \qquad (7-10)$$

where the symbol log is defined as $\log_{10}$. Plots of phase angle and magnitude in decibels of Eq. (7–9) and (7–10) versus ω are Bode plots.

Both magnitude and phase angle plots may be plotted on semilog paper. In a magnitude plot, the ordinate is 20 log $G(j\omega)H(j\omega)$ plotted on a linear scale, while the abscissa is ω plotted on a logarithmic scale. Similarly, in a phase angle plot, the ordinate is the phase angle $\theta(\omega)$ in degrees plotted on equally spaced divisions, while the abscissa is ω plotted on a logarithmic scale. The number of logarithmic cycles required on the ω axis is determined by the range of frequency over which the system performance is to be investigated. The advantage of using semilog paper and plotting the Bode diagrams on the same paper will be obvious when we perform the stability analysis using Bode plots.

7–5 CONSTRUCTING BODE PLOTS

A close examination of the magnitude equation $(7-10)$ suggests that 20 log $G(j\omega)H(j\omega)$ is composed of a number of factors; hence the composite magnitude plot is the superimposition of the contributions of each of these factors. Each factor exhibits certain characteristics when its logarithm is plotted separately versus ω. The most commonly found factors of $G(j\omega)H(j\omega)$ are $\pm K$, $(j\omega)^{\pm 1}$, $(j\omega t + 1)^{\pm 1}$, and $[(j\omega)^2 + 2\zeta\omega_n(j\omega) + \omega_n^2]^{\pm 1}$. With a knowledge of the behavior of each factor, the construction of Bode plots may be simplified and speeded. In addition, it is possible to construct approximate Bode plots by recognizing the behavior of each factor. The approximate diagrams are the asymptotic representation of the factors of $G(j\omega)H(j\omega)$. It is possible, as will be seen, to draw *actual Bode plots* which are more accurate than the asymptotic representations. Actual Bode plots are preferred over asymptotic plots, especially if the system is marginally or conditionally stable.

The first step in the construction of Bode plots is obviously to analyze the behavior of each of the basic factors. For the sake of clarity, we will draw both asymptotic and actual Bode plots for each of the basic factors mentioned above.

Bode Plots of a Constant: ±K

The constant K in the loop transfer function $G(j\omega(H(j\omega)$ is the gain or product of gains associated with one or more block transfer functions of the system. In fact, the gain K may be thought of as a complex number with zero imaginary part, and hence may be a positive or negative real number.

EXAMPLE 7–1

Draw Bode plots for the following loop transfer function:

$$G(s)H(s) = K$$

if (a) $K = 10$; (b) $K = -10$; (c) $K = 0.1$.

SOLUTION   First we substitute $s = j\omega$ in the given transfer function and then obtain the magnitude and phase angle equations.

$$G(j\omega)H(j\omega) = K$$

Hence

$$20 \log_{10} |G(j\omega)H(j\omega)| = 20 \log_{10} K$$

or

$$M(\omega) \text{ dB} = 20 \log_{10} K \qquad (7-11a)$$

where

$$M(\omega) \text{ dB} = 20 \log_{10} |G(j\omega)H(j\omega)|$$

and

$$\theta(\omega) = \tan^{-1} \frac{0}{K} \qquad (7-11b)$$

(a) Substituting $K = 10$ in Eqs. (7-11a) and (7-11b), we get

$$M(\omega) \text{ dB} = 20 \text{ dB} \qquad \text{and} \qquad \theta(\omega) = \tan^{-1} \frac{0}{10} = 0°$$

Thus the magnitude plot is a horizontal line of 20 dB and the phase angle plot is also a horizontal line of 0°, as shown in Fig. E7–1a. Generally, Bode diagrams are constructed on a five-cycle or six-cycle semilog graph papers. Figure E7–1 uses five-cycle semilog paper. The horizontal log axis is used for angular frequency ω, which is in rad/s. The vertical axis on the left of ω axis is used for the magnitude in decibels and the vertical axis on the right-hand side of the ω axis is used for the phase angle of the system. The vertical magnitude axis is marked in decibels and uses linear scale. Similarly, the vertical phase angle axis is marked in degrees and also uses a linear scale (Fig. E7–1).

(b) Next substitute $K = -10$ in Eqs. (7-11a) and (7-11b). Thus

$$M(\omega) \text{ dB} = 20 \text{ dB}$$

and

$$\theta(\omega) = \tan^{-1} \frac{0}{-10} = -180°$$

The magnitude and phase angle plots are shown in Fig. E7–1b.

(c) Finally, let us substitute $K = 0.1$ in Eqs. (7-11a) and (7-11b). We have

$$M(\omega) \text{ dB} = 20 \log_{10} 0.1 = 20 \log_{10} \frac{1}{10}$$

$$= -20 \text{ dB}$$

and

$$\theta(\omega) = \tan^{-1} \frac{0}{0.1} = 0°$$

The magnitude and phase angle plots are shown in Fig. E7–1c.

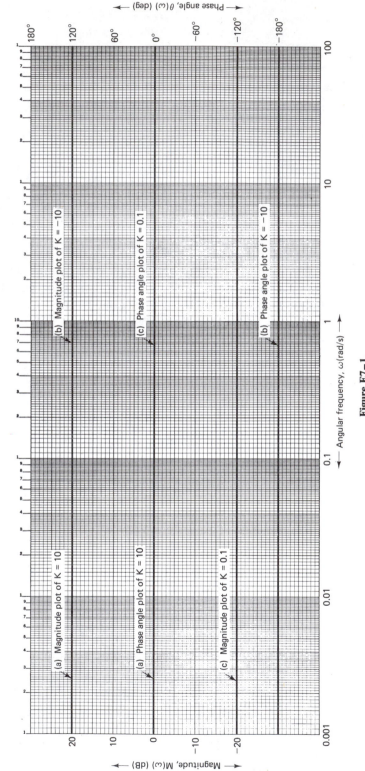

Figure E7–1.

From the discussion above it is obvious that the magnitude and phase angle plots of a constant are horizontal lines, since the constant K is independent of operating angular frequency ω. However, the phase angle depends on the sign of the gain constant K and is either 0 or 180°. Remember that the straight-line asymptotic plots and the Bode plots are the same for the proportional block since the gain constant K is not a function of angular frequency ω.

Bode Plots of Pole–Zero at the Origin of the s plane: $(s)^{\pm 1}$

Next let us consider the loop transfer function

$$G(s(H(s) = (s)^{\pm 1}$$

Examples 7–2 and 7–3 illustrate the construction of Bode plots for the loop transfer function with a zero and a pole respectively, at the origin of the s plane.

EXAMPLE 7–2

Draw Bode plots for the following loop transfer function:

$$G(s)H(s) = s$$

SOLUTION To obtain the magnitude and phase angle equations, first substitute $s = j\omega$. Hence

$$G(j\omega)H(j\omega) = j\omega$$

and

$$20 \log_{10} |G(j\omega)H(j\omega)| = 20 \log_{10} |j\omega|$$

or

$$M(\omega) \text{ dB} = 20 \log_{10} \omega \qquad (7-12a)$$

Also,

$$\theta(\omega) = \tan^{-1} \frac{\omega}{0} = 90° \qquad (7-12b)$$

Table E7–2 is constructed using Eqs. (7–12a) and (7–12b) for various values of ω. It is obvious from Eq. (7–12a) that the magnitude increases with increase in the value of ω. For example:

At $\omega = 1$: $M(\omega)$ dB $= 0$ dB.
At $\omega = 2$: $M(\omega)$ dB $= 6.02$ dB.
At $\omega = 4$: $M(\omega)$ dB $= 12.04$ dB.
At $\omega = 40$: $M(\omega)$ dB $= 32.04$ dB.

In short, every time ω is doubled: that is, ω is increased from 1 to 2, or 2 to 4, or 4 to 8, there is 6-dB increase in the value of $M(\omega)$ dB. Therefore, the magnitude

---

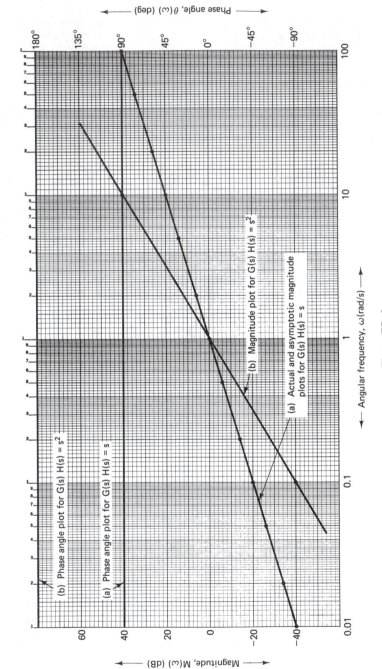

(b) Phase angle plot for G(s) H(s) = s²

(a) Phase angle plot for G(s) H(s) = s

(b) Magnitude plot for G(s) H(s) = s²

(a) Actual and asymptotic magnitude plots for G(s) H(s) = s

Phase angle, θ(ω) (deg)

Magnitude, M(ω) (dB)

Angular frequency, ω(rad/s)

Figure E7–2.

Angular frequency, ω (rad/s)	Magnitude, M(ω) (dB)	Phase angle, θ(ω) (deg)
0.01	−40	90
0.02	−33.98	90
0.05	−26.02	90
0.1	−20	90
0.2	−13.98	90
0.5	−6.02	90
1.0	0.0	90
2	6.02	90
5	13.98	90
10	20	90
20	26.02	90
50	33.98	90
100	40	90

is said to increase at the rate of 6 dB/octave, where octave means a twofold increase in the value of ω. Similarly, if ω is increased by a decade (tenfold increase), the magnitude $M(\omega)$ dB will increase by 20 dB. In other words, the magnitude $M(\omega)$ dB increases at the rate of 20 dB/decade every time ω is increased by tenfold. The data from Table E7–2 are used to construct the Bode magnitude and phase angle plots shown in Fig. E7–2a. Note that the magnitude plot $M(\omega)$ dB crosses 0 dB at $\omega = 1$ rad/s, which is said to be a crossover frequency for a *zero* at $s = 0$. The significance of the crossover frequency is that a magnitude plot for $G(s)H(s) = s$ is obtained if a straight line is drawn through this frequency at the slope of approximately 6 dB/octave.

EXAMPLE 7–3

Draw Bode plots for the loop-transfer function

$$G(s)H(s) = \frac{1}{s}$$

SOLUTION   The given transfer function has only one pole, which is at the origin of the s plane. Substituting $s = j\omega$, we have

$$G(j\omega)H(j\omega) = \frac{1}{j\omega}$$

Hence the magnitude and phase angle equations are

$$20 \log |G(j\omega)H(j\omega)| = 20 \log \left| \frac{1}{j\omega} \right|$$

or

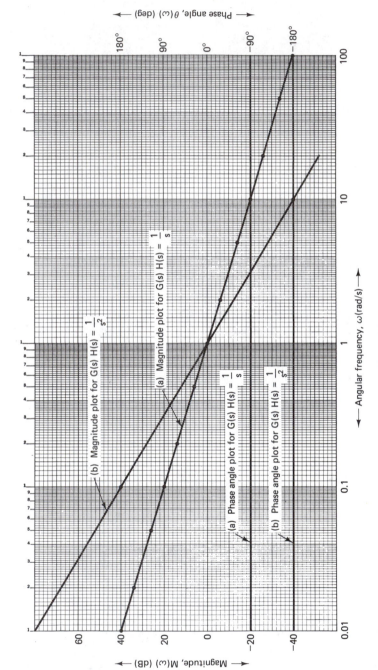

Figure E7–3.

TABLE E7–3 DATA FOR EXAMPLE 7–3

Angular frequency, ω (rad/s)	Magnitude, $M(\omega)$ (dB)	Phase angle, $\theta(\omega)$ (deg)
0.01	40	−90
0.02	33.98	−90
0.05	26.02	−90
0.1	20	−90
0.2	13.98	−90
0.5	6.02	−90
1.0	0.0	−90
2	−6.02	−90
5	−13.98	−90
10	−20	−90
20	−26.02	−90
50	−33.98	−90
100	−40	−90

$$M(\omega)\ dB = -20 \log \omega \qquad (7\text{–}13a)$$

and

$$\theta(\omega) = \tan^{-1}\frac{0}{1} - \tan^{-1}\frac{\omega}{0}$$
$$= -90° \qquad (7\text{–}13b)$$

From Eq. (7–13b) it is obvious that the phase angle is fixed at −90°. Hence the phase angle plot is a straight line at −90°. However, the magnitude in decibels increases in the negative direction as the value of ω is increased [Eq. (7–13a)]. Table E7–3 includes the values of magnitude $M(\omega)$ dB and phase angle $\theta(\omega)$ as ω is varied from 0.01 rad/s to 100 rad/s. The data in Table E7–3 are used to construct Bode plots of Fig. E7–3a. Again note that the magnitude plot $M(\omega)$ dB crosses 0 dB at $\omega = 1$ rad/s. The value of ω at which $M(\omega)$ dB = 0 dB is referred to as the *crossover frequency*. Thus the crossover frequency for a pole at the origin is 1 rad/s.

The only difference in the plots of Fig. E7–2a and E7–3a is that in Fig. E7–2a the magnitude increases with increase in the angular frequency ω, whereas in Fig. E7–3a it decreases with the increase in ω. Thus the rate of *increase* of 20 dB/decade in the magnitude of the transfer function (Bode plot) is the characteristic of a zero at the origin, whereas the *decrease of* 20 dB/decade in the magnitude of the transfer function is the characteristic of a pole at the origin of the s plane.

The concepts above can easily be extended to a multiple zero or pole at the origin. For example, let us assume that there are N zeros at the origin. Therefore, the slope of the magnitude plot will be $20N$ dB/decade. Similarly, for N poles at the origin, the slope of the magnitude plot will be $-20N$ dB/decade. Figures E7–2b and E7–3b show the magnitude of the phase angle plots for $(s)^{\pm 2}$ where $N = 2$.

Bode Plots of Finite Zero or Pole: $(s\tau + 1)^{\pm 1}$

A *zero* or *pole* on the real axis of the s plane is said to be a finite zero or pole. Most of the practical systems' transfer functions contain finite pole or zero or both. In Examples 7–4 and 7–5 we will study transfer functions with pole and zero on the negative real axis of the s plane.

EXAMPLE 7–4

Construct Bode plots for the following loop transfer function:

$$G(s)H(s) = s\tau + 1$$

where τ = time constant = 2 s.

SOLUTION The transfer function above has a single time constant and is commonly known as the *simple lead* (see Chapter 4). To obtain data for the magnitude plot, we need

$$20 \log |j\omega\tau + 1| = 20 \log \sqrt{(\omega\tau)^2 + 1}$$

or

$$M(\omega) \, \text{dB} = 20 \log \sqrt{(2\omega)^2 + 1} \qquad (7\text{–}14a)$$

Similarly, the data for the phase angle plot are obtained from

$$\theta(\omega) = \tan^{-1} \frac{\omega\tau}{1} = \tan^{-1} \frac{2\omega}{1} \qquad (7\text{–}14b)$$

TABLE E7–4 DATA FOR EXAMPLE 7–4

Angular frequency, ω (rad/s)	$\omega\tau$	Magnitude, $M(\omega)$ (dB)	Phase angle, $\theta(\omega)$ (deg)
0.001	0.002	0	$0.11 \simeq 0$
0.005	0.01	0	$0.57 \simeq 0$
0.01	0.02	0	1.14
0.02	0.04	0	2.29
0.07	0.14	$\simeq 0$	7.97
0.1	0.2	0.17	11.3
0.2	0.4	0.64	21.8
0.5	1.0	3.01	45
1	2	6.99	63.4
2	4	12.3	75.96
5	10	20.04	84.3
10	20	26.03	87.14
20	40	32.04	88.57
50	100	40	89.4
100	200	46.02	89.7

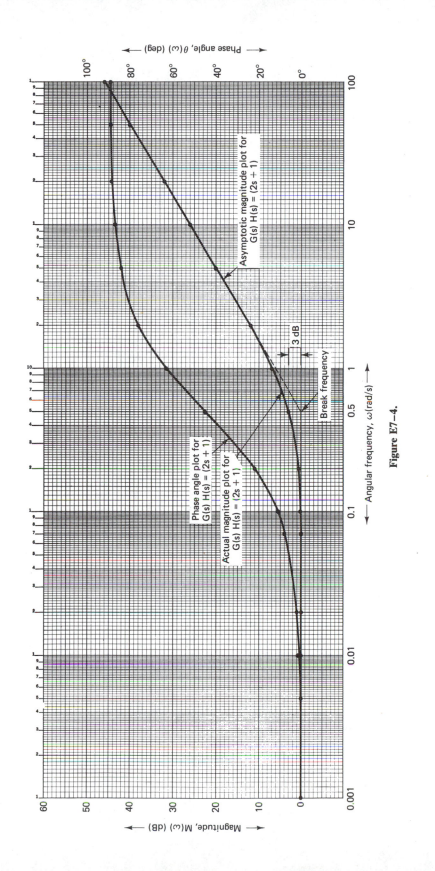

Figure E7–4.

The data in Table E7–4 are obtained by substituting various values of ω in Eqs. (7–14a) and (7–14b). Note that at low frequencies where $\omega\tau \ll 1$,

$$M(\omega) \text{ dB} \simeq 0 \text{ dB} \qquad \text{and} \qquad \theta(\omega) \simeq 0°$$

For example, at $\omega = 0.025$ rad/s, $\omega\tau = 0.025(2) = 0.05 \ll 1$, and

$$M(\omega) \text{ dB} = 20 \log \sqrt{(0.05)^2 + 1} \simeq 0 \text{ dB}$$

and

$$\theta(\omega) = \tan^{-1} 0.05 = 2.86°$$

However, if the value of ω is such that $\omega\tau = 1$ (or $\omega = 1/\tau$), the magnitude at this frequency is

$$M(\omega) \text{ dB} = 20 \log \sqrt{(1)^2 + 1} = 3.01 \text{ dB}$$

$$\theta(\omega) = \tan^{-1} 1 = 45°$$

This frequency $\omega = 1/\tau$ is called the *break frequency, 3-dB frequency*, or *corner frequency*. All these terms are equivalent and can be used interchangeably. The frequency $\omega = 1/\tau$ is called the *break frequency* because at this frequency there is a break in the slope of the magnitude curve as shown in Fig. E7–4. The name *3-dB frequency* arises simply because at this frequency, $\omega = 1/\tau$, the magnitude changes by 3 dB from its initial value. Finally, $\omega = 1/\tau$ is referred to as a *corner frequency* because in an asymptotic magnitude plot the frequency $\omega = 1/\tau$ is at the corner of the two straight lines (Fig. E7–4).

At high frequencies where $\omega\tau \gg 1$, $M(\omega)$ dB is governed by the value of ω (Table E7–4). The magnitude and phase angle plots of Figure E7–4 are constructed using the data of Table E7–4. In Fig. E7–4 the magnitude and phase angles corresponding to the values of ω are denoted by the small dots. Beyond the corner frequency $\omega = 1/\tau$, the magnitude $M(\omega)$ dB increases at the rate of 20 dB/decade with the increase in the value ω.

EXAMPLE 7–5

Construct Bode plots for the transfer function

$$G(s)H(s) = \frac{1}{s\tau + 1}$$

where τ = time constant = 5 s.

SOLUTION   The given transfer function possesses the characteristics of a simple lag, in that it has a finite pole on the negative real axis of the s plane. To construct Bode plots, we need

$$G(s)H(s) = \frac{1}{j\omega\tau + 1}$$

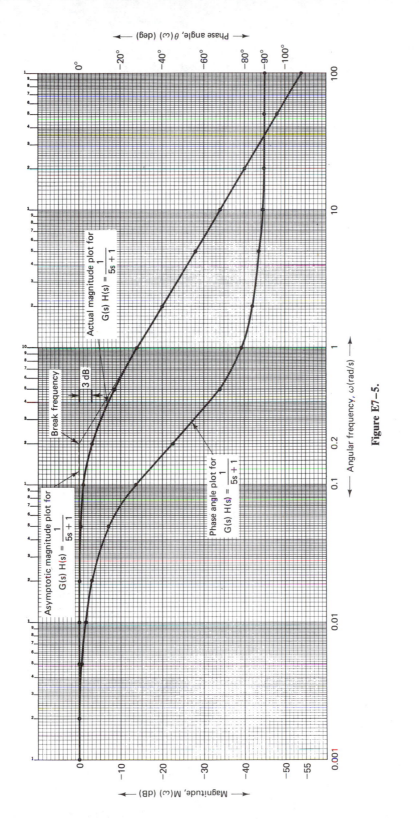

Figure E7–5.

or

$$20 \log |G(j\omega)H(j\omega)| = 20 \log \left| \frac{1}{j\omega\tau + 1} \right|$$

$$M(\omega) \text{ dB} = -20 \log \sqrt{(\omega\tau)^2 + 1}$$
$$= -20 \log \sqrt{(5\omega)^2 + 1} \qquad (7\text{–}15a)$$

and

$$\theta(\omega) = \tan^{-1}\frac{0}{1} - \tan^{-1}\frac{\omega\tau}{1}$$

$$= -\tan^{-1}\frac{5\omega}{1} \qquad (7\text{–}15b)$$

Again from Eq. (7–15a) the break frequency is $\omega = 1/5$ rad/s at which the magnitude $M(\omega)$ dB $= -3.01$ dB (Table E7–5). The data in Table E7–5 are obtained by substituting various values of ω in Eqs. (7–15a) and (7–15b). Note that the important region of the frequency response (magnitude and phase angle plots) is normally within ± 2 decades of the break frequency. Therefore, if possible, select values of ω two decades above and two decades below the corner frequencies. Also, select several values of ω within a decade such that enough data are available to maintain the accuracy of the magnitude and phase angle plots. The data in Table E7–5 are plotted in Fig. E7–5.

TABLE E7–5 DATA FOR EXAMPLE 7–5

Angular frequency, ω (rad/s)	$\omega\tau$	Magnitude $M(\omega)$ (dB)	Phase angle, $\theta(\omega)$ (deg)
0.001	0.005	0	$-0.29 \approx 0$
0.002	0.010	0	$-0.57 \approx 0$
0.005	0.025	0	-1.4
0.01	0.05	≈ 0	-2.9
0.02	0.10	≈ 0	-5.7
0.05	0.25	$-0.26 \approx 0$	-14.0
0.1	0.5	-0.97	-26.6
0.2	1.0	-3.01	-45 (break frequency)
0.5	2.5	-8.6	-68.2
1.0	5.0	-14.15	-78.7
2	10	-20.04	-84.3
5	25	-27.97	-87.71
10	50	-33.98	-88.9
20	100	-40	-89.4
50	250	-47.96	-89.8
100	500	-53.98	-89.9

Bode Plots for a Quadratic: $[s^2 + 2\zeta\omega_n s + \omega_n^2]^{\pm 1}$

Another important factor that appears quite frequently in transfer functions is the quadratic, especially in the denominator as a quadratic lag. Let us consider a loop transfer function with a quadratic lag:

$$G(s)H(s) = \frac{1}{(s^2 + 2\zeta\omega_n s + \omega_n^2)}$$

As we studied in Chapter 4, the roots of the quadratic depend on the value of ζ.

1. If $\zeta > 1$, the roots of a quadratic are real and simple, and the response due to the quadratic is overdamped.
2. If $\zeta = 1$, the roots of the quadratic are repeated, and the response due to the quadratic is critically damped.
3. If $\zeta < 1$, the roots of the quadratic are complex conjugates, and the response due to the quadratic is underdamped.
4. If $\zeta = 0$, the roots of the quadratic are pure imaginary, and the response due to the quadratic is oscillatory.

Examples 7–6 to 7–9 illustrate how to construct Bode plots for each of the foregoing cases.

EXAMPLE 7–6

Construct Bode plots for the transfer function

$$G(s)H(s) = \frac{1}{s^2 + 3s + 2}$$

SOLUTION   The given transfer function has a denominator quadratic; comparing it with the standard quadratic $s^2 + 2\zeta\omega_n s + \omega_n^2$, we get

$$2\zeta\omega_n = 3$$

$$\omega_n^2 = 2$$

$$\zeta = \frac{3}{2\sqrt{2}} = 1.06$$

This means that the transfer function has real and simple poles. The poles are at $s = -2$ and $s = -1$. Therefore,

$$G(s)H(s) = \frac{1}{(s + 2)(s + 1)}$$

$$= \frac{0.5}{(0.5s + 1)(s + 1)} \tag{7–16}$$

Substituting $s = j\omega$, we obtain

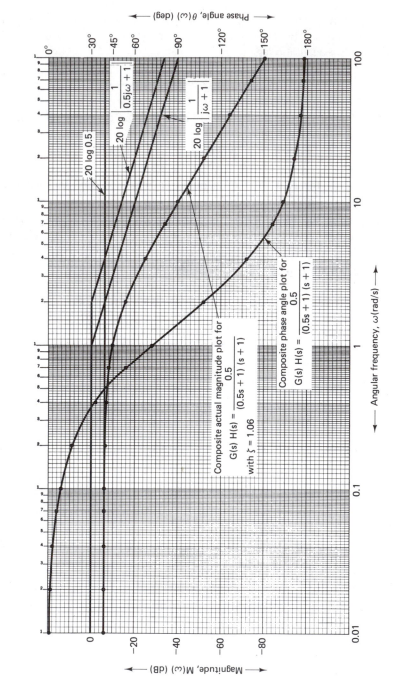

Figure E7-6.

$$G(s)H(s) = \frac{0.5}{(0.5j\omega + 1)(j\omega + 1)}$$

or

$$M(\omega) \text{ dB} = 20 \log 0.5 - 20 \log \sqrt{(0.5\omega)^2 + 1} - 20 \log \sqrt{\omega^2 + 1} \qquad (7-17a)$$

and

$$\theta(\omega) = -\tan^{-1} \frac{0.5\omega}{1} - \tan^{-} \frac{\omega}{1} \qquad (7-17b)$$

The break frequencies are at $\omega = 2$ and $\omega = 1$. Therefore, we will assign the values for ω from 0.01 to 100 rad/s, since it covers the frequency range approximately two decades above and two decades below the break frequencies. The values of magnitude and phase angles corresponding to the desired values of ω are given in the Table E7–6. The data of Table E7–6 are used to plot the magnitude and the phase angle plots of Fig. E7–6. As expected, the phase angle approaches $-180°$ as ω increases toward infinity. This is because the maximum contribution due to one pole is $-90°$ and there are two poles.

TABLE E7–6 DATA FOR EXAMPLE 7–6

Angular frequency, ω (rad/s)	Magnitude, $M(\omega)$ (dB)	Phase angle, $\theta(\omega)$ (deg)
0.01	−6.02	−0.86
0.02	−6.02	−1.72
0.04	−6.02	−3.44
0.07	−6.02	−6.0
0.1	−6.07	−8.6
0.2	−6.2	−17.0
0.4	−6.83	−33.11
0.7	−8.25	−54.3
1.0	−10.0	−71.6
2	−16.02	−108.4
4	−25.31	−139.4
7	−34.23	−155.9
10	−40.21	−163.0
20	−52.09	−171.4
40	−64.09	−175.71
70	−73.8	−177.5
100	−80	−178.3

EXAMPLE 7–7

Construct Bode plots for the following transfer function:

$$G(s)H(s) = \frac{1}{(s + 0.1)^2}$$

SOLUTION The given transfer function has repeated poles at $s = 0.1$, which means that $\zeta = 1$, and time-domain-system response must be critically damped. We will rewrite the given transfer function as

$$G(s)H(s) = \frac{100}{(10s + 1)^2}$$

The transfer function above, in which the poles are expressed in the form of $(s\tau + 1)$, is convenient to obtain the magnitude and phase angle data for Bode plots.

There are two poles on the negative real axis at $s = -0.1$, and the double break frequency is at $\omega = 0.1$. Therefore, an asymptotic magnitude plot remains constant at 40 dB until the break frequency $\omega = 0.1$ rad/s. However, after $\omega = 0.1$ it decreases at the rate of 40 dB/decade because of the double pole at $s = -0.1$ (Fig. E7–7).

However, for the actual Bode plots, we need magnitude and phase angle equations. Substituting $s = j\omega$, we have

$$G(j\omega)H(j\omega) = \frac{100}{(10j\omega + 1)^2}$$

and the magnitude equation is

$$M(\omega) \text{ dB} = 20 \log (100) - 40 \log \sqrt{(10\omega)^2 + 1} \qquad (7\text{–}18a)$$

Similarly, the phase angle equation is

$$\theta(\omega) = -2 \tan^{-1} \frac{10\omega}{1} \qquad (7\text{–}18b)$$

The data for the magnitude and the phase angle plots are given in Table E7–7. The Bode plots in Fig. E7–7b are constructed using the data in Table E7–7.

TABLE E7–7 DATA FOR EXAMPLE 7–7

Angular frequency, ω (rad/s)	Magnitude, $M(\omega)$ (dB)	Phase angle, $\theta(\omega)$ (deg)
0.001	40	−1.15
0.002	40	−2.29
0.005	≃40	−5.72
0.010	≃40	−11.42
0.020	39.66	−22.62
0.050	38.06	−53.13
0.10	33.98	−90
0.2	26.02	−126.87
0.5	11.70	−157.38
1.0	−0.09 ≈ 0	−168.58
2	−12.06	−174.28
5	−27.96	−177.71
10	−40.0	−178.85
20	−52.04	−179.43
50	−67.96	−179.77
100	−80.0	−179.89

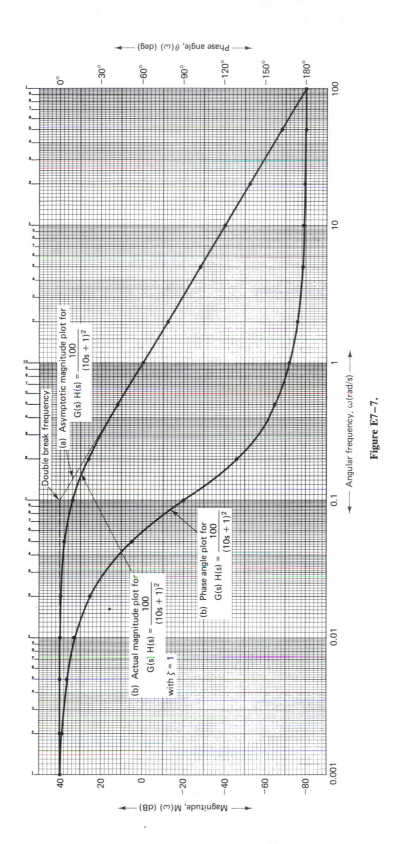

Figure E7-7.

EXAMPLE 7-8

Construct Bode plots for the following transfer function:

$$G(s)H(s) = \frac{1}{s^2 + 2s + 2}$$

SOLUTION The given transfer function has complex conjugate roots and $\zeta = 0.707$. Therefore,

$$G(s)H(s) = \frac{1}{(s + j1 + 1)(s - j1 + 1)}$$

and the magnitude equation is

$$M(\omega) \text{ dB} = -20 \log \sqrt{(\omega + 1)^2 + 1} - 20 \log \sqrt{(\omega - 1)^2 + 1} \qquad (7-19a)$$

and the phase angle equation is

$$\theta(\omega) = -\tan^{-1}\frac{\omega + 1}{1} - \tan^{-1}\frac{\omega - 1}{1} \qquad (7-19b)$$

Substituting the values of ω from 0.01 through 100 rad/s in Eqs. (7-19a) and (7-19b), we obtain the data in Table E7-8.

TABLE E7-8 DATA FOR EXAMPLE 7-8

Angular frequency, ω (rad/s)	Magnitude, $M(\omega)$ (dB)	Phase angle, $\theta(\omega)$ (deg)
0.01	-6.02	-0.57
0.02	-6.02	-1.15
0.05	-5.99	-2.87
0.1	-5.98	-5.74
0.2	-6.02	-11.53
0.5	-6.09	-29.74
1.0	-6.99	-63.43
2.0	-13.01	-116.57
5	-27.98	-156.50
10	-40	-168.47
20	-52.04	-174.26
50	-67.96	-177.71
100	-80.01	-178.85

The data in Table E7-8 are used to construct the Bode plots of Fig. E7-8.

EXAMPLE 7-9

Construct Bode plots for the following transfer function:

$$G(s)H(s) = \frac{1}{s^2 + 4}$$

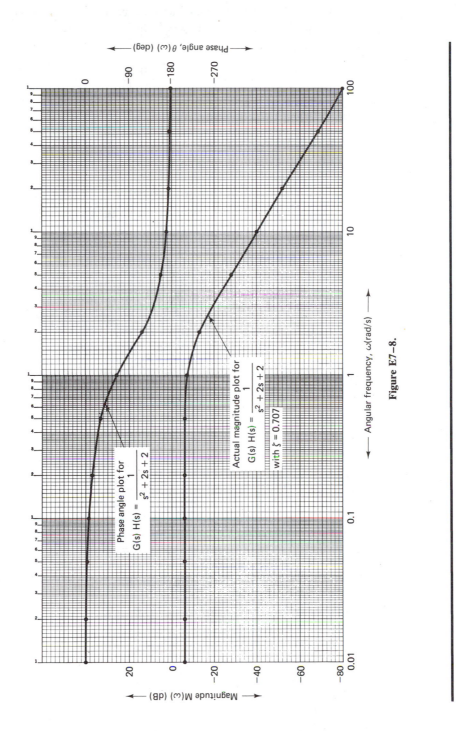

Figure E7–8.

SOLUTION   First we rearrange the transfer function, in that we reduce the constant term in the denominator to unity, which simplifies the magnitude and phase angle calculations. Hence

$$G(s)H(s) = \frac{0.25}{(0.5s + j1)(0.5s - j1)}$$

The magnitude and phase angle equations are

$$M(\omega) \text{ dB} = 20 \log 0.25 - 20 \log (0.5\omega + 1) - 20 \log (0.5\omega - 1) \qquad (7-20a)$$
$$\theta(\omega) = -180° \qquad (7-20b)$$

Obviously, the phase angle is constant at $-180°$ and the magnitude changes with change in ω (Table E7−9).

TABLE E7−9   DATA FOR EXAMPLE 7−9

Angular frequency, ω (rad/s)	Magnitude, $M(\omega)$ (dB)	Phase angle, $\theta(\omega)$ (deg)
0.1	−12.01	−180
0.2	−11.95	−180
0.5	−11.48	−180
1.0	−9.54	−180
1.4	−6.18	−180
1.6	−3.17	−180
1.8	+2.38	−180
1.9	+8.18	−180
2.0	∞	−180 (break frequency)
2.1	+7.74	−180
2.2	+1.52	−180
2.3	−2.21	−180
2.5	−7.04	−180
2.6	−8.81	−180
3.0	−13.98	−180
3.4	−17.66	−180
3.8	−20.37	−180
4	−21.58	−180
4.4	−23.72	−180
5	−26.44	−180
7	−33.05	−180
10	−39.64	−180
20	−51.95	−180
50	−67.94	−180
100	−79.99	−180

The data of Table E7−9 are used to construct the Bode diagrams of Fig. E7−9. Note that the break frequency is $\omega = 2$ rad/s. At this break frequency the magnitude of the transfer function is infinity, as shown in Fig. E7−9.

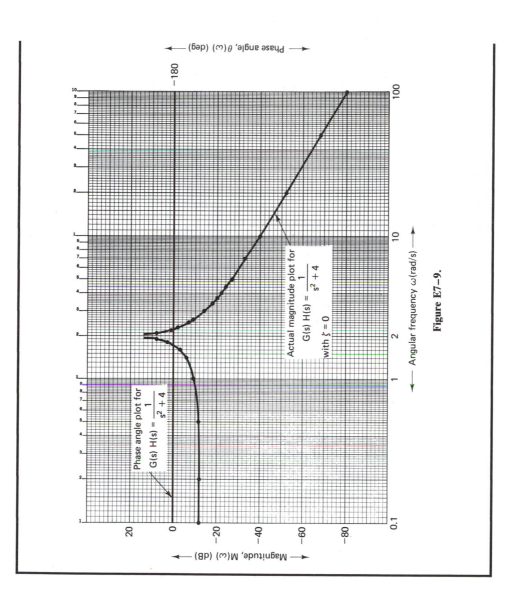

Figure E7–9.

In Examples 7-6 to 7-9 we learned how to construct Bode plots for a loop transfer function with a quadratic lag. The results obtained in these examples are summarized in Table 7-2, and should serve as a quick reference.

TABLE 7-2 SUMMARY OF BODE DIAGRAMS OF A QUADRATIC LAG

Damping ratio	$\zeta = 1.06$	$\zeta = 1$	$\zeta = 0.707$	$\zeta = 0$
Given transfer function	$\dfrac{1}{s^2 + 3s + 2}$	$\dfrac{1}{(s + 0.1)^2}$	$\dfrac{1}{s + 2s + 2}$	$\dfrac{1}{s^2 + 4}$
Transfer function suitable for Bode diagrams	$\dfrac{0.5}{(0.5s + 1)(s + 1)}$	$\dfrac{100}{(10s + 1)^2}$	$\dfrac{1}{(s + j1 + 1)(s - j1 + 1)}$	$\dfrac{0.25}{(0.5s + j1)(0.5s - j1)}$
Break frequency	$\omega = 1$ and $\omega = 2$	$\omega = 0.1$	$\omega = 1$	$\omega = 2$

7-6 STEPS FOR PLOTTING BODE PLOTS

In Section 7-5 we studied how to construct Bode plots of four basic transfer functions: $\pm K$, $(s)^{\pm 1}$, $(s\tau + 1)^{\pm 1}$, and $(s^2 + 2\zeta\omega_n s + \omega_n^2)$. In practice, often we have systems with combinations of one or more of the basic transfer functions above. The construction of Bode plots of such systems is made easier because of the additive nature of plots on logarithmic coordinates. To facilitate the construction of Bode diagrams for a given system, the following steps should be used.

1. Express the transfer function $G(s)H(s)$ in factored form so that these factors may take any of the four basic forms discussed in Section 7-5.
2. Rearrange the factors such that the constant terms associated with simple lead, simple lag, and quadratic terms are unity.
3. Substitute $s = j\omega$ in the factored transfer function.
4. Identify the break frequencies due to simple lead, simple lag, and quadratic factors.
5. Obtain the magnitude and phase angle equations for the given transfer function.
6. Calculate the data for the magnitude and phase angle plots by substituting various values of ω in the magnitude and phase angle equations. The values chosen for ω depend on the break frequencies and preferably should cover frequencies two decades below and two decades above the minimum and maximum break frequencies, respectively.
7. Plot the magnitude and the phase angle plots using the data of step 6.

Before digital calculators and computers were available, approximate magnitude (asymptotic) and phase angle plots were used to establish the general nature of the plots. These approximate plots obviously required a minimum of actual calculations and were less accurate. However, at present, calculators and even computers are readily available, so actual data for Bode diagrams can be obtained very easily. In fact, digital computers can even produce actual Bode diagrams. Thus actual plots are more accurate than approximate plots and hence are widely used.

7–7 STABILITY ANALYSIS USING BODE PLOTS

In the time domain, a system is said to be stable if its output response reaches a steady-state value in finite time in response to a given input. On the contrary, a system is said to be unstable if its output increases with increase in time. In such systems the output keeps on increasing until the system breaks down. However, in the s domain, such as in a pole-zero diagram, a system is said to be stable if its open-loop transfer function has no poles on the right-half of the s plane. Also, a system that is open-loop stable is also closed-loop stable if the feedback used is negative.

In the frequency domain, that is, using Bode plots, the stability of a system is determined from both magnitude and phase angle plots by calculating the phase angle at 0 dB and the magnitude at $-180°$.

1. *Stable system*. A system is said to be stable if the phase angle is less than $-180°$ when the magnitude is 0 dB, and the magnitude $M(\omega)$ dB is negative decibels when the phase angle is $-180°$. In some systems, especially first order, the phase angle never exceeds $-180°$; for such systems $M(\omega)$ dB at $-180°$ is undefined. However, these systems are stable because the phase angle is always less than $-180°$ when the magnitude $M(\omega)$ dB is 0 dB.

2. *Marginally stable system*. A system is said to be marginally stable if the magnitude is 0 dB when the phase angle is $-180°$. A marginally stable system is also referred to as a conditionally stable system.

3. *Unstable system*. A system is unstable if the phase angle is more negative than $-180°$ at the frequency corresponding to 0 dB magnitude, or if the magnitude is positive decibels at the frequency corresponding to $-180°$ phase shift.

In short, from Bode plots not only can we determine the stability of a system but we can also learn what modifications should be made to keep it stable or make it stable.

The terms *gain margin* and *phase margin* are synonymous with Bode plots and are used to indicate the stability of the system. These terms are defined below.

Gain Margin

On Bode plots the *gain margin* is defined as the magnitude in decibels of the magnitude plot when the phase angle is $-180°$. In equation form

$$\text{gain margin} = M(\omega) \text{ dB} \Big|_{\text{phase angle} = -180°} \qquad (7-21)$$

This means that the magnitude in decibels corresponding to a phase angle of $-180°$ may be positive or negative. Thus, the gain margin may be positive or negative decibels. A positive gain margin implies that the system is unstable. However, for a marginally stable system the gain margin is 0 dB and for the stable system it is negative. A positive or negative gain margin indicates the amount by which the gain of the system may be increased or decreased, respectively, to achieve the marginal stability. For systems whose phase angle is less than $-180°$, the gain margin is undefined. Therefore, systems with a phase shift of less than $\pm 180°$ are generally stable. For example, almost all first-order systems are stable because the phase angle never exceeds $\pm 90°$.

Phase Margin

The *phase margin* is an angle defined as $-180°$ minus the phase angle at the frequency corresponding to 0 dB magnitude. In equation form,

$$\text{phase margin} = -180° - (\text{phase angle}) \Big|_{M(\omega)\text{dB}=0\text{dB}} \qquad (7-22)$$

Equation (7–22) indicates that the phase margin may be a positive or a negative angle. A positive phase margin indicates that the system is unstable, whereas a negative phase margin indicates that it is stable. However, a zero phase margin means that the system is marginally stable. Often the magnitude plot may not cross 0 dB (the magnitude is never 0 dB); in such cases the phase margin is undefined.

In short, to determine the stability of a given system, both the gain margin and phase margin must be known. However, as mentioned earlier, sometimes the gain margin or phase may be undefined. In such situations, the known parameter (gain margin or phase margin) should be used. In general, a system is stable if its gain margin is negative decibels and the phase margin is a negative angle. Also, a system is said to be unstable if its gain margin is a positive decibels or the phase margin is a positive angle. Similarly, a system is said to be marginally or conditionally stable if its gain margin and phase margin are both zero.

7–8 SYSTEM ANALYSIS USING BODE PLOTS

In preceding sections we studied how to construct Bode plots and determine the stability of a system by computing its gain margin and phase margin. These concepts and procedures may be applied in several ways in the design and analysis of practical control systems. We consider next specific examples that illustrate the use of Bode plots in the analysis and design of control systems.

EXAMPLE 7–10

Plot Bode diagrams for the phase-lag network whose transfer function is

$$G(s)H(s) = \frac{1.5(s + 1)}{2s + 1}$$

From the Bode plots determine the gain margin and the phase margin and then establish the stability of the system.

SOLUTION   The given transfer function has a zero at $s = -1$ and a pole at $s = -1/2$; Therefore, the break frequencies are $\omega = 1$ rad/s and $\omega = 0.5$ rad/s. The magnitude and the phase angle equations are

$$M(\omega) \text{ dB} = 20 \log 1.5 + 20 \log \sqrt{\omega^2 + 1} - 20 \log \sqrt{(2\omega)^2 + 1} \qquad (7\text{–}23a)$$

$$\theta(\omega) = \tan^{-1}\frac{0}{1.5} + \tan^{-1}\frac{\omega}{1} - \tan^{-1}\frac{2\omega}{1} \qquad (7\text{–}23b)$$

Substituting various values of ω, we can obtain the corresponding magnitudes and phase angles as shown in Table E7–10.

TABLE E7–10   DATA FOR EXAMPLE 7–10

Angular frequency, ω (rad/s)	Magnitude, $M(\omega)$ (dB)	Phase angle, $\theta(\omega)$ (deg)
0.001	3.52	$-0.057 \approx 0$
0.002	3.52	$-0.11 \approx 0$
0.005	3.52	$-0.29 \approx 0$
0.01	3.52	$-0.57 \approx 0$
0.02	3.52	-1.14
0.05	3.49	-2.85
0.1	3.39	-5.60
0.2	3.05	-10.49
0.5	1.48	-18.43
1.0	-0.46	-18.43
2.0	-1.79	-12.53
5.0	-2.37	-5.60
10	-2.47	-2.85
20	-2.49	-1.43
500	-2.49	-0.57
100	-2.5	-0.29

The data of Table E7–10 are used to construct the Bode plots of Fig. E7–10. From these plots

$$\text{phase margin} = -180° - (-18°)$$
$$= -162°$$

However, the gain margin is undefined because the phase angle never exceeds $-180°$. Thus the system stability is to be determined from the phase margin alone. Since the phase margin is negative ($-162°$), the given system is absolutely stable.

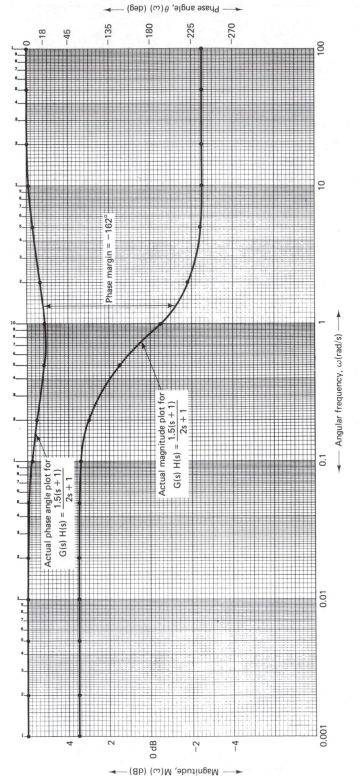

Figure E7–10.

K_{max} for a Marginally Stable System

Using the concepts of gain margin and phase margin, we can determine the range of values for the system gain K which is needed to maintain the stability of the system. A system whose stability is based on system gain may be referred to as a *conditionally stable system*. The characteristic of a conditionally stable system is that its gain margin is 0 dB and its phase margin is 0°.

EXAMPLE 7–11

Draw Bode plots for the following transfer function:

$$G(s)H(s) \frac{K}{(s + 10)(s + 5)(s + 0.5)(s + 20)}$$

Using the Bode plots, establish the maximum gain K that just maintains the stability of the system.

SOLUTION The given transfer function can be rewritten as follows:

$$G(s)H(s) = \frac{K/500}{(0.1s + 1)(2s + 1)(0.2s + 1)(0.05s + 1)}$$

$$G_N(s)H_N(s) = \frac{1}{(0.1s + 1)(2s + 1)(0.2s + 1)(0.05s + 1)}$$

Substituting $s = j\omega$ and obtaining the magnitude and phase angle equations, we have

$$M_N(\omega) \text{ dB} = -20 \log \sqrt{(0.1\omega)^2 + 1} - 20 \log \sqrt{(2\omega)^2 + 1}$$
$$-20 \log \sqrt{(0.2\omega)^2 + 1} - 20 \log \sqrt{(0.05\omega)^2 + 1}$$

$$(7-24a)$$

TABLE E7–11 DATA FOR EXAMPLE 7–11

Angular frequency, ω (rad/s)	Magnitude, $M_N(\omega)$ (dB)	Phase angle, $\theta(\omega)$ (deg)
0.01	0	−1.3
0.02	0	−2.7
0.05	−0.04	−6.7
0.1	−0.17	−13.3
0.2	−0.64	−25.8
0.5	−3.1	−55
1.0	−7.2	−83.3
2.0	−13.15	−114.8
5	−24.28	−169.9
10	−37	−222.1
20	−54.34	−273
50	−82.79	−320.6
100	−106.24	−339.8
200	−130.15	−349.9
500	−161.95	−355.9
1000	−186.02	−357.9

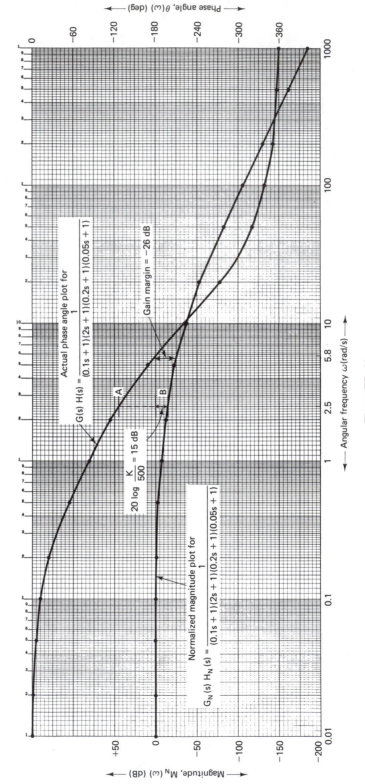

Figure E7–11.

Note that $M_N(\omega)$ dB indicates normalized magnitude in decibels, that is, the magnitude in decibels of the given transfer function with $(K/500) = 1$.

$$\theta(\omega) = -\tan^{-1}\frac{0.1\omega}{1} - \tan^{-1}\frac{2\omega}{1} - \tan^{-1}\frac{0.2\omega}{1}$$

$$-\tan^{-1}\frac{0.05\omega}{1} \qquad\qquad (7-24b)$$

Since the break frequencies are $\omega = 0.5$, 5, 10, and 20 rad/s, we will vary the value of ω from 0.91 to 1000 rad/s. The corresponding values of magnitudes and phase angles are given in Table E7–11. Note that the normalized phase angle $\theta_N(\omega) = \theta(\omega)$.

Finally, the data of Table E7–11 are used to plot the Bode diagrams of Fig. E7–11. From the Bode diagrams we can see that the magnitude is -26 dB when the phase angle is $-180°$ or the gain margin is -26 dB. Thus for the system to be marginally stable, we have to shaft the normalized magnitude plot up by 26 dB so that the resultant gain and phase margins are zero. In short

$$20\log\frac{K}{500} = 26\text{ dB}$$

or

$$\frac{K}{500} = 10^{1.3}$$

$$K = 9976.3$$

The frequency at which the magnitude is 0 dB and the phase angle is $-180°$ is 5.8 rad/s. Thus the system is stable if $K < 9976.3$. However, it is marginally stable if $K = 9976.3$ and unstable if $k > 9976.3$.

K for a Desired Phase Margin

Example 7–11 illustrated that the stability of a system may depend on the system gain K. Therefore, we have to limit the value of K to maintain the system's stability. Often, we have to adjust the gain K to ensure a desired phase margin for the system. This is because the phase margin is a function of system gain K.

EXAMPLE 7–12

Adjust the system gain K in the transfer function of Example 7–11 so that its phase margin is $-54°$.

SOLUTION First we determine the phase angle that will result in a phase margin of $-54°$. Therefore, using Eq. (7–22), we have

$$\text{phase margin} = -180° - [\text{phase angle at } M(\omega) = 0\text{ dB}]$$

$$-54° = -180° - [\text{phase angle at } M(\omega)\text{ dB} = 0\text{ dB}]$$

or

$$[\text{phase angle at } M(\omega) \text{ dB} = 0 \text{ dB}] = -180° + 54°$$
$$= -126°$$

Next, we can readily use the Bode plots of Fig. E7–11 to obtain the gain K for a phase angle of $-126°$.

First on the phase angle plot we will locate the $-126°$ point. This angle is denoted as point A. Point A is then projected onto the normalized magnitude curve, which intersects the magnitude curve at point B. The distance between point B and the 0-dB axis is 15 dB, which is the required gain in decibels to give us a desired phase margin of $-54°$. Thus

$$20 \log \frac{K}{500} = 15 \text{ dB}$$

or

$$K = 2811.7$$

In other words, with $K = 2811.7$ the given transfer function becomes

$$G(s)H(s) = \frac{5.623}{(0.05s + 1)(0.1s + 1)(0.2s + 1)(2s + 1)}$$

If we plot Bode diagrams for the transfer function above, the magnitude $M(\omega)$ should be 0 dB when the phase angle is approximately $-126°$ or the phase margin is $-54°$. Thus we can adjust the K for a desired value of the phase margin.

Effect of Delay on System Stability

There are many control systems that have a time delay. A *time delay* is defined as the time elapsed since the input is applied and until the time output is produced. A time delay occurs especially in mechanical and hydraulic systems which have a movement of a substance that requires a finite time to pass from an input to an output. For example, in a home water system, there is a finite time delay between the valve adjustment and the water output through the sink taps. In this example, the time delay is a function of the length of water pipe between the valve and the sink taps, its diameter, and the pressure of the water. The delay time can be reduced by decreasing the length of pipe between the valve and the sink taps, and increasing the water pressure. Another system in which time delay occurs is a toilet. The delay time is equal to the time it takes a toilet to refill to its initial level after it is flushed. If the toilet is flushed again while it is refilling, the amount of water discharged is not as much as it would be if the toilet were refilled. In short, the operation of a system is affected by its delay time. For digital circuits (ICs) a delay time is commonly referred to as a *propagation time*. Sometimes a delay time is also called a *dead time* because no output is produced during this time.

A time delay without attenuation is called a *pure time delay* and is given by the following transfer function:

$$G_D(s) = e^{-SD} \qquad (7-25a)$$

where D is delay or dead time. When s is replaced by $j\omega$, we have

$$G_D(s) = e^{-j\omega D}$$
$$= \cos \omega D - j \sin \omega D \qquad (7-25b)$$

To plot Bode diagrams we need the magnitude and the phase angle equations for the time-delay term. Hence

$$|G_D(j\omega)| = \sqrt{\cos^2 \omega D + \sin^2 \omega D}$$
$$= 1$$

or

$$20 \log |G_D(j\omega)| = 20 \log 1 = 0 \qquad (7-25c)$$

and

$$\theta_D(\omega) = \tan^{-1} \frac{-\sin \omega D}{\cos \omega D}$$
$$= -\omega D \qquad (7-25d)$$

In short, a time delay, e^{-SD}, introduces a phase lag but no magnitude change. Example 7–13 shows how a delay time affects the stability of a control system.

EXAMPLE 7–13

The chemical concentration control system of Fig. E7–13a is to maintain a constant composition of the output mixture by adjusting the feed-flow valve.[*] The transport time D of the feed along the conveyor is 2 s.

(a) Draw the Bode diagrams for the given system.
(b) Determine the stability of the system by computing the gain margin and the phase margin.

SOLUTION (a) The loop transfer function for the block diagram of Fig. E7–13a is

$$G(s)H(s) = \frac{5(s + 1)e^{-2S}}{s(5s + 1)}$$

Substituting $s = j\omega$, we have

$$G(j\omega)H(j\omega) = \frac{5(j\omega + 1)e^{-2j\omega}}{(j\omega)(5j\omega + 1)}$$

Hence the magnitude and the phase angle equations are

[*] D. Y. Etchart, "Forecasting and compensating the effects of deadtime on a commonly applied chemical pacing control loop," *ISA Transactions*, 16(4): 59–67, 1977.

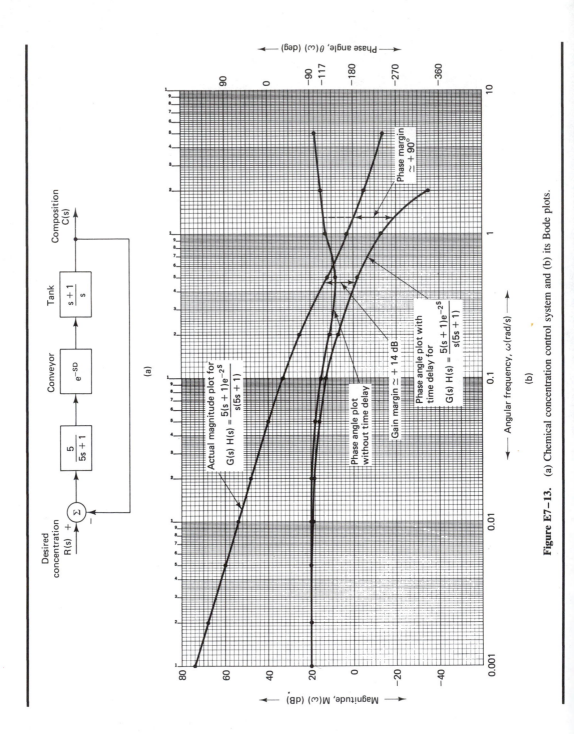

Figure E7–13. (a) Chemical concentration control system and (b) its Bode plots.

$$20 \log \left| G(j\omega)H(j\omega) \right| = 20 \log \left| \frac{5(j\omega + 1)e^{-2j\omega}}{j\omega(5j\omega + 1)} \right|$$

or

$$M(\omega) \text{ dB} = 20 \log 5 + 20 \log \sqrt{\omega^2 + 1} - 20 \log \omega$$
$$- 20 \log \sqrt{(5\omega)^2 + 1} \qquad (7-26a)$$

and

$$\theta(\omega) = \tan^{-1}\frac{\omega}{1} - 2\omega - \tan^{-1}\frac{\omega}{0} - \tan^{-1}\frac{5\omega}{1}$$

or

$$\theta(\omega) = -90° - 2\omega - \tan^{-1}\frac{5\omega}{1} + \tan^{-1}\frac{\omega}{1} \qquad (7-26b)$$

Next we will obtain the data in Table E7–13 by substituting the values of ω from 0.001 to 5 rad/sec in Eqs. (7–26a) and (7–26b).

TABLE E7–13 DATA FOR EXAMPLE 7–13

Angular frequency, ω (rad/s)	Magnitude, $M(\omega)$ (dB)	Phase angle, $\theta(\omega)$ (deg)	
		Without time delay	With time delay
0.001	73.98	−90.23	−90.35
0.002	67.96	−90.46	−90.69
0.005	60	−91.14	−91.71
0.01	53.97	−92.29	−93.44
0.02	47.92	−94.56	−96.86
0.05	39.75	−101.2	−106.9
0.1	33.1	−110.86	−122.32
0.2	25.12	−123.69	−146.59
0.5	12.37	−131.63	−188.93
1.0	2.84	−123.69	−238.29
2.0	−5.09	−110.86	−340.06
5	−13.82	−99.02	−672.02

The data of Table E7–13 are used to construct the magnitude and phase angle plots of Fig. E7–13b. Note that the phase angle plot without the time delay is also drawn to illustrate the effect of time delay on the stability of a system.

(b) From the Bode plots of Fig. E7–13b we can see that

$$\text{gain margin} \simeq 14 \text{ dB}$$
$$\text{phase margin} \simeq +90°$$

Hence the system is unstable. However, if we ignore the time delay, the same system becomes stable becuase the phase margin is −63° and the gain margin is undefined.

From Example 7–13 it is obvious that the time delay affects the stability of a control system. Although time delay cannot be completely eliminated from some control systems, it may certainly be reduced. In addition, the system loop gain may be adjusted until the system is stable.

7–9 TIME-DOMAIN VERSUS FREQUENCY-DOMAIN ANALYSIS

In most control systems, the state of the output and its variation with time are usually of major interest. In such systems an input signal is applied to a system and its output response is studied in the time domain. The time-domain response of a control system is composed of two parts: the *transient response* and the *steady-state response*.

In frequency-domain analysis of a control system, a sinusoidal input with variable frequency is used. The frequency of the input is varied over a desired frequency range so that the magnitude and phase angle variations can be measured. Thus the frequency response is defined as the steady-state response of a system subjected to a sinusoidal input of a fixed amplitude and variable frequency. Bode diagrams are one of the most commonly used and convenient means of obtaining the steady-state response for a system subjected to a sinusoidal input.

Time-domain analysis of a control system is performed by using such analytical techniques as Laplace transform and inverse Laplace transform. On the other hand, frequency-domain analysis is performed by using such graphical techniques as Bode diagrams.

SUMMARY

1. The frequency response is defined as the steady-state response of a system subjected to a sinusoidal input of fixed amplitude and variable frequency. Bode diagrams are the graphical method most commonly used to study the frequency response of control systems. Bode diagrams are composed of magnitude versus frequency and phase angle versus frequency plots.

2. The frequency response analysis of a given system using Bode plots requires its loop transfer function. The loop transfer function of all practical control systems may contain one or more of the following factors: $\pm K$, $(s)^{\pm 1}$, $(s\tau + 1)^{\pm 1}$, or $(s^2 + 2\zeta\omega_n s + \omega_n^2)^{\pm 1}$. An asymptotic magnitude plot is an approximate plot because it uses straight-line approximation. Actual Bode plots are more accurate than asymptotic plots and hence are commonly used.

3. The stability of a system is determined from its Bode diagrams by computing the gain margin and the phase margin. A gain margin is defined as the magnitude in decibels corresponding to a phase angle of $-180°$. On the other hand, a phase margin is defined as the phase difference between $-180°$ and the phase angle at $M(\omega)$ dB $= 0$. A system is said to be stable if its gain margin and phase margin are both negative. However, for some systems the gain margin is undefined; therefore, phase margin is used to define its stability, or vice versa. A marginally stable system has zero phase margin and zero gain margin.

4. A system with time delay may cause it to be unstable because the time delay contributes a negative phase angle, which increases with an increase in angular frequency ω.

5. The major difference between time-domain and frequency-domain analysis is that the first is performed using such analytical techniques as Laplace transform, whereas the latter is performed using such graphical techniques as Bode diagrams. Time-domain analysis of a system is composed of transient and steady-state response, whereas frequency-domain analysis (i.e., Bode diagrams) give the steady-state response of the system.

QUESTIONS

7–1. What is the difference between time-domain and frequency-domain analysis? Explain.

7–2. What is a frequency response?

7–3. In Bode diagrams, angular frequency ω is assigned a log scale. Why?

7–4. Why should the magnitude and phase angle plots be drawn on the same sheet of semilog paper?

7–5. Compare and contrast asymptotic and actual Bode magnitude plots.

7–6. What is the difference between open-loop, closed-loop, and loop transfer functions for a typical closed-loop system?

7–7. List the factors found most commonly in the loop transfer function of a system.

7–8. Is 6 dB/octave the same as 20 dB/decade? Explain.

7–9. What is the difference between a break frequency and a crossover frequency?

7–10. Define a gain margin and a phase margin with reference to Bode diagrams.

7–11. What is the difference between a stable, an unstable, and a marginally stable system in terms of gain and phase margins?

7–12. Explain the effect of loop gain K on the stability of a system.

7–13. Explain the significance of time delay and its effect on system stability. Give a practical system example to support your explanation.

PROBLEMS

7–1. Make polar plots for networks with the following transfer functions:

(a) $\dfrac{V_o(s)}{V_i(s)} = \dfrac{-2}{s}$

(b) $\dfrac{V_o(s)}{V_i(s)} = 10 + \dfrac{5}{s}$

7–2. The transfer functions for the ideal and practical integrators of Fig. P7–2 are given in parts (a) and (b), respectively. Draw Bode diagrams for these networks, and determine the break and crossover frequencies in each case.

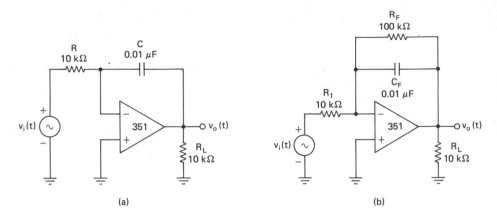

Figure P7–2. (a) Ideal integrator; (b) practical integrator of Problem 7–2.

(a) $\dfrac{V_o(s)}{V_i(s)} = \dfrac{-10^4}{s}$

(b) $\dfrac{V_o(s)}{V_i(s)} = \dfrac{-10}{0.001s + 1}$

Compare and contrast the Bode plots obtained in parts (a) and (b).

7–3. The transfer function of the ac amplifier of Fig. P7–3 is

$$\frac{V_o(s)}{V_i(s)} = \frac{0.0047s}{0.0001s + 1}$$

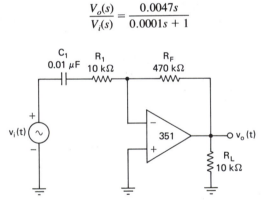

Figure P7–3. Ac inverting amplifier of Problem 7–3.

(a) Draw the asymptotic magnitude plot for the amplifier.
(b) Draw the actual Bode plots for the amplifier.
(c) From the actual magnitude plot determine the crossover frequency.

7–4. The transfer functions for the ideal and practical differentiation amplifiers of Fig. P7–4 are given in parts (a) and (b), respectively. Draw Bode plots for these networks, and in each case determine the break and crossover frequencies, if applicable.

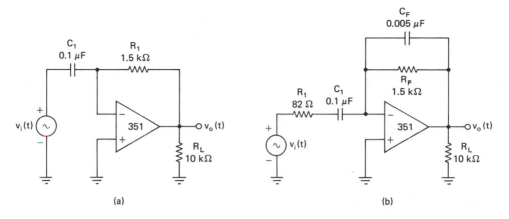

(a)

(b)

Figure P7–4. (a) Ideal differentiation amplifier; (b) practical differentiation amplifier of Problem 7–4.

(a) $\dfrac{V_o(s)}{V_i(s)} = -0.00015s$

(b) $\dfrac{V_o(s)}{V_i(s)} = \dfrac{-(24.4)\,(10^5)s}{[s + 1.3(10^5)]\,[s + 1.2(10^5)]}$

7–5. The transfer function of the multiple-feedback narrow bandpass filter of Fig. P7–5 is as follows:

$$\frac{V_o(s)}{V_i(s)} = \frac{s/R_1 C_1}{s^2 + [(C_1 + C_2)/R_3 C_1 C_2]\,s + (R_1 + R_2)/R_1 R_2 R_3 C_1 C_2}$$

$$= \frac{21.3(10^3)s}{s^2 + 2(10^3)s + 3.7(10^7)}$$

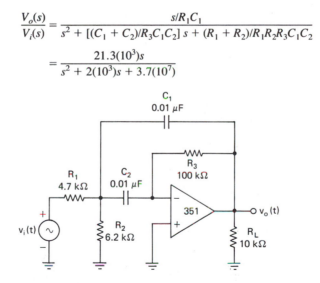

Figure P7–5. Multiple-feedback narrow-bandpass filter of Problem 7–5.

Draw actual Bode plots for the transfer function and then determine the crossover frequency.

7–6. Draw the actual Bode diagrams for the second-order high-pass Butterworth filter shown in Fig. P7–6. The transfer function of the filter is

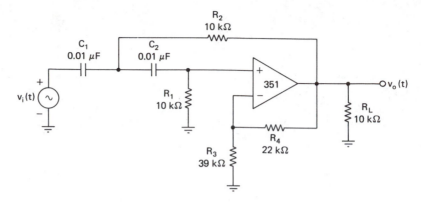

Figure P7–6. Second-order high-pass Butterworth filter of Problem 7–6.

$$\frac{V_o(s)}{V_i(s)} = \frac{A_F s^2}{s^2 + [(3 - A_F)/RC] s + 1/R^2C^2}$$

$$= \frac{(1.6)s^2}{s^2 + 1.4(10^4)s + 10^8}$$

7–7. A second-order band-reject filter using a twin-T network has the following transfer function (see Fig. P7–7):

$$\frac{V_o(s)}{V_i(s)} = \frac{[(R_1 + R_2)/R_1] (s^2 + 1/R^2C^2)}{s^2 + \{[2 - 2(R_2/R_1)]/RC\} s + 1/R^2C_2}$$

$$= \frac{1.3[s^2 + 14(10^4)]}{s^2 + 528s + 14(10^4)}$$

Draw the actual Bode diagrams.

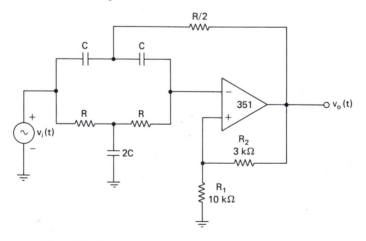

Figure P7–7. Second-order band-reject filter of Problem 7–7.

7–8. The asymptotic logarithmic magnitude plot for a certain transfer function is shown in Fig. P7–8. Determine the transfer function for the system using dc gain and break frequencies.

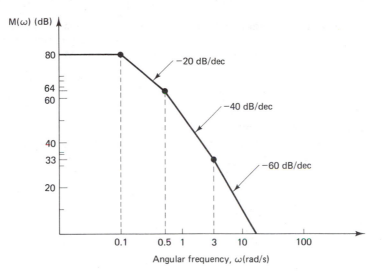

Figure P7-8. Asymptotic magnitude plot for Problem 7–8.

7–9. From the Bode plots obtained in Problem 7–2:
   (a) Calculate the gain and phase margins.
   (b) Determine the stability of the systems in each part.

7–10. From the Bode plots obtained in Problem 7–3, determine the stability of the system.

7–11. Determine the stability of the networks in Problem 7–4 by computing gain and phase margins. Use the Bode plots obtained in Problem 7–4.

7–12. From the Bode plots obtained in Problem 7–5:
   (a) Calculate the gain and phase margins.
   (b) Determine the stability of the system.

7–13. From the Bode plots obtained in Problem 7–6:
   (a) Calculate the gain and phase margins.
   (b) Determine the stability of the system.

7–14. Determine the stability of the system in Problem 7–7 by computing the gain and phase margins. Use the Bode plots obtained in Problem 7–7.

7–15. The loop transfer function of a speed control system is

$$G(s)H(s) = \frac{K}{(s + 100)(s + 25)}$$

Use Bode diagrams to determine the loop gain K that will result in a phase margin of $-45°$.

7–16. The loop transfer function of a certain control system is

$$G(s)H(s) = \frac{100e^{-SD}}{(s + 10)(s + 2)}$$

Using Bode diagrams, establish the amount of time delay D in seconds that will result in a marginally stable system.

7–17. For the system of Problem 7–16, determine the maximum amount of time delay D if the desired phase margin is $-30°$.

7–18. Write a program in BASIC for determining the magnitude in decibels and the phase angle in degrees for the system of Problem 7–3. Run the program for ω = 1 k, 2 k, 10 k, 20 k, 50 k, 100 k, 500 k, 1 M, 2 M, 5 M, and 10 M rad/s. Compare the results with those obtained in Problem 7–3.

LAB EXPERIMENT 7–1
FREQUENCY RESPONSE OF A FIRST-ORDER
LOW-PASS FILTER

Objective. To verify the frequency response of a first-order low-pass filter. At the end of this experiment, you should be able to:

1. Obtain the transfer function of a given first-order filter.
2. Compute the theoretical data for Bode diagrams of a first-order filter.
3. Obtain the experimental data for frequency response plots of the low-pass filter.
4. Construct Bode plots for the low-pass filter from experimental data.
5. Discuss the difference between theoretical and experimental Bode diagrams of a first-order low-pass filter.

Equipment

Dual-trace oscilloscope
Audio-signal generator
±15-V power supply

Materials

351 or equivalent op amp
3.3-kΩ resistor
Three 10-kΩ resistors
0.047-μF capacitor
Semilog paper

Procedure

1. Using the Laplace transform methods, obtain the transfer function $V_o(s)/V_i(s)$ for the circuit of Fig. LE7–1.
2. Find the magnitude in decibels and phase angle equations for the filter using the transfer function obtained in step 1.

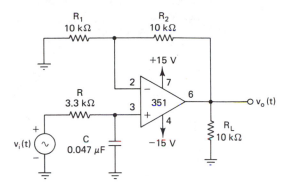

Figure LE7–1. First-order low-pass filter (pin numbers refer to 8-pin mini DIP).

3. Calculate the frequency response data $M(f)$ dB and $\theta(f)$ in degrees at various values of input frequency f. Complete Table LE7–1.1.

TABLE LE7–1.1 THEORETICAL DATA FOR
BODE DIAGRAMS OF FIRST-ORDER
LOW-PASS FILTER OF FIG. LE7–1

Input frequency, f (Hz)	Magnitude, $M(f)$ (dB)	Phase angle, $\theta(f)$ (deg)
1. 10		
2. 20		
3. 50		
4. 100		
5. 200		
6. 500		
7. 1 k		
8. 2 k		
9. 5 k		
10. 10 k		
11. 20 k		
12. 50 k		
13. 100 k		

4. Construct the first-order filter of Fig. LE7–1.

5. Apply sinusoidal input $v_i(t)$ of amplitude 1-V peak to peak at 10 Hz. Use a dual-channel oscilloscope to measure the input and output signals. Record these values in Table LE7–1.2. Vary the frequency of operation from 10 Hz through 100 kHz, measuring the input and output amplitudes as well as the phase shift between them at each frequency setting. Enter the measured values in Table LE7–1.2. Make sure that the oscilliscope is used in an appropriate triggering mode. Complete Table LE7–1.2.

TABLE LE7–1.2 EXPERIMENTAL DATA FOR BODE PLOTS OF
FIRST-ORDER LOW-PASS FILTER OF FIG. LE7–1

Input frequency, f (Hz)	$v_i(t)$ p-p (V)	$v_o(t)$ p-p (V)	$\dfrac{v_o}{v_i}(t)$	$20 \log \dfrac{v_o}{v_i}$	Phase shift, $\theta(\omega)$, between $v_o(t)$ and $v_i(t)$ (deg)
10	1				
20	1				
50	1				
100	1				
200	1				
500	1				
1 k	1				
2 k	1				
10 k	1				
20 k	1				
50 k	1				
100 k	1				

6. On semilog paper construct the Bode diagrams using the theoretical data of Table LE7–1.1.

7. Construct Bode diagrams on the semilog paper using the experiment data of Table LE7–1.2.

Test data

Evaluation. The responses to the following questions and/or conclusions must be supported whenever possible by the applicable theory.

1. What is the difference between theoretical and experimental Bode diagrams? Explain your answer.
2. How does the theoretical break frequency compare to the corresponding experimental break frequency? If they differ by more than 10%, suggest causes.
3. Is the filter of Fig. LE7–1 absolutely stable, marginally stable, or unstable? Explain.

EXPERIMENT 7–2
FREQUENCY RESPONSE OF A PEAKING AMPLIFIER

Objective. To verify the frequency response of a peaking amplifier. At the end of this experiment, you should be able to:

1. Calculate the transfer function of a peaking amplifier.
2. Compute the theoretical data for the Bode plots of a peaking amplifier.
3. Obtain the experimental data for the Bode plots of a given peaking amplifier.

4. Construct Bode diagrams for the peaking amplifier using the experimental data.

5. Explain the difference between theoretical and experimental Bode plots of a peaking amplifier.

Equipment

Dual-trace oscilloscope
Audio-signal generator
± 15-V power supply

Materials

351 or equivalent op amp
Two 1-kΩ resistors
10-kΩ resistor
100-kΩ resistor
10-mH inductor with $R_L \leq 30\ \Omega$
0.01-μF capacitor
Semilog paper

Procedure

1. Using the Laplace transform methods, obtain the transfer function $V_o(s)/V_i(s)$ for the peaking amplifier of Fig. LE7–2.

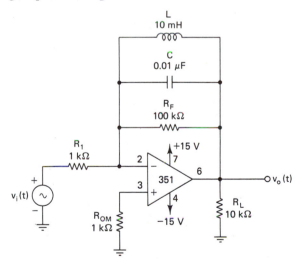

Figure LE7–2. Peaking amplifier (pin numbers refer to 8-pin mini DIP).

2. Obtain the magnitude in decibels and the phase angle equations for the peaking amplifier from the transfer function obtained in step 1.

3. Calculate the frequency response data $M(f)$ dB and $\theta(f)$ in degrees at the frequencies listed in Table LE7–2.1. Complete the table.

TABLE LE7–2.1 THEORETICAL DATA FOR BODE DIAGRAMS OF THE PEAKING AMPLIFIER OF FIG. LE7–2

Input frequency, f (Hz)	Magnitude, $M(f)$ (dB)	Phase angle, $\theta(f)$ (deg)
1. 100		
2. 200		
3. 500		
4. 1000		
5. 2 k		
6. 5 k		
7. 10 k		
8. 20 k		
9. 50 k		
10. 100 k		
11. 200 k		
12. 500 k		
13. 1 M		
14. 2 M		
15. 5 M		
16. 10 M		

4. Construct the peaking amplifier of Fig. LE7–2.

5. Apply sinusoidal input $v_i(t)$ of 1 V peak to peak at 100 Hz. Use a dual-channel oscilloscope to measure the input and output signals. Record these values in Table

TABLE LE7–2.2 EXPERIMENTAL DATA FOR BODE PLOTS OF THE PEAKING AMPLIFIER OF FIG. LE7–2

Input frequency, f (Hz)	$v_i(t)$ p-p (V)	$v_o(t)$ p-p (V)	$\dfrac{v_o}{v_i}(t)$	$20 \log \dfrac{v_o}{v_i}$	Phase shift, $\theta(\omega)$, between $v_o(t)$ and $v_i(t)$ (deg)
1. 100	1				
2. 200	1				
3. 500	1				
4. 1000	1				
5. 2 k	1				
6. 5 k	1				
7. 10 k	1				
8. 20 k	1				
9. 50 k	1				
10. 100 k	1				
11. 200 k	1				
12. 500 k	1				
13. 1 M	1				
14. 2 M	1				
15. 5 M	1				
16. 10 M	1				

LE7–2.2. Vary the frequency of input from 100 Hz through 10 MHz, measuring the input and output amplitudes and the phase shift between them at each frequency setting. Enter the measured values in Table LE7–2.2. Make sure that the oscilloscope is used in an appropriate triggering mode. Complete Table LE7–2.2.

6. On semilog paper construct Bode diagrams using the theoretical data of Table LE7–2.1.

7. Using the experimental data of Table LE7–2.2, construct the Bode diagrams on the semilog paper.

Test data

Evaluation. The responses to the following questions and/or conclusions must be supported whenever possible by the applicable theory.

1. What is the difference between the theoretical and experimental Bode diagrams? Explain your answer.
2. At what frequency does the amplitude of the output peak? (Use an experimental Bode magnitude plot.)
3. What is the amplitude of the output signal at peak frequency?
4. Is the peaking amplifier of Fig. LE7–2. stable? Explain your answer.

ANALOG CONTROL SYSTEMS

chapter 8 ——————————

8–1 INTRODUCTION

Often it is said that we live in an analog world, where changes in the physical properties of our environment generally occur in a continuous manner. Light, heat, sound, and motion are examples of energy forms that we experience in our everyday life. Conversion of energy to an electrical form is required in applications where an electrical system processing the energy responds only to voltages or currents. A microphone, for example, converts the acoustical energy to an electrical signal. This signal may be amplified and then converted back to sound waves by means of the speaker. The microphone and the speaker are examples of a transducer, which converts energy from one form to another. The output of the microphone is an example of analog information, which is continuous, as opposed to the digital information, which is in discrete form. The analog signal is also an exact electrical replica of a physical entity. If the acoustical energy incident on the surface of the microphone and the voltage at the output of the microphone were both plotted as a function of time, the two waveforms would be indistinguishable.

A control system is a combination of amplifiers, transducers, and actuators which collectively act on a process to maintain some condition at a required value. A home heating system where the thermostat is the temperature sensing element is an example of a temperature control system. Control systems generally use negative feedback, which helps to improve system performance. A typical feedback path begins at the system's output, which may be temperature, speed of a motor, or an angular position of the motor's shaft, and end at the input to the summing junction, which is an electrical amplifier requiring the presence of voltage at that point. The transducer must then be in the feedback path converting the temperature or the speed of the motor or the angle of the motor's shaft to an electrical voltage which is suitable as an input to the summing amplifier. The transducer thus plays an indispensable role in control systems, as it provides a bridge between the electrical world and the world of physical quantities, such as heat, speed, position, and so on.

Although the home heating system is an example of a practical workable system, its mathematical treatment at best is not an easy one. The reason for this is the nonlinear operation of this system. The nonlinear character is due primarily to the switching action of various components within the system, such as the solenoid zone control valves, which are turned on and off by the thermostat. In short, this is an on–off type of control system which is incapable of incremental sensing and control.

In this chapter attention is focused only on linear analog control systems, and a mathematical treatment is provided for such systems. In a typical linear system, each component must be linear. The inference is that it is possible to represent each linear component by a linear block with a transfer function, and the output of the linear block is linearly related to its input. From the mathematical view, each linear component must satisfy the superposition principle.

Four control systems are presented in this chapter: speed and position control using a dc PM motor, temperature control, and liquid-level control. There are three first-order systems, and the position control is a second-order system. Being very different from the other three, the position control system is presented separately in its own section. The first-order systems are also discussed in separate sections; however, due to their similarity, the performance characteristics of the three first-order systems are combined in a single section. This chapter also includes, as a special feature, several practical control systems that were constructed and tested in the laboratory.

8–2 SPEED CONTROL SYSTEM

An open-loop speed control system is shown in Fig. 8–1. The input voltage V_i is first amplified by $-R_f/R$ and then by the voltage gain $-K_1$ of the power amplifier. The power amplifier also generates the necessary current to drive the motor. The armature voltage V_a establishes the operating speed ω_m of the motor. The load that the motor drives consists of inertia and viscous friction. The direction of rotation may be reversed by reversing the polarity of V_i.

The transfer function of the open-loop speed control system is obtained from the open-loop system block diagram as the product of the motor transfer function and the gain $K_a = (-R_f/R)(-K_1)$. Hence

$$\frac{\Omega_m(s)}{V_i(s)} = \frac{K_a K_m}{\tau_m s + 1} \tag{8-1}$$

The reader is referred to Chapter 3 for the dc motor operation, motor models, and parameters of the motor.

An open-loop system is extremely sensitive to (1) changes in load, (2) changes in parameter values such as the amplifier gain, and (3) external disturbances. The inability

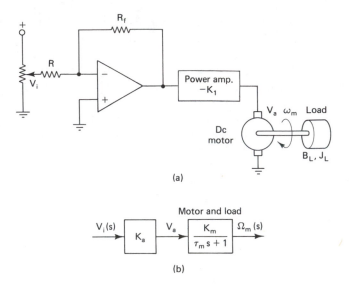

Figure 8–1. Open-loop control system: (a) system diagram; (b) block diagram.

of an open-loop system to compensate for such effects results in a variation in the operating speed of the motor, and therefore of the load. This is a serious disadvantage which is typical of open-loop operation. The use of negative feedback in a closed-loop speed control system, which is considered next, solves these problems.

As shown in the configuration of Fig. 8–2, a generator whose shaft is mechanically coupled to the shaft of the motor serves as the feedback component, providing negative feedback. Other components may also be used to provide negative feedback; the optical disk, OID, FVC speed-sensing configuration shown in Fig. 2–31 is one such example.

The feedback voltage V_f from the generator represents the *actual* speed of the motor, and the input voltage V_i represents the *desired* motor speed. The difference $V_i - V_f$ is amplified by $(-R_f/R)$ $(-K_1)$ and becomes the motor input voltage V_a. The motor responds to this input and changes its speed in the direction to reduce the difference signal $V_i - V_f$ from the summing junction to a value that is close to zero but not equal to zero. This is a consequence of negative feedback. Positive feedback, on the other hand, would cause an increase with time of the summing junction output resulting in an unstable operation. Under steady-state conditions the difference between the desired motor speed represented by V_i and the actual motor speed is the steady-state error, which is discussed later.

A change in the speed of the motor caused by a load variation or amplifier gain change (which may be due to aging or unstable parameters or power supply change) or by external disturbances causes the difference signal from the summing junction to change, which after amplification acts on the motor, forcing it to return to its original speed. The exact original speed (before the change takes place) may never be recovered. One reason for this, as will be shown later, is the increase in steady-state error due to the increase in load. In any case, the closed-loop system senses the change in the operating speed and provides compensation for various load and disturbance effects

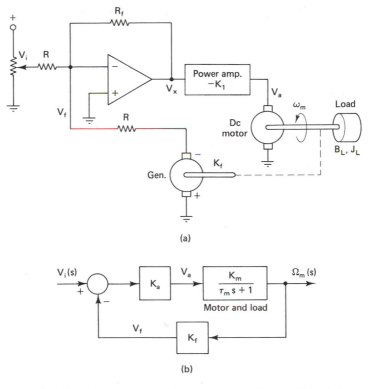

(a)

(b)

Figure 8–2. Closed-loop speed control system: (a) system diagram; (b) block diagram.

that are responsible for the speed change. This is not possible to accomplish with the open-loop system, which has no provision to sense and make a comparison between the desired and the actual speeds. An ideal open-loop system, on the other hand, is unconditionally stable, whereas it is possible that a system with feedback may become unstable. Closed-loop operation also reduces the system gain, due to the use of negative feedback. The many advantages of closed-loop operation outweigh by far these few disadvantages.

The determination of the differential equation for a closed-loop speed control system may begin with Eq. (3–22), which is a statement of the motor armature current. Restating this equation, we have

$$I_a = \frac{V_a - V_b}{R_a}$$

and the motor developed torque may be expressed by Eq. (3–23) as

$$T_m = K_i I_a = K_t(V_a - K_b \omega_m) \tag{8–2}$$

As we have here a specific system, V_a has a somewhat different meaning than it had in Chapter 3, where the motor was all by itself. Turning to the block diagram of the system in Fig. 8–2b, V_a may be expressed in terms of the closed-loop system parameters as

$$V_a = (V_i - K_f \omega_m) K_a \tag{8–3}$$

Substituting this expression for V_a in Eq. (8–2) and replacing T_m by T_L, the load torque, which in the given system consists of the inertial and viscous friction torques, the following result is obtained:

$$K_t[(V_i - K_f\omega_m)K_a - K_b\omega_m] = (J_L + J_m)\dot{\omega}_m + B_L\omega_m \tag{8–4}$$

Rearranging terms yields

$$(J_L + J_M)\dot{\omega}_m + (B_L + B_m + K_aK_fK_t)\omega_m = K_aK_tV_i(t) \tag{8–5}$$

Dividing both sides of Eq. (8–5) by $(B_L + B_m)$ and using the definitions of K_m and τ_m from Chapter 3, we have

$$\tau_m\dot{\omega}_m + (1 + K_aK_mK_f)\omega_m = K_aK_mV_i(t) \tag{8–6}$$

Dividing both sides of Eq (8–6) by $(1 + K_aK_mK_f)$ gives us

$$\tau_s\dot{\omega}_m + \omega_m(t) = K_sV_i(t) \tag{8–7}$$

where

$$K_s = \frac{G_o}{1 + G_oK_f} \quad \text{(rpm/V or rad/s/V)} \qquad \text{closed-loop system speed constant}$$

$$\tau_s = \frac{\tau_m}{1 + G_oK_f} \quad \text{(s)} \qquad \text{closed-loop system time constant}$$

$$G_o = K_aK_m \,(\text{rpm/V or rad/s/V}) \qquad \text{open-loop gain, the forward-path transfer function}$$

$$G_oK_f = \text{loop gain}$$

Equation (8–7) is the differential equation model of the closed-loop speed control system. The transfer function model may be obtained by taking the Laplace transform of both sides in Eq. (8–7) and then solving for $\Omega_m(s)/V_i(s)$, or it may also be obtained from the system block diagram, as follows:

$$\frac{\Omega_m(s)}{V_i(s)} = \frac{K_s}{\tau_s s + 1} \tag{8–8}$$

As usual, all initial conditions must be set to zero when using the transfer function in the analysis.

8–3 POSITION CONTROL SYSTEM

As was described in the preceding section, the operation of the speed control system is preoccupied primarily with the task of maintaining the speed of the load at a constant value. The position control system, on the other hand, advances the load through a desired angular displacement and then comes to a stop. A position control system using a dc motor cannot be operated open loop as was the speed control system. The reason, of course, is that an applied voltage will cause the motor shaft to spin without stopping.

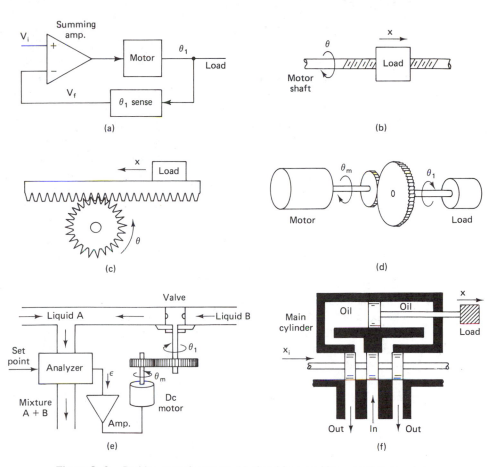

Figure 8-3. Position control systems: (a) closed-loop position control servo using a dc motor and providing rotational displacement to the load; (b) load configuration using lead screw arrangement which imparts translational motion to the load; (c) rack-and-pinion arrangement which provides translational motion to the load; (d) high-resolution rotational system; (e) position control of valve in a chemical plant; (f) hydraulic position control system.

Open-loop position control systems are possible, but then a stepper motor must be used. Feedback is often used even with the stepper motor positioning systems to guarantee that the next excitation input is applied only after the feedback circuits sense that a step was made by the motor. This applies particularly to critical positioning applications where no missed steps are permitted.

A conventional position control system is shown in a simple block diagram of Fig. 8-3a. This type of system will be used in analysis later. The lead screw and the rack-and-pinion configuration shown in Fig. 8-3b convert the rotational motion of the motor to the translational motion of the load. The rack-and-pinion structure shown in Fig. 8-3c was also discussed in Chapter 3 as one of possible motor load configurations. The use of gears as shown in Fig. 8-3d reduces the angular motion of the load and thus improves the resolution; for gear teeth ratio of 50:1, a 10° displacement of motor shaft results in 0.2° motion of the load, a displacement that may be difficult to accomplish without the gears. In a chemical process control application shown in Fig. 8–3e, the position control system adjusts the valve opening and thus regulates the flow of chemical B, which mixes with chemical A. When the desired concentration of B in the mixture is attained, the error signal from the analyzer is reduced to zero and the valve is thus adjusted to the proper opening.

A hydraulic positional control system is illustrated in Fig. 8–3f. The externally applied pressure results in a pressure differential across the piston in the oil-filled (or filled with another incompressible liquid) main cylinder. The product of the pressure differential and the cross-section area of the piston results in a force that drives the cylinder in the direction of lower pressure, thus providing translational motion to the load, which is coupled to the shaft of the cylinder. The hydraulic system may be configured as a closed-loop system by sensing the load position and applying the resulting feedback signal together with the desired position signal to a summing junction. The amplified error signal is then applied to the motor driving a rack and pinion, where the rack is coupled to the shaft of the lower cylinder, which controls the pressure application to the main cylinder. The hydraulic systems are generally capable of positioning large loads; a hydraulic car lift is one example.

The position control system shown in Fig. 8–4 is used for the analysis. This system uses a dc PM motor as the load position actuator. Once again the feedback component plays an important role, as it provides the feedback voltage proportional to the actual position of the motor shaft. In this application a potentiometer whose shaft is mechanically coupled to the shaft of the motor is used as the position feedback component. The use of the potentiometer as the angle-to-voltage transducer and the definition of the pot constant K_p were discussed in Chapter 2. The potentiometer used in a servo application such as this is very different in construction from a general-purpose potentiometer which is intended for circuit application, and it is also more expensive. Its shaft is generally mounted on ball bearings to minimize the load on the motor as well as the wearing of its shaft, and the wiper arm, which is frequently subjected to friction, is made of a special material which can better withstand the wear and prolong the useful life of the potentiometer. Some servo pots are multiturn; 10-turn pots, for example, are commercially available. The construction of such pots includes a helix structure which allows the wiper arm to rotate continuously more than one turn.

Another approach to position sensing, which is probably less expensive and adaptable for direct interfacing to a microprocessor, is optical position sensing, shown in Fig. 2–32. This configuration includes an optical disk, two OIDs, a D flip-flop, an 8-bit binary counter, and a DAC. The two OIDs which are phased in quadrature, provide signals for sensing the direction of shaft's rotation, and together with the D flip-flop control the up/down input of the binary counter. This is important, as the counter, whose contents must accurately represent the instantaneous position of the motor shaft, is incapable of recognizing the direction of rotation on the basis of the incoming pulse train.

Returning to the position control system shown in Fig. 8–4, where the potentiometer is used, for the sake of simplicity, as a feedback component, the system operation may be described by considering the two inputs to the summing op amp. The feedback

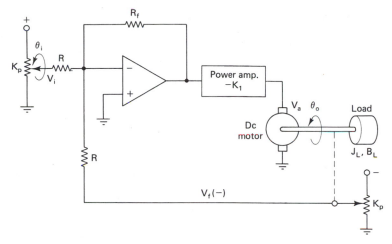

(a)

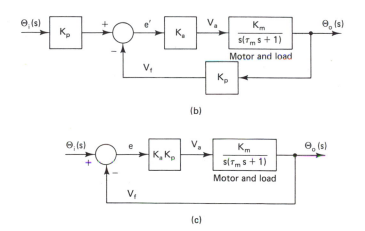

(b)

(c)

Figure 8–4. Position control system: (a) system diagram; (b) block diagram; (c) simplified block diagram.

potentiometer produces V_f, which represents the actual shaft position. The input V_i, representing θ_i, is the desired shaft position. Thus the input θ_i may be regarded as the desired output. The difference signal e', after being amplified by $(-R_f/R)(-K_1)$, becomes the V_a input to the motor. In response to V_a, the motor rotates in the direction to reduce e' as a consequence of negative feedback. When $e' = 0$, the motor stops, as its drive is cut off. This must necessarily occur when $\theta_i = \theta_o$ because both potentiometers have the same constant K_p. The result is that the motor shaft with the load has advanced an angular increment corresponding to the command input θ_i, and in the process position control was accomplished.

System Models

The differential equation model for the position control system is derived by first duplicating Eq. (8–2) after replacing ω_m by $\dot{\theta}_o$, the controlled variable for the position control system. Hence

$$T_m = T_L = (J_m + J_L)\ddot{\theta}_o + B_L\dot{\theta}_l = K_t(V_a - K_b\dot{\theta}_o) \qquad (8\text{–}9)$$

where the load torque includes the inertial and viscous friction effects. The simplified block diagram shown in Fig. 8–4c is obtained from part (b) by moving the summing junction ahead of the input block K_p. From the simplified diagram,

$$V_a = (\theta_i - \theta_o)K_aK_p$$

Substituting this expression for V_a in Eq. (8–9) yields

$$(J_m + J_L)\ddot{\theta}_o + B_L\dot{\theta}_o = K_t[(\theta_i - \theta_o)K_aK_p - K_b\dot{\theta}_o]$$

Rearranging terms and substituting $K_bK_t = B_m$, we obtain

$$(J_m + J_L)\ddot{\theta}_o + (B_L + B_m)\dot{\theta}_o + K_aK_pK_t\theta_o = K_aK_pK_t\theta_i(t) \qquad (8\text{–}10)$$

Dividing both sides of Eq. (8–10) by $(B_m + B_L)$ and using the definitions for τ_m and K_m from Chapter 3 gives us

$$\tau_m\ddot{\theta}_o + \dot{\theta}_o + G_o\theta_o(t) = G_o\theta_i(t) \qquad (8\text{–}11)$$

where $G_o = K_aK_mK_p$ is the dc value of the forward-path transfer function (open-loop gain). The system transfer function is obtained by taking the Laplace transform of both sides of Eq. (8–11), setting all initial conditions to zero, and solving for $\theta_o(s)/\theta_i(s)$; hence

$$\frac{\Theta_o(s)}{\Theta_i(s)} = \frac{G_o/\tau_m}{s^2 + (1/\tau_m)s + G_o/\tau_m} \qquad (8\text{–}12)$$

The transfer function (8–12) may also be derived directly from the block diagram of Fig. 8–4c.

The transfer function for a general second-order system is of the form

$$\frac{C(s)}{R(s)} = \frac{\omega_n^2}{s^2 + (2\zeta\omega_n)s + \omega_n^2} \qquad (8\text{–}13)$$

where ω_n is the natural resonant frequency and ζ is the damping ratio, which is related to energy losses within the system. The closed-loop position control system may therefore be classified as the second-order system on the basis of the form of the system's differential equation (8–11) and the form of its transfer function (8–12) as compared to the form of the transfer function of the general second-order system (8–13). The parameters of the position control system are related to ζ and ω_n through comparison of the two forms as follows:

$$\omega_n = \left(\frac{G_o}{\tau_m}\right)^{1/2} = \left(\frac{K_aK_pK_t}{J_L + J_m}\right)^{1/2} \qquad (8\text{–}14)$$

Solving for ζ from $2\zeta\omega_n = 1/\tau_m = (B_m + B_L)/(J_m + J_L)$, we have

$$\zeta = \frac{1}{2(G_o\tau_m)^{1/2}} = \frac{B_L + B_m}{2[K_aK_pK_t(J_L + J_m)]^{1/2}} \tag{8-15}$$

System Transient Response to Step and Ramp Inputs

A detailed derivation of the general second-order-system response to the step and ramp inputs is given in Chapter 6. As the position control system is also a second-order system, the results of Chapter 6 with minor adjustment in symbols are used here, and the reader is referred to Chapter 6 for details. To apply the results of Chapter 6, the parameters of the position control system are related to ζ and ω_n by Eqs. (8-14) and (8-15). Also, the general output $C(s)$ is replaced by $\Theta_o(s)$, the angular position of the motor shaft, and $R(s)$ is replaced by $\Theta_i(s)$, the position input to the system (in the time domain) $c(t)$ is replaced by $\theta_o(t)$ and $r(t)$ is replaced by $\theta_i(t)$.

The responses of the position control system to step and ramp inputs for various conditions of damping are presented in Tables 8-1 and 8-2, respectively. The scaled responses are included together with the real-time responses. The step response magnitudes are scaled by E (the input step height), and the ramp response magnitudes are scaled by E/ω_n (where E is the slope of the input ramp). In both cases the real time and the scaled time are related by $t_n = \omega_n t$. The use of scaling simplifies the response form and reduces the number of parameters. The scaled responses are all a function of ζ. Each of these responses may then be plotted to scale for a given value of ζ without requiring the values of E and ω_n.

The absence of damping implies that there are no energy losses within the system, which is not possible in a practical sense. Therefore, the case $\zeta = 0$ is included for completeness and must be viewed in a purely mathematical sense. As shown in Tables 8-1 and 8-2, the undamped responses oscillate with constant amplitude about the input that serves as the dc or average value of the response.

Step response (Table 8-1).   The underdamped response ($\zeta < 1$) is characterized by damped oscillations. It consists of the steady-state term E (which is same as the input) and the transient term. Under steady-state conditions, when the transient term is reduced to zero, the output is equal to the input. We may therefore conclude that the control system positions the load in response to a step input, without error. The expressions θ_k and T_k are used to find the value of the kth response peak and the corresponding time at which this peak occurs. The maximum difference between the response and the input that occurs at time T_1 of the first peak is characterized by the percent overshoot (POT). The overshoot decreases exponentially with increasing ζ. Time T_1 and also T_r, on the other hand, behave in the opposite manner (i.e., they increase with ζ). This is illustrated graphically in Fig. 6-8. It follows then, that a fast system (T_r is small) must be accompanied by a large value of POT, and vice versa. In practical applications a compromise must be reached between system speed and overshoot. The settling time T_s is the time required by the system to reach the steady state. From a mathematical standpoint, this takes infinite time, but in a practical sense the response whose peaks decay exponentially requires $3\tau_s$ to be within 5% of the steady-state value.

TABLE 8-1 SECOND ORDER SYSTEM RESPONSE TO STEP INPUT $\theta_i(t) = Eu(t)$

Damping	Response equations and parameters	Response waveforms
$\zeta = 0$	$\theta_o(t) = E(1 - \cos \omega_n t)$ Scaled response: $\theta_o(t_n) = 1 - \cos t_n$	
$\zeta < 1$	$\theta_o(t) = E\left[1 - \dfrac{e^{-\zeta\omega_n t}}{\beta} \sin(\omega_n \beta t + \phi)\right]$ Scaled response: $\theta_o(t_n) = 1 - \dfrac{e^{-\zeta t_n}}{\beta} \sin(\beta t_n + \phi)$ kth peak time $T_k = \dfrac{k\pi}{\omega_n \beta}$ kth peak value $\theta_k = 1 + (-1)^{k+1} e^{-k\pi\zeta/\beta}$ POT $= 100 e^{-\pi\zeta/\beta} = 100[(\theta_1 - E)/E]$ Rise time $T_r = \dfrac{\pi - \phi}{\omega_n \beta}$; $T_d = 2\pi/(\omega_n)$ Settling time $T_s = 3\tau_s = 3/(\zeta\omega_n) = 6\tau_m$ where τ_s is the system time constant	
$\zeta = 1$	$\theta_o(t) = E[1 - e^{-\omega_n t}(\omega_n t + 1)]$ Scaled response: $\theta_o(t_n) = 1 - e^{-t_n}(t_n + 1)$ $\tau_s = 1/\omega_n$; $T_r = 3.35/\omega_n$	
$\zeta > 1$	$\theta_o(t) = E(1 - C_1 e^{-t/\tau_1} + C_2 e^{-t/\tau_2})$ Scaled response: $\theta_o(t_n) = 1 - C_1 e^{-(\zeta - \alpha)t_n} + C_2 e^{-(\zeta + \alpha)t_n}$ $C_{1,2} = (\zeta \pm \alpha)/2\alpha$; $\tau_{2,1} = \dfrac{1}{\omega_n(\zeta \pm \alpha)}$ $C_1 - C_2 = 1$; $\tau_1 \tau_2 = 1/\omega_n^2$; $C_2/C_1 = \tau_2/\tau_1 = (\zeta - \alpha)^2$	

Magnitude scale factor = E, time scale $t_n = \omega_n t$.
Equations (8-14) and (8-15) relate the position
control system parameters to ω_n and ζ.

$\phi = \cos^{-1} \zeta$

$\zeta < 1 \quad \zeta^2 + \beta^2 = 1$
$\zeta > 1 \quad \zeta^2 - 1 = \alpha^2$

The time at which the response enters for the first time the band of width $0.1E$ (this is $\pm 5\%$) and centered on E is the settling time T_s. The system is then in the steady state for times greater than T_s. As the criterion for establishing the value of T_s depends largely on the specifications or the performance requirements of a given system, the $3\tau_s$ value used here may be $4\tau_s$ or $5\tau_s$ in other cases. Regardless of the method used to determine T_s, T_s marks the point at which the system steady state occurs, and what is very important, the point at which the next step input may be applied.

The critically damped ($\zeta = 1$) and overdamped ($\zeta > 1$) responses are characterized by an exponential rise toward the steady state. The critically damped system is the fastest, that is, its response time is the lowest as compared to the overdamped system, whose response time increases with the value of ζ. The overdamped system includes two transient terms, one with the time constant τ_1 and the other with a smaller time constant, τ_2. Generally, the transient response is dominated by larger of the two time constants. The conditions under which the transient term with the smaller time constant has a negligible effect have been derived in Chapter 6 and illustrated graphically in Fig. 6–10.

Ramp response (Table 8–2) The underdamped ramp response ($\zeta < 1$) is also characterized by damped oscillations. It includes the steady-state part and the transient term, which decays exponentially with time. In some respects the step and ramp responses are similar. Their transient terms are similar in form having the same time constant $1/\zeta\omega_n$ and having the same damped frequency of oscillation $\omega_n\beta$. There are significant differences between the two responses. First, the steady-state part consists of two terms $Et - 2\zeta E/\omega_n$, where Et is the input ramp. Under steady-state conditions the transient term vanishes and the steady-state value of the output may be expressed as $\theta_{oss} = E(t - 2\zeta/\omega_n)$. Mathematically, this means that the output is equal to the input ramp after the ramp has been translated along the positive time scale by $2\zeta/\omega_n$, which represents the response constant time lag under steady-state conditions. The difference between the input and the output (i.e., $Et - \theta_{oss}$) is also constant and is equal to $2\zeta E/\omega_n$. This suggests that in contrast to the step response, which has no steady-state error term, the ramp response does have a steady-state error, whose value is $2\zeta E/\omega_n$. This is the reason why in the case of the step response the damped oscillations are centered on the step input level E, which is the average value of these oscillations, suggesting that after a sufficiently long time the response becomes the same as the input E, resulting in no error, and in the case of the underdamped ramp response the damped oscillations are centered on the line that is parallel to and displaced from the ramp input by $2\zeta E/\omega_n$, suggesting that after a sufficiently long time (under steady-state conditions) the response becomes the same as the parallel line, resulting in a steady-state error of $2\zeta E/\omega_n$. As the value of ζ is decreased, the steady-state error decreases and the amplitude of the oscillations increases, and vice versa.

These observations coincide with the steady-state error results summarized in Table 5–1. The position control system is a type 1 system, and according to Table 5–1, the steady-state error is 0 for step input and E/HK_1 for ramp input. In the case of the position control system $H = 1$, and $K_1 = \omega_n/2\zeta$, resulting in the steady-state error as deduced above (i.e., $2\zeta E/\omega_n$). The steady-state error is related to the position control system parameters by using Eqs. (8–14) and (8–15) as follows:

$$e_{ss} = \frac{2\zeta E}{\omega_n} = \frac{E}{K_a K_m K_p} = \frac{E(B_L + B_m)}{K_a K_p K_t} \tag{8–16}$$

The critically damped and overdamped responses ($\zeta \geq 1$) are characterized, as in the case for the step response, by the exponential rise toward the steady-state value. The steady-state error appears here also in the same form and has the same meaning as it did with the step response. The critically damped ramp response attains the steady

TABLE 8–2 SECOND ORDER SYSTEM RESPONSE TO RAMP INPUT $\theta_i(t) = Etu(t)$

Damping	Response equations and parameters	Response waveforms
$\zeta = 0$	$\theta_o(t) = \dfrac{E}{\omega_n}(\omega_n t - \sin \omega_n t)$ Scaled response: $\theta_o(t_n) = t_n - \sin t_n$	
$\zeta < 1$	$\theta_o(t) = Et - \dfrac{2\zeta E}{\omega_n} + \dfrac{Ee^{-\zeta\omega_n t}}{\beta\omega_n}\sin(\omega_n \beta t + 2\phi)$ Scaled response: $\theta_o(t_n) = t_n - 2\zeta + \dfrac{e^{-\zeta t_n}}{\beta}\sin(\beta t_n + 2\phi)$ Syst. time const. $\tau_s = 1/\zeta\omega_n$	
$\zeta = 1$	$\theta_o(t) = Et - \dfrac{2E}{\omega_n} + \dfrac{Ee^{-\omega_n t}}{\omega_n}(\omega_n t + 2)$ Scaled response: $\qquad \tau_s = 1/\omega_n$ $\theta_o(t_n) = t_n - 2 + e^{-t_n}(t_n + 2)$	
$\zeta > 1$	$\theta_o(t) = Et - \dfrac{2\zeta E}{\omega_n} + C_3 e^{-t/\tau_1} - C_4 e^{-t/\tau_2}$ Scaled response: $\theta_o(t_n) = t_n - 2\zeta + \dfrac{(\zeta + \alpha)^2 e^{-(\zeta - \alpha)t_n}}{2\alpha} - \dfrac{(\zeta - \alpha)^2 e^{-(\zeta + \alpha)t_n}}{2\alpha}$ $C_{3,4} = E[(\zeta \pm \alpha)^2/2\omega_n\alpha]$ $\tau_{2,1} = \dfrac{1}{\omega_n(\zeta \pm \alpha)} \qquad C_4/C_3 = (\tau_2/\tau_1)^2 = (\zeta - \alpha)^4$	

Magnitude scale factor = E/ω_n, time scale $t_n = \omega_n t$.
Equations (8-14) and (8-15) relate the position control system parameters to ω_n and ζ. The steady-state error is $2\zeta E/\omega_n$. The scaled steady-state error is 2ζ.

$\phi = \cos^{-1}\zeta$
$\zeta < 1 \quad \zeta^2 + \beta^2 = 1$
$\zeta > 1 \quad \zeta^2 - 1 = \alpha^2$

state in shortest time with the time constant $1/\omega_n$. As in the case of the step response, the transient part of the ramp response consists of two terms, and once the response is in progress, the term with the larger time constant (which is $\tau_1 = 1/[\omega_n(\zeta - \alpha)]$) dominates the response. It can easily be verified that for any value of $\zeta \geq 1$, $\tau_1 \geq 1/\omega_n$. Figure 6–10 shows that for values of $\zeta > 1.2$ and for times greater than $1t_n$, the transient term with the smaller time constant contributes less than 5% to the overall response and may therefore be ignored. For values of $\zeta \geq 1$, the critically damped response has the lowest steady-state error since the steady-state error is a function of ζ. As shown in Table 8–2, the steady-state error increases for larger values of ζ.

414 Analog Control Systems Chap. 8

Effect of load on system response.   The effect of load on the position control system response may be deduced from Eqs. (8–14) and 8–15). An increase in the viscous friction B_L with J_L unchanged, increases ζ and therefore increases the product $\zeta\omega_n$ since ω_n remains constant. This results in a lower system time constant τ_s and a lower value of the settling time T_s. Both the step and ramp responses are therefore faster with lower value of the overshoot. It may seem strange that an increase in viscous friction makes the system faster. We have seen, however, a similar situation with the stepper motors, where an external resistor in series with the phase winding was used to reduce the time constant L/R and thereby lower the rise time of the phase current, making the system in the process faster and capable of high-speed operation. This is not surprising, as the character of R and B_L is very similar. Within their respective system each is responsible for energy losses. The reverse effect to that described above results when B_L is lowered and J_L unchanged.

The negative effect, according to Eq. (8–16), of increasing B_L is to increase the value of e_{ss} and thus lower the system accuracy. A reduction in B_L improves the accuracy but unfortunately, as described above, makes the system slower and the overshoot larger.

An increase in load inertia J_L with B_L unchanged lowers both ω_n and ζ. This results in a lower product $\zeta\omega_n$, and therefore in a larger system time constant and a larger settling time T_s. As intuition suggests, a system with a higher inertial load is slower and less responsive to the step or ramp inputs. The increased inertia has no effect on system accuracy, as in the steady state the inertia has no effect on the response. This conclusion may also be deduced from Eq. (8–16), where ζ/ω_n is independent of inertia.

Effect of amplifier gain on system response and system accuracy.   Suppose that the amplifier gain changes from K_{a1} to K_{a2}, and the gain ratio is represented by $K_{a2}/K_{a1} = X^2$. Corresponding to K_{a1}, the damping ratio is ζ_1 from Eq. (8–15), the natural resonant frequency is ω_{n1} from Eq. (8–14), the time of the first response peak T_1 is expressed by Eq. (6–28), the damped frequency is $\omega_{d1} = \omega_{n1}\beta_1$, and the steady-state error is e_{ss1} from Eq. (8–16). The gain K_{a2} determines the values of ζ_2, ω_{n2}, $(T_1)_2$, ω_{d2}, and e_{ss2}. Assuming that the load and all remaining parameters in these equations do not change, the following results, which are obtained by taking ratios, should be verified by the reader:

$$\frac{\zeta_2}{\zeta_1} = \frac{1}{X} \qquad \text{and} \qquad \frac{\omega_{n2}}{\omega_{n1}} = X \qquad\qquad (8–17)$$

and the product

$$\frac{\zeta_2}{\zeta_1}\frac{\omega_{n2}}{\omega_{n1}} = 1$$

Therefore,

$$\zeta_1\omega_{n1} = \zeta_2\omega_{n2}$$

and consequently,

$$\tau_{s1} = \tau_{s2} \qquad \text{(system time constant)}$$

$$T_{s1} = T_{s2} \qquad \text{(settling time)}$$

Also,

$$\frac{\omega_{d2}}{\omega_{d1}} = \frac{(T_1)_1}{(T_1)_2} = X \left| \frac{1 - \zeta_2^2}{1 - \zeta_1^2} \right|^{1/2} \tag{8-18}$$

and

$$\frac{e_{ss2}}{e_{ss1}} = \frac{1}{X^2} \tag{8-19}$$

If the amplifier gain increases so that $K_{a2} > K_{a1}$, ζ decreases and ω_n increases but the product $\zeta\omega_n$ remains constant, and therefore the system time constant and the settling time remain unchanged. Also, ω_d increases, T_1 decreases, and e_{ss} decreases. From this we can draw a conclusion that the system speed and its accuracy are improved when the amplifier gain is increased. However, the increased gain causes ζ to decrease, which results in a larger overshoot according to Eq. (6–29). A decrease in the amplifier gain will result in a lower overshoot (due to the increase in ζ), but the system speed and the accuracy will both be degraded. Thus the increase in gain does not provide a simultaneous reduction in the overshoot and an improvement in system speed. The observation was also made in Chapter 6 and illustrated in Fig. 6–8. To achieve independent control over the overshoot and the speed of the system, independent control of system damping is required. This is possible through the use of velocity feedback, to be discussed later.

EXAMPLE 8–1

Consider the position control system in Fig. 8–4. Its step response is underdamped with 37% overshoot. It is required to reduce the overshoot to 20%. (a) Calculate the required percent change in amplifier gain to meet this requirement. Also calculate the percent change in (b) ω_d, (c) T_1, and (d) e_{ss} which occur due to the change in gain.

SOLUTION (a) From Eq. (6–29), $\zeta_1 = 0.3$ for $POT_1 = 37\%$, and $\zeta_2 = 0.45$ for $POT_2 = 20\%$. Then, from Eq. (8–17),

$$\left(\frac{\zeta_1}{\zeta_2} \right)^2 = X^2 = \frac{K_{a2}}{K_{a1}}$$

$$\frac{K_{a2}}{K_{a1}} = \left(\frac{0.3}{0.45} \right)^2 = 0.44$$

The amplifier gain must therefore be lowered by 56%.
(b), (c) From Eq. (8–18),

$$\left(\frac{0.3}{0.45} \right) \frac{1 - 0.45^2}{1 - 0.3^2} = 0.62$$

Therefore,

$$\frac{\omega_{d2}}{\omega_{d1}} = 0.62 \qquad \text{and} \qquad \frac{(T_1)_2}{(T_1)_1} = \frac{1}{0.62} = 1.6$$

The damped resonant frequency decreases by 38%, and the T_1 peak time increases 60%.

(d) From Eq. (8–19),

$$\frac{e_{ss2}}{e_{ss1}} = \left(\frac{\zeta_2}{\zeta_1}\right)^2 = \left(\frac{0.45}{0.3}\right)^2 = 2.25$$

The system accuracy is therefore degraded by 125%.

Derivative Feedback

The oscillatory character of the underdamped system response is objectionable in many practical applications. First, the next input step or pulse may not be applied until the system has reached the steady state, where the oscillations are reduced to a tolerable level. This limits the frequency of the applied pulses. Second, the oscillatory response presents a real possibility that when subjected to severe oscillations, the load may fail catastrophically. Finally, the oscillatory response may be objectionable on the basis of comfort. An automobile suspension system is one such example. An automobile suspension, which may be represented by a simple model consisting of mass of the car, spring, and the shock absorber (damping), is a second-order system in which the vertical displacement of the car is the system response. Suppose that due to a badly worn shock absorber the system is underdamped. The step input, which may be nicely simulated by a pothole, may cause the automobile to oscillate up and down for the next several hundred feet down the road.

As noted earlier, any attempts to lower the overshoot by raising the gain result in a degradation of system speed and accuracy. Thus another control of system damping is required which is independent of the gain. The velocity feedback provides exactly this type of damping control.

Figure 8–5a shows the position control system with an additional feedback block whose transfer function is sK_f. Consequently, the output of this block in the s-domain is $s\theta_oK_f$, but in the time domain it is ω_oK_f. Recalling that multiplication by s in the s domain is equivalent to performing the differentiation d/dt in the time domain, we may conclude that the additional block in the feedback path provides a feedback voltage proportional to speed of the motor's shaft, and for that reason it is called the velocity feedback or derivative feedback. The other block in the feedback path, whose transfer function is K_p, provides feedback proportional to the angular position of the motor's shaft (i.e., $K_p\theta_o$). When the summing junction is moved ahead of the input K_p block and the two feedback blocks are combined, the resulting feedback transfer function, shown in Fig. 8–5b, is $1 + (K_f/K_p)s$.

We may determine next the expression for the system transfer function with velocity feedback, and from the resulting expression deduce the effect that the velocity feedback has on the system damping ratio. From Fig. 8–5b, $H(s) = 1 + K_f s/K_p$, and $G(s) = K_aK_pK_m/[s(\tau_m s + 1)]$. Substituting G and H in the general expression $G/(1 + GH)$ for

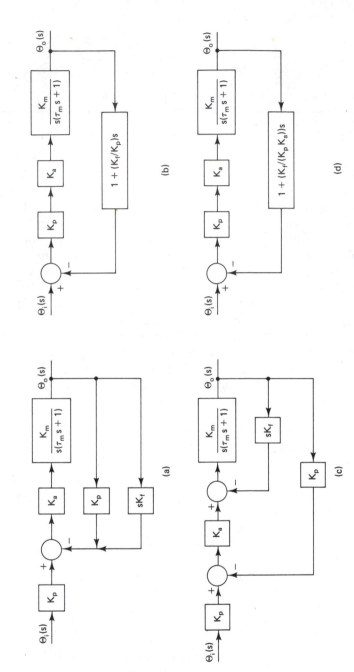

Figure 8–5. Position control system with velocity feedback. Overall feedback: (a) block diagram; (b) simplified block diagram. Local feedback: (c) block diagram; (d) simplified block diagram.

the closed-loop system transfer function with negative feedback, the following expression is obtained after rearrangement of terms:

$$\frac{\theta_o(s)}{\theta_i(s)} = \frac{G_o/\tau_m}{s^2 + (1/\tau_m)(1 + G_oK_f/K_p)s + G_o/\tau_m}$$ (8–20)

where $G_o = K_aK_mK_p$. Comparing transfer function (8–20) with the general second-order system transfer function, we have

$$\omega_n = \left(\frac{G_o}{\tau_m}\right)^{1/2}$$

and

$$2\zeta\omega_n = \frac{1 + G_oK_f/K_p}{\tau_m}$$

Solving for ζ and substituting for G_o, the following result is obtained:

$$\zeta = \frac{1 + K_aK_mK_f}{2(K_aK_mK_p\tau_m)^{1/2}} = \frac{B_L + B_m + K_aK_tK_f}{2[K_aK_pK_t(J_L + J_m)]^{1/2}} \qquad \text{(overall feedback)}$$ (8–21)

Comparing Eqs. (8–15) and (8–21), the latter has the additional term $K_aK_tK_f$ in the numerator, which is associated with the velocity feedback. As K_f may be adjustable, it therefore provides an independent means of adjusting the value of the system damping.

In the configuration where the velocity feedback is used locally around the motor, as shown in Fig. 8–5c, the effect of velocity feedback is slightly different. Comparing the forms of the feedback transfer functions in the simplified block diagrams of Fig. 8–5b and d, that in part (d) may be obtained from part (b) after K_f is replaced by K_f/K_a. This relationship is maintained throughout the derivation (which is exactly the same as above) of the damping ratio with local derivative feedback. The final form of the damping ratio for the local derivative feedback is obtained from Eq. (8–21), replacing K_f by K_f/K_a; hence

$$\zeta = \frac{1 + K_mK_f}{2(K_aK_pK_m\tau_m)^{1/2}} = \frac{B_L + B_m + K_tK_f}{2[K_aK_pK_t(J_L + J_m)]^{1/2}} \qquad \text{(local feedback)}$$ (8–22)

The main difference between Eqs. (8–21) and (8–22) is K_a, which appears in the numerator of the former and does not in the latter. If K_a and K_f are both adjusted in a given situation, the net effect on the damping ratio may be deduced by removing $(K_a)^{1/2}$ from the denominator and applying it to the numerator in Eq. (8–21). Equation (8–21) may be rewritten as

$$\zeta = \frac{(K_a)^{1/2}[(B_L + B_m)/K_a + K_mK_f]}{2(K_pK_m\tau_m)^{1/2}}$$ (8–23)

If the $B_L + B_M$ term in Eq. (8–23) is dominant and K_a is reduced, for example, to one-fourth of its original value, the $B_L + B_m$ term is doubled in value and the K_mK_f term is halved. If the K_mK_f term is dominant (or of the order of magnitude of the $B_L + B_m$ value), the adjustment in K_f will provide the necessary change in the system damping ratio. In a situation where the value of K_a is too low, the desired value of the damping

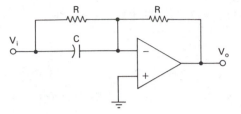

Figure 8–6. Proportional plus derivative feedback circuit.

ratio will depend on the choice of K_f. If $(K_a)^{1/2}$ is factored out in Eq. (8–22) and applied to the numerator, the net effect on the damping ratio is that $(K_a)^{1/2}$ will equally divide each of the terms in the numerator. In contrast to the overall derivative feedback, the local feedback is not amplified by K_a, and therefore it has less of an effect on the damping ratio.

The derivative feedback may be achieved in several ways. A dc generator whose shaft is coupled to the shaft of the motor, may be used. Of course, its constant $K_g = K_f$ is not adjustable, and if adjustment in K_f is required, an amplifier with an adjustable gain may be used in series with the generator to provide the required adjustment in K_f. Another approach, which is cheaper, depends on the use of a differentiator. Consider the circuit shown in Fig. 8–6. Routine analysis shows that its transfer function is

$$V_o(s) = -(1 + RCs)V_i(s)$$

Suppose that V_i is provided by the wiper arm of the feedback potentiometer of the position control system; then $V_i = + K_p\theta_o$, assuming that positive supply is used on the feedback pot. Substituting for V_i in the expression for V_o above, we have

$$V_o(s) = - [K_p + (K_pRC)s]\Theta_o(s) \tag{8–24a}$$

The combined feedback in Fig. 8–5a may be expressed as

$$V_f(s) = -(K_p + K_fs)\Theta_o(s) \tag{8–24b}$$

Comparing Eqs. (8–24a) and (8–24b), we may conclude that

$$K_f = K_pRC \tag{8–25}$$

If the position feedback potentiometer provides the input to the circuit in Fig. 8–6, its output consists of two terms: the $K_p\theta_o$ component, which is associated with the angular position of the motor shaft, and the $(K_pRC)\omega_o$ term, which provides feedback proportional to the speed of the motor. The circuit thus provides *proportional plus derivative feedback*. To produce a proportional plus derivative type of feedback, K_f must be replaced by K_pRC in Eqs. (8–21) and (8–22).

Adaptive Position Control

It was shown in the preceding section that velocity feedback provides an effective means of controlling the damping ratio. An adaptive control system is a system that adapts its operation to meet and maintain a required level of performance. The system must adapt

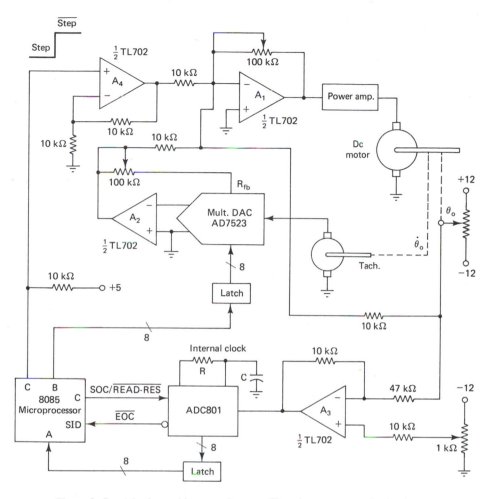

Figure 8–7. Adaptive position control system. The microprocessor maintains the system critically damped in the presence of load and other parameter fluctuations.

automatically whenever its performance falls outside the prescribed limits. The adaptive feature is generally implemented with additional circuitry, which may include a microprocessor. To be more specific, in the present case the system under consideration is a position control system of the type shown in Fig. 8–4. In its basic form the system is underdamped even with the amplifier gain set to unity. Suppose that it is required that this system be critically damped. Velocity feedback may be used to adjust the system damping to unity, and thus solve the problem. However, variations in the load may cause the system to deviate from the critical damping. This may be good reason to implement the adaptive feature into the system, so that any deviations in damping from the desired value are corrected automatically.

An adaptive position control system using an 8085 microprocessor is shown in Fig. 8–7. The control software is shown in Fig. 8–8. This system was constructed and tested in the lab. The oscilloscope display of the generator output was used as a

```
"8085"

*

          POSCON — POSITION CONTROL PROGRAM
*
*         8085 BASED MICROCOMPUTER

*         OUTPUT STEP GOES TO PORT C BIT 0
*         OUTPUT READ/RESET SIGNAL GOES TO PORT C BIT 4
*         INPUT EOC BAR GOES TO SID 8085

          ORG     1400H
          MVI     A,4EH     ;CONFIGURE 8155 FOR PORTA-INPUT
          OUT     CSR       ;PORTB-OUTPUT — PORTC-OUTPUT
          MVI     D,01H     ;SET MULTIPLICATION FACTOR OF 1
STUFF     MOV     A,D       ;SEND TO AD7523 MDAC
          OUT     PORTB
          MVI     A,00H     ;STEP — RESET A/D
          OUT     PORTC
          CALL    DELAY     ;WAIT FOR POINT WHERE OVERSHOOT
          MVI     A,10H     ;STEP — SOC                OCCURS
          OUT     PORTC                          THE GREATEST
NOTYET    RIM
          RAL               ;WAIT FOR EOC BAR
          JC      NOTYET
          MVI     A,00H     ;STEP — READ/RESET A/D
          OUT     PORTC
          IN      PORTA     ;GET DC VALUE OF POSITION
          MOV     E,A       ;PUT IN REG.E
          CALL    STEP
          CALL    DELAY1    ;ALLOW TO SETTLE
          MVI     A,01H     ;NO-STEP — HOLD A/D
          OUT     PORTC
          CALL    DELAY1    ;ALLOW TO SETTLE
          JMP     STUFF     ;PLAY IT AGAIN SAM

STEP      CPI     81H       ;CHECK IF OVER DAMPED
          JNC     TOOLOW
          CPI     7FH       ;CHECK IF UNDER DAMPED
          JC      TOOHGH
          RET               ;NO CHANGE
TOOHGH    INR     D         ;INCREASE MULTIPLICATION FACTOR
          RET
TOOLOW    DCR     D         ;DECREASE MULTIPLICATION FACTOR
          RET

*         DELAY TO WAIT FOR POINT WHERE OVERSHOOT IS THE GREATEST

DELAY     LXI     B,03FFFH  ;60-80 MILLISECONDS
STUFF1    DCR     C
          JNZ     STUFF1
          DCR     B
          JNZ     STUFF1
          RET

*         DELAY TO ALLOW SYSTEM TO SETTLE OUT

DELAY1    LXI     B,0FFFFH  ;320-360 MILLISECONDS
STUFF2    DCR     C
          JNZ     STUFF2
          DCR     B
          JNZ     STUFF2
          RET

CSR       EQU     08H
PORTA     EQU     09H
PORTB     EQU     0AH
PORTC     EQU     0BH
```

Figure 8–8. Software for the adoptive position control system in Fig. 8–7.

measure of the system's response. The initial 60% overshoot was almost totally eliminated by the adaptive control, resulting in a critically damped response.

The multiplying DAC plays an important role in the adaptive control. One of its inputs is the velocity feedback from the generator, and the other input is a binary count which comes from port B of the microprocessor. For an 8-bit multiplying DAC (MDAC), the range of the count is from 0 to 255. In initializing the ports, port A is assigned as the input port for the position data from the ADC, port B as an output port of the incremental control count for the MDAC, and port C is configured as the STEP and ADC control port: LSB controls the STEP, and bit 4 controls the SOC/READ-RESET ADC input.

At the beginning of the operation, the microprocessor outputs a STEP, causing the motor to respond in an underdamped fashion. The microprocessor places itself in a wait mode by executing the first of the two delays, which is 60 to 80 ms in duration. This is the approximate time that it takes the system to reach the first peak, and is stored in register pair BC as 3FFFH of the first delay subroutine (this value, which is determined experimentally, will be different for another system).

After the first delay, the microprocessor outputs the SOC to ADC and goes into a looping mode, constantly monitoring its SID input pin with the RIM instruction. Upon completing the conversion, the ADC outputs the EOC signal (LOW), which is detected by the RIM instruction. The microprocessor outputs 00H on port C, which transfers the position data from the ADC latch to the accumulator by way of port A and resets the ADC (and outputs the STEP, thus maintaining the STEP input condition to the position control system).

Next, the microprocessor compares the position data against the dead band, which is two counts wide (7F to 81)H and centered on 80H. The value 80H is the midpoint of the MADC, and it also corresponds to the step input and the steady-state value of the response. If the result of the comparison is outside the dead band, the microprocessor increments (thus increasing the velocity feedback) or decrements (decreasing the velocity feedback) the MDAC as the case may be, and if the position data are within the dead band, the microprocessor executes the second delay (which for this particular system is 320 to 360 ms, corresponding to the FFFFH contents of the BC register pair), allowing the system to reach the steady state. The microprocessor then applies the $\overline{\text{STEP}}$ to the position control system, causing it to respond in the opposite direction, and repeats the program.

System Dynamics

Critically damped system (step input).   When a critically damped position control system responds to a step input, its position (angular position of the motor's shaft), velocity, and acceleration vary in a specific manner. As the velocity attains a maximum value at some point in time, it gives rise to the maximum viscous friction torque. Maximum acceleration, on the other hand, gives rise to the maximum inertial torque. This type of dynamic response of the system provides information on the amount of torque that the motor must generate at a specific time during the response to overcome the viscous friction and the inertial reaction torques due to the load.

The velocity and the acceleration are derived by differentiating the position $\theta_o(t)$

with respect to t and then differentiating the velocity to get the acceleration. The final results in the normalized form are

$$\theta_n(t_n) = 1 - e^{-t_n}(1 + t_n) \qquad \text{position} \qquad (8-26)$$

$$\dot{\theta}_n(t_n) = t_n e^{-t_n} \qquad \text{velocity} \qquad (8-27)$$

$$\ddot{\theta}_n(t_n) = e^{-t_n}(1 - t_n) \qquad \text{acceleration} \qquad (8-28)$$

where the magnitude scale factors for position, velocity, and acceleration are E, $\omega_n E$, and $\omega_n^2 E$, respectively, and the time scale is defined by $t_n = \omega_n t$. Equations (8-26) to (8-28) are represented graphically in Fig. 8-9.

The total load reaction torque, which is expressed by the following general form, must include the inertial torque porportional to acceleration, the viscous friction torque proportional to velocity, and a constant torque which is independent of velocity or acceleration:

$$T_L = T_m = J_{\text{tot}}\ddot{\theta}_o + B_{\text{tot}}\dot{\theta}_o + T_c \qquad (8-29)$$

Several important deductions may be made from the graph of velocity and acceleration in Fig. 8-9. At $t_n = 0$, the acceleration is maximum and velocity is zero, resulting in maximum inertial torque and zero viscous friction torque at the beginning of motion. At $t_n = 1$ (the displacement of motor shaft is 26% of its final value), the acceleration is reduced to zero and the velocity is maximum. At this time the system experiences no inertial torque, and maximum viscous friction torque. As the motion continues the acceleration reverses polarity, resulting in a negative inertial torque reaching a maximum negative value at $t_n = 2$, a point in time where the motor shaft has advanced to 60% of its final value. As the motor shaft approaches the steady-state value, where it comes to a stop, both the velocity and the acceleration decay to zero.

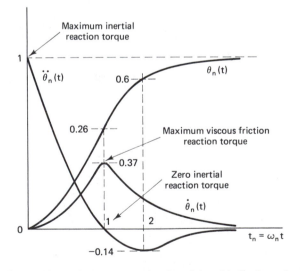

Figure 8-9. Position, velocity, and acceleration of the critically damped system step response.

Underdamped system (step input). In contrast to the critically damped system, the step response of the underdamped system is characterized by sinusoidal variations. Thus the position, velocity, and acceleration vary in a damped sinusoidal manner. The expressions for velocity and acceleration are determined in the usual way [i.e., through successive differentiation of position (θ_o in Table 8–1)]. The results in normalized form are as follows:

$$\theta_n(t_n) = 1 - \frac{e^{-\zeta t_n}}{\beta} \sin (\beta t_n + \phi) \qquad \text{position} \qquad (8-30)$$

$$\dot{\theta}_n(t_n) = \frac{e^{-\zeta t_n}}{\beta} \sin \beta t_n \qquad \text{velocity} \qquad (8-31)$$

$$\ddot{\theta}_n(t_n) = -\frac{e^{-\zeta t_n}}{\beta} \sin (\beta t_n - \phi) \qquad \text{acceleration} \qquad (8-32)$$

where the magnitude scale factors for position, velocity, and acceleration are E, $\omega_n E$, and $\omega_n^2 E$, respectively, and the time scale is $t_n = \omega_n t$.

As in the case of the critically damped response, the motor must generate sufficient torque to overcome the viscous friction and the inertial reaction torques which are due to the load, and thus advance the load to its final position. Equation (8–29), which accounts for the load reaction torque, applies in this case as well.

Referring to Fig. 8–10, the velocity $\dot{\theta}_o$ is the derivative of position θ_o (the velocity is the slope at any point on the position response), and the acceleration $\ddot{\theta}_o$ is the derivative of velocity (acceleration is the slope at any point on the velocity response). Consequently, the value of velocity must be zero at each peak of the position response, and the value of acceleration must be zero at each peak of the velocity response. The peak times expressions are shown in Table 8–3.

The peak values (minima and maxima) of the position response and the zeros of the velocity response occur at times $t_n = k\pi/\beta$. Evaluating the position expression at these times gives us an equation (Table 8–3) which may be used to calculate any peak value of the position response. The velocity peaks and the acceleration zeros coincide, occurring at times $T_k = (\phi + k\pi)/\beta$. The expression for the velocity peaks is obtained by evaluating Eq. (8–31) at $t_n = T_k$. Although the expression for $\ddot{\theta}_o$ is not included here, the reader can readily verify that the acceleration peaks occur at times $T_k = (2\phi + k\pi)/\beta$. The expression for the acceleration peak value is obtained by evaluating Eq. (8–32) at $t_n = T_k = (2\phi + k\pi)/\beta$. The peak times and various peak values are summarized in Table 8–3.

As the underdamped position control system begins its response to a step input from rest, the velocity is zero and the acceleration is at its maximum value. Therefore, there is no viscous friction torque initially, and practically all of the motor-developed torque is applied to accelerating the load inertia. Subsequent acceleration peaks gradually decrease, requiring the motor to develop less torque for driving the inertial load. At points in time where the velocity peaks, the acceleration is zero, resulting in zero inertial torque and maximum viscous friction torque. At these times, therefore, the system experiences no inertial reaction torque. As was the case with the acceleration, the velocity peaks decrease in time, requiring less and less of the motor-developed torque for overcoming the viscous friction.

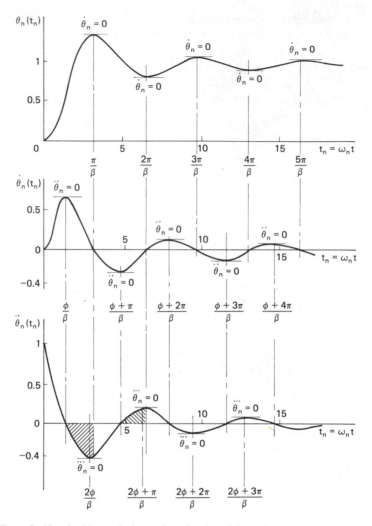

Figure 8–10. Position, velocity, and acceleration of the underdamped position control system step response. The response curves are to scale for $\zeta = 0.3$. ϕ and β have not been evaluated at $\zeta = 0.3$ to show explicitly their relationship to t_n.

TABLE 8–3 PEAK VALUES OF RESPONSES IN FIG. 8–8

	kth peak time, T_k	Value of kth peak	Index
Position	$\dfrac{k\pi}{\beta}$	$1 - \dfrac{(-1)^{k+1}e^{-\zeta T_k}}{\beta}$	$k = 1, 2, 3, \ldots$
Velocity	$\dfrac{\phi + k\pi}{\beta}$	$\dfrac{(-1)^k e^{-\zeta T_k}}{\beta}$	$k = 0, 1, 2, \ldots$
Acceleration	$\dfrac{2\phi + k\pi}{\beta}$	$\dfrac{(-1)^{k+1}e^{-\zeta T_k}}{\beta}$	$k = 0, 1, 2, \ldots$

[*] T_k represents the value of the normalized time t_n at the kth peak.

There are instances during the response when the inertia provides a breaking effect. This occurs at times when the velocity is yet positive (negative) and the acceleration has reversed its direction to negative (positive). For example, in the time interval ($\phi/\beta \leq t_n \leq \pi/\beta$) the velocity is positive but the acceleration is negative, and in the time interval ($\phi + \pi)/\beta \leq t_n \leq 2\pi/\beta$ the velocity remains negative when the acceleration becomes positive. At each of the velocity peaks, the reversal of acceleration (i.e., the acceleration is opposite to the direction of motion) is responsible for the breaking action, which reduces the velocity to zero before the motion (velocity) reverses and assumes the direction of the acceleration.

There is a parallel to the foregoing observations in electrical circuits and in physics. Consider the parallel LC circuit, with C initially charged to E. The resistance in the circuit is due to the coil. The damped oscillations are characterized by the voltage $V = L\, dI/dt$ across the parallel LC circuit and the current I which flows through L and C. The slope of the current is thus proportional to the voltage V. Consider next a pendulum which is initially at one extreme of its motion, where the potential energy (PE) is maximum and the kinetic energy (KE) = 0. As the pendulum moves through its lowest point, PE = 0 and KE is maximum. As the pendulum passes the lowest point (maximum velocity), the acceleration reverses direction as velocity is reduced to zero and the pendulum comes to a stop at the opposite extreme of its motion. Note that the acceleration of the pendulum past the point of maximum velocity reverses direction, just as the voltage across the capacitor (in the parallel LC circuit) reverses polarity past the point of maximum current through the inductor. The reader should be able to verify as well as justify that the $\dot{\theta}_o$ graph in Fig. 8–10 may represent equally well the current I in the LC circuit or the KE of the pendulum, and the $\ddot{\theta}_o$ graph may also represent the capacitor voltage in the LC circuit, or the acceleration (or PE) of the pendulum.

The period of oscillation of the position, velocity, and the acceleration responses in Fig. 8–10 is $2\pi/\beta$. In simple harmonic motion (undamped motion), the position and the acceleration peaks coincide in time. As can be seen in Fig. 8–10, these peaks do not coincide in time. The reason these peaks do not coincide is due to damping. The position response peaks depend on β, and the acceleration peaks depend on both β and ϕ. Both ϕ and β, in turn, depend on the value of the damping ζ. As $\zeta \rightarrow 0$, $\beta = 1$ and $\phi = \cos^{-1}\zeta = \pi/2$. Clearly, under such conditions the first position and acceleration peaks coincide at $t_n = \pi$, and the remaining peaks occur at integer multiples of π.

The largest possible velocity for the entire transient interval motion occurs at the first peak. In Table 8–3 the first velocity peak occurs for $k = 0$, at $T_k = \phi/\beta$. At this time acceleration is zero. Therefore, the torque that the motor must develop at this time may be expressed as follows using Table 8–3 and Eq. (8–29) and neglecting the Coulomb friction torque T_c:

$$T_m(T_0) = B_t\dot{\theta}_o(T_0) = B_t\omega_n E \dot{\theta}_n \frac{\phi}{\beta} = \frac{B_t\omega_n E}{\beta} e^{-\zeta\phi/\beta} \tag{8–33}$$

where $T_0 = \phi/\beta$ is the normalized time of the first velocity peak corresponding to $k = 0$, and the velocity is denormalized through multiplication by $\omega_n E$. The largest acceleration occurs at the start of motion $t_n = 0$. This is the absolute maximum acceleration. Its normalized value of 1 is much greater than the other acceleration peaks, as shown in Fig. 8–10. The torque that the motor must develop to accelerate the inertial load is the product of the total inertia and the value of the initial acceleration from Eq. (8–32); hence

$$T_m(0) = J_t \ddot{\theta}_n(0) = J_t \omega_n^2 E = K_a K_p K_t E \qquad (8\text{--}34)$$

Use has been made of Eq. (8–14) to relate T_m to the position control system parameters, and the Coulomb friction T_c was neglected. Equations (8–14) and (8–15) may also be used to express Eq. (8–33) in terms of the position control system parameters.

EXAMPLE 8–2

In the position control system of Fig. 8–4, $J_L = 0.03$ oz-in.-s^2, $J_m = 0.02$ oz-in.-s^2, $B_L = 0.3$ oz-in.-s, $B_m = 0.2$ oz-in.-s, $K_t = 5$ oz-in./V, $K_p = 1$ V/rad, $K_a = 2$, $T_a = 2$ oz-in., and $\theta_i = 40u(t)$ degrees. Calculate (a) the motor-developed torque at $t = 60$ ms; (b) the time and value of the first velocity peak; (c) the percent change in the system response time due to the doubling of the load inertia.

SOLUTION (a) To calculate the motor developed torque at 60 ms, the velocity and acceleration must first be calculated at that time. The values of various parameters are $K_m = K_t/B_t = 5/(0.2 + 0.3) = 10$ oz-in./V, $G_o = K_a K_m K_p = 20$, $\omega_n = (G_o/\tau_m)^{1/2} = 14.1$ rad/s, $\zeta = 1/[2(G_o\tau_m)^{1/2}] = 1/[2(2)^{1/2}] = 0.35$, $\phi = \cos^{-1} 0.35 = 1.2$ rad, $= [1 - (0.35)^2]^{1/2} = 0.94$, and $\theta_i = 40° = 0.67$ rad. From Eq. (8–31) after denormalizing (multiplying by $\omega_n E$) and using $t_n = \omega_n t = 14.1(0.06) = 0.85$, we have

$$\dot{\theta}_o(0.06) = \frac{1}{0.94}(14.1)(0.67)[e^{-(0.35)(0.85)}] \sin[0.94(0.85)] = 5.35 \text{ rad/s}$$

The acceleration is evaluated from Eq. (8–32) after denormalization as

$$\ddot{\theta}_o(0.06) = \frac{1}{0.94}(14.1)^2(0.67)[e^{-(0.35)(0.85)}] \sin[0.94(0.85)$$

$$- (1.2)] = +41.1 \text{ rad/s}^2$$

From Eq. (8–29)

$$T_m(0.06) = 0.05(41.1) + 0.3(5.35) + 2 = 5.66 \text{ oz-in.}$$

(b) From Table 8–3, the first velocity peak occurs at $T_0 = \phi/\beta$ for $k = 0$; therefore, $T_0 = 1.2/0.94 = 1.28$, and the real time of the first velocity peak is $t_0 = T_0/\omega_n = 1.28/14.1 = 90$ ms. The value of the first velocity peak from Table 8–3 for $k = 0$ after denormalization,

$$\dot{\theta}_o(T_0) = \frac{\omega_n E}{\beta} e^{-\zeta\phi/\beta} = \frac{1}{0.94}(14.1)(0.67)e^{-0.35(1.2)/0.94} = 6.43 \text{ rad/s}$$

(c) If the time T_1, the first peak of the position response, is to be used as an indicator of system response change due to the doubling of the inertia, the original value of T_1 is $T_{1i} = \pi/\omega_n\beta = \pi/14.1(0.94)] = 237$ ms. Assuming that the load inertia now doubles, $\omega_n = (G_o/\tau_m)^{1/2} = (K_a K_p K_m/\tau_m)^{1/2} = [2(10)(0.5)/0.08]^{1/2} = 11.2$ rad/s, $\zeta = 1/[(2)(K_a K_p K_m \tau_m)^{1/2}] = 0.5/[10(2)(0.08)/0.5]^{1/2} = 0.28$ and $\beta = (1 - \zeta^2)^{1/2} = 0.96$. The new value, $T_{1f} = \pi/\omega_n\beta = \pi/11.2(0.96)] = 292$ ms. The ratio $292/237 = 1.23$ indicates that the system response is degraded by 23% due to the doubling of load inertia.

8–4 PRACTICAL SERVOMECHANISM SYSTEMS

Speed Control System

The system configuration shown in Fig. 8–11 has been constructed and tested in the laboratory. It is different from the speed control system shown in Fig. 8–2. The main difference is in the form of the feedback. Whereas the system in Fig. 8–2 uses a dc generator as a source of velocity feedback, the system shown depends on optical speed sensing. The optical disk, which has 120 holes, and the OID combination produces a pulse train at the OID output. According to Eq. (2–12), the frequency of pulses is 2ω, where ω is at the motor speed in rpm. The pulse frequency is converted to a dc voltage by the 9400 frequency-to-voltage converter (the 9400 chip is manufactured by Teledyne Semiconductor, Mountain View, Calif.). The circuit that follows the FVC is a unity-gain amplifier which filters the ripple at the output of FVC. (This circuit is recommended by Teledyne in their *1984 Data Acquisition Design Handbook* as a filter of FVC output ripple.) The ac component of the FVC output is applied to the + and − inputs of the OA_4. If the gains in both paths are identical, the ripple is completely canceled. The amount of mismatch between the gains of the two paths determines the degree of ripple cancellation; for example, if the gains are matched within 2%, the incoming ripple is then reduced by 1/50. The dc component of the FVC output is applied to the + input of OA_4 and is given a gain of +1.

The filtered dc level at the output of OA_4 is given additional gain by the OA_1 inverting amplifier, whose gain is adjustable by the 10-kΩ potentiometer in the feedback path. The negative feedback voltage at the output of OA_1 represents the actual speed of the motor. As negative feedback must be maintained, this negative voltage corresponds to a positive input V_i. For a negative input V_i, the manual switch, S_1, is used to disconnect the amplifier OA_1 and connect in series with OA_4 the noninverting amplifier OA_7, whose gain is same as that of OA_1. This results in the positive feedback voltage required for negative feedback.

OA_2 is the summer with unity gain and with a polarity inversion. The output of OA_1 is the difference signal $-(V_i - V_f)$, which is amplified without the polarity inversion by OA_3. The output of OA_3 drives the power amplifier. As noted earlier, the power amplifier generates sufficient current (up to several amperes in this case) to drive the motor. The power stage consists of p-type and n-type Darlingtons (each Darlington is commercially available on a chip) connected in a complementary-symmetry arrangement. The power amplifier provides an additional polarity inversion. The complementary symmetry allows reversal of the motor current and hence the direction of rotation. The polarity of V_i determines the direction of rotation.

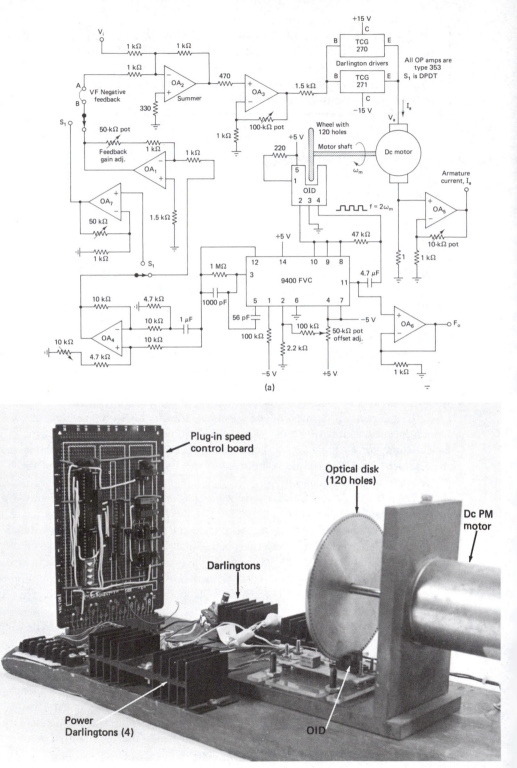

Figure 8–11. Practical speed control system using optical speed sensing: (a) block diagram; (b) laboratory prototype; (c) power supply, ±20 V, 3 A.

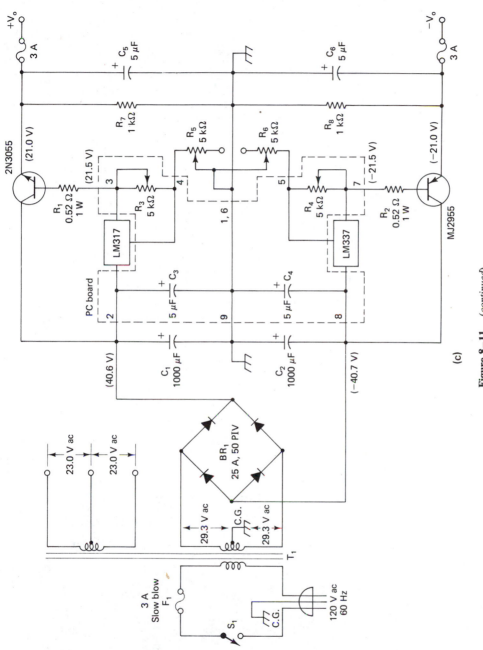

Figure 8–11. (*continued*)

431

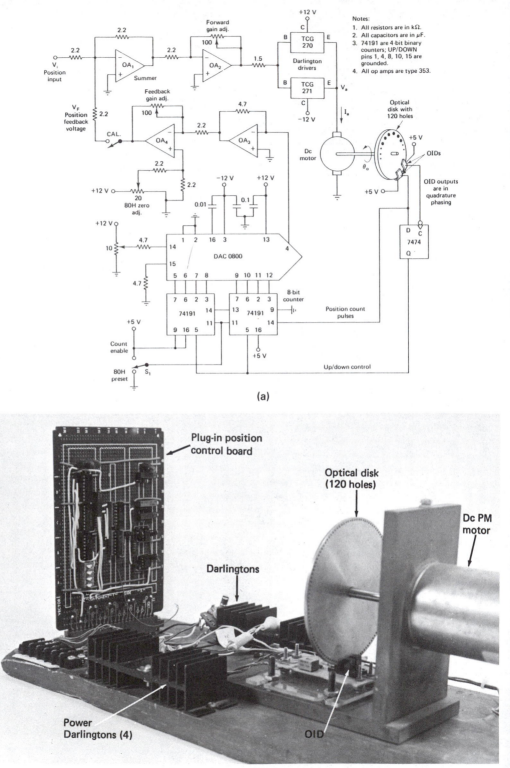

Figure 8–12. Practical position control system with optical position sensing: (a) system diagram; (b) laboratory prototype (same as Fig. 8–11b); (c) power supply (same as Fig. 8–11c).

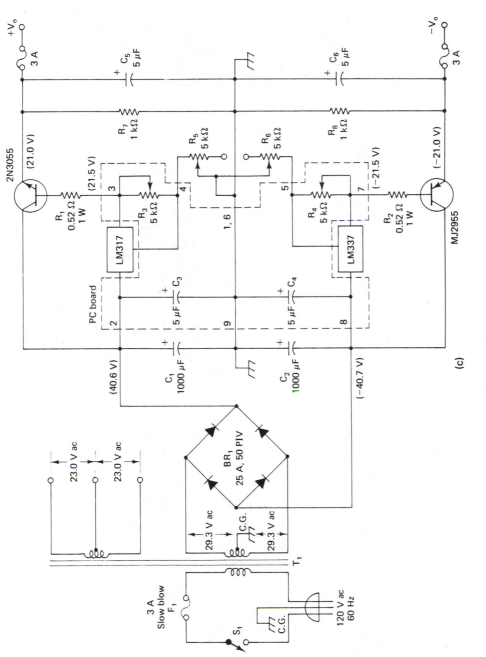

Figure 8–12. (*continued*)

The manual switching of feedback polarity is satisfactory for most applications where reversals in the direction of rotation do not occur too frequently. In cases where they do, the manual switch, S_1, must be replaced with a relay or an electronic switch. If an electronic switch is used, it may be triggered by a direction-sensing circuit of the type used in the position control system of Fig. 8–12.

The OA_5 amplifier monitors the motor armature current, which is equal to the voltage across the 1-Ω resistor. As the voltage is small, the amplifier gain scales the current so that the output voltage, which is displayed on a DVM, is in a low-voltage range.

The OA_6 buffer provides the motor speed test point. The frequency at the output of OA_6, which in this case is numerically equal to twice the motor speed, is displayed on the frequency counter.

Position Control System

The practical position control system shown in Fig. 8–12 uses an optical disk with 120 holes, the same disk that was used by the speed control system in Fig. 8–11, except in this case the function of the disk is to sense the angular position of the motor's shaft. As the shaft advances from some reference position, each hole on the optical disk produces a pulse at the output of the OID. Since the holes on the disk are spaced at 3° intervals, each pulse must represent a 3° angular displacement of the motor's shaft. The shaft moves $3N$ degrees for N pulses at the output of the OID. These pulses are counted by an 8-bit binary up/down counter which is constructed using two 74191 chips, each chip being a 4-bit counter. The 8-bit binary output from the counter is a digital representation of the actual position of the motor shaft. It may be applied to an 8-bit latch whose output may be directly interfaced to an 8-bit microprocessor.

There are instances when the shaft oscilates back and forth for some time before reaching the steady state. This occurs when an underdamped system responds to a step input. The counter in this case will continue to count up (or down) due to the pulses from the OID, despite the fact that the shaft is merely moving back and forth, not advancing in one direction. The result is that the contents of the counter no longer represent accurately the angular position of the shaft.

To resolve this dilemma, a direction-sensing circuit consisting of two OIDs and a D flip-flop is used. The OID outputs in quadrature phasing are applied to the D flip-flop, whose output is HIGH for one direction of rotation and LOW for the opposite direction. The D flip-flop output is used as the up/down control for the counter. The direction-sensing circuit is discussed in Chapter 2 with the configuration and the waveforms shown in Fig. 2–32. With the help of the direction-sensing circuit, the counter is able to track the shaft rotation even if the shaft should reverse its rotation.

The counter output is applied to the 8-bit DAC, which converts the digital shaft position information to the analog form. The output of the current amplifier OA_3 is a dc level which uniquely represents the angular position of the shaft. This voltage level is further amplified in OA_4 and applied as the position feedback voltage to the summing amplifier OA_1 together with the position input. The position input voltage is the desired position of the motor shaft. As we know from the theory, when the two inputs to the summer become equal, the shaft would have advanced to the position as dictated by the position input command.

The counter is preset to 80H, the midpoint of its full-scale count, to allow for the bidirectional operation of the system. This is accomplished by momentarily grounding pin 11 on both chips through switch S_1, and then returning the switch to +5 V, the count enable position. The DAC output due to the 80H input from the counter is zeroed out by the 80H ZERO ADJ. control, producing zero volts at the output of OA_4. The position input may now be positive or negative, depending on the desired direction of rotation. The complex position feedback structure beginning with the optical disk and terminating with OA_4 replaces the simple potentiometer used in Fig. 8–4. Despite the complexity, the component chips that constitute the position feedback are inexpensive and not subject to wear as is the potentiometer, and in addition this type of feedback structure can easily be interfaced to the microprocessor. On the other hand, the optical feedback structure lacks the fine resolution offered by the potentiometer. The optical disk used in the illustration can resolve the shaft position only within 3°. The resolution may be improved by increasing the number of holes.

The feedback voltage at the output of OA_4 also varies in a discrete manner. For an 8-bit DAC that uses a 12-V reference, each step is approximately 47 mV. The size of the step may be reduced by using a DAC with a greater number of bits; the size of the step is 3 mV for a 12-bit DAC using a 12-V reference.

The value of the position feedback voltage V_f is adjustable by the 100-kΩ feedback gain adjust potentiometer. The ratio V_f/θ_o is equivalent to K_p in Fig. 8–4b. Similarly dividing the position input voltage V_i, which represents the desired position of the motor shaft by the ratio $V_f/\theta_o = K_p$, results in θ_i, the equivalent angular position input shown in Fig. 8–4b.

OA_1 is a unity-gain summing amplifier, and the forward gain is adjustable by the 100-kΩ pot used with OA_2. The Darlington drivers perform the function of the power amplifier with bipolar motor drive capability. TCG270 responds to a position voltage on its base and supplies the motor current in the direction shown, causing the motor to advance in one direction. TCG271, on the other hand, responds to a negative voltage at its base and supplies the motor current in the direction opposite to that shown, causing the motor to reverse its direction of rotation.

The equivalent circuit shown in Fig. 8–4c applies to this system if K_a is equated to the combined voltage gain of OA_2 and the Darlington drivers, and K_p is equated to the value of V_f/θ_o. The dynamic and static tests performed on this system closely correlate to the theory on position control system developed in this chapter and in Chapter 6.

8–5 TEMPERATURE CONTROL SYSTEM

The object of the temperature control system shown in Fig. 8–13 is to maintain the temperature within the chamber at a value corresponding to the set point. The set point is a voltage level representing the desired chamber temperature T_0.

The temperature sensor inside the chamber produces a voltage proportional to temperature T_0. It is amplified by K_1 and is applied to the summing amplifier together

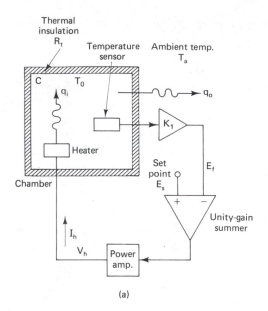

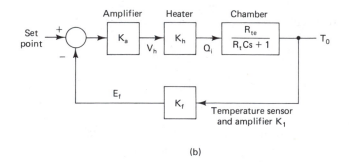

Figure 8–13. Temperature control system: (a) system diagram; (b) block diagram.

with the set point voltage E_s. The difference signal from the summing amplifier is applied to the power amplifier (a low-level amplifier may be used before the power amp), which supplies the current to drive the heater element inside the chamber. The input heat q_i causes the temperature T_0 as well as the feedback voltage E_f to increase, and the difference signal at the output of the summing amplifier, to decrease (a consequence of negative feedback). As the difference signal approaches the zero value, the drive to the power amp is cut off, and the supply of heat to the chamber is also cut off. As a result, the temperature inside the chamber corresponds to the set point E_s.

The thermal equilibrium within the chamber thus established does not persist indefinitely. The reason is that the chamber walls are not made of a perfect insulator, and therefore some heat is lost through the walls. The amount of heat flow q_0 through the walls depends on the temperature differential $(T_0 - T_a)$ that exists across the walls, and the thermal resistance of the wall material.

It is a matter of time, then, before sufficient heat is lost through the walls, causing the temperature within the chamber to drop. When this happens, E_f will decrease, and the increased difference signal from the summing amplifier will act on the power amp, which will, in turn, supply current to the heater element. The heater element will continue to supply heat to the chamber until its original temperature is restored, thus causing the output of the summing amplifier to become zero again and the heater to be cut off.

Through the use of negative feedback, the system provides automatic control of the chamber temperature. The ambient temperature T_a, which is external to the chamber, has the effect of a load on the system. Any variations in T_a result in a fluctuating load which is attempting to change the chamber temperature. Also, any variation in the power amp gain or the power supply voltage is also making an attempt to change the chamber temperature. As noted above, through the use of negative feedback, the system senses these effects and maintains the chamber temperature at the set point or at least close to the set point, differing only by the steady-state error. The quantitative aspects of the system performance are described in a later section.

System Models

The mathematical model for the system may be derived by considering the law of thermodynamics, which governs the heat flow. Accordingly, the difference between the heat supplied and heat lost through the walls must be equal to the heat rise inside the chamber. The heat accumulation within the chamber is proportional to the rate at which the chamber temperature is changing, with C_t the thermal capacitance of the medium inside the chamber, being the constant of proportionality. Stated mathematically, we have

$$q_i - q_o = C_t \frac{dT_0}{dt} \qquad \text{Btu/s} \qquad (8-35)$$

where $q_i(t)$ = heat supplied to the chamber (Btu/s)
$q_o = (T_0 - T_a)/R_t$ = heat flow (Btu/s) through the chamber walls
R_t = thermal resistance of the wall material (°F, sec/Btu)
C_t = thermal capacitance of the medium inside the chamber (Btu/°F)

The expression for q_o is substituted in Eq. (8–35), and after rearranging terms the following first-order differential equation is obtained for the chamber:

$$\tau_t \dot{T}_0 + T_0(t) = R_t q_i(t) + T_a \qquad (8-36)$$

To represent the chamber as a linear block in the system block diagram, we must determine the transfer function for the chamber. This may be done in a routine manner by taking the Laplace transform of both sides of Eq. (8–36), and then solving for the ratio of output $T_0(s)$ to the input $Q_i(s)$. However, there is a problem using this approach. Even if the initial condition $T_0(0)$ is equated to zero, the additional term T_a on the right-hand side of Eq. (8–36) makes it difficult to solve for the transfer function. This difficulty may be resolved through the definition of the equivalent or the effective thermal resistance R_{te} of the chamber walls as follows:

$$R_{te} = \frac{(R_t q_i + T_a)}{q_i} \qquad (8-37)$$

This type of mathematical adjustment may be done easily in situations where the heat q_i is applied at a constant rate and where T_a is also a constant. R_{te} may then be treated as another system parameter, which includes the effect of the load (the external chamber temperature T_a may be regarded as a load on the system). The reader may recall that a similar type of adjustment was done in Eq. (3–38), where the constant friction load T_c was included as part of k_{te}. Using R_{te}, the chamber differential equation (3–38) may be expressed as

$$\tau_t \dot{T}_0 + T_0(t) = R_{te} q_i(t) \tag{8–38}$$

The transfer function for the chamber may now be obtained from Eq. (3–38) by taking the Laplace transform of both sides of the equation, equating the initial condition $T_0(0)$ to zero, and solving for the ratio of the output $T_0(s)$ to the input $Q_i(s)$ as follows:

$$\frac{T_0(s)}{Q_i(s)} = \frac{R_{te}}{\tau_t s + 1} \tag{8–39}$$

where $\tau_t = R_t C_t$ is the thermal time constant of the chamber. As shown in Fig. 8–13b, this transfer function is used to represent the chamber by a linear block in the block diagram of the temperature control system.

Next, we must determine the mathematical model for the closed-loop temperature control system. From the system block diagram

$$q_i = (E_s - K_f T_0) K_a K_h \tag{8–40}$$

Substituting this expression for q_i in Eq. (8–38) and rearranging terms, the following result is obtained:

$$\tau_s \dot{T}_0 + T_0(t) = K_s E_s(t) \tag{8–41}$$

where

$$\tau_s = \frac{\tau_t}{1 + G_o K_f} \quad \text{closed-loop system thermal time constant (s)}$$

$$G_o = R_{te} K_a K_h \quad \text{dc value of the forward-path transfer function (°F/V)}$$

$$K_s = \frac{G_o}{1 + G_o K_f} \quad \text{closed-loop system temperature constant (°F/V)}$$

Equation (8–41) is the differential equation model for the closed-loop temperature control system. The system transfer function model may be obtained in the usual way by Laplace transforming Eq. (8–41), equating the initial condition $T_0(0)$ to zero, and solving for the ratio of the output $T_0(s)$ to the input $E_s(s)$ with the final result

$$\frac{T_0(s)}{E_s(s)} = \frac{K_s}{\tau_s s + 1} \tag{8–42}$$

The mathematical models of the system, the differential equation and the transfer function equations (8–41) and (8–42), are used later to characterize the performance of the closed-loop temperature control system. The temperature control system shown in Fig.

8–13 regulates the temperature of the chamber only if it drops below the set point. There is no provision in this configuration to control a rising temperature within the chamber. Such control would require bipolar amplifiers and a controllable heat exchanger for removing the excess heat from the chamber.

An experimental temperature control system shown in Fig. 8–14 was constructed and tested in the laboratory. The chamber is made of wood approximately 9-in.[3] and well insulated. The feedback component used in this system is a commercially available temperature sensor chip LM334 (made by National Semiconductor Corporation, Santa Clara, Calif.). It is an adjustable current source whose current is established with a single external resistor. The voltage V_s measured across the 1-kΩ resistor exhibits a linear temperature dependence with a TC of 0.33%/°C over the temperature range 0 to 70°C.

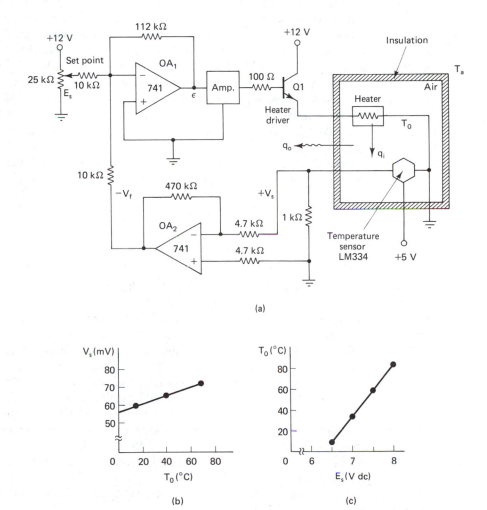

(a)

(b)

(c)

Figure 8–14. Practical temperature control system: (a) system diagram; (b) temperature sensor's transfer characteristics; (c) closed-loop system transfer characteristics.

The voltage V_s is amplified in OA_2, and then it is applied together with the set point input to the summing amplifier OA_1. The resulting difference signal is first brought to a power level by the intermediate power amplifier, which is sufficient to drive the power transistor Q_1. The current required by the resistive heater is in the range 1 to 10 A. The transient response tests done on the system show that the system time constant under open-loop conditions is approximately 650 s, and 140 s under closed-loop conditions.

8–6 LIQUID-LEVEL CONTROL SYSTEM

A mechanical liquid-level control system is shown in Fig. 8–15. In a typical application a system such as this may be required to feed liquid from the tank to several remote points through one or more outlet valves. The fluctuating demand will generally result in a varying flow rate through the outlet valves. This will cause the liquid level inside the tank to fluctuate as well. To avoid an overflow or possibly the tank being emptied during periods of peak demand, it is required that the liquid level inside the tank be maintained by a control system at a constant value corresponding to the set point.

The control system in this case is purely mechanical. It uses a float as the liquid-level sensor. The float translates in a one-to-one correspondence the changes in the liquid level x_0 to the displacement x_f of link L_3, which is attached to the float. The seesaw arrangement of link L_1–L_2, which is connected to L_3, makes possible the reversal of motion. An upward displacement x_f produces a downward displacement y, which acts on the inlet control valve. The relationship between x_f and Y may be deduced from the similar triangles in Fig. 8–15b as

$$y = -\frac{L_1}{L_2} x_f \tag{8–43}$$

The set-point input represents the desired value of the liquid level x_0. As shown in the illustration, under equilibrium or steady-state conditions, the liquid level is approximately equal to the set point (i.e., $x_0 \doteq x_s$). To input a larger set-point value, the set-point adjust screw is loosened and link L_3 is raised to the desired set-point value as indicated by the pointer on the scale. After tightening the setscrew and releasing link L_3, the weight of the float forces the inlet control valve to open, thereby increasing the inlet flow q_i. When the liquid level reaches the new set point, the upward motion of the float closes the inlet valve to the extent necessary to maintain the new steady-state liquid level. The forces necessary to overcome the friction in the linkage and the control valve, and to produce the required opening of the control valve, are implicit in the design of the float.

As shown in Fig. 8–16a, the upward-acting buoyant force due to the liquid displaced by the float (shaded area) is just balanced by its weight W. In Fig. 8–16b the downward-acting force pushed the float deeper, displacing an additional volume ΔV, which produces an additional buoyant force:

$$\Delta F_n = \rho \, \Delta V \tag{8–44}$$

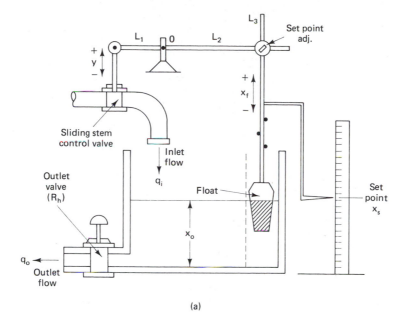

(a)

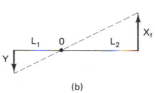

(b)

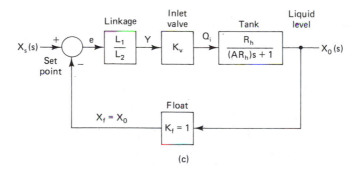

(c)

Figure 8–15. Mechanical liquid-level control system: (a) system diagram; (b) displacement diagram relating X_f and Y; (c) system block diagram.

where ρ is the density of liquid being displaced by the float (the fresh water, for instance, has a density of 62.4 lb/ft^3). The float thus submerges to the extent necessary to develop the additional buoyant force F_b, as expressed by Eq. (8–44), to counteract F_L. In Fig. 8–16c the upward-acting external force lifts the float slightly above the liquid level, reducing the volume of the displaced liquid by ΔV. By Eq. (8–44), the buoyant force

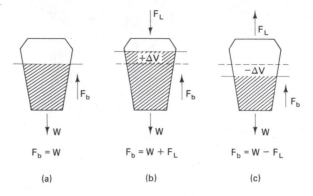

$$F_b = W \qquad\qquad F_b = W + F_L \qquad\qquad F_b = W - F_L$$

(a) (b) (c)

Figure 8–16. Float dynamics: (a) no external force; (b) increase in the displaced liquid produces an additional buoyant force to balance the external downward load; (c) upward-acting load produces a decrease in the displaced volume of liquid, thus reducing the buoyant force by the amount of the load.

is reduced also to the extent necessary to develop the new equilibrium position of the float, where the reduced buoyant force and F_L just balance the downward-acting weight of the float.

The float design, that is, its volume and its weight, obviously play a critical role in the dynamics of the opening and the closing of the sliding stem inlet control valve. As the liquid level rises and falls due to the fluctuating demand in the outlet flow, the float motion acts on the inlet control valve to maintain the liquid level at the value dictated by the set point. Negative feedback is clearly at work in this design, since a rising liquid level results in the closing of the inlet valve, and vice versa.

A more succinct description of the system is shown in the block diagram of Fig. 8–15c. As shown, the difference signal $x_s - x_f$ acts on the linkage, producing the motion y, which acts on the inlet control valve, which in turn produces the inlet flow q_i. The transfer function of the tank is shown as being first order. It will be derived next.

System Models

Tank. The mathematical model of the tank is derived by invoking the law of conservation of mass. Accordingly, the difference between the inlet and outlet flow rates must be equal to the rate at which the volume of liquid inside the tank is increasing or decreasing. The volume inside the tank remains constant in the case where the inlet and outlet flow rates are equal. This is translated to the following mathematical form:

$$q_i - q_o = A \frac{dx_0}{dt} \qquad \text{ft}^3/\text{s} \qquad\qquad (8\text{–}45)$$

where A is the cross-sectional area of the tank in ft^2, and the RHS, which represents the increase in liquid volume inside the tank, is expressed in terms of the liquid level x_0, the controlled variable of the system. An assumption is made at this time regarding the manner in which the liquid flows through the outlet pipe. A viscous flow is assumed to be "laminar" if the value of the Reynolds number R is 2000 or less (the Reynolds number is defined as $R = \rho VD/\eta$, where ρ, V, and η are liquid's density, velocity and viscosity, respectively, and D is the diameter of the pipe). Accordingly, for laminar flow, the hydraulic resistance R_h is defined as the ratio of the "head" (liquid level height x_0) to the outlet flow q_o; hence (consult Appendix B)

$$R_h = \frac{x_0}{q_o} \quad \text{s/ft}^2 \tag{8-46}$$

In applications where the head is constant, resulting in a constant flow, R_h, being a constant also, may be treated as a parameter. Substituting Eq. (8-46) in Eq. (8-45) yields

$$q_i - \frac{x_0}{R_h} = A \frac{dx_0}{dt} \tag{8-47}$$

After rearranging the terms, the following first-order differential equation, the mathematical model of the tank, is obtained:

$$\tau_h \dot{x}_0 + x_0(t) = R_h q_i(t) \tag{8-48}$$

where $\tau_h = AR_h$ is the hydraulic time constant of the tank.

The transfer function for the tank is obtained by taking the Laplace transform of both sides of Eq. (8-48), equating the initial conditions to zero, and finally solving for the ratio of output to input as follows:

$$\frac{X_0(s)}{Q_i(s)} = \frac{R_h}{\tau_h s + 1} \tag{8-49}$$

This transfer function represents the tank as a linear block in the block diagram of the system shown in Fig. 8-15c.

Closed-loop system. From the block diagram of Fig. 8-15c,

$$q_i = (x_s - x_0) \frac{L_1}{L_2} K_v$$

This expression for q_i is substituted in Eq. (8-47), and after algebraic manipulation and simplification, the following first-order differential equation, which represents the closed-loop hydraulic or level control system, is obtained:

$$\tau_s \dot{x}_0 + x_0(t) = K_s x_s(t) \tag{8-50}$$

where

$$\tau_s = \frac{\tau_h}{1 + G_o} \qquad \text{closed-loop system time constant (s)}$$

$$\tau_h = AR_h \qquad \text{tank and open-loop system time constant (s)}$$

$$K_s = \frac{G_o}{1 + G_o} \qquad \text{closed-loop system level constant}$$

$$G_o = R_h K_v \frac{L_1}{L_2} \qquad \text{dc value of the forward gain and the system loop gain } (K_f = 1)$$

The transfer function mathematical model of the closed-loop level control system is derived from the differential equation using the Laplace transformation in the usual way. The final result is as follows:

$$\frac{X_0(s)}{X_s(s)} = \frac{K_s}{\tau_s s + 1} \tag{8-51}$$

Electromechanical Liquid-Level Control System

Another version of the liquid-level control system is shown in Fig. 8–17. There are significant differences between this system and the purely mechanical system of the preceding section. They are similar in that both are based on the use of negative feedback

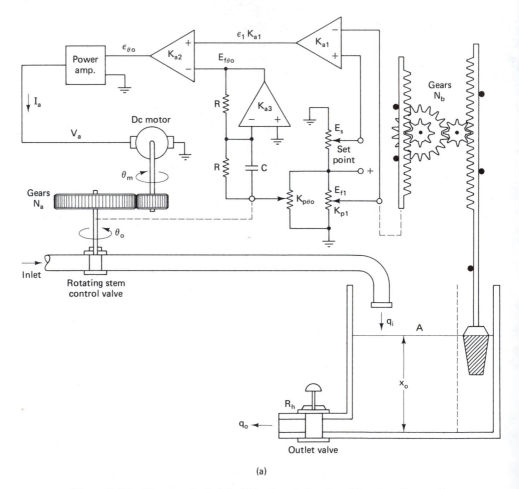

Figure 8–17. Electromechanical liquid-level control system: (a) system diagram (A, cross-sectional area of the tank; R_h, hydraulic resistance of the outlet valve); (b) block diagram.

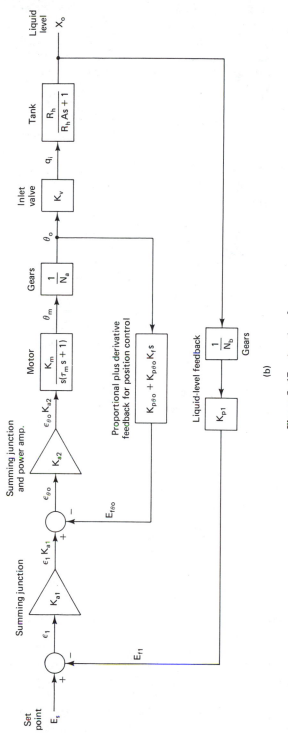

Figure 8-17. (*continued*)

and both use the float as the liquid-level sensor. The similarities end here, however, as the electromechanical system uses multiple feedback loops, resulting in a second-order system, whereas the mehcanical system uses a single feedback loop and is a first-order system. The use of amplifiers by the electromechanical system makes it possible to achieve large loop gain, and therefore make the system more accurate and faster in its response.

As shown in the system diagram, the motion of the float is transformed by the gears N_b to the potentiometer K_{p1}, with the wiper arm of K_{p1} being mechanically coupled to the output gear (the straight gear) of N_b. Gears N_b provide distance scaling by matching the maximum distance traveled by the float to that of the K_{p1} wiper arm. The analog signal at the wiper arm of K_{p1} represents the actual position of the liquid level, and the set point E_s represents the desired position of the liquid level. The two signals are applied to the summing amplifier K_{a1}, whose output is proportional to the difference between the set point and the actual liquid level, and therefore it is also proportional to the error in the actual liquid level.

The dc motor is configured as a position control actuator. The rotation of the motor shaft is stepped down by gears N_a, with the shaft of the driver gear mechanically coupled to the motor's shaft, while the shaft of the driven gear is mechanically coupled to the shaft of the inlet control valve and to the wiper arm of the potentiometer $K_{p\theta_o}$. The motor and the reduction gears provide an incremental control of liquid flow into the tank by controlling the stem position of the inlet value. The voltage at the wiper arm of $K_{p\theta_o}$ represents the stem position of the inlet valve and therefore the inlet flow rate q_i. This voltage is applied to the proportional plus derivative network, consisting of R, C, and K_{a3}. The output of K_{a3} provides the proportional plus derivative type of feedback voltage for the inner loop, which is the position control servo loop. The derivative feedback provides additional damping to ensure that the response of the motor is critically damped. As the motor is the position actuator for the inlet valve, it is important that its response be critically damped or perhaps slightly overdamped so that the motion of the motor's shaft is smooth and nonoscillatory. The position feedback signal $E_{f\theta_o}$ and the signal $\epsilon_1 K_{a1}$, which represents the error in the liquid level, are applied to the second summing amplifier K_{a2}. The output of K_{a2} is amplified in the power amplifier, and the resulting signal drives the motor.

In a typical application, the tank may be required to provide liquid to remote locations through several outlet valves. The liquid level inside the tank will remain constant as long as the total outlet flow is equal to the inlet flow q_i. There may be times, however, when the liquid level may vary due to the fluctuating demand in outlet flow. Any deviation in liquid level from the set point will result in an error signal at the output of K_{a1} due to the motion of the float. The error is then applied to K_{a2}, the summing amplifier for the position actuating servo, which causes the motor act on the inlet control valve and adjust the inlet flow rate.

A fully automatic control of this system by a microprocessor from a remote location is also possible after some modification. If the microprocessor is in the same general area as the tank, then the signal E_{f1}, which represents the actual liquid level, must be converted to a digital form by an ADC and applied to the microprocessor. The set point input from the microprocessor, after conversion from the digital to the analog form by the DAC, is applied to K_{a1}. The outlet valve may be modified to include the motor position actuator, thus providing a means of controlling the outlet flow by the microprocessor. The microprocessor will thus have the capability of monitoring the liquid level, making adjustments in the liquid level, and provide total control of the outlet flow rate. In a typical process control environment, the microprocessor may be required to control other process parameters in addition to liquid level.

EXAMPLE 8–3

The block diagram shown in Fig. E8–3a applies to the system shown in Fig. 8–15. The height of the tank is 25 ft and the transfer characteristic of the inlet valve is as shown. Calculate (a) the value of the set point x_s; (b) the diameters of the tank and the outlet; (c) the outlet flow under the given conditions; (d) the time that it takes the liquid level to attain 17.25 ft if the input x_s were suddenly increased by 1.5 ft. (e) What happens to the steady-state operation of the system if the outlet valve is adjusted to increase the outlet flow by 25%? Explain.

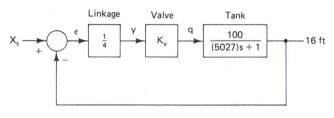

(a)

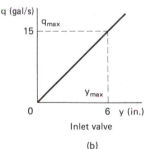

(b)

Figure E8–3.

SOLUTION (a) The value of K_v may be obtained as the slope of the given inlet valve flow characteristics; hence

$$K_v = \frac{15}{7.48(6/12)} = 4 \text{ ft}^2/\text{s}$$

where 7.48 constant converts from gal to ft³. From the given block diagram

$$(x_s - 16)(\tfrac{1}{4})(4)(100) = 16$$

Solving for X_s yields

$$x_s = 16.16 \text{ ft}$$

(b) From Eq. (B–10) in Appendix B, assuming a circular cross section of the outlet pipe,

$$R_h = \frac{1}{\pi D_2^2/4} \left(\frac{16}{64.4}\right)^{1/2} = 100$$

Solving for the outlet diameter gives us

$$D_2 = 0.08 \text{ ft} = 0.96 \text{ in.}$$

The diameter of the tank may be calculated from the given tank time constant. Since $\tau_h = AR_h$, and assuming a circular cross section, we have

$$5027 = \frac{\pi D_t^2}{4} (100)$$

Solving for D_t yields

$$D_t = 8 \text{ ft}$$

(c) From Eq. (B–5),

$$q_o = A_2(2gh)^{1/2} = \frac{\pi(0.08)^2}{4} [64.4(16)]^{1/2} = 0.16 \text{ ft}^3/\text{s} = 1.2 \text{ gal/s}$$

Under steady-state conditions, we would expect that $q_o = q_i$. In checking, we find

$$q_i = \frac{x_0}{R_h} = \frac{16}{100} = 0.16 \text{ ft}^3/\text{s}$$

(d) From Table 8–3,

$$G_o = \frac{R_h K_v L_1}{L_2} = 100(4)(\tfrac{1}{4}) = 100$$

$$K_s = \frac{G_o}{1 + G_o} = \frac{100}{101}$$

$$\tau_s = \frac{\tau_h}{1 + G_o} = \frac{5027}{101} = 49.8 \text{ s}$$

then

$$x_0(t) = x_0(0) + EK_s(1 - e^{-t/\tau_s})$$

$$17.52 = 16 + (1.5)\frac{100}{101}(1 - e^{-t/49.8})$$

Solving for t gives us

$$t = 91.8 \text{ s}$$

(e) A 25% increase in outlet flow results in an increase in the outlet flow from 0.16 to 0.2. Under steady-state conditions the inlet flow q_i is also equal to 0.2 ft³/s. From the block diagram $\epsilon = q_i/[K_v(L_1/L_2)] = 0.2/1 = 0.2$ ft. Therefore, the new value of the liquid level, assuming that the input remains at 16.16 ft, is $16.16 - 0.2 - 15.96$ ft, and the new value of the hydraulic resistance is obtained by dividing the new liquid level by the new flow; hence $R_h = 15.96/0.2 = 79.8$ s/ft².

The reduction in hydraulic resistance causes a reduction in loop gain from 100 to 79.8. The reduction in loop gain in turn causes an increase in the steady-state error. The original error is $16.6 - 16 = 0.16$ ft $= 1.92$ in. The new value of error is corresponding to loop gain of 79.8 is $16.16 - 15.96 = 0.2$ ft $= 2.4$ in. The increased outlet flow may be regarded as an increased load on the system, causing an increase in the steady-state error, as a result of which the liquid level drops an additional 0.04 ft or 0.48 in.

8–7 PERFORMANCE CHARACTERISTICS OF FIRST-ORDER SYSTEMS

The speed control, temperature control, and liquid-level control are the three first-order systems presented in this chapter. The three systems (and there are many other systems of this type) belong to a family of systems generally referred to as first-order systems. They are so called because they are modeled mathematically by a first-order differential equation. Because of this, the analysis, as well as the results of the analysis that characterize system performance, are identical for each of the three control systems. Furthermore, we shall not undertake at this point the task of detailed analysis, since the analytical treatment of a general first-order system is presented in sufficient detail in Chapters 5 and 6. We shall therefore apply the results from these chapters to the three first-order systems here. The reader is reminded that each of the three systems is an automatic control type of system using negative feedback. (A simple low-pass filter using one resistor and one capacitor is also a first-order system; however, it is not a control system, nor does it use feedback.)

The important characteristics of the three systems are summarized in Table 8–3. Descriptions of the input, output, and type of feedback are included for each system. The open-loop gain, also referred to as the dc value of the forward-path transfer function, and appearing in several equations, is defined for each system. The open-loop and closed-loop time constants and the closed-loop gain K_s are also included for each system. As mentioned earlier, the three systems are described by the same differential equation and transfer function models. The transfer function and the differential equation are included in the table and are expressed in the general form in terms of K_s and τ_s. To use one of these equations, one simply needs to substitute the value of K_s and τ_s for the given system.

TABLE 8–3 PERFORMANCE CHARACTERISTICS SUMMARY OF THREE FIRST-ORDER CONTROL SYSTEMS

Characteristic/variable or parameter	Speed control	Temperature control	Liquid-level control
Controlled variable, $C(s)$	Motor speed, $\Omega_m(s)$	Chamber temperature, $T_0(s)$	Liquid level, $X_0(s)$
Reference or set-point input, $R(s)$	Voltage input, $V_i(s)$ (desired motor speed)	Voltage input, $E_s(s)$ (desired chamber temp)	Set-point displacement input, $X_s(s)$
Feedback, $H(s)$	Speed sensing (tach. or optical), K_f	Temperature sensor (chip), K_f	Liquid-level sensor (float), $K_f = 1$
Open-loop-system time constant	$\tau_m = \dfrac{J_L + J_m}{B_L + B_m}$	$\tau_t = R_t C$	$\tau_h = A R_h$
Closed-loop-system time constant, τ_s (Loop gain, $G_o K_f$)	$\dfrac{\tau_m}{1 + G_o K_f}$	$\dfrac{\tau_t}{1 + G_o K_f}$	$\dfrac{\tau_h}{1 + G_o}$
Open-loop dc gain, G_o	$K_a K_m$	$R_{te} K_a K_h$	$\dfrac{R_h K_v L_1}{L_2}$
Closed-loop-system gain, K_s		$\dfrac{G_o}{1 + G_o K_f}$	

Characteristic/variable or parameter		
Closed-loop-system transfer function	$T(s) = \dfrac{C(s)}{R(s)} = \dfrac{K_s}{\tau_s s + 1}$	Eq. (5–1)
Closed-loop-system differential equation	$\tau_s \dot{c} + c(t) = K_s r(t)$ $c(t)$ = controlled variable $r(t)$ = reference input	Eq. (6–4)
Steady-state error (step input, $Eu(t)$)	$e_{ss} = c_{des} - c_{ss} = \dfrac{E}{K_f} - EK_s = \dfrac{E}{K_f(1 + G_o K_f)}$	Eq. (5–10) Eq. (5–11)
Sensitivity	$S_G = \dfrac{\partial T/T}{\partial G_o/G_o} = \dfrac{1}{1 + G_o K_f}$ (small changes only)	Eq. (5–20)
	$S_G = \dfrac{\Delta T/T}{\Delta G_o/G_o} = \dfrac{1}{1 + G_o K_f(1 + \Delta G_o/G_o)}$	Eq. (5–25)
Step response	$c(t) = c(0) + EK_s(1 - e^{-t/\tau_s})$ where $\begin{aligned} c(0) &= r(0)K_s \\ r(0) &= \text{initial value of input} \\ c(0) &= \text{initial value of output} \end{aligned}$	(Eq. 6–9)
Disturbance rejection	$DR = G_1 \dfrac{R}{D}$	Eq. (5–29) Fig. 5–7

Accuracy

The accuracy of a system is generally characterized by the value of the steady-state error e_{ss}. It was found in Chapter 5 that the steady-state error depends on system type value and on the form of the input (step, ramp, parabolic). Each of the three systems considered here is a type 0 system, and the steady-state error for a type 0 system is infinite when the input is parabolic or ramp. The steady-state error is finite and nonzero for a step input. It is repeated here from eqs. (5–10) and (5–11) and included in Table 8–3 as

$$e_{ss} = c_{des} - c_{ss} = \frac{E}{K_f} - EK_s = \frac{E}{K_f(1 + G_oK_f)} \tag{8-52}$$

The steady-state error as expressed above is the difference between the desired value and the actual steady-state value of the system's controlled variable. In Eq. (8–52), E represents the value of the step input.

Sensitivity

The sensitivity factor expresses the extent to which the parameter changes within the system have an effect on the system's output. The sensitivity of the system with respect to parameter changes in the forward path $[G(s)]$ is expressed by Eq. (5–20) for differential (small) changes in G_o as

$$S_G = \frac{\partial T}{T} \bigg/ \frac{\partial G_o}{G_o} = \frac{1}{1 + G_oK_f} \tag{8-53}$$

Assuming that the input to the system remains constant, the ratio $\partial T/T$ reduces to percent change in the output or $\partial c/c$. Equation (8–53) applies for small changes in G_o. The sensitivity factor derived by the use of the increment method is defined by Eq. (5–25) as follows:

$$S_G = \frac{\Delta T}{T} \bigg/ \frac{\Delta G_o}{G_o} = \frac{1}{1 + G_oK_f(1 + \Delta G_o/G_o)} \tag{8-54}$$

and applies for small or large changes in G_o.

Disturbance Rejection

The ability of a closed-loop control system to reject external disturbances is considered in Section 5–5. The disturbance rejection is expressed as the ratio C_r/C_d, where C_r is the system's output due to the useful input R alone, and C_d is the unwanted output of the system due to the disturbance D alone. The rejection ratio is expressed by Eq. (5–29) as

$$DR = \frac{C_r}{C_d} = G_1\frac{R}{D} \tag{8-55}$$

Clearly, the larger DR, the better is the disturbance rejection. DR is increased by increasing the ratio of R/D, which may be difficult to achieve in a practical sense since the value of the disturbance is unknown in most cases. DR may also be increased by increasing the value of G_1. The G_1 block shown in Fig. 5–7, is the first block after the summing junction. For the first-order systems considered here, this block is typically the lower-power amplifier, whose gain is adjustable. It was discovered in Chapter 5 that DRs for the open- and closed-loop system is exactly the same. It was shown toward the end of Section 5–5 that under certain conditions it is possible to realize a better rejection ratio for the closed-loop system.

Step Response

The response of the first-order system to the step input $Eu(t)$ as derived in Eq. (6–9) gives the final form of the response, and Fig. 6–4 shows the input and response graphically. In Eq. (6–10), x_0 and $y(0)$ are the initial values of the input and the response, respectively, and K, which is expressed by K_s in Table 8–3, is the closed-loop system gain. If the system is in the steady state at the time when the step input is applied, then $y(0) = x_0 K_s$. This makes possible some simplification of Eq. (6–9), with the final result as follows:

$$y(t) = y(0) + EK_s(1 - e^{-t/\tau_s}) \qquad (8-56)$$

Equation (8–56) shows that the steady-state value of the output is

$$y_{ss} = (x_0 + E)K_s = y(0) + EK_s \qquad (8-57)$$

which results when $t \to \infty$ in Eq. (8–56). The steady state is reached as the transient term $EK_s e^{-t/\tau_s}$ approaches zero with the system time constant τ_s.

General Remarks

It is evident from the preceding results that the loop gain $G_o K_f$ plays an important role in the performance of first-order systems. This is a consequence of using negative feedback of magnitude K_f. The use of feedback offers many advantages and one relatively minor disadvantage. The use of negative feedback always contributes to a reduction in closed-loop system gain. In the present case K_s represents the closed-loop gain of each of the three systems considered, and the value of K_s decreases as the loop gain $G_o K_f$ is increased. When $G_o K_f \gg 1$, $K_s \doteq 1/K_f$, which shows that the closed-loop gain depends inversely on the value of the feedback and is independent of the forward system parameters. The reduction in closed-loop gain due to feedback may be regarded as a disadvantage, but at the same time a large loop gain offers many advantages.

One such advantage is a reduction in the system sensitivity factor S_G. The system is thus less sensitive to variation in the forward-path parameters. In a practical sense this means that cheaper and less stable components used in the forward path G_o do not significantly degrade system performance. Not mentioned here, but derived in Eq. (5–22), is the sensitivity factor S_H, which expresses the effect on the output due to variation in the value of feedback. For large loop gains $S_H \doteq 1$, showing clearly that the feedback components must be very stable. It is largely due to the fact that S_G is small that the complex internal structure of the operational amplifier with all its variations has no effect on the amplifier gain, which is determined only by the external resistors.

Other advantages of using a large loop gain lie in reduction in the steady-state error e_{ss} and in reduction of the system time constant τ_s. From the operational standpoint, this means that the system becomes more accurate and faster as loop gain is increased. For large loop gain the speed control system time constant may be approximated by $\tau_s = \tau_m/G_o K_f = (J_L + J_m)/K_a K_f K_t$ being independent of the term $(B_L + B_m)$ in the denominator, which is canceled with the identical term in the numerator. The same occurs in the case of thermal and liquid-level-system time constants, where the thermal resistance R_t almost cancels R_{te}, and the hydraulic resistance R_h completely cancels out. It may therefore be concluded that under the condition of large loop gain, load parameters such as viscious friction coefficient B_L, thermal resistance R_t of the chamber walls, and hydraulic resistance R_h of the tank outlet have little or no effect on the system time constant, and therefore little or no effect on the speed of the system.

DC Motor Dynamics

We must return briefly to Chapter 3, where the dc motor is discussed, and complete some unfinished business. Specifically, we wish to consider the viscous friction and inertial torque components of motor-developed torque as well as the motor-developed power and the resulting armature current during the transient phase of the step response. This was not possible to do at that time because the transient analysis is done in a later chapter.

The form of the differential equation models for the dc motor [Eq. (3–39)] and the closed-loop speed control system [Eq. (8–7)] are exactly the same. Consequently, the solution of a differential equation for a general first-order system as expressed by Eq. (8–56) also aplies to the dc motor. After substituting appropriate parameters [Eq. (8–56)], the speed of the motor is as follows:

$$\omega_m(t) = \omega_m(0) + EK_m(1 - e^{-t/\tau_m}) \tag{8–58}$$

where E is the amplitude of the voltage step input to the armature circuit, K_m is the motor speed constant Eq. (3–39), the initial motor speed is $\omega_m(0) = K_m V_a(0)$, $V_a(0)$ is the armature voltage at the time when the step E is applied, and τ_m is the motor time constant, including the load [see Eq. (3–39)]. The viscous friction load torque is the product of B_L and the motor speed; hence

$$T_B(t) = B_L \omega_m(t) = B_L[\omega_m(0) + EK_m(1 - e^{-t/\tau_m})] \tag{8–59}$$

The total inertial torque is the product of $(J_L + J_m)$ and the acceleration $\dot{\omega}_m(t)$. Using Eq. (8–58), the final form of the inertial torque is

$$T_J(t) = (J_L + J_m)\dot{\omega}_m = \frac{EK_m}{\tau_m} e^{-t/\tau_m} = EK_{te} e^{-t/\tau_m} \tag{8–60}$$

where K_{te} is defined by Eq. (3–38) ($K_{te} = K_t$ when the Coulomb friction $T_c = 0$). The sum of the viscous friction and the inertial torques must be the motor-developed torque, which must at all times be equal to $K_i I_a$; hence

$$K_i I_a(t) = T_J(t) + T_B(t) \tag{8–61}$$

As shown in Fig. 8–18, the inertial torque $T_J(t)$ is maximum initially (when the step input is applied). It decays exponentially with the time constant τ_m and reaches a zero

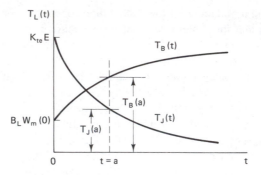

Figure 8–18. Variation of the inertial and viscous friction load torques during the transient phase of the step response of the first-order system. At $t = a$, for example, the motor-developed torque $T_m(a) = T_J(a) + T_B(a)$.

value in the steady state, where the velocity is constant and the acceleration is zero. The viscous friction torque, on the other hand, is normally equal to zero unless the initial velocity is nonzero, in which case the initial value of viscous friction torque is $B_L\omega_m(0)$, as shown in the diagram, and attains its maximum value in the steady state.

Because the speed control system is also a first order system, Eqs. (8–58) to (8–60) apply after some minor parameter changes. Therefore, for the basic speed control system shown in Fig. 8–2, the speed, the acceleration, and the torques may be expressed as follows:

$$\omega_m(t) = \omega_m(0) + EK_s(1 - e^{-t/\tau_s}) \tag{8–62}$$

$$T_J(t) = (J_L + J_m)\dot{\omega}_m = EK_aK_t e^{-t/\tau_s} \tag{8–63}$$

$$T_B(t) = B_m[\omega_m(0) + EK_s(1 - e^{-t/\tau_s})] \tag{8–64}$$

Equation (8–61), which states one of the fundamental properties of the armature-controlled dc motor—that the armature current I_a is converted to the torque required by the load—applies also to the motor in the closed-loop speed control system configurations. The output power of the motor, which is equal to the load power, is restated here from Eq. (2–63). The mechanical power being the product of speed and torque may be expressed in horsepower from Eq. (2–63) as

$$P = \frac{T\omega}{5252} \qquad \text{hp} \tag{8–65}$$

where T is in ft-lb. and ω in rpm. It may also be expressed in watts or ft-lb/s or several other forms of units. Some of the power conversion formulas are included in Eqs. (2–54 and 2–55).

EXAMPLE 8–4

Given the closed-loop speed control system shown in Fig. 8–2 and $K_f = 10$ V/1000 rpm, $J_L = 0.03$ oz-in.-s^2, $J_m = 0.02$ oz-in.-s^2, $B_L = 30$ oz-in.-s, and $R_a = 2\ \Omega$; Calculate (a) the value of K_a for $\omega_{ss} = 900$ rpm and $V_i = 12u(t)$ V; (b)

Analog Control Systems   Chap. 8

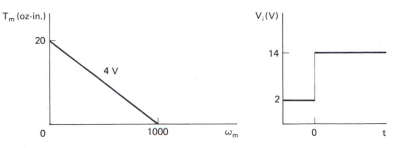

Figure E8–4. Motor speed–torque characteristics.

the time $t > 0$ such that $\omega_m(t) = 800$ rpm, V_i as shown in Fig. E8–4, and $K_a = 3$; (c) $I_a(0)$, I_{ass}, and $I_a(0.05)$ (use V_i from Fig. E8–4, and $K_a = 2$); (d) the percent change in motor speed if K_a changes from 5 to 4.5 (use the sensitivity factor); (e) the value of K_a such that the steady-state error is 40 rpm [$V_i = 12\,u(t)$].

SOLUTION (a) From the given speed–torque characteristics, $K_t = 20/4 = 5$ oz-in./V and the slope $B_m = 20/1 = 20$ oz-in./1000 rpm; therefore, $K_m = K_t/(B_L + B_m) = 5/(20 + 30) = 100$ rpm/V, and the closed-loop speed constant $K_s = K_a K_m/(1 + K_a K_m K_f) = 100 K_a/(1 + K_a)$; hence the steady-state speed $\omega_{ss} = EK_s = 900 = 1200 K_a/(1 + K_a)$. Solving for K_a, its value is found as $K_a = 3$.

(b) From part (a), $K_s = 300/4 = 75$ rpm/V. Since $V_i(0) = 2$ V, the initial speed is $\omega_m(0) = V_i(0)K_s = 2(75) = 150$ rpm, where $V_i(0)$ is given on the graph. The motor time constant $\tau_m = (J_L + J_m)/(B_L + B_m) = 0.05/[0.05(30/\pi)] = 104.7$ ms, where $B_L + B_m = (20 + 30)/1000$ oz-in./rpm $= 0.05$ (oz-in./rpm)$[(60/(2\pi)$rpm/rad/s]. The closed-loop time constant $\tau_s = \tau_m/(1 + K_a K_m K_f)$; hence $\tau_s = 104.7/(1 + 3) = 26.2$ ms. The closed-loop speed $\omega_m(t) = \omega_m(0) + EK_s(1 - e^{-t/\tau_s})$, and the value of the step input from the graph is $E = 14 - 2 = 12$ V. Substituting for E and other values yields

$$800 = 150 + 75(12)(1 - e^{-t/0.0262})$$

Solving for t from the equation above gives us

$$t = 33.6 \text{ ms}$$

(c) At $t = 0$, from part (b), $\omega_m(0) = 150$ rpm and $K_i = R_a K_t = (2)(5) = 10$ oz-in./amp; then $T_B(0) = B_L \omega_m(0) = 0.03(150) = 4.5$ oz-in., and the inertial torque from Eq. (8–63) $T_J(0) = 12(2)(5) = 120$ oz-in. Using Eq. (8–61), $120 + 4.5 = 10 I_a(0)$; therefore, $I_a(0) = 124.5/10 = 12.45$ A.

(d) The sensitivity factor, which is defined as $S_G = (\partial T/T)/(\partial G/G) = (\partial \omega/\omega)/(\partial K_a/K_a) = 1/(1 + K_a K_m K_f)$, where the percent change in T is equal to percent change in ω assuming that V_i is held constant, and the percent change in $G = K_a K_m$ is equal to percent change in K_a assuming that K_m does not change. Since $\partial K_a/K_a = (4.5 - 5)/5 = -0.1$, then $S_G = (\partial \omega_m/\omega_m)/(0.1) = 1/(1 + 5) = \frac{1}{6}$ or $\partial \omega_m/\omega_m = -0.1/6 = -1.67\%$. Thus a 10% decrease in K_a results in a 1.67% drop in the motor speed. The change in speed is less for a larger value of loop gain (i.e., for a larger value of the nominal value of K_a).

(e) The steady-state error for a step input is defined as $e_{ss} = E/[K_f(1 + K_aK_mK_f)]$. Substituting values, $e_{ss} = 40 = 12/[(0.01)(1 + K_a)] = 1200/(1 + K_a)$; therefore, $K_a = 29$. Thus the actual speed is less than the desired speed by 40 rpm (i.e., $\omega_{ss} = 12/0.01 - 40 = 1160$ rpm). The steady-state error is reduced by increasing the value of K_a, which also means that the loop gain must be increased.

This example illustrates, among other things, the effect that the loop gain has on the performance of the closed-loop speed control system, and in general on any feedback system. By increasing the loop gain through adjustment of K_a, it is possible to reduce the closed-loop system time constant, steady-state error, and the value of the sensitivity factor, and thus make the system faster, more accurate, and less sensitive to variations in the values of the forward-path parameters.

SUMMARY

1. Three first-order systems are presented in this chapter: speed control, temperature control, and liquid-level control. Each of the systems presented use negative feedback and may therefore be characterized in general as being automatic control systems. The speed control system regulates the speed of the motor, the temperature control system regulates the temperature within the chamber, and the liquid-level control system maintains the height of the liquid within the tank close to the set point.

2. The reader is referred to Table 8–3, where the important properties of the three first-order control systems are summarized. As can be seen from the table, the three systems are similar in many respects, and the reason for that is that they are first-order systems. The differential equation and the transfer function are the mathematical models for these systems. The performance of each of these systems may be described by the transient response to the step input; by the steady-state error, which expresses system accuracy; by the sensitivity factor, which expresses the sensitivity of the system to variation in system's forward parameters; and by the disturbance rejection ratio, which expresses the ability of the system to reject external disturbances. This type of system performance information and other first-order system characteristics are included in Table 8–3.

3. The fourth control system presented in this chapter is a position control system using a dc motor, with a potentiometer as a feedback component. This is a second-order system whose mathematical models, the differential equation and the transfer function, are expressed by Eqs. (8–11) and (8–12). Being a second-order system, its response to the step and ramp inputs may be underdamped, critically damped, or overdamped, depending on the value of the damping ratio ζ. The step and ramp responses are summarized in Tables 8–1 and 8–2. The steady-state error is zero for a step input, and the error due to the ramp input is expressed by Eq. (8–16). The error due to the ramp input is directly proportional to the load viscous friction.

4. The use of derivative feedback in the position control system increases system damping and thus reduces POT. Local and overall velocity feedback schemes are shown

in Fig. 8–5. The simple differentiating circuit shown in Fig. 8–6 accepts position signal at its input and generates a proportional plus derivative type of feedback at its output.

5. This chapter features several practical systems. They are important because they use present-day technology components, and in that respect they deviate from the traditional approach. They are also important because they were constructed and tested in the laboratory with excellent correlation between measurements and theory. The first of these systems is the speed control system shown in Fig. 8–11, which replaces the traditional taco-generator type of feedback with optical speed-sensing feedback. The position control system shown in Fig. 8–12 replaces the traditional potentiometer with optical position-sensing feedback. Both systems share some of the components, including the disk and the OIDs.

6. Another practical system featured in this chapter is the adoptive position control system shown in Fig. 8–7. This system has also been constructed and tested in the laboratory. The software for the system is shown in Fig. 8–8. The heart of the system is the 8085 microprocessor, which senses position and outputs a control signal to the multiplying DAC to maintain system damping, and hence POT, at a desired value. Growing emphasis on the use of microprocessors in today's industrial control environment is the primary reason for including this type of system in this chapter.

QUESTIONS

8–1. Describe the operation of the closed-loop speed control system using a dc motor and a taco generator.

8–2. Suppose that the optical motor speed-sensing approach is used to display the speed of the motor digitally using four eight-segment display chips. Assume that the disk has 180 holes. Explain, using values whenever possible, how this can be accomplished. Draw the block diagram.

8–3. Describe the operation of the closed-loop temperature control system to maintain a chamber at a temperature corresponding to the set point.

8–4. Describe the operation of a purely mechanical liquid-level control system using negative feedback. In particular, explain exactly how negative feedback is achieved.

8–5. Considered the three first-order systems presented in this chapter, and for each system express K_s in terms of the individual system parameters. What is the practical meaning of K_s? In the case of the speed control system, how does K_s differ from K_m? Is K_s related to system gain? Explain.

8–6. Explain the effect of changes in loop gain on the performance of a first-order system.

8–7. Determine the expression for the open-loop gain, loop gain, and closed-loop gain for each of the first-order systems in this chapter. What is the difference between the open-loop gain, loop gain, and closed-loop gain? Explain.

8–8. Suppose that a constant input is applied to a first-order closed-loop system. Explain the difference in system performance that you would expect for values of the sensitivity factor S_G of 0.05 and 0.5.

8–9. Describe the operation of the position control system using a dc motor and a potentiometer feedback.

8–10. Explain the operation of an optical position-sensing circuit whose output is a position feedback voltage. Draw a block diagram.

8–11. What effect does the system damping ratio have on the operation of the position control system? Explain.

8–12. Sketch the response of an underdamped position control to step and ramp inputs. Indicate on the diagram the steady-state error for each response.

8–13. Suppose that an oscilloscope displays the response of an underdamped position control system to a step input. What measurements must be made to determine the values of POT, the system time constant, and ω_n?

8–14. Suppose that the settling time for the step response of an underdamped position control system is defined as the time that it takes the response to be bounded for the first time by a band whose width is 30% of the step input E and centered on E. Derive the expression of the settling time that satisfies this definition.

8–15. Show by means of appropriate waveforms how the derivative feedback used with an underdamped position control system reduces the overshoot.

8–16. Draw the flowchart for the software in Fig. 8–8 which is used with adaptive position control, and explain the operation of the system in Fig. 8–7.

8–17. Explain the operation of the electromechanical liquid-level control system in Fig. 8–17.

8–18. Suppose that the electromechanical liquid-level control system in Fig. 8–17 is to be reconfigured so that the feedback circuits shown are replaced by the microprocessor, which senses the liquid level and controls the opening of the inlet valve. Determine the new block diagram configuration for this system with the microprocessor, and the necessary interface circuits.

PROBLEMS

For Problems 8–1 to 8–10, consider the closed-loop speed control system shown in Fig. 8–2, and the following parameter values: $B_L = B_m = 40$ oz-in./1000 rpm, $K_b = 8$ V/1000 rpm, $R_a = 1.6$ Ω, $J_m = 0.02$ oz-in.-s^2, $J_L = 0.03$ oz-in.-s^2, $K_f = 10$ V/1000 rpm, and $T_c = 0$. Assume that the given load is applied unless otherwise required.

8–1. If $V_i = 20u(t)$ and $K_a = 10$, calculate the steady-state speed with and without the load. Calculate the desired speed ω_d, and compare the two speeds calculated with ω_d.

8–2. The input is $V_i = 20u(t)$ and $K_a = 10$. Calculate the steady-state error with and without the load. Do the steady-state error values calculated here verify the differences between the desired and the actual speeds in Problem 8–1? If not, explain.

8–3. The input $V_i = 10u(t)$ V and $K_a = 5$. Calculate the armature current at $t = 0$, $t = 20$ ms, and in the steady state. Assume $\omega(0) = 0$.

8–4. The input $V_i = 20u(t)$ V, $K_a = 10$, and $\omega_m(0) = 400$ rpm. Suppose that additional viscous friction has been added to the load, increasing B_L to some value B_x. When the given input is applied, the motor speed (closed loop) is measured as 1268 rpm at 8 ms. Calculate the amount by which B_L is increased.

8–5. Suppose that $K_a = 10$, and V_i is as shown in Fig. P8–5. Calculate ω_m at $t = 1.02$ s, and at $t = 1.1$ s.

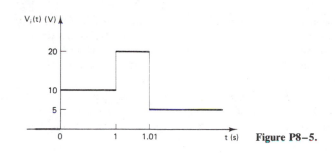

Figure P8–5.

8–6. When the step input $V_i = Eu(t)$ is applied, the initial value of armature current is measured as 15.5 A, and the steady-state value of armature current is measured as 1.25 A. Assuming that $K_a = 2$, calculate E, $\omega_m(0)$, and ω_m steady state.

8–7. The input is $V_i = 18u(t)$ V. Calculate the value of K_a so that the steady-state error is 50 rpm.

8–8. Suppose that the steady-state speed is 1000 rpm and the steady-state error is 40 rpm. Assuming that the input is not given and that $\omega_m(0) = 0$, calculate the value of K_a.

8–9. Calculate the time at which the error in Problem 8–2 is 800 rpm.

8–10. Suppose that the input is $V_i(t) = 20u(t)$ V and that $K_a = 10$. Calculate the efficiency with which the motor operates in the steady state.

For Problems 8–11 to 8–20, consider the position control system shown in Fig. 8–4, and the following parameter values: $B_m = 40$ oz-in./1000 rpm, $B_L = 10$ oz-in./1000 rpm, $J_L \doteq 0.03$ oz-in.-s², $J_m = 0.02$ oz-in.-s², $K_b = 8$ V/1000 rpm, $R_a = 1.6\ \Omega$, $\theta_{\text{active}} = 344°$, $E = 12$ V for both potentiometers, and the Coulomb friction $T_c = 0$. Assume that the given load is applied unless otherwise required.

8–11. The step input $\theta_i(t) = 60u(t)$ degrees and $K_a = 10$. From Fig. P8–11, calculate **(a)** t_2 and t_4; **(b)** t_1 and t_3; **(c)** $\theta_1 - \theta_2$; **(d)** the settling time; **(e)** the system time constant τ_s.

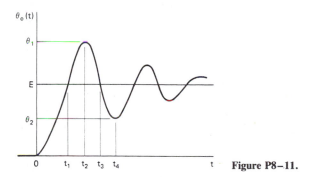

Figure P8–11.

8–12. Suppose that $\theta_1 = 90°$ in Fig. P8–11. Assuming that the input is a step of amplitude $E = 60°$, calculate the value of K_a.

8–13. In Fig. P8–11, the input $E = 60°$, $\theta_2 = 49.7$ ms and $t_3 = 169.8$ ms. Determine the position control system transfer function and its differential equation with all parameter values included.

8–14. The input is a step of height $E = 60°$, and $K_a = 8$. Calculate the armature current at the instant when **(a)** the acceleration is zero for the first time after the application of the step input; **(b)** the motor shaft position is at its first maximum peak.

8–15. Suppose that the input terminal of the circuit shown in Fig. 8–6 is connected to the wiper arm of the feedback pot, and its output is applied to the input summing junction of the position control system in Fig. 8–4. If $R = 100$ kΩ and $K_a = 5$, calculate the value of C for which the position control system's response to a step input is critically damped. Recalculate the value of C for local derivative feedback.

8–16. The input to the system is a ramp increasing at a constant rate of 1 deg/ms.
    (a) Calculate the value of K_a for the steady-state error of 1.5°.
    (b) Assuming that $K_a = 10$, calculate the value of the maximum error and the time at which it occurs.

8–17. Suppose that the settling time of the given position control system is defined as the time required for the displacement θ_o to be bounded for the first time by the band whose width is 30% of the step input E. Derive the equation for the settling time in terms of the system time constant.

8–18. Derive the time-domain response to the parabolic input $\theta_i(t) = \frac{1}{2}Et^2u(t)$ applied to a position control system. Using the expression derived, the given position control system parameter values, and $E = 1000°/\text{s}^2$, calculate the error at the time equal to 3 system time constants after application of the parabolic input.

8–19. The transfer function for some position control system (the given position control system parameter values do not apply) is $\theta_o(s)/\theta_i(s) = 625/(s^2 + 12.5s + 625)$. Calculate for this system the values of POT and T_1, assuming the step input.

8–20. The ramp response of some position control system (the given position control system parameter values do not apply) is $\theta_o(t) = 1000t - 8 + 40e^{-C_1t}\sin(C_2t + C_3)$ degrees. Calculate the values of ζ and ω_n for this system.

For Problems 8–21 to 8–26, consider the system shown in Fig. P8–21.

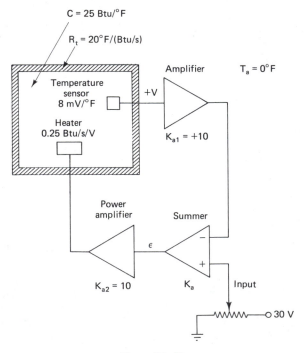

Figure P8–21.

8–21. The input is 15 V. Calculate the value of K_a for a steady-state chamber temperature of 180°F.

8–22. Assuming that the input is 16 V, it is required that the chamber temperature not deviate from the desired value by more than 2°F. Calculate the minimum value of K_a to meet this requirement.

8–23. The steady-state chamber temperature must not decrease by more than 1% when the summer amplifier gain K_a decreases by 40%. Calculate the minimum value of the initial amplifier gain K_a to meet this requirement through the use of the sensitivity factor.

8–24. The input is $16u(t)$ V, $K_a = 10$, and the initial chamber temperature is 50°F. Assuming that the 16 V step is applied above the input level required to maintain the initial temperature, calculate **(a)** the initial values of the input heat flow and the heat flow (loss) through the chamber walls; **(b)** the time required by the chamber to attain 100°F; **(c)** the time at which the heat flow (loss) through the chamber walls is 80% of the input heat flow.

8–25. Suppose that R_t triples. Answer all questions for the new value of R_t and compare the performance of the system for the two values of R_t in Problem 8–24.

8–26. Consider a chamber whose walls have perfect insulation ($R_t = \infty$). The heat applied to the chamber increases at a constant rate of 1 Btu/s². If the chamber $C = 1000$ Btu/°F and the initial chamber temperature is -50°F, calculate the time that it takes the chamber to reach 200°F after the heat is applied.

For Problems 8–27 and 8–28, consider the system shown in Fig. P8–27 and the following system specifications: cylindrical float having radius = 5 in., height = 9 in., weight = 5 lb; tank having $A = 20$ ft², capacity = 1200 gal; inlet valve of Sliding stem type, having valve const = 0.25 ft²/s, maximum stem display = 3 in., force to open/close valve = 0.5 lb; linkage: $L_1 + L_2 = 3$ ft; liquid: $\rho = 100$ lb/ft³; outlet valve $R_h = 100$ s/ft².

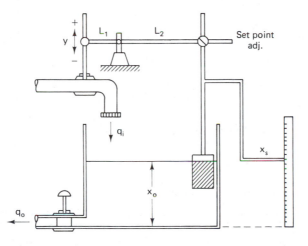

Figure P8–27.

8–27. Consider the liquid-level control system shown in Fig. P8–27. The length $L_2 = 1$ ft, $x_0(0) = 4$ ft, and $q_o \neq 0$. At $t = 0$ the set point is adjusted to 5 ft. Calculate **(a)** the initial inlet flow; **(b)** the initial position of the inlet valve stem; **(c)** the time required for the liquid inside the tank to increase by 50 gal; **(d)** the time required for the liquid level to reach 4.5 ft; **(e)** $x_{o_{ss}}$; **(f)** the steady-state error; **(g)** the increase in the inlet flow under steady-state conditions; **(h)** the outlet flow under steady-state conditions; **(i)** the velocity of the outlet flow in in./s assuming that the diameter of the outlet pipe is 2 in.; **(j)** the minimum change in x_0 to initiate the opening of the inlet valve.

8-28. Suppose that the system in Fig. P8-27 is in the steady state with initial conditions $x_0(0)$ = 4 ft and L_2 = 1 ft. At t = 0 the outlet flow is suddenly increased by 30%.

(a) Did the original 4-ft liquid level change when the steady-state conditions are attained again? If so, how much?

(b) Calculate the inlet flow under the steady-state conditions.

(c) How long does it take for the inlet flow to be within 10% of its final value?

LAB EXPERIMENT 8-1
OPEN-LOOP SPEED CONTROL SYSTEM

Objective. To evaluate the performance of an open-loop speed control system. The performance characteristics to be considered include the system's step response, its sensitivity to changing gain, and the effect of the inertial load on its response time.

Equipment

Digital voltmeter

Frequency counter

Dual-trace oscilloscope

Pulse generator

Materials. As shown in the test circuit.

Test system. The test system used in this experiment is the speed control system with optical speed sensing shown in Fig. 8-11. This system is configured as an open-loop system by removing the connection between points A and B, grounding point A, and feeding the output of OA_4 to the input of OA_1. It is important that all offset voltages (if present) of the operational amplifiers must be nulled out.

Procedure

1. Apply a 50% duty cycle, 5-V_{peak} (0 to +5 V), 1.0-Hz square wave to the input V_i. Adjust the gain of OA_3 for motor speed of 2000 rpm as measured on the frequency counter at the output of OA_6, and adjust the gain of OA_1 for 5-V peak value of the response waveform at the output of OA_1. Measure and record the following: (a) the OA_1 and OA_3 gains; (b) the response waveform at the output of OA_1. Use suitable graph paper showing in sufficient detail the transient portion of the response.

2. Apply V_i = 5 V dc, and adjust the gains of OA_1 and OA_3 to the same values as in step 1, resulting in the same motor speed of 2000 rpm and the same OA_1 output of 5 V as obtained in step 1. Measure and record the values of the output voltages of OA_1, OA_2, OA_3 and the value of V_a.

3. Adjust the gain of OA_3 to 5. Apply a dc voltage to the input V_i and adjust its value for the motor speed of 2000 rpm as measured by the frequency counter at the output of OA_6. Record the value of V_i.

4. Adjust the gain of OA_3 to 4 and apply the value of V_i used in step 3. Measure and record the motor speed.

5. Apply the inertial load to the motor's shaft and repeat steps 1 and 2. The inertial load is a solid brass cylinder whose inertia is approximately equal to the inertia of motor's armature.

6. Repeat step 5 but use an inertial load whose inertia is approximately twice the inertia of motor's shaft.

7. Repeat step 5 but use an inertial load whose inertia is approximately three times the inertia of motor's shaft.

Test data.   The format for the test data and the calculated results, as well as the report format, must comply with and meet the requirements of your lab instructor.

Evaluation.   As the ultimate objective of any lab experiment is to verify the theory-based predictions through measurements, significant differences between theory and measurements (generally, differences over 10%) must be resolved through additional experimentation, investigation, or retest.

1. The open-loop speed control system used in this experiment is represented by a block diagram as shown in Fig. LE8–1. Determine the values of the following constants in this block diagram: (a) K_{OL}, the speed constant for the open-loop speed control system; (b) τ_{OL}, the time constant for the open-loop speed control system; (c) K_{OS}, the transfer function of the optical speed-sensing circuit. Use the results of experiments in Chapters 2 and 3 and the test data of this experiment to determine the values of these constants.

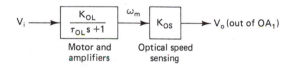

Figure LE8–1.

2. In this step we are concerned with comparing and verifying the measured values by the theory-based predictions. The transient and steady-state portions of the system's step response are considered here. (a) Using the linear block model of the system in step 1 with all values of constants included, determine the transient response equation $W_m(t)$. Using this equation, calculate the time that it takes the system to reach 1000 rpm (this is 50% of the steady-state speed, which is 2000 rpm). Compare this value with the measured value obtained in step 1 of the Procedure. (b) Using the system speed constant K_{OL}, calculate the steady-state speed for $V_i = 5$ V. How does this value compare with the 2000 rpm used in this experiment?

3. Using the data from steps 3 and 4 of the Procedure, calculate the system's sensitivity factor S_G, which is expressed as the ratio of percent change in speed to percent change in the transfer function value. Compare this experimentally obtained value of S_G against the theoretical value.

4. For each of the inertia values in steps 5 to 7 of the Procedure, calculate the time taken by the system to reach 1000 rpm as suggested in step 2a, compare these values with those obtained experimentally.

5. What effect did the inertial loading have on the transient part of the system's step response? on the system's steady-state response? Explain.

6. What conclusions can you reach regarding the sensitivity of the open-loop speed control system to changing gain? Explain. Suggest other effects that can influence the operation (i.e., system speed) of the system in this experiment.

LAB EXPERIMENT 8–2
CLOSED-LOOP SPEED CONTROL SYSTEM

Objective.   To evaluate the performance of a closed-loop speed control system. The performance characteristics to be considered include the system's step response, its sensitivity to variation in the forward-path and feedback gains, system accuracy, and the effect of loop gain on system's accuracy, sensitivity, and transient response.

Equipment

Digital voltmeter
Frequency counter
Dual-trace oscilloscope
Pulse generator

Materials.   As shown in the test circuit.

Test system.   The test system used in this experiment is the speed control system with optical speed sensing shown in Fig. 8–11. In this experiment the system is operated as a closed-loop system. Prior to testing, all offset voltages of the operational amplifiers must be nulled out.

Procedure

1. Adjust the gain of OA_3 to 2 and that of OA_1 to 20.

2. Apply a 50% duty cycle, 10-V_{peak} (0 to 10.0 V), 0.5-Hz square wave to the input V_i. Monitor the motor speed response waveform at the output of OA_1 and V_i on a dual-trace oscilloscope. Measure and record the following: (a) the speed response waveform on suitable graph paper, showing in sufficient detail the transient portion of the response; (b) t_{50}, the time required by the motor to attain 50% of its steady-state speed.

3. Apply $V_i = 10.0$ V dc. The gains remain at the same value as set in step 1. Measure and record the following: (a) the output of OA_2 (this is the difference voltage between the input and the feedback voltages); (b) ω_{ss}, the steady-state speed measured on the frequency counter at the output of OA_6.

4. Repeat steps 2 and 3 for each of the following OA_3 gains: 4 to 20 in steps of 2.

5. Apply $V_i = 10$ V dc, set the gain of OA_3 to 10. Measure and record values as specified in step 3 for the following OA_1 gains: 14, 16, and 18.

6. Apply the inertial load to the motor's shaft and repeat steps 1 to 4. The inertial load is a brass cylinder whose inertia is approximately twice the inertia of the motor's armature.

7. Repeat step 6 except use an inertial load whose inertia is approximately three times the inertia of motor's shaft.

Test data.   The format for the test data and the calculated results, as well as the report format, must comply with and meet the requirements of your lab instructor.

Evaluation.   As the ultimate objective of any lab experiment is to verify the theory-based predictions through measurements, significant differences between theory and measurements (generally, differences over 10%) must be resolved through additional experimentation, investigation, or retest.

1. Plot t_{50} versus loop gain K_fG_o. There must be three curves: the J_m curve, corresponding to steps 2 and 4 of the Procedure, and the J_2 and J_3 curves, corresponding to steps 6 and 7, respectively. The loop gain $K_fG_o = K_aK_mK_f$, where K_a includes the gain of OA_3 and the gain of the Darlington driver, which is available from Lab Experiment 8–1. K_f (the transfer function of the feedback circuits) may be determined from the data taken in step 3a of the Procedure, and the speed constant of the motor K_m is available from Lab Experiment 3–1. Plot on the same graph the J_m, J_2, and the J_3 curves, which are based only on theory.

2. Plot on suitable graph paper the measured ω_{ss} and ω_{ss} based on theory as a function of loop gain K_fG_o, with the gain of OA_1 being held at 20. Include on the graph the desired ω_{ss} as predicted by theory.

3. Plot on suitable graph paper the steady-state error e_{ss} as a function of loop gain K_fG_o with OA_1 gain held at 20. The steady-state error is, by definition, the difference between the desired ω_{ss} (E/K_f) and actual ω_{ss}. Include on the graph the steady-state error as predicted by theory.

4. Plot on suitable graph paper the sensitivity factor S_G based on measurements as a function of loop gain K_fG_o with OA_1 gain held at 20. S_G may be calculated for the OA_1 gain increment between 2 and 4 and plotted as a point corresponding to the value of loop gain calculated for OA_1 gain of 2; then the increment in OA_1 gain of 4 to 6 is considered, then 6 to 8, and so on. Include on the graph the S_G curve based on theory.

5. Calculate the sensitivity factor S_H corresponding to the three data points in step 5 of the Procedure, and compare these values with S_H values based on theory.

The responses to the following questions and/or conclusions must be supported whenever possible by the appropriate theory. Draw conclusions from the graphical experimental results and explain.

6. The effect of loop gain on the transient respnose.
7. The effect of loop gain on steady-state speed.
8. The effect of loop gain system accuracy.
9. System sensitivity to variation in the forward-path gain.
10. System sensitivity to variation in feedback gain.
11. The effect of inertial load on the system's speed of response.
12. The effect of inertial load on the steady-state speed.
13. The effect of feedback gain on system accuracy.
14. What effect, if any, might there be on system performance if the number of holes on the optical disk is increased? decreased? (No data are taken in this experiment.)
15. Suppose that the response time of FVC was equal or greater than that of the motor. How would this affect system performance? (No data are taken in this experiment.)
16. Can the output of OA_2 be called the error voltage? Explain.
17. What problem results if the polarity of the input voltage is suddenly reversed? What must be done to the system configuration to avoid such a problem?
18. Compare and contrast the following operational characteristics of the open-loop system in Lab Experiment 8–1 and the closed-loop system in this experiment: (a) speed of response and the effect of the inertial load on the speed of response; (b) system sensitivity to variation in the forward-path gain.
19. What advantage (if any) would there be to using local negative feedback in the system configuration of this experiment, between the output of the darlington driver and the input to the summing amplifier OA_2? Explain.

LAB EXPERIMENT 8–3
POSITION CONTROL SYSTEM

Objective.   To evaluate the response of a position control system with optical position sensing to a step input.

Equipment

Dual-trace oscilloscope
Digital voltmeter
Pulse generator

Materials.   As shown in the test system.

Test system. The test system used in this experiment is the position control system, which uses an optical sensing technique for the angular position of the motor's shaft, as shown in Fig. 8–12. As shown, this is a closed-loop system that uses negative feedback. Prior to testing, all offset voltages (if present) of the operational amplifiers must be nulled out.

Procedure

1. Clear the counter [store $(0)_2$] and measure the output voltage at pin 4 of the DAC. It should read 0.00 V dc. If not, adjust the zero control until it does.

2. Open the CAL switch at the output of OA_4, and adjust the 80H ZERO ADJ. potentiometer for 0 V dc at the (+) input to OA_4. With 0 V dc at the output of the DAC (pin 4), the output of OA_4 should read 0 V dc. If not, check the offset voltages of OA_3 and OA_4, and adjust if necessary.

3. Set the feedback gain so that the output of OA_4 is -12 V dc when the optical disk completes one full revolution. This is done as follows. With the CAL switch open, store $(120)_2$ in the counter (this corresponds to one full revolution of the optical disk), and adjust the FEEDBACK GAIN ADJ. potentiometer for -12 V dc as measured by the digital voltmeter at the output of OA_4.

4. With the CAL switch open, preset the counter to 80H with the switch S_1. Adjust the 80H ZERO ADJ. potentiometer for 0.00 V dc at the output of OA_4. Close the CAL switch. The system is now calibrated and ready to receive an input at V_i. If the optical disk exhibits any motion with $V_i = 0$ V dc, repeat steps 1 to 4.

5. Set the gain of OA_2 to 20 (by the FORWARD GAIN ADJ. potentiometer), and apply to the input V_i a 50% duty cycle, 2.5-V_{peak} (0 to 2.50 V dc), 1-Hz square wave. Monitor the position response waveform at the output of OA_4 and V_i on a dual-trace oscilloscope. Measure and record (a) the response waveform at the output of OA_4, which includes accurately measured values of voltages and times corresponding to the first five peaks (three positive and two negative); (b) the settling time based on the 10% band; (c) the time at which the response is equal to the input for the first time.

6. Repeat step 5 for each of OA_2 gains: 16, 12, 8, and 4.

7. Apply the inertial load to the motor's shaft and repeat step 5. The inertial is a brass cylinder whose inertia is approximately twice the inertia of motor's armature.

8. Repeat step 7 but use the inertial load whose inertia is approximately three times that of the motor's shaft.

Test data. The format for the test data and the calculated results, as well as the report format, must comply with and meet the requirements of your lab instructor.

Evaluation. As the ultimate objective of any lab experiment is to verify the theory-based predictions through measurements, significant difference between theory and measurements (generally, differences over 10%) must be resolved through additional experimentation, investigation, or retest.

1. Plot on graph paper the response θ_o (deduce θ_o values from the output voltage of OA_4) as a function of time using the test data for the gain of OA_2 of 20 and no inertial load. Show the values of the five response peaks and the values of times at which the peaks occur. Include on the same graph the input θ_i, which may be deduced from V_i. Include on the graph the response curve, which is based on theory (i.e., the values of the response peaks and times are obtained from theory based equations).

2. Plot on the graph paper the percent overshoot POT based on measurements as a function of loop gain $G_o = K_a K_m K_f$. (The value of motor's K_m is available from Lab Experiment 3–1, and the value of K_f may be determined from the data in this experiment or from question 6 of the Evaluation in Lab Experiment 2–2, and K_a includes the gains of OA_2 and the Darlington driver.) There must be three POT curves, which are based on measurements: the J_m curve, corresponding to no externally applied inertial load, and the J_2 and J_3 curves, corresponding to steps 7 and 8 of the Procedure, respectively. Include on the graph three POT curves based on theory.

3. Repeat step 2 but instead of plotting POT, plot the three T_1 (time of the first peak) curves.

4. Repeat step 2, replacing POT by the rise time t_r (the time at which the response and the input are equal for the first time).

5. Repeat step 2, replacing POT by the settling time T_s.

6. Based on measurements of step 5 of the Procedure, calculate the damped resonant frequency W_d and the system time constant which governs the decay of the response oscillations. Compare these two values against the corresponding theory-based values.

The responses to the following questions and/or conclusions must be supported whenever possible by the applicable theory.

7. Draw conclusions from data and graphical results and explain the following: (a) the effect of forward gain on POT; (b) the effect of forward gain on t_r, T_1, and T_s; (c) the effect of load inertia in parts (a) and (b).

8. Which performance parameters (if any) would be affected if the number of holes on the optical disk is to be increased? How?

9. Why is the system initially preset to 80H?

DIGITAL CONTROL SYSTEMS

chapter 9 _____

9–1 INTRODUCTION

In this chapter you will study some of the most widely used digital control systems. These are systems that employ digital techniques to control the output variable. The type of controlled variable is generally used to identify the control system. For example, in a digital temperature control system, temperature is the controlled variable. The choice of components (analog or digital) obviously depends on accuracy, simplicity, performance, and the cost of the system.

Advancements in digital technology and efficient fabrication techniques have led not only to more complex and powerful ICs, but also to more reliable and economical ICs. The use of such ICs in modern control systems makes them more accurate, efficient, simpler, and yet inexpensive. In this chapter we examine some typical digital control systems.

9–2 DIGITAL TEMPERATURE CONTROL SYSTEM

In this section we design and analyze a temperature control system. An example of the most commonly used temperature control system is a home heating and cooling system. Although the system that will be discussed here is not an exact replica of a home heating and cooling system, the concepts presented here should be applicable to all temperature control systems. Recently, with the advent of integrated-circuit technology, digital temperature control systems have become more popular. In such systems digital techniques are used to process and control the temperature.

System Specifications

Whenever any system is to be designed or evaluated, it is necessary to know what the design criteria or specifications are. Therefore, for the temperature control system, let us use the following specifications.

1. Use digital techniques to process and control a desired temperature.
2. Temperature will be maintained at a desired value as long as needed.
3. Use of LEDs to indicate whether the heater or the air conditioner is on.
4. Performance, accuracy, and cost are the important factors and must be considered in the design of the system.

Block Diagram

The first step in the design of a system is to draw a block diagram. The block diagram of a temperature control system is shown in Fig. 9–1. Let us examine briefly the function of each block. The block labeled *desired temperature setting* enables the user to select a desired temperature and also to change it as needed. Since we wish to design a digital temperature control system, the output of this block would be a digital

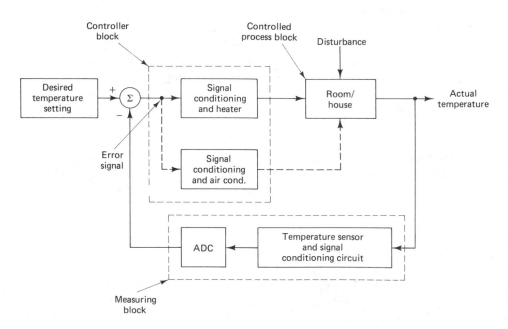

Figure 9–1. Digital temperature control system block diagram.

signal. To maintain the temperature at a fixed value, the desired temperature must be compared with the actual temperature and the error signal must be used to drive the actual temperature toward its desired value. In Fig. 9–1 an actual temperature in a house is sensed by a temperature sensor. The output of the temperature sensor is an analog signal that is converted into a digital signal by an analog-to-digital converter (ADC). Thus the output of the ADC represents the actual temperature. The temperature sensor and the ADC together perform the function of measuring an actual temperature and converting it into an appropriate form so that it can be compared with the input temperature setting. For this reason the temperature sensor and the ADC blocks are often combined into one and are referred to as a *measuring block*.

The output of the ADC is compared with the output of the desired temperature-setting block by using an error detector. Based on the difference between the desired and actual temperatures, the output of the error detector turns on either a heater or an air conditioner. Specifically, if the desired temperature is higher than the actual temperature, the heater is turned on. On the other hand, if the desired temperature is lower than the actual temperature, the air conditioner is turned on. However, if the desired temperature is equal to the actual temperature, both the heater and the air conditioner remain off. The blocks that contain the heater and the air conditioner also include the necessary signal conditioning circuitry. The heater and the air-conditioner blocks may be replaced by a single block, called a *controller block* because it performs the function of controlling the temperature.

The output of the heater or air conditioner is an input to the controlled process block. The *controlled process block* handles the temperature-controlling process. In a practical home heating system a house performs the function of a controlled process block. The process of heating or cooling a house has to be controlled by closing doors and windows and having adequate insulation in the house. In addition, variations in the outside temperature also affect the temperature inside the house. The factors that affect the temperature inside the house are represented by a *disturbance signal* to the controlled process block. The actual temperature is an output signal of the controlled process block.

System Design

Next, we design a circuit for each of the blocks such that it satisfies the requirement(s) discussed in the preceding section.

Input microswitches. We use two 4-bit switches to select an input temperature. These microswitches may be connected as shown in Fig. 9–2. Since there are two switches, any binary number from 00_2 to $1111\ 1111_2$ may be selected using these switches. We will consider switch settings after we design the *measuring block* circuitry.

Temperature transducer. We need a transducer to sense a temperature. There are a variety of temperature transducers, from a simple diode to a specialized IC. We will use LM334, a three-terminal adjustable current source, as a temperature sensor

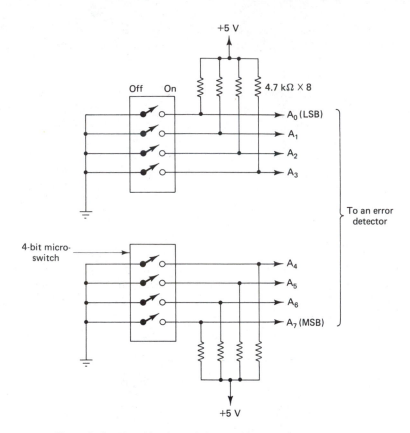

Figure 9–2. Four-bit microswitches used to set an input temperature.

because it is reliable, economical, and compact in size. The electrical specifications of a LM334 are as follows:

Operates from 1 to 40 V

10,000:1 (1 μA to 10 mA) operating current range

0.02% per volt current regulation

Guaranteed over a temperature range of 0 to 70°C

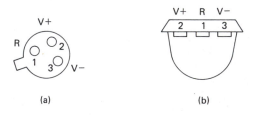

(a)   (b)

Figure 9–3. LM334: (a) bottom view of TO-46 metal can package (pin 3 is electrically connected to case); (b) bottom view of TO-92 plastic package. (Courtesy of National Semiconductor, Inc.)

The LM334 is available in TO-46 hermetic and TO-92 plastic packages. The bottom views of plastic and metal can packages are shown in Fig. 9–3.

The output current I_{set} of the LM334 is directly proportional to absolute temperature in kelvin (K) and is given by

$$I_{set} = \frac{(227 \ \mu V/1 \ K)(T)}{R_{set}} \qquad (9-1)$$

where I_{set} = output current of the LM334

R_{set} = resistor that determines I_{set} and is connected between the R and V terminals of LM334

T = temperature in kelvin

Using the connection diagram of the LM334 shown in Fig. 9–4 and Eq. (9–1), we may express the output voltage V_o of the LM334 in terms of I_{set} and R_L. In other words,

$$\begin{aligned} V_o &= I_{set}R_L \\ &\approx 10 \ mV/K \end{aligned} \qquad (9-2)$$

This means that the output V_o changes at the rate of 10 mV per 1 K change in temperature. In other words, the output voltage of the sensor will vary from a minimum of 2730 mV to a maximum of 3430 mV as the temperature varies from 0 to 70°C.

The output voltage of the sensor is an analog voltage. We will use an analog-to-digital converter to process this voltage into a digital signal. We will use National Semiconductor's ADC0804, an 8-bit ADC that is easy to use and less expensive. A glance at ADC0804 data sheet reveals that its analog input voltage range is 0 to 5 V. This means that the output voltage of the sensor, 2.73 to 3.43 V, must be transformed into 0 to 5 V.

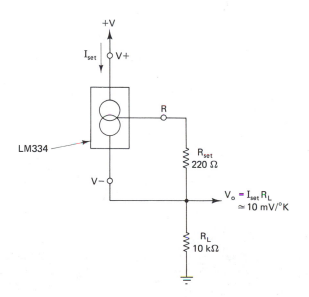

Figure 9–4. LM334 connection diagram. (Courtesy of National Semiconductor, Inc.)

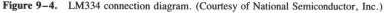

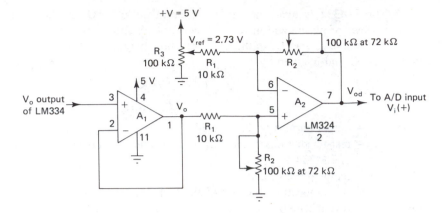

Figure 9–5. Signal conditioning through voltage follower and differential amplifier.

Level translator stage. There are a variety of ways that we can condition the output signal of the temperature transducer to make it suitable for the ADC0804. One of the easiest ways is to use a voltage follower and a differential amplifier, as shown in Fig. 9–5. The voltage follower eliminates "loading" and the differential amplifier provides the gain necessary to make the signal suitable for the ADC0804. We need a differential amplifier with an adjustable input offset voltage of 2.73 V and a gain of

$$A_F = \frac{5 \text{ V}}{3.43 - 2.73} \simeq 7.2 \tag{9-3}$$

This is accomplished with the help of resistor R_1 and potentiometer R_2, as shown in Fig. 9–5. Thus the output of the differential amplifier at 0°C is

$$V_{od} = A_F(V_o - V_{ref}) = 0 \text{ V} \tag{9-4a}$$

where V_{od} = output voltage of the differential amplifier
$\quad A_F$ = closed-loop gain of the differential amplifier
$\quad V_o$ = output voltage of the sensor
$\quad V_{ref}$ = reference voltage of the differential amplifier = 2.73 V
Similarly, the output voltage of the differential amplifier at 70°C is

$$\begin{aligned} V_{od} &= A_F(V_o - V_{ref}) \\ &= \frac{R_2}{R_1}(3.43 - 2.73) \\ &= 7.2(0.7) \\ &= 5 \text{ V} \end{aligned} \tag{9-4b}$$

Analog-to-digital converter. Next we consider how to configure an analog-to-digital converter so that an analog signal is converted into an appropriate digital signal. An ADC0804 is a CMOS 8-bit successive approximation ADC that uses a differential potentiometric ladder. The pin diagram of the ADC0804 is shown in Fig. 9–6. The ADC0804 has an on-chip clock generator which requires the use of an *RC* network

474 Digital Control Systems   Chap. 9

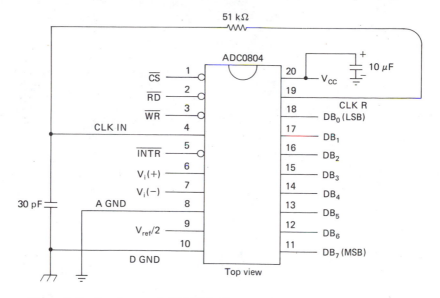

Figure 9-6. Pin diagram of ADC0804. (Courtesy of National Semiconductor, Inc.)

between pins 4 and 19, as shown in Fig. 9-6. The clock frequency as a function of R and C is given by

$$f_{\text{clk}} \simeq \frac{0.91}{RC} \tag{9-5}$$

The ADC has a total error of ± 1 LSB and a conversion time of 100 μs. The digital output can be decoded by dividing 8 bits into two hexadecimal (hex) characters, the

TABLE 9-1 DIGITAL OUTPUT DECODING OF ADC0804

Hex	Binary	Fractional binary value		Output voltage	
		MS group	LS group	VMS group	VLS group
F	1111	15/16	15/256	4.80	0.30
E	1110	7/8	7/128	4.48	0.28
D	1101	13/16	13/256	4.16	0.26
C	1100	3/4	3/64	3.84	0.24
B	1011	11/16	11/256	3.52	0.22
A	1010	5/8	5/128	3.20	0.20
9	1001	9/16	9/256	2.88	0.18
8	1000	1/2	1/32	2.56	0.16
7	0111	7/16	7/256	2.24	0.14
6	0110	3/8	3/128	1.92	0.12
5	0101	5/16	5/256	1.60	0.10
4	0100	1/4	1/64	1.28	0.08
3	0011	3/16	3/256	0.96	0.06
2	0010	1/8	1/128	0.64	0.04
1	0001	1/16	1/256	0.32	0.02
0	0000	0	0	0	0

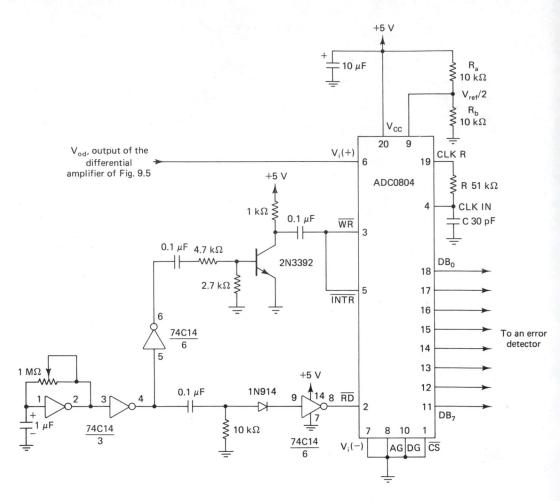

Figure 9-7. Connection diagram of ADC0804.

four most significant (MS) and the four least significant (LS). Table 9–1 shows the fractional binary equivalent of these two 4-bit groups and the corresponding output voltages.

We will connect the output of the differential amplifier of Fig. 9–5 to the $V_i(+)$, pin 6. Also, to obtain the clock for the analog-to-digital conversion, we will connect RC combination between pins 19 and 4 as shown in Fig. 9–7, which results in a clock frequency of

$$f_{\text{clk}} \simeq \frac{0.91}{(51 \text{ k}\Omega)(30 \text{ pF})} \simeq 357 \text{ kHz}$$

A $V_{\text{ref}}/2 \simeq 2.5$ V is derived from the supply voltage of 5 V. Finally, the read ($\overline{\text{RD}}$), write ($\overline{\text{WR}}$), and interrupt ($\overline{\text{INTR}}$) signals are derived by using 74C14 hex inverters. Specifically, 74C14/3 is configured as a clock generator, the frequency of which is given by

TABLE 9–2 INPUT MICROSWITCH SETTINGS

Temperature (°C)	Transducer output (V)	Input to ADC (V)	Input microswitch settings in binary
0	2.73	0	0000 0000
10	2.83	0.72	0010 0100
20	2.93	1.44	0100 1000
30	3.03	2.16	0110 1100
40	3.13	2.88	1001 0000
50	3.23	3.60	1011 1110
60	3.33	4.32	1101 1000
70	3.43	5.04	1111 1100

$$f_{RW} \simeq \frac{1}{1.7RC} \qquad (9-6)$$

The frequency of the clock is varied by adjusting 1-MΩ potentiometer. The pot is adjusted until the ADC functions satisfactorily. The $\overline{WR}$ and $\overline{RD}$ signals are derived from the 74C14/3 clock such that they are opposite in phase; when one is HIGH, the other is LOW, and vice versa. The analog-to-digital signal conversion takes place each time the $\overline{WR}$ and $\overline{INTR}$ signals are active LOW.

The 8-bit digital output appears at pins 11 through 18. This output must be compared with the binary "desired temperature setting" so that the actual temperature can be driven toward the desired temperature. The outputs of the ADC and the microswitches must be compatible in order to be compared. Hence the input microswitches should be set according to the output of the ADC. Table 9–2 shows the input microswitch settings corresponding to the desired temperature values. The microswitch settings for intermediate temperatures may be calculated by using Tables 9–1 and 9–2.

Error detector. The function of an error detector is to compare the binary number representing an input temperature setting to the binary number at the output of the ADC, which corresponds to an actual temperature. Based on the difference between the two binary numbers, the output of the error detector may turn on either the heater or the air conditioner. A 4-bit magnitude comparator 74LS85 will perform the function of an error detector because it can compare two 4-bit numbers to produce three separate signals: less than (<), larger than (>), or equal to (=). The pin diagram and the function table of a 74LS85 4-bit magnitude comparator are shown in Fig. 9–8.

Remember that we have two sets of 8-bit numbers to compare; hence we need two 74LS85 4-bit magnitude comparators. These comparators should be connected as shown in Fig. 9–9. In this figure A represents the binary number that corresponds to the desired temperature and B represents the binary number that corresponds to the actual temperature. The output $A > B$ will be used to drive the heater, and the output $A < B$ will be used to drive the air conditioner. However, $A = B$ output of the MSD 4-bit comparator is not used, because if the desired temperature is equal to the actual temperature, both the heater and the air conditioner should remain off.

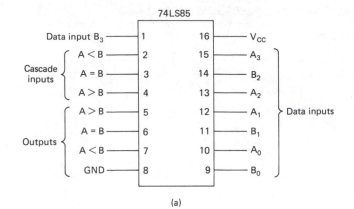

74LS85

Pin			Pin	
Data input B₃	1		16	V_CC

Data input B_3 —— 1   16 —— V_{CC}

Cascade inputs:
$A < B$ —— 2   15 —— A_3
$A = B$ —— 3   14 —— B_2
$A > B$ —— 4   13 —— A_2

Outputs:
$A > B$ —— 5   12 —— A_1 } Data inputs
$A = B$ —— 6   11 —— B_1
$A < B$ —— 7   10 —— A_0
GND —— 8   9 —— B_0

(a)

Comparing inputs				Cascading inputs			Outputs		
A_3, B_3	A_2, B_2	A_1, B_1	A_0, B_0	$A>B$	$A<B$	$A=B$	$A>B$	$A<B$	$A=B$
$A_3 > B_3$	X	X	X	X	X	X	H	L	L
$A_3 < B_3$	X	X	X	X	X	X	L	H	L
$A_3 = B_3$	$A_2 > B_2$	X	X	X	X	X	H	L	L
$A_3 = B_3$	$A_2 < B_2$	X	X	X	X	X	L	H	L
$A_3 = B_2$	$A_2 = B_2$	$A_1 > B_1$	X	X	X	X	H	L	L
$A_3 = B_3$	$A_2 = B_2$	$A_1 < B_1$	X	X	X	X	L	H	L
$A_3 = B_3$	$A_2 = B_2$	$A_1 = B_1$	$A_0 > B_0$	X	X	X	H	L	L
$A_3 = B_3$	$A_2 = B_2$	$A_1 = B_1$	$A_0 < B_0$	X	X	X	L	H	L
$A_3 = B_3$	$A_2 = B_2$	$A_1 = B_1$	$A_0 = B_0$	H	L	L	H	L	L
$A_3 = B_3$	$A_2 = B_2$	$A_1 = B_1$	$A_0 = B_0$	L	H	L	L	H	L
$A_3 = B_3$	$A_2 = B_2$	$A_1 = B_1$	$A_0 = B_0$	L	L	H	L	L	H
$A_3 = B_3$	$A_2 = B_2$	$A_1 = B_1$	$A_0 = B_0$	X	X	H	L	L	H
$A_3 = B_3$	$A_2 = B_2$	$A_1 = B_1$	$A_0 = B_0$	H	H	L	L	L	L
$A_3 = B_3$	$A_2 = B_2$	$A_1 = B_1$	$A_0 = B_0$	L	L	L	H	H	L

(b)

Figure 9–8. 74LS85: (a) pin configuration; (b) function table. (Courtesy of Texas Instruments, Inc.)

Heater and air-conditioner control circuitry. The next circuit we need to design is the control circuitry for the heater and air conditioner. If the desired temperature is higher than the actual temperature, the output $A > B$ at pin 5 of the 74LS85 is HIGH (Fig. 9–9). When this happens we wish to turn the heater on. To accomplish this, we need to condition the $A > B$ signal such that it can drive the heater. Also, we need to isolate the high-voltage ac signal from the low-voltage $A > B$ digital signal. An optically isolated triac driver MOC3010 will satisfy this requirement. The pin diagram and the

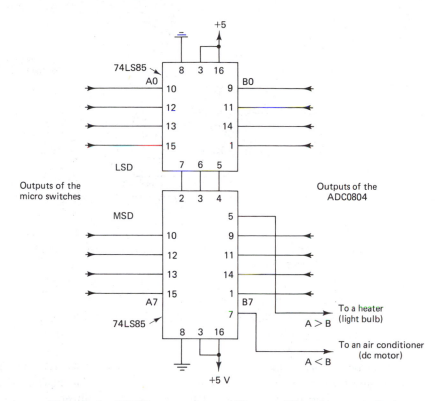

Figure 9–9. 74LS85s as error detector. (Courtesy of Texas Instruments, Inc.)

internal structure of a MOC3010 is shown in Fig. 9–10. The electrical ratings of a MOC3010 are as follows:

Forward current (continuous) I_F = 50 mA.
Forward voltage V_F = 1.5 V maximum at I_F = 10 mA.
Reverse voltage V_R = 3 V maximum.
Off-state output terminal voltage = 250 V.
Peak nonrepetitive surge current = 1.2 A.

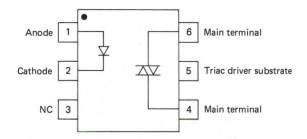

Figure 9–10. MOC3010 internal structure and pin diagram. (Courtesy of Motorola Semiconductor, Inc., Phoenix, Ariz.)

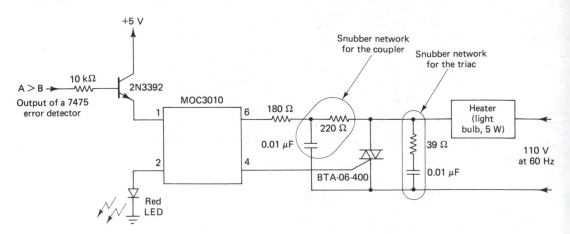

Figure 9–11. Drive circuit for the heater (light bulb). (Courtesy of Motorola Semiconductor, Inc., Phoenix, Ariz.)

Isolation surge voltage = 7500 V ac.

In addition, the holding current in either direction is 100 μA, and the LED trigger current is typically 8 mA.

To increase the drive capability of a MOC3010, an additional triac may be connected at the output of the MOC3010. The selection of this extra triac depends on the electrical ratings of the load it drives. In addition to the triac, a snubber network to protect the drive circuit may also be considered.

Since the $A > B$ signal output of the 74LS85 is not capable of driving the MOC3010 directly, we need to use a current amplifier. This amplifier may be formed by using an

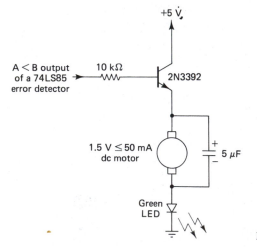

Figure 9–12. Drive circuit for a dc motor.

Digital Control Systems   Chap. 9

npn transistor and a resistor. Thus the input to the current amplifier is the $A > B$ signal, and its output current drives the MOC3010 (Fig. 9–11).

The schematic diagram of a current amplifier, an optically isolated triac driver, a triac, and snubber networks are shown in Fig. 9–11. If input $A > B$ is HIGH, transistor 2N3392 is ON, which in turn triggers the MOC3010 diac. Once this happens, the triac BTA–06–400 is turned ON and hence the light bulb. As long as $A > B$ is HIGH, the bulb remains ON. However, when $A > B$ switches LOW, the bulb turns off. A red LED in the emitter circuit of the current amplifier is used to indicate the status of the load.

On the other hand, if $A < B$ is HIGH, the air conditioner should be ON. Again, to accomplish this we need a voltage-to-current amplifier. Note that the operation of an air conditioner is simulated by a dc motor. The drive circuit for the motor is shown in Fig. 9–12. When the output $A < B$ of the MSD 74LS85 is HIGH, the motor is ON, and when it is LOW, the motor is OFF. A green LED is used in the emitter circuit of the voltage-to-current amplifier to indicate the status of the motor.

Schematic Diagram

The complete schematic diagram of the temperature control system is shown in Fig. 9–13. In this figure the output of the temperature transducer is calibrated for degrees Celsius. However, it can be calibrated for degrees Fahrenheit instead. The digital temperature control system of Fig. 9–13 is fairly simple, accurate, and inexpensive. The concepts illustrated in the figure can easily be applied to any temperature control system.

EXAMPLE 9–1

In the circuit of Fig. 9–13, determine (a) the input voltage $V_i(+)$ of the ADC0804; (b) the output voltage of the ADC if the temperature in the room is 15°C and the input switches are set at $0000\ 1111_2$. Briefly describe the operation of the circuit as a result of this action.

SOLUTION (a) The output of the LM334 at 15°C will be $(273 + 15) \times 10$ mV $= 2.88$ V. Since the gain of the differential amplifier is 7.2, the input voltage $V_i(+)$ of the ADC will be $V_i(+) = (2.88 - 2.73) \times 7.2 = 1.08$ V.

(b) An LSB value of the ADC is approximately 20 mV. Therefore, the digital output corresponding to an input of 1.08 V will be 1.08 V/20 mV $\simeq$ 54 $= 0011\ 0110_2$. Thus the output of the ADC is $0011\ 0110_2$. The output can also be verified from Table 9–1. Referring to Table 9–1, we see that the binary value for MS group, 0011_2, corresponds to an output voltage of 0.96 V. Similarly, the binary value of the LS group, 0110_2, corresponds to an output voltage of 0.12 V. Since 0.96 V + 0.12 V = 1.08 V, the binary output is $0011\ 0110_2 = 54_{10}$, as expected.

The binary 8-bit microswitch is set at $0000\ 1111_2$, which is less than the output $0011\ 0110_2$ of the ADC; therefore, the air conditioner (a dc motor) should turn on (Fig. 9–13).

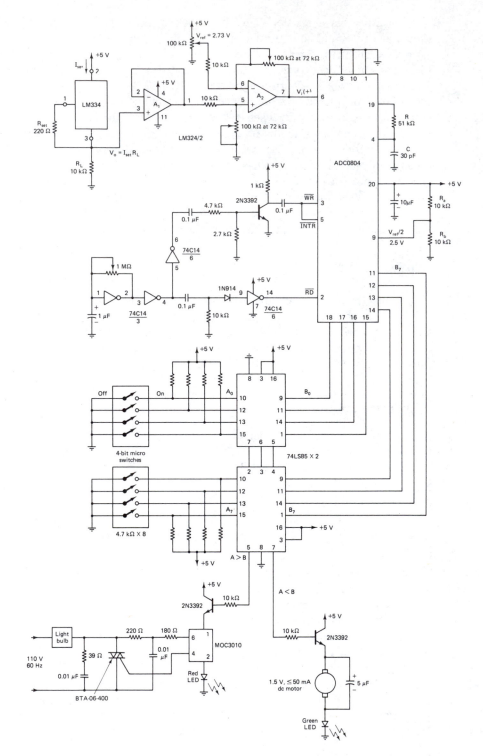

Figure 9-13. Digital temperature control system.

In this section we study the concepts involved in the digital speed control system using a dc motor. The concepts illustrated in the digital control system will then be used in the design of the microprocessor-based speed control system in Chapter 10.

System Specifications

Let us begin the design of the digital dc motor speed control system with its performance specifications. These specifications are as follows:

1. Use digital techniques to process and maintain the desired speed.
2. The speed of the dc motor should be maintained at a constant value as long as necessary.
3. The speed may be changed with the help of microswitches if desired, within the minimum and maximum limits offered by the system.
4. Performance, accuracy, and cost are the important factors and must be considered in the design of the system.

Block Diagram

Once we know the system specifications, the next step in the design of the system is to construct the block diagram. Let us begin with the input, the desired speed setting. The desired speed may be selected using an eight-position microswitch. This will allow the user to select any speed, from a minimum to a maximum rpm, simply by selecting the appropriate switch positions.

The binary speed setting must be converted into an analog input so that it can be compared to the signal representing the actual speed of the motor. Therefore, a digital-to-analog converter (DAC) should be used to convert the input binary number into an equivalent analog signal. This analog signal is then compared to the actual motor speed signal using an error detector. The output of the error detector is applied to the *controller block*, which transforms and processes the error signal such that it results in controlling the speed of the motor. The *controlled process block* contains the drive circuitry and also a dc motor. The output of the controlled process block is the actual speed of the motor shaft in rpm.

Finally, the actual speed must be sensed and converted into an appropriate signal by a measuring block so that it can be compared to the desired speed setting signal. The difference between the desired and the actual speed signals is then used to drive the actual speed toward the desired speed.

Thus the digital dc motor speed control system should consist of the following blocks: a speed setting block, a DAC block, an error detector, a controller block, a controlled process block, and a measuring block. A block diagram that includes these blocks is shown in Fig. 9–14.

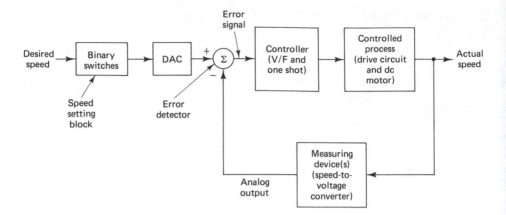

Figure 9–14. Block diagram of a digital dc motor speed control system.

System Design

Once we know the block diagram of the digital dc motor speed control system, the next step is to design the circuit for each block.

Eight-position microswitch: input speed setting. An eight-position microswitch is used to select the desired motor speed. Therefore, the speed of the dc motor may be changed by altering the switch positions. To accomplish this function, the binary switches may be configured as shown in Fig. 9–15. As the switch setting is

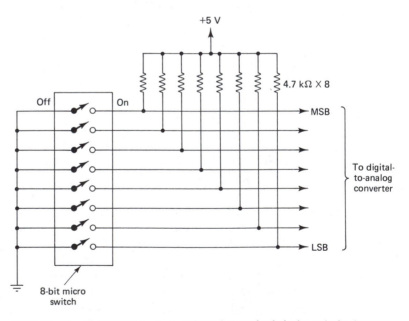

Figure 9–15. Eight-position microswitch used to set the desired speed of a dc motor.

changed successively from 0000 0001$_2$ to 1111 1111$_2$, the speed of the motor should increase from a minimum to a maximum, or vice versa. Thus, ideally, the speed of the motor may be changed through 256 different steps.

Digital-to-analog converter. A desired speed of a motor is in binary and is selected by the use of an 8-bit microswitch. However, to compare the desired speed with the actual speed, the binary number representing it must be converted into an analog signal. This function can be performed by a digital-to-analog converter (DAC). We will use a MC1408 DAC to convert the binary number into an analog signal because it is readily available, inexpensive, and commonly used. Some of the important electrical specifications of the MC1408 are as follows:

Full-scale output current, I_o settling time: 300 ns

Relative accuracy (error relative to full scale I_o) of better than $\pm 0.19\%$

Power supply currents independent of bit codes

Interfaces directly with TTL, DTL, or CMOS logic levels, and is a direct replacement for the DAC0808

Standard supply voltages: $+5$ V and -5 to -15 V

The pin configuration and connection diagram for a MC1408 are shown in Fig. 9–16.
    For our purposes we will use $+V_{CC} = 5$ V, $-V_{EE} = -12$ V, $+V_{ref} = +5$ V, $-V_{ref} = 0$ V, $R_1 = 5$-kΩ pot, $R_2 = 4.7$-kΩ pot, and $C_1 = 0.01$ μF. For the circuit configuration shown in Fig. 9–16b, the output current I_o is given by

$$I_o = \frac{+V_{ref}}{R_1} \left(\frac{A_1}{2} + \frac{A_2}{4} + \frac{A_3}{8} + \frac{A_4}{16} + \frac{A_5}{32} + \frac{A_6}{64} + \frac{A_7}{128} + \frac{A_8}{256} \right)$$

or

$$I_o = \frac{+V_{ref}}{R_1} \frac{N}{256} \qquad (9\text{--}7\text{a})$$

where $+V_{ref}$ = positive reference voltage
       R_1 = resistor connected to pin 14 and $+V_{ref}$
       N = input binary number in decimal
The output current I_o must be converted into a voltage so that it can be compared to the voltage (feedback), which represents the motor speed. Therefore, we need to use a current-to-voltage converter at the output of the DAC as shown in Fig. 9–16b. Thus the output voltage V_o is

$$V_o = I_o R_3$$
$$= \frac{+V_{ref}}{R_1} \frac{N}{256} R_3 \qquad (9\text{--}7\text{b})$$

where R_3 is the load resistor. Let us choose $R_3 = 4.7$ kΩ. A capacitor C_3 across R_3 is used to minimize the overshoot and ringing.
    In Fig. 9–16b, with all the binary inputs logic 1, the pot R_1 should be adjusted

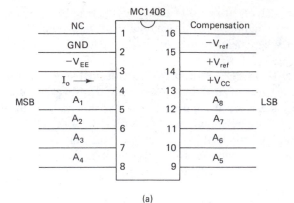

(a)

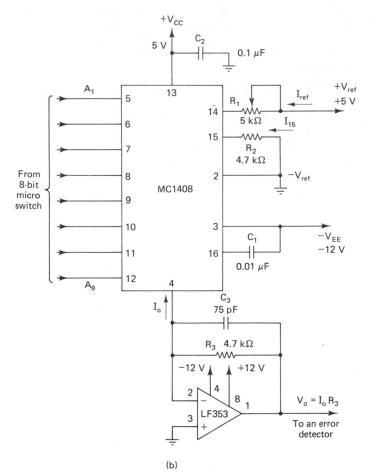

(b)

Figure 9–16. MC1408: (a) pin configuration; (b) connection diagram. (Courtesy of Motorola Semiconductor, Inc., Phoenix, Ariz.)

until $V_o = 5V$. Therefore, the full-scale output of the DAC is +5 V, which results in the resolution of 5 V/256 = 19.5 mV. Thus the output of the DAC will vary from 19.5 mV to 5V as the input binary number is changed from $0000\ 0001_2$ to $1111\ 1111_2$.

Error detector circuit: differential amplifier. In Fig. 9–14 a desired speed is converted into a voltage by the DAC 1408. Similarly, an actual speed will be converted into a voltage by a speed-to-voltage converter. To maintain the speed of a dc motor at a desired value, these two voltages must be compared and the difference should be used to drive the actual speed toward the desired speed. An error detector compares the two voltages to produce an error signal. The most commonly used error detector is the differential amplifier. A differential amplifier can be configured as shown in Fig. 9–17a, the output voltage of which is

$$V_e = (V_o - V_F) \tag{9-8}$$

where V_o = output of the DAC
$\quad V_F$ = feedback voltage proportional to the actual motor speed
$\quad V_e$ = error voltage
Initially, the feedback voltage is almost zero because the speed of the motor is zero. This means that error voltage V_e is approximately equal to the output voltage V_o of the DAC, which causes the motor to move faster. However, as the motor moves faster, the feedback voltage V_F increases toward V_o until V_e attains a constant value. If V_e is constant, V_o and V_F must also be constant [Eq. (9–8)]. Often, an error detector is referred to as a summing junction and is represented by a small circle, as shown in Fig. 9–17b.

Controller block: voltage-to-frequency converter plus the one-shot. In Fig. 9–14 the output of the error detector is an input to the controller block. The function of the controller is two-fold. First, it has to process the error signal so that digital techniques can be used to control the speed of the motor. Second, the controller must produce an output signal that drives the controlled process block such that the larger the input binary number, the stronger the output signal and the higher the motor speed.

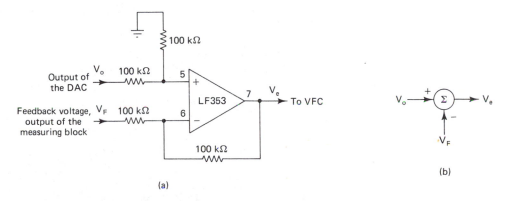

(a)

(b)

Figure 9–17. Error detector: (a) schematic diagram; (b) block diagram representation.

The first functional requirement of the controller can be met by converting the error signal (analog) into a digital signal, namely the pulse waveform. One of the simplest methods to convert an analog voltage into a pulse waveform is to use a *voltage-to-frequency converter* (VFC) such as the NE/SE 566, LM331, or Teledyne 9400 series. We choose the LM331 VFC mainly because it is readily available and simple to use.

The second functional requirement of the controller is that it should provide a pulse waveform such that its width (duty cycle) increases with an increase in the input error voltage. This is needed because for a given pulse rate, wider pulse-width waveforms will keep the motor ON for a longer duration and hence will increase its speed. One such circuit is a *monostable multivibrator*, commonly referred to as a *one-shot*. We

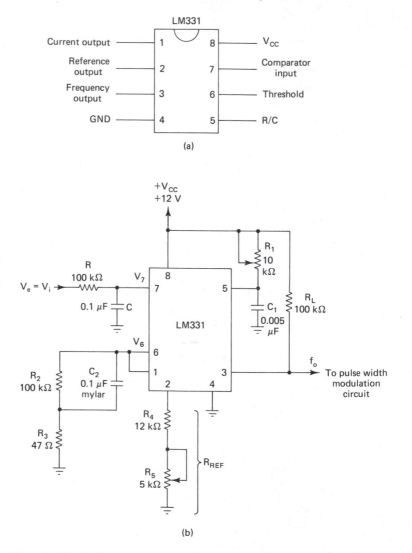

Figure 9–18. LM331 VFC: (a) pin configuration; (b) connection diagram. (Courtesy of National Semiconductor, Inc.)

will choose a CD4098B COS/MOS dual monostable multivibrator because it is simple to use.

Let us begin with the VFC LM331. Figure 9–18a shows the pin configuration of a LM331. Some of the important features of the LM331 are as follows:

Wide frequency range: 1 Hz to 100 kHz

Excellent linearity and temperature stability: 0.01% maximum and ±50 ppm/°C maximum, respectively

Pulse output compatible with all logic types

Operates on split or single supply

Absolute maximum supply voltage V_{CC} of 40 V and maximum input voltage range of −0.2 V to + V_{CC}

A typical connection diagram of the VFC using LM331 is shown in Fig. 9–18b.

The frequency output, f_o, in addition to the input voltage V_i, is a function of R_1, R_2, R_{ref}, and C_1. In equation form it is expressed as

$$f_o = \frac{V_i}{2.09\text{ V}} \frac{R_{ref}}{R_2} \frac{1}{R_1 C_1} \tag{9–9}$$

Let us assume that we want the full-scale frequency output f_o = 10 kHz at V_i = 5 V. If we choose R_2 = 100 kΩ and R_{ref} = (10 kΩ + 5-kΩ pot), then from Eq. (9–9),

$$R_1 C_1 = \frac{5\text{ V}}{2.09\text{ V}} \frac{10\text{ k}\Omega}{100\text{ k}\Omega} \frac{1}{10\text{ kHz}}$$

$$\simeq 24\ \mu s$$

Now let C_1 = 0.01 μF. Therefore, $R_1 \simeq 2.4$ kΩ. Note that in the calculations we used R_{ref} = 10 kΩ instead of 15 kΩ because 5-kΩ pot can be used to fine tune the frequency output to 10 kHz.

The final VFC circuit with the component values is shown in Fig. 9–18b. In this figure when $V_i \simeq 20$ mV, which corresponds to a binary input of 0000 0001$_2$, the frequency output from Eq. (9–9) is

$$f_o \simeq 40\text{ Hz}$$

Thus the frequency output f_o of the VFC will vary from 40 Hz to 10 kHz as the input is varied from 20 mV to 5 V, respectively.

The other half of the controller is the monostable multivibrator circuit. Let us study the CD4098B multivibrator circuit next.

Figure 9–19 shows the pin configuration and the circuit diagram of a CD4098B. Some of the important features of a CD4098B are:

Triggering from the leading or trailing edge

Wide range of output pulse width

Power supply voltage range: +5 to +15 V

Retriggerable/resettable capability

Q and $\overline{Q}$ buffered outputs available

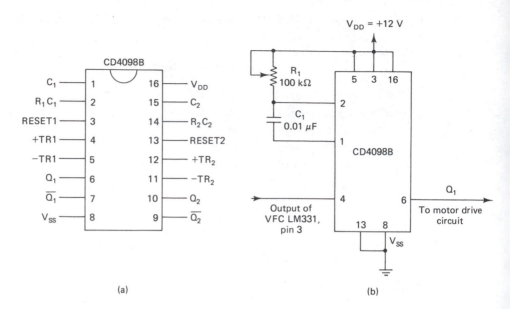

Figure 9–19. CD4098B: (a) pin configuration, (b) connection diagram. (Courtesy of RCA, Inc.)

A CD4098B is a dual monostable multivibrator and has two identical multivibrators. Therefore, we will examine only one section.

A monostable multivibrator requires two external components, R_1 and C_1, to control its operation. The time period of the multivibrator can be approximated by

$$T = \frac{R_1 C_1}{2} \tag{9-10}$$

for $C_1 \geq 0.01$ μF. The maximum value of $C_1 = 100$ μF, and the minimum value of R_1 is 5 kΩ.

Next, let us choose C_1 and R_1 such that the monostable multivibrator performs satisfactorily over the frequency range 40 Hz to 10 kHz. Let us choose $C_1 = 0.01$ μF and $R_1 = 100$-kΩ pot. These values are within the specified limits for C_1 and R_1.

When the circuit of Fig. 9–19b is connected it should be calibrated by adjusting the 100-kΩ pot such that the *desired output pulse width* is maintained over an entire input frequency range of 40 Hz to 10 kHz. For calibration a square or pulse input of amplitude ≤ 12 V p-p may be used. Here the *desired pulse width* is defined as the pulse width that is necessary to keep the motor running when the input frequency is varied over the range 40 Hz to 10 kHz.

Controlled process: motor power drive circuit and motor. The function of the motor drive circuit is to process the output of the controller block such that a desired motor speed is maintained. In addition, it should control the direction of rotation of the motor shaft.

To achieve efficient operation and provide direction control, we may connect four

Digital Control Systems   Chap. 9

power transistors in a Wheatstone bridge arrangement, two *pnp*s and two *npn*s. For clockwise (CW) rotation of the motor shaft, the transistors in the opposite arms of the bridge should be turned ON; for counterclockwise (CCW) rotation, the transistors in the remaining two arms of the bridge should be turned ON. This means that we must use complementary power transistor pairs in the opposite arms of the bridge. In addition, we need an appropriate logic circuitry to control the operation of these transistors based on the desired direction of rotation.

Let us begin with the basic power drive circuit. The design of the power drive circuit depends on the motor specifications. Let us choose a TRW405A100−1 reversible motor with the following specifications.

The motor is rated at 12 V with a no-load current of 150 mA.

The rated torque is 0.5 oz-in. at 300 mA.

The armature resistance is 15.3 Ω with the torque constant of 2.6 oz-in./A.

The no-load speed is 4800 rpm with $(1.5)(10^{-4})$ oz-in.-s^2 inertia.

The complementary transistor pairs that are used for the motor power drive circuit must have higher current and voltage ratings than the motor they drive. Therefore, we

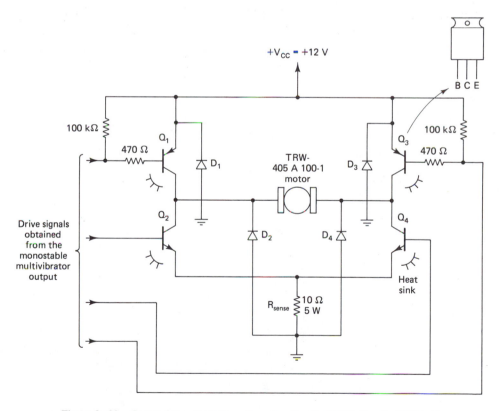

Figure 9–20. Power drive circuit for a dc motor. Q_1, Q_3: TIP30; Q_2, Q_4: TIP29; D_1 to D_4: 1N4004.

will use Texas Instrument's TIP29 (*npn*) and TIP30 (*pnp*) complementary transistor pair, which has the following electrical specifications. For the TIP29:

Collector–emitter breakdown voltage $V_{CEO}(BR) = 40$ V min

Continuous collector current $I_C = 1$ A

Peak collector current 3 A

Continuous device dissipation at $\leq 25°C$, 30 W maximum

Base–emitter voltage $V_{BE} = 1.3$ V

Collector–emitter saturation voltage $V_{CE(sat)} = 0.7$ V

Small-signal $h_{fe} = 20$ at $f = 1$ kHz, $V_{CE} = 10$ V, and $I_C = 0.2$ A

The basic motor drive circuit using power transistors TIP29 and TIP30 and 405A100–1 motor is shown in Fig. 9–20. In this figure diodes D_1 through D_4 are used to protect the collector–emitter junctions of transistors from inductive kicks and switching spikes. Also, the emitter resistor R_{sense} is used to detect the speed of the motor. Based on the

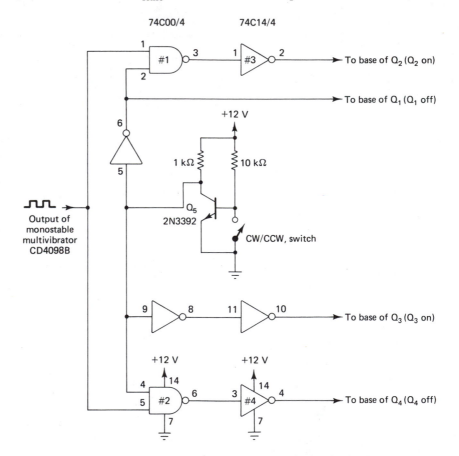

Figure 9–21. Direction control and power drive logic circuitry for the circuit of Fig. 9–20.

Digital Control Systems   Chap. 9

speed and duration of the on-time of the motor, heat sinks may be used for the power transistors as shown in Fig. 9–20. The complementary pair Q_1–Q_4 and Q_2–Q_3 must be turned on simultaneously. Moreover, when Q_1–Q_4 is ON, Q_2–Q_3 is OFF, and vice versa.

Next, let us consider the logic circuit which should turn transistors Q_2 and Q_3 ON for counterclockwise (CCW) motion and transistors Q_1 and Q_4 ON for clockwise (CW) motion of the motor shaft. The output of the monostable multivibrator and the CW/CCW direction control signal are the inputs to the logic circuit. One of the simplest circuits that will satisfy these requirements is shown in Fig. 9–21. The circuit uses a 74C00 quad two-input NAND gate and a 74C14 hex Schmitt trigger. The pin numbers of 74C00 and 74C14 are included in Fig. 9–21 for convenience.

In Fig. 9–21 a transistor circuit with a switch is used to control the direction of rotation. When the CW/CCW switch is open, Q_5 is ON, and hence the transistor Q_3 is ON also. In addition, the output of a *one-shot* is applied to the base of transistor Q_2. This action causes the motor to turn in the counterclockwise direction. On the other hand, when the CW/CCW switch is closed, the transistor pair Q_1–Q_4 is ON and the motor turns in the clockwise direction (Fig. 9–20).

Measuring block: speed-to-voltage converter. The function of a measuring block is to convert the speed of a motor into an equivalent voltage. This voltage is then compared to the desired speed setting voltage. The difference voltage, in turn, is used to drive the motor toward its desired speed.

Remember that the current drawn by the motor is directly proportional to its speed. This current is sensed by connecting a resistor R_{sense} from the emitter junction of Q_2–Q_4 to ground (Fig. 9–20). The voltage developed across this resistor may be amplified so that it can be compared to the output of the DAC (desired speed setting). Figure 9–22 shows a noninverting amplifier that is used to amplify the voltage across the resistor R_{sense}. The output of this amplifier is applied to the inverting input of the differential amplifier (error detector) of Fig. 9–17a. The gain of the amplifier depends

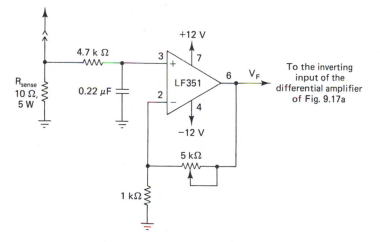

Figure 9–22. Speed-to-voltage converter.

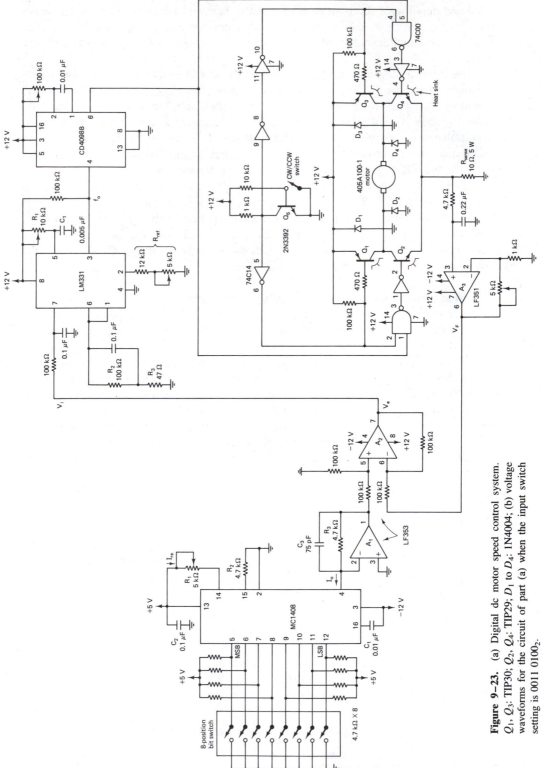

Figure 9–23. (a) Digital dc motor speed control system. Q_1, Q_3: TIP30; Q_2, Q_4: TIP29; D_1 to D_4: 1N4004; (b) voltage waveforms for the circuit of part (a) when the input switch setting is 0011 0100$_2$.

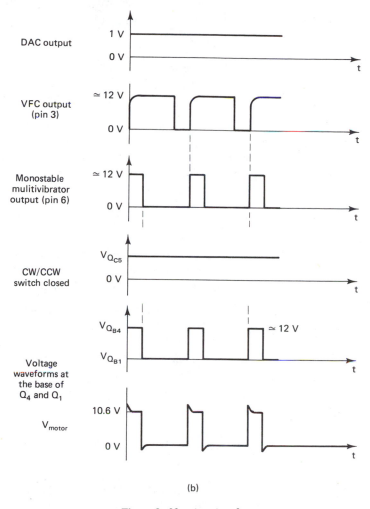

(b)

Figure 9–23. (*continued*)

on the current rating of the motor and the output voltage range of the DAC. For the maximum output voltage of the DAC, the gain of the amplifier must be adjusted until the speed of the motor reaches a maximum steady-state value. For example, in the circuit of Fig. 9–22, when all the inputs are set HIGH (maximum speed setting), the output of the DAC is 5 V. The motor should attain a maximum speed at this setting. If the motor used is TRW405A100–1, it will draw 300 mA at its maximum speed. This means that the voltage across R_{sense} will be $10(0.3) = 3$ V. Since the feedback voltage V_F and the DAC output voltage V_o are to be approximately equal [see Eq. (9–8)], the gain of the speed-to-voltage converter is

$$A_{F3} = \frac{V_{o(\text{max})}}{(R_{\text{sense}})(I_{m(\text{max})})}$$

$$= \frac{5 \text{ V}}{3 \text{ V}} \approx 1.67 \qquad (9-11)$$

Schematic Diagram

The complete schematic diagram of the speed control system is shown in Fig. 9–23a. When the circuit is assembled, to ensure proper operation, each block must be tested individually. The necessary waveforms for the circuit at key points are shown in Fig. 9–23b when the input binary setting is 0011 0100$_2$.

EXAMPLE 9–2

For the circuit of Fig. 9–23a, calculate the output voltage of the DAC if the input binary number is 0011 0100$_2$. Also, draw the output waveforms for the VFC, the monostable multivibrator, and across the dc motor.

SOLUTION Remember that 0011 0100$_2$ = 52$_{10}$. Therefore, using Eq. (9–7b), the output voltage V_o of the DAC is

$$V_o = (1.06 \text{ m})\left(\frac{52}{256}\right)(4.7 \text{ k}\Omega) \approx 1.0 \text{ V}$$

If we assume that initially the feedback voltage $V_F = 0$ V, the output frequency f_o of the VFC is given by Eq. (9–9):

$$f_o = \frac{V_i}{2.09} \frac{R_{\text{ref}}}{R_2} \frac{1}{R_1 C_1}$$

$$= \left(\frac{1.0}{2.09}\right)\left(\frac{12 \text{ k}\Omega}{100 \text{ k}\Omega}\right)\left(\frac{1}{(5.7 \text{ k}\Omega)(0.005 \text{ μF})}\right)$$

$$= 2.0 \text{ kHz}$$

Recall that the monostable multivibrator pulse width has to be adjusted such that it is sufficient to sustain a steady speed at the minimum and maximum speed limits of the motor. Let us assume that the pulse width of the monostable multivibrator output is 0.09 ms when the input frequency is 1 kHz. Therefore, the pulse width of the monostable multivibrator at 2 kHz is

$$\frac{2 \text{ kHz}}{1 \text{ kHz}}(0.09 \text{ m}) \approx 0.18 \text{ ms}$$

Also, let us assume that the direction of rotation of the motor shaft is clockwise (CW/CCW switch *closed*). This means that transistors Q_1 and Q_4 are ON (Fig. 9–23a). Since $V_{CE(\text{sat})} = 0.7$ V for each transistor, the voltage across the motor is ≈ 10.6 V. Thus the output waveforms of the VFC, monostable multivibrator, and across the dc motor are as shown in Fig. 9–23b.

EXAMPLE 9-3

In the circuit of Fig. 9–23a, what changes need to be made if 12-V, 0.2-A, and 4800-rpm dc motor is used?

SOLUTION Recall that the circuit of Fig. 9–23a is calibrated for a maximum motor speed of 10,000 rpm. Therefore, the motor with a maximum rpm of 4800 should work satisfactory in it. However, the speed of the motor may not always correspond to the input binary switch settings. Specifically, the switch setting of greater than 120_{10} or greater than $(4.8k\Omega/10k\Omega)(5\ V) = 2.4\ V$ will not result in an increase in speed. Therefore, if we need to maintain 256 speed steps (increments), we must recalibrate the circuit. Namely, we need to recalibrate the controller block (VFC and monostable multivibrator) and the speed-to-voltage converter (noninverting amplifier).

The VFC should be recalibrated such that its output is 4.8 kHz at a 5-V input signal. The monostable multivibrator output pulse width may also be readjusted so that the motor operates satisfactory over the frequency range 20 Hz to 4.8 kHz. In addition, the gain of the noninverting amplifier should be adjusted to $5\ V/10(0.2) = 2.5$.

9–4 STEPPER MOTOR POSITION CONTROL SYSTEM

In this section we study a position control system using a stepper motor. We will examine the operation of a translator/driver IC and an electronic drive used for a position control system. Position control systems using stepper motor(s) are commonly found in mainframe computers, printers, and commercial printing cameras, where precise positioning is important.

System Specifications

Before we design a stepper motor speed/position control system, it is important to establish some system specifications.

1. Design a position control system so that any stepper motor rated up to 2 A (per winding) at 12 V may be used.
2. The user should be able to select and change the speed/position of a load as desired.
3. The user should also be able to control the motion of the load in a clockwise (forward) or counterclockwise (backward) direction.
4. The system should be stand-alone and easy to operate.
5. Accuracy, performance, and cost are the most important considerations in the evaluation of the system.

Block Diagram

The first step in the design of a given system is its block diagram. Let us begin with a stepper position selector block. This block may simply be a 8-bit microswitch which allows the user a choice of altering the stepper position as necessary. The next block should be a translator block because whenever any changes are made in the position selector block, it should also affect the output of the translator block. This action, in turn, should change the position of the stepper motor accordingly. The block after the translator should contain a power drive circuit, which acts on the output of the translator and helps position the stepper. Often, to improve the performance as well as the holding torque, a high-voltage supply is used for a stepper motor system. Therefore, we will use a block for high voltage. We need a block for the stepper motor. The operation of this block is controlled by the outputs of the stepper power drive circuitry and the high-voltage supply. The last block is the load, which is driven by the stepper motor. Based on the type of application, the load may require double-ended shaft or integral lead screw shaft.

The above-mentioned blocks and the relationship between them are shown in the block diagram of Fig. 9–24. This is an open-loop block diagram because the actual position of the stepper is not compared with the desired position. A device such as an optical encoder or Hall effect transducer may be used to configure the system as a closed loop. The function of these devices is to count the revolutions (rpm) of the motor shaft. The actual revolutions are then compared to the desired revolutions and corrective action is generated to reduce the error signal to zero. The system described above with feedback is often implemented by microprocessors.

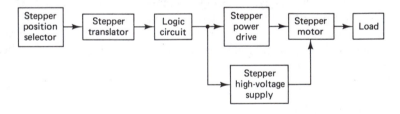

Figure 9–24. Block diagram of a stepper position control system.

System Design

In the preceding section we studied the block diagram of a stepper motor position control system. The next step is to design a circuit for each of the blocks so that overall performance criteria can be met.

It is important to consider the stepper translator first even before the position selector block because the characteristics of the former influence the latter. There are several stepper translators, such as UCN4202A, PMM8713, and SA1027. Basically all these specialized ICs can operate in a full stepping (four-step) mode as a function of the input clock. However, there are some important differentiating features that make these ICs unique.

Stepper translator: UCN4202A.   The UCN4202A is a stepper motor translator and a driver manufactured by Sprague Electric Company. It is specifically designed for driving small-to-medium permanent-magnet (PM) stepper motors rated to 500 mA and 15 V. However, with an external power drive and appropriate logic circuitry, it may be used to drive stepper motors rated for >500 mA and 15 V.

The UCN4202A uses a bipolar TTL technology and is TTL compatible. As a translator it can be operated in a four-step (full-stepping) or eight-step (double-stepping) sequence. It has open-collector output power transistors that can handle currents up to 600-mA maximum and hence may be used to drive a stepper motor directly. Thus the UCN4202A can be used as a translator and also as a driver.

The functional schematic diagram and the pin configuration for the UCN4202A are shown in Fig. 9–25a. The IC is available as a 16-pin DIP package. Let us briefly examine the function of each pin.

Stepper Drive Signals.   The drive signals for the stepper motor appear at pins 3, 4, 5, and 6 and are labeled as outputs A, B, C, and D. These are open-collector output signals and each has a current sink capacity of 600 mA at 15 V maximum. Based on the status of pin 11, the output signals A, B, C, and D can provide a full-step or a double-step mode of operation.

In the *full-step mode*, one pulse at the step input causes the stepper motor shaft to rotate its normal step distance, whereas in the *double-step mode*, the same pulse rate causes the motor shaft to step twice the specified angular increment. The advantage of the double-step mode is that it provides improved torque characteristics for the stepper motor. Outputs A, B, C, and D are also functions of the DIRECTION CONTROL input (pin 12) and can provide clockwise or counterclockwise rotation for the stepper motor. Another output, labeled DRIVER OUTPUT, appears at pin 2, which provides an additional uncommitted open-collector driver.

Inductive Load Suppression.   To protect the internal circuitry, especially the output drive transistors, the UCN4202A uses diodes to suppress the inductive spikes generated by the stepper motor. The anodes of these protective diodes are connected to the collectors of the output drive transistors and the cathodes are connected together and are brought out to pin 7. This pin is marked ''K'' for the common cathodes of the five protective diodes (Fig. 9–25a). To protect the output power transistors from inductive spikes, pin 7 must be connected to the power supply of the drive transistors.

Output Enable.   Pin 1, labeled OUTPUT ENABLE, controls outputs A through D. When this input is logic HIGH, it inhibits all four outputs. Under certain emergency conditions it is desirable to stop the motor immediately by disabling its power drive. A function, such as "emergency stop," is accomplished by pulling pin 1 logic HIGH. On the other hand, when pin 1 is logic LOW, all the outputs are active or enabled. Therefore, for normal motor operation the OUTPUT ENABLE input must be logic LOW.

In addition to the OUTPUT ENABLE input there is another input marked DRIVER INPUT, pin 15, which is specifically used to control the DRIVER OUTPUT at pin 2. When DRIVER INPUT is logic LOW, the power transistor connected to the DRIVER OUTPUT, pin 2, is ON, or vice versa. If DRIVER OUTPUT is not used, the DRIVER INPUT need not be used either (Fig. 9–25a).

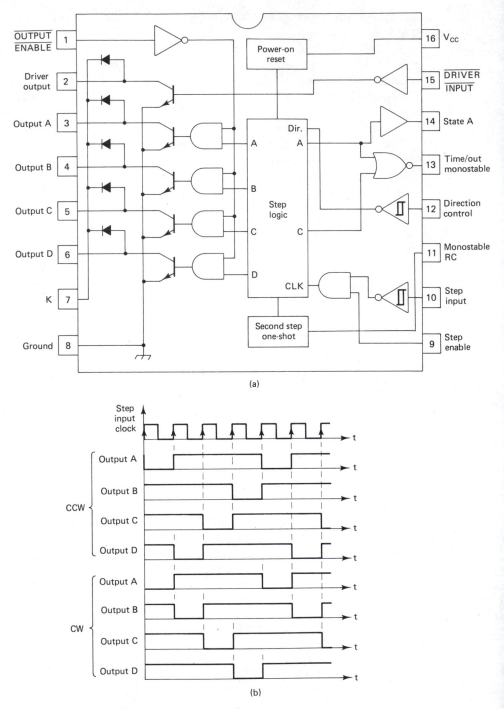

Figure 9-25. UCN4202A: (a) functional block diagram and pin configuration; (b) output waveforms for fullstep mode. CW, clockwise; CCW, counterclockwise. (Courtesy of Sprague Electric Company.)

The status of outputs *A* through *D* is monitored by an output marked as STATE *A* at pin 14. This output is the same as output *A* except may be in amplitude.

Power Supply. A supply voltage of $+5$ V typical is applied to V_{CC} input pin 16. With the initial application of V_{CC}, an internal *RS* flip-flop sets output *A* ON. However, once the flip-flop is reset, the device functions according to the logic input conditions. Note that in Fig. 9–25a the V_{CC}, pin 16, is connected to the POWER-ON RESET CIRCUIT. All the external voltages, including the supply voltage V_{CC}, are measured with respect to pin 8. This pin is marked GROUND.

Monostable RC and OUT. If the MONOSTABLE RC (pin 11) is tied to V_{CC}, the device operates in the *full-step mode*. In this mode, outputs *B* and *D* are in stationary states. In fact, to move through each of the four output states a separate input pulse is applied to STEP INPUT (Fig. 9–25b). On the other hand, in the *double-step mode*, a series *RC* network junction is connected to the MONOSTABLE RC (pin 11). In this mode, states *B* and *D* are transition states whose duration is determined by the MONOSTABLE RC timing.

The timing signal, which is a function of *RC* components connected to pin 11, appears at pin 13. It is an output and is labeled TIME/OUT MONOSTABLE (Fig. 9–25a).

Direction Control. The UCN4202A provides a single-input direction control, pin 12, which determines the direction of rotation of the motor shaft. If the DIRECTION CONTROL input is logic HIGH, the rotation is counterclockwise ($A-D-C-B$); if logic LOW, the rotation is clockwise ($A-B-C-D$). The DIRECTION CONTROL is a Schmitt trigger input and NMOS and CMOS compatible.

Step Input and Enable. The UCN4202A uses the integral step logic to control the PM stepper motors. To step the motor from one position to the next, the STEP INPUT, pin 10, is used. Normally, it is logic HIGH. When logic LOW for ≥ 1 μS, the step logic will advance one position on the positive transition by activating one of four output sink drivers. Thus the position of the motor is a function of the clock frequency applied to STEP INPUT.

Pin 9 is the STEP ENABLE input, which must be held logic HIGH to enable the STEP INPUT pulses for advancing the motor. When this input is pulled LOW, it inhibits the translator logic. In almost all applications it is necessary to "hold" the motor in a known position when not stepping, to maintain the current position of the load. Obviously, this function can be accomplished by pulling the STEP ENABLE low.

So far we studied the characteristics and electrical specifications of the UCN4202A stepper motor translator and driver. Next, let us configure it to control the operation of the motor and, in turn, the position of the load. First, we shall consider the source for STEP INPUT.

Stepper position selector. We know that on each positive transition of the pulse applied to the STEP INPUT, the load advances from one position to the next. Therefore, the frequency and length of time the pulse source of the translator is active determines the distance traveled by the load. Remember that the load is driven by the stepper motor. Based on the desired application, the pulse source may be a microprocessor, a signal generator, or an oscillator. The main function of the pulse source is to provide

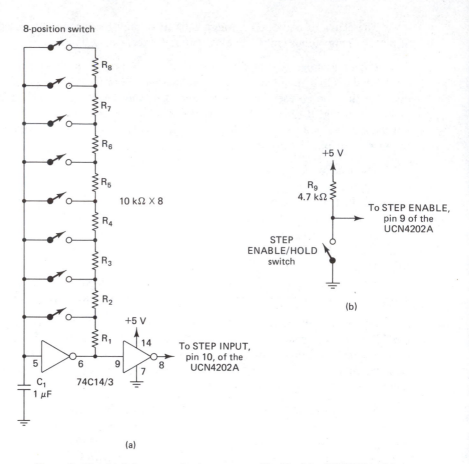

Figure 9–26. (a) Pulse source for the STEP INPUT Pin 10 of the UCN4202A. (b) STEP ENABLE/HOLD switch.

the desired number of pulses at a known rate so that the load can travel a given distance.

Since we need a stand-alone stepper motor position control system, let us use an oscillator for the pulse source. In addition, let us use an eight-position bit switch to vary the frequency of oscillation.

One of the ways to design an oscillator is to use the 74C14 hex Schmitt trigger shown in Fig. 9–26a. The frequency of oscillation is varied by changing the value of resistor. The various resistor values are, however, switched in and out of the circuit using an eight-position bit switch.

The frequency of oscillation is given by Eq. (9–6)

$$f_o = \frac{0.59}{RC_1}$$

Therefore, if $C_1 = 1$ μF and R is varied from 10 kΩ to 1.25 kΩ using bit switches, f_o will change from 59 Hz to 472 Hz, respectively.

Note that the frequency range for the pulse source is determined based on the speed versus torque characteristics of the motor and the intended application. Also, the STEP ENABLE, pin 9, can be used as *timer*, in that, after a desired distance is traveled by a load, the STEP ENABLE input may be used to *hold* the load in that position. Both the ENABLE and HOLD functions may be accomplished with a ON/OFF switch, as shown in Fig. 9–26b. When the switch is open, the stepper motor is enabled; when it is closed, the motor is in the HOLD state.

We will also use a switch to control the direction of motion of the motor shaft. The switch will be connected to the DIRECTION CONTROL, pin 12, as shown in Fig. 9–27. When the switch is open, the motor moves clockwise, and when it is closed, the shaft rotation is counterclockwise.

Because of improved torque and better motor stability for high step rates, we will use a double-step mode of operation. To do this we need to connect a series *RC* circuit to the MONOSTABLE RC pin 11 as shown in Fig. 9–27. Recall that in the double-step mode states *B* and *D* are transition states with duration determined by the *RC* timing. More important, for the stepper logic of 4202A to work properly, the STEP INPUT must remain low for ≥ 1 µs. This means that MONOSTABLE RC timing must be ≥ 1 µS. Let us assume that it is 5.6 ms. This means that in the half-step mode the output *B* and *D* will remain low for approximately 5.6 ms. In equation form the MONOSTABLE RC timing is expressed as

$$t_{DS} = 1.12 R_{10} C_2 \tag{9–12}$$

where t_{DS} is the pulse width when *B* and *D* outputs are low. If we assume that $C_2 = 0.05$ µF, then from Eq. (9–12), $R_{10} \simeq 100$ kΩ.

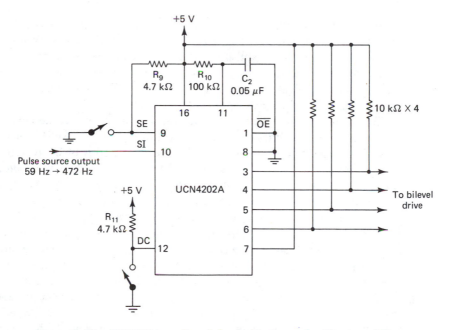

Figure 9–27. UCN4202A configured for double-step mode. (Courtesy of Sprague Electric Company.)

Recall that we may drive directly any PM stepper motor rated up to 500 mA at 12 V using the configuration of Fig. 9–27 if we replace the 10-kΩ resistors by a stepper motor and the +5-V supply by a +12-V supply. However, we wish to design a stepper drive so that it can be used for motors rated up to 2 A per winding at 12 V. With this in mind, let us consider the power drive circuit next. Once we design the power drive circuit, we will design the logic circuit needed to match the translator outputs to the power drive circuit.

Stepper power drive.   We know that the stepper translator 4202A performs the function of providing electrical signals for the stepper. However, an electronic drive is the key in the overall performance of a stepper motor and hence is often referred to as the heart of the stepper position control system. Most commonly used electronic drives include:

1. L/R unipolar
2. L/R bipolar
3. Bilevel
4. Reactive
5. Chopper

The choice of an electronic drive depends on the characteristics of the load and the desired torque versus speed performance of the system. The bilevel electronic drive yields the most efficient operation and hence is commonly used for high-speed operations. Therefore, we will discuss only the bilevel drive.

Bilevel electronic drive.   As its name indicates, the bilevel electronic drive uses two power supplies. The power drive circuit is configured such that initially the high voltage is applied to establish a current required to cause the motor to step. Once the required current level is established, the low voltage is applied to the motor to sustain it. Then the high-voltage supply is turned off. Thus, when the high-voltage supply is ON, the low voltage supply is OFF, or vice versa.

We will use a four-phase stepper motor. Therefore, one of the configurations for such a motor using the bilevel drive is shown in Fig. 9–28. As shown in the figure, +12 V, 5 A is the high-voltage supply, V_H, and +5 V, 10 A is the low-voltage supply, V_L. The voltage and the current ratings of the high-voltage supply are dictated by the torque and speed specifications of the motor. Obviously, a high-speed, large-torque motor requires a high-voltage, high-current supply. On the other hand, a low-voltage, high-current supply is necessary to sustain the necessary torque.

In Fig. 9–28 transistors Q_1 and Q_2 are configured as drive transistors for switch (or pass) transistors Q_3 and Q_4. In addition, transistors Q_5, Q_6 and Q_7, Q_8 are configured as current sink transistors. The stepper motor windings, ϕ_1 through ϕ_4, are connected between the collectors of switch transistors Q_3 and Q_4 and the collectors of sink transistors Q_5, Q_6, and Q_7, Q_8. Also, outputs A, B, C, and D of the translator are applied to the bases of transistors Q_5, Q_7, Q_6, and Q_8, respectively. Note that the low voltage, V_L, is applied to the junctions of windings ϕ_1–ϕ_3 and ϕ_2–ϕ_4 through the high-current diodes D_1 and D_2.

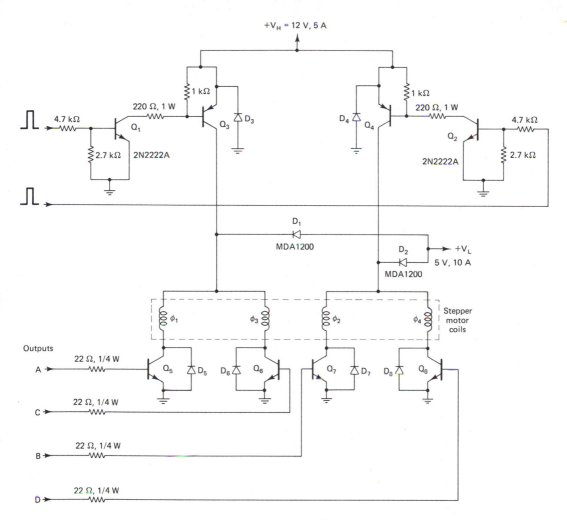

Figure 9–28. Bilevel electronic drive; Q_3, Q_4: TIP125; Q_5 to Q_8: TIP120; D_3 to D_8: MR820.

In the normal single-supply operation, the state of transistors Q_5 to Q_8 is controlled by inputs A to D. When these inputs are high, the corresponding transistor is turned on, which in turn energizes the respective motor winding. However, in bilevel operation a high voltage, V_H, is applied to the motor windings for a short period of time when the motor drive signals A, B, C, and D are logic HIGH. The function above can be accomplished by adding *drive* and *switch* transistor pairs Q_1–Q_2 and Q_3–Q_4 to the basic single-supply drive circuitry as shown in Fig. 9–28.

A pulse waveform that is synchronized with the rising edge of the STEP INPUT clock is applied to the drive transistors Q_1 and Q_2 when drive signals A, B, C, or D are logic HIGH. A high voltage, V_H, is applied to the motor windings each time the pulse waveform is logic HIGH. However, when the pulse waveform is logic LOW, the low voltage, V_L, is applied to the motor windings. This action maintains the current in

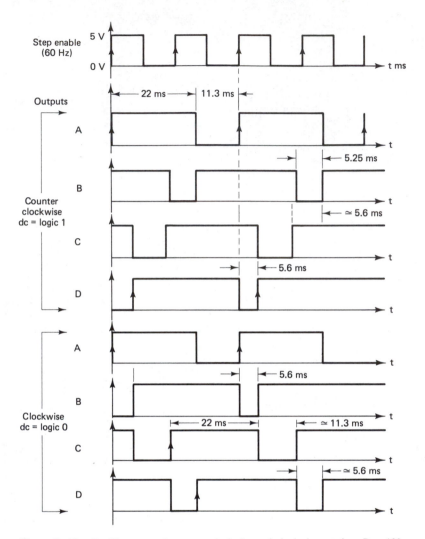

Figure 9–29. Double-step mode: counterclockwise and clockwise rotation. $R = 100$ kΩ and $C = 0.05$ μF, $t_p = 1.12(R)(C) \approx 5.6$ ms. (Courtesy of Sprague Electric Company.)

the stepper coils that was initially established by the high-voltage supply and causes the motor to step. The important waveforms for the bilevel stepper motor drive are shown in Fig. 9–29.

In the circuit of Fig. 9–28, TIP120 and TIP125 is a complementary Darlington power transistor pair. The electrical specifications of the TIP120 are as follows:

$$I_{C(\text{max})} = 5 \text{ A continuous}$$

$$V_{CE(\text{max})} = 60 \text{ V}$$

$$P_{D(\text{max})} = 65 \text{ W at 25°C case temperature}$$

$$I_{B(\text{max})} = 100 \text{ mA continuous}$$

$$V_{CE(\text{sat})} = 2 \text{ V at } I_C = 3 \text{ A and } I_B = 12 \text{ mA}$$

$$h_{FE} = 1000 \text{ at } I_C = 3 \text{ A and } V_{CE} = 3 \text{ V}$$

Also, two 2N2222As are used as drive transistors. These are small-signal transistors with following electrical specifications:

$$I_{C(\text{max})} = 800 \text{ mA}$$

$$V_{CE(\text{max})} = 40 \text{ V}$$

$$f_T = 300 \text{ MHz minimum at } I_C = 20 \text{ mA}$$

$$V_{CE(\text{sat})} = 0.5 \text{ V}$$

$$h_{fe} = 300 \text{ maximum at } I_C = 150 \text{ mA}$$

It is available as a metal transistor in a TO-18 package.

Finally, Motorola's bridge rectifier MDA1200 is used for diodes D_1 and D_2, which are rated at 12 A, and a V_{PIV} of 50 V. Also, fast-recovery diodes MR820 are used for D_3 to D_8, which protect transistors Q_3–Q_8, respectively, against negative inductive spikes. The MR820 is available in a plastic package and has the following electrical specifications.

Average rectified forward current $I_o = 5$ A

Peak inverse voltage $V_{PIV} = 50$ V

Maximum forward current $I_F = 300$ A

Logic circuit. To interface the output signals of the translator to the bilevel electronic drive, we need a logic circuit. The translator outputs A, C, B, and D are to be used to drive transistors Q_5, Q_6, Q_7, and Q_8, respectively. To provide sufficient current drive for transistors Q_5 to Q_8, we will use high-current source drivers such as UDN2585A. The schematic diagram of UDN2585A is shown in Fig. 9–30a. Some of the important features of the 2585A are

TTL and CMOS compatible

High output current up to 120 mA

Internal diode transient suppression

18-pin dual-in-line plastic package

Output-sustaining voltage $= 15$ V

Output $V_{CE(\text{sat})} = 1.2$ V at $I_o = 120$ mA

Input voltage $V_{i(\text{on})} = -4.6$ V

The input of a 2585A is a *pnp* stage; hence we need to invert the outputs of the translator before they are applied to the 2585A. We also need to provide adequate drive for the 2585A. Both the signal inversion and necessary current drive are provided by the use of 74C00 CMOS NAND gate (Fig. 9–30c).

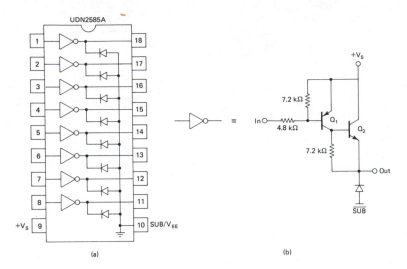

(a)

(b)

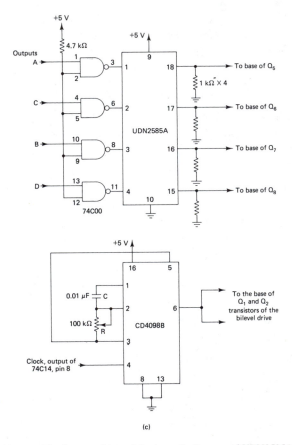

(c)

Figure 9–30. (a) Pin diagram; (b) partial schematic diagram of UDN2585A; (c) logic circuit for the stepper motor position control system. [(a) and (b) Courtesy of Sprague Electric Company.]

Similarly, to provide the pulse waveform for the bilevel electronic drive, we will use a monostable multivibrator such as CD4098B. Recall that the CD4098B was discussed in Section 9–3. In fact, we can use the same circuit configuration as that used in Fig. 9–19b.

In Fig. 9–30c the trigger input to the monostable multivibrator is the output of the pulse source. The output of the monostable is used to drive transistors Q_1 and Q_2 so that the high voltage, V_H, is applied to the motor winding(s) when one or more of the stepper drive signals A, B, C, and D are active. The pulse width of the monostable output waveform is, however, dependent on the electrical specifications of the stepper motor and may be adjusted using a 100-kΩ potentiometer until the motor performs satisfactorily.

Thus the logic circuit of Fig. 9–30c is composed of high-current AND-gate source drivers and a monostable multivibrator. The next block to consider is the stepper motor itself.

Stepper motor. Remember that according to the design specifications we need to design a stand-alone position control system in which we can use a stepper motor rated up to 2 A per winding at 12 V. Eastern Air Devices, Sigma Instruments, North American Philips, and Superior Electric are some of the manufacturers of stepper motors. The reader may refer to the stepper motor discussion of Chapter 3 when selecting a stepper motor.

A Superior Electric M111-FD-310 motor, which has the following specifications, should perform satisfactorily in the position control system of Fig. 9–31:

Rated at 1.25 A per winding, 10 V dc

Inductance per phase = 45 mH and resistance = 7 Ω

Motor step angle = 1.8° or 200 steps/rev

Number of leads = 6

Nominal rated torque = 425 oz-in.

Holding torque = 625 oz-in.

Maximum overhang load = 25 lb

Approximate weight = 8 lb with a shaft diameter of 0.375 in.

Load. In the block diagram of Fig. 9–24, the last block is the load. Based on the loads driven, the stepping motors are chosen from a wide variety of special configurations, including double-ended shaft, and integral lead screw shaft. Also, flats, keyways, tapers, holes, threads, and splines are some of the important special shaft configurations.

Generally, the load is coupled to the stepping motor through a lead screw or gear train. For example, in a PC printer, stepping motors are used to move the paper by controlling the motion of the load, a take-up reel. On the other hand, in applications such as litho camera, relatively heavy loads are attached to the lead screws which are driven by the stepper motors. Also, typically the integral lead screw shaft steppers are used for positioning the magnetic pickup in floppy disk applications. In short, the type of load driven by the stepping motor dictates its size and configuration.

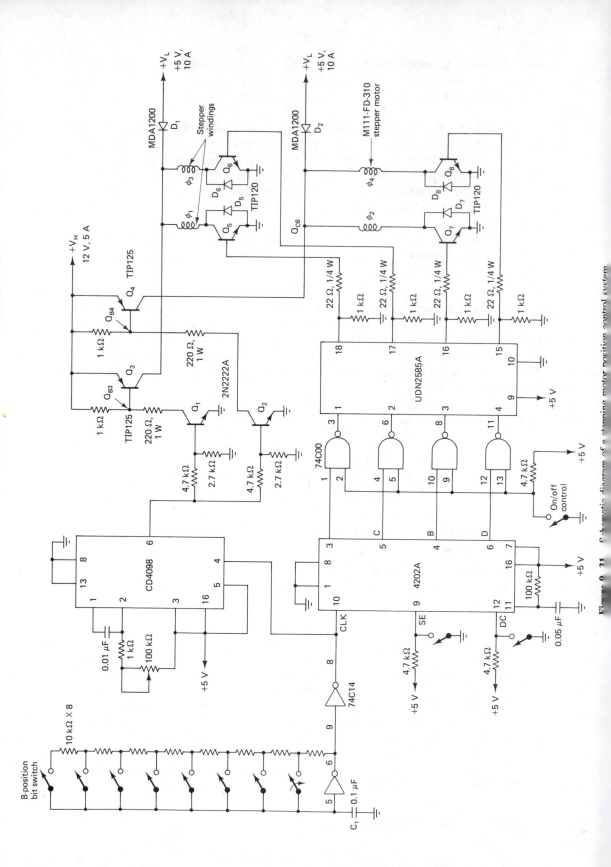

Figure 9-31. Schematic diagram of a stepping motor position control system.

Schematic Diagram

Figure 9–31 shows the complete schematic diagram of a stepping motor position control system. The circuit is stand-alone and can be used for stepping motors rated up to 2 A per winding at 12 V. More important, it is flexible, in that it can easily be upgraded for larger and faster stepping motors. For example, for high-speed stepping motors the pulse source may be upgraded by changing the resistor values in Fig. 9–31. Also, for larger motors with higher current and voltage ratings (>2 A and 12 V) the bilevel electronic drive may be upgraded by replacing Q_3 to Q_8 and D_1 and D_2 by higher-rated power transistors and diodes, respectively.

EXAMPLE 9–4

In a machine tool application such as a drilling machine, stepper motors are used to position an x–y table. For a certain drilling machine the x–y table must be positioned at a rate of 5 in./min. The table is positioned by a three-pitch ball nut lead screw.

(a) What is the required speed in steps/s to meet this requirement if a 1.8°, 200-step/rev motor is used?

(b) If the motor speed is 100 steps/s, how far will the table move in 2 s?

SOLUTION (a) Note that a three-pitch ball nut screw has three threads per inch. Also, the relationship between the linear speed, steps/s, steps/rev, and the lead screw pitch p in threads/inch is given by

$$\frac{\text{in.}}{\text{s}} = \frac{\text{steps/s}}{(\text{steps/rev})\,(p)} \tag{9–13}$$

where p is the lead screw pitch in threads/inch. Substituting the given values, we get

$$\frac{5}{60} = \frac{\text{steps/s}}{200(3)}$$

or 50 steps/s.

(b) If the motor speed is 100 steps/s, according to Eq. (9–13) the table will move 1/3 in. in 2 s.

9–5 APPLIANCE TIMER

The most commonly used digital circuit in appliances and entertainment games is a timer. A timer is used in appliances and consumer machines such as a coin-operated car wash, a vacuum cleaner, a washer, a dryer, video games, and a TV set. There are a variety of circuit configurations that provide timer operation. However, the counters and divide-by-N networks are the most basic building blocks in these circuits. In this section we use a *down counter* as a timing device to determine the on time of an appliance.

System Specifications

We wish to design a digital appliance timer with the following specifications:

1. The user should be able to select and change the on-time of an appliance as desired.
2. The input BCD switches may be used to select the desired on-time from approximately 10 s to 10 min.
3. The system should be stand-alone and easy to operate.
4. Accuracy, performance, and cost are equally important factors in the evaluation of the system.

Block Diagram

The first step in the design of a system is the block diagram. Let us begin with an *on time selector block*. To allow 10 min of on-time, we may use two BCD switches. Thus the function of this block is to allow the user to select any BCD number between 01 and 99. The output of this block should be processed such that it determines the on-time of an appliance. One of the easiest options is to jam the number selected into a counter and then decrement the counter until its content is reduced to zero. The difference between the time the counter is loaded and the time its content is reduced to zero is the on-time. However, this means that the counter needs a clock to decrement its content. Thus the second block is a down counter with an external clock. The number loaded into the counter and the frequency of the clock applied to it determine the time required to decrement the content to zero.

Next, the output of the counter block should be conditioned such that it turns the appliance on and off. Thus the third block should be a control circuit that controls the appliance operation based on the output of the counter. The last block will be an appliance itself. This appliance may be operated off a dc or ac voltage. The block diagram of an appliance timer that matches this description is shown in Fig. 9–32.

System Design

The next step is to design the circuit for each block such that the function of each block as well as the functional relationship between the blocks is satisfied.

Time selector block: two BCD switches.   The function of this block is to select a desired on-time. The two BCD switches will satisfy this function. These switches

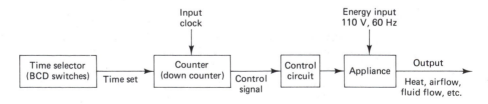

Figure 9–32.   Block diagram of an appliance timer.

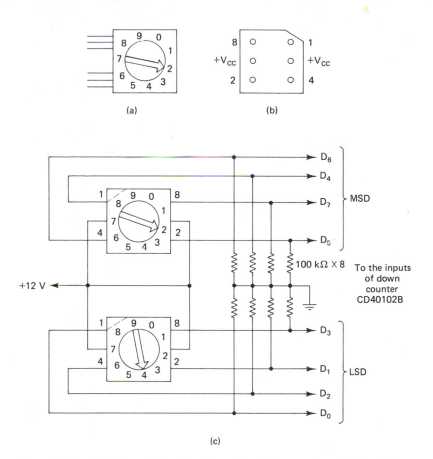

(a)

(b)

(c)

Figure 9–33. BCD micro-DIP switch: (a) top view; (b) bottom view. (c) Two BCD switches configured to select an on-time for an appliance.

may be configured as shown in Fig. 9–33. The output of the BCD switches is an input to the counter. The positions of these switches determine the number that is loaded into the counter. Since we are using two BCD switches, the maximum number that can be loaded is 99.

Down counter: CD40102B. The second block in Fig. 9–32 is a counter. The function of the counter is to provide an on-time for an appliance based on the number that is set on the BCD switches. We will use a down counter CD40102B. The CD40102B is a COS/MOS eight-stage presettable synchronous down counter, which is easy to use and is readily available. It is configured as two cascaded 4-bit BCD counters with a single output that is active LOW when the internal count is zero. The pin diagram of a CD40102B is shown in Fig. 9–34a. Some of the important features of the CD40102B are as follows:

Synchronous or asynchronous operation
Cascadable to increase the count down time

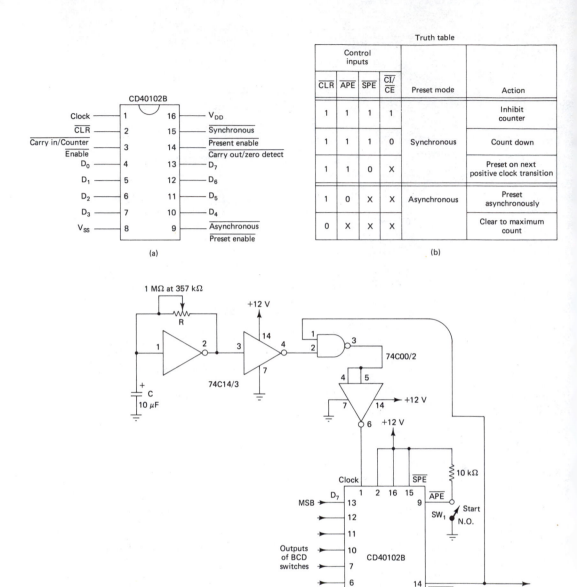

Figure 9–34. CD40102B: (a) pin diagram; (b) truth table (0, low level; 1, high level; x, don't care). (c) connection diagram. (Courtesy of RCA, Inc.)

Wide operating voltage range of 5 to 15 V

Maximum clock input frequency of 4.8 MHz at $V_{DD} = 15$ V

Figure 9–34b shows the truth table for the CD40102B down counter. According to this table, there are four control inputs: clear, asynchronous preset enable, synchronous preset enable, and carry in/counter enable. For a given mode—synchronous or asynchronous—the status of these control inputs determine the operation of the down counter.

Let us briefly examine the function of each pin. An external clock input is applied to pin 1. The counter is decremented by one count on each positive transition of the clock. However, counting is inhibited when the CARRY-IN/COUNTER ENABLE ($\overline{\text{CI/CE}}$) input is high. Pin 3 is a $\overline{\text{CI/CE}}$ control input. The CLEAR ($\overline{\text{CLR}}$) input is pin 2. When it is LOW, the counter is asynchronously cleared to its maximum count, 99_{10}, regardless of the state of any other control input.

Inputs D_0 through D_7 represent two 4-bit BCD words for the CD40102B and are applied to pins 4, 5, 6, 7, 10, 11, 12, and 13 respectively. Bits D_0 to D_3 form LSD and bits D_4 to D_7 form MSD. The SYNCHRONOUS PRESET-ENABLE ($\overline{\text{SPE}}$) input is applied to pin 15. When this input is low, data at inputs D_0 to D_7 are clocked into the counter on the next positive clock transition, regardless of the state of the $\overline{\text{CI/CE}}$ control input (Fig. 9–34b). On the other hand, pin 9 is the ASYNCHRONOUS PRESET-ENABLE ($\overline{\text{APE}}$) input. When it is LOW, data at inputs D_0 to D_7 are asynchronously forced into the counter regardless of the state of the $\overline{\text{CI/CE}}$, $\overline{\text{SPE}}$, or CLOCK inputs.

A power supply voltage V_{DD} is applied to pin 16, and ground, V_{SS}, is pin 8. Pin 14 is the CARRY-OUTPUT/ZERO-DETECT ($\overline{\text{CO/ZD}}$). The output goes LOW when the count reaches zero. In addition, the output remains LOW for one full clock period if the $\overline{\text{CI/CE}}$ input is LOW. Also, at the time of zero count, if all control inputs are HIGH, the CD40102B jumps to the maximum count, 99_{10}.

Once we know the pin functions of the CD40102B down counter, the next step is to configure it as a timer. First, let us consider the clock input. Since the maximum on-time for the appliance is 10 min and the maximum number that can be loaded into the CD40102B is 99_{10}, the time period of the clock input must be (10 min/99) = 0.101 min or 0.165 Hz.

The clock may be derived by using inverters such as 74C14 and an appropriate RC combination. For example, if we use a 74C14 hex Schmitt trigger with RC combination, as shown in Fig. 9–34c, the frequency of oscillation is given by Eq. (9–6). Let us assume that $C = 10$ μF. Then for $f_o = 0.165$ Hz, the value of R from Eq. (9–6) is

$$R = \frac{0.59}{0.165(10^{-5})} = 357.6 \text{ k}\Omega$$

we will use $R = 1$-MΩ pot.

To obtain clean, *spike*-free waveform, we may run the oscillator output through an additional Schmitt trigger inverter as shown in Fig. 9–34c. Before the oscillator output is applied to the clock input (pin 1) of the CD40102B, the potentiometer R is adjusted for an output frequency of 0.165 Hz.

The next step is to decide how we wish to configure the four control inputs: $\overline{\text{CI/CE}}$, $\overline{\text{CLR}}$, $\overline{\text{SPE}}$, and $\overline{\text{APE}}$. To enable a CD40102B, we will connect $\overline{\text{CI/CE}}$ to ground.

Also, we will connect $\overline{\text{CLR}}$ input to $+V_{DD}$ so that the counter is not asynchronously cleared to its maximum count, 99_{10}. We will connect the $\overline{\text{SPE}}$ input to $+V_{DD}$ because we do not wish to use the synchronous preset mode to load data at inputs D_0 to D_7. However, we will use $\overline{\text{APE}}$ control input to load the data asynchronously into the counter. To load data asynchronously, we will use switch SW_1. The switch is marked "start" because it initiates counter operation, see Fig. 9–34c. Each time an appliance is turned ON for a specific time, the START switch is momentarily closed to load the desired BCD number asynchronously into the counter.

To ensure that the counter is inhibited after its content is reduced to zero, the $\overline{\text{CO/ZD}}$ output will be ANDed with the clock input. Therefore, the output of the AND gate formed by using a 74C00 quad NAND gate will drive the clock input of the CD40102B (Fig. 9–34c). We will use a supply voltage of $+12$ V and connect it to pin 16. Finally, the data inputs D_0 to D_7 will be connected to the outputs of the BCD switches. The BCD switches and, in turn, the data inputs may be set to either 0 or $+12$ V, based on the required on time. Thus the complete connection diagram of a CD40102B counter is shown in Fig. 9–34c.

Control circuit. The function of the control circuit block is to process the $\overline{\text{CO/ZD}}$ output of the down counter such that an appliance is ON when the $\overline{\text{CO/ZD}}$ is logic 1 and OFF when it is logic 0. Since an appliance is operated off 110 V AC at 60 Hz, the control circuit should also provide adequate isolation between the high-voltage ac signals and the low-voltage digital signals. A MOC3010 optically isolated triac driver should provide this needed isolation. However, to drive an infrared LED of a MOC3010, we need a current amplifier. In fact, we will use a control circuit similar to that in Fig. 9–11. The control circuit to process $\overline{\text{CO/ZD}}$ output is shown in Fig. 9–35. It consists of a voltage-to-current amplifier, an optically isolated triac driver, and an extra triac to match the power requirements of the given appliance. The only difference between the circuits of Figs. 9–11 and 9–35 is the operating voltage for the voltage to current amplifier.

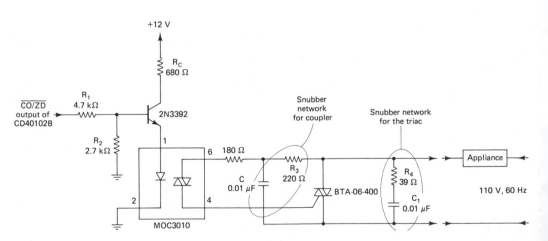

Figure 9–35. Control circuit for the appliance timer. (Courtesy of Motorola Semiconductor, Inc., Phoenix, Ariz.)

Digital Control Systems   Chap. 9

Thus, in Fig. 9–35 when the $\overline{CO/ZD}$ output is HIGH, the transistor 2N3392 is ON, which in turn triggers the MOC3010 ON. This action causes the triac to turn ON, which results in turning the appliance ON. On the other hand, when $\overline{CO/ZD}$ is LOW, the 2N3392 is OFF and so is the MOC3010. As a end result the appliance is also OFF.

Appliance.   The last block in the appliance timer system is an appliance. The appliance is connected to and driven by the control circuit, as shown in Fig. 9–35. Any appliance that operates on 110 V AC and draws ≤ 6 A current may be used in the circuit of Fig. 9–35. Obviously, the size of the appliance used is dictated by the current rating of an external triac.

Schematic Diagram

The complete schematic diagram of an appliance timer is shown in Fig. 9–36. When the circuit is breadboarded, each block must be checked individually to ensure proper circuit operation.

In the circuit of Fig. 9–36, the maximum time an appliance can be ON is 10 minutes. The on-time can be extended either by decreasing the frequency of the clock input or by cascading CD40102B down counters, whichever is economical and applicable to a given application.

As you may have noticed, the appliance timer is an open-loop control system; hence it is less accurate than the equivalent closed-loop system. For example, if the appliance is a dryer, there is no guarantee that the clothing inside the dryer will be completely dry after a cycle. In fact, almost all consumer control systems, including a washer, a dryer, a dishwasher, and a bread toaster, are *calibrated open-loop systems* in which the outputs are calibrated for desired input settings. Generally, an open-loop control system is cheaper and simpler than the corresponding closed-loop system.

The major drawback in the appliance timer of Fig. 9–36 is that initially the output $\overline{CO/ZD}$ of the counter may not be logic 0. One way to solve this problem is to load 00_{10} before a number corresponding to the desired on time is loaded into the counter. Another method is to connect the appliance to the control circuit at the same time the timer is loaded. Recall that the timer is loaded by depressing switch SW_1. However, the procedure described above needs to be exercised initially just once.

EXAMPLE 9–5

Referring to the appliance timer of Fig. 9–36, determine (a) the on time of the appliance if the input BCD switches are set to 35_{10}; (b) the BCD switch settings if the appliance is to remain on for approximately 61 s.

SOLUTION   (a) The frequency of the clock in Fig. 9–36 is 6.1 s. On each positive transition of the clock, the number in the down counter is decremented by one count. Hence, to decrement a count of 35, will take $35(6.1) = 213.5$ s or 3.56 min. Thus the appliance will remain on for approximately 3.56 min.

(b) To keep an appliance on for 61 s, the input BCD switches must be set to $61/6.1 = 10$.

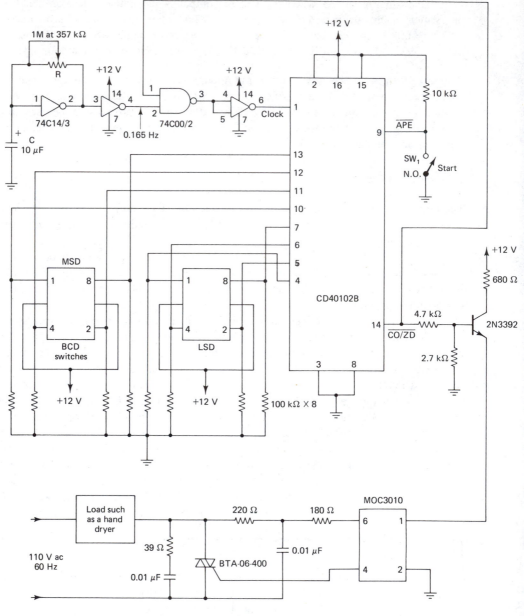

Figure 9–36. Timer for a hand dryer.

9–6 DIGITAL LIGHT-INTENSITY CONTROLLER

Often, there is a need to control an intensity of light at a known value in lux for a certain duration, especially in film-exposing applications. Based on where and how it is used, a digital light intensity controller may also be referred to as a precision light dimmer or a light integrator.

The intensity of light is varied by changing the voltage applied to the light source. If the light source is ac operated and is ON for the entire 360° of the ac cycle, we may change its intensity by controlling its on-time. In this section we will see how we can control the conduction angle of the triac to control the intensity of a light source.

System Specifications

We wish to design a light-intensity controller system according to the following specifications:

1. The system should be able to provide various intensity levels.
2. It should be simple and convenient for the user to select a desired intensity level.
3. The user should be able to maintain the desired intensity or change it as required.
4. The system is operated off of ac voltage and should be stand-alone.

Block Diagram

It is convenient and easier to design a system from its block diagram. Therefore, let us consider the block diagram first.

We will use BCD switches to select a desired intensity of light. The use of switches should make it simpler and more convenient for the user to select and change a desired intensity. Thus the first block should be an intensity selector block.

We shall utilize a light bulb as a load and assume that ac line voltage is used to operate it. We will use a down counter and load it with BCD numbers (0 to 99) so that its output will switch to logic LOW at different times with reference to the ac signal. However, to maintain a constant intensity corresponding to a certain BCD number, we need to obtain a signal that is synchronized with a 110-V AC, 60-Hz input. This synchronizing signal is used to load the number into the counter. Thus we need a down counter and a *sync signal blocks*.

We need another block which is used to process the output of the down counter and makes it suitable to drive a light source. This block is referred to as a *control circuit block*.

Finally, we need a block that represents a light bulb. The output of the control circuit block governs the operation of a light bulb by regulating the energy input (110 V AC, 60 Hz) applied to it.

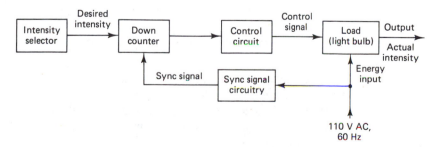

Figure 9–37. Block diagram of a light-intensity controller.

Thus the light intensity controller should be composed of the following blocks: intensity selector, down counter, control circuit, sync signal circuit, and load. The resulting block diagram is shown in Fig. 9–37.

System Design

Let us begin the design of a light intensity controller with an intensity selector block.

Intensity selector block: two BCD switches. The function of this block is to offer flexibility to the user in selecting a desired intensity. The intensity that can be set with two BCD switches can be varied from 00 to 99, where 00 represents bright and 99 represents dim light, respectively. The reason for this is that when 00 is loaded into the counter, its output is logic 0, which allows the maximum voltage to be applied to the light bulb and hence the light is bright. Exactly the opposite action takes place when 99 is loaded into the counter. It takes a maximum allowable time for the counter to reduce its content to zero, which in turn applies the least voltage to the light bulb. The end result is dim or no light. Thus the two BCD switches may be configured as shown in Fig. 9–33.

Down counter: CD40102B. As we learned in Section 9–5, there are two important considerations when we use a CD40102B down counter. The first is the input clock frequency and the second is the mode of operation.

Let us consider an input clock frequency first. We will assume that an optically isolated triac driver and an additional triac are used to control the operation of a light bulb. Since a triac consists of two identical SCRs connected in series opposing, we need to consider the operation of a triac over a half-cycle of 110-V AC, 60 Hz sine wave only. Therefore, the time it takes the counter to decrement its content to zero is ≤ 8.33 ms (or 180°). Since the maximum BCD number to be loaded into the counter is 99 and maximum allowed time is 8.33 ms, the required frequency of the clock is

$$\frac{8.33 \text{ ms}}{99} = 84.2 \text{ } \mu s$$

$$\approx 11.9 \text{ kHz}$$

Again, to obtain a clock frequency of 11.9 kHz, we will use 74C14 hex Schmitt trigger inverters. Let us assume that $C = 0.01$ μF. Therefore, for an f_o of 11.9 kHz, the value of R from Eq. (9–6) is

$$R = \frac{0.59}{(11.9 \text{ kHz}) (0.01 \text{ } \mu F)} = 4.96 \text{ k}\Omega$$

Let R be a 10-kΩ potentiometer. To obtain spike-free clock input, we will use an extra inverter, as shown in Fig. 9–38.

Thus, with a clock input frequency of 11.9 kHz, it will take the counter approximately 8.33 ms to reduce BCD 99 to zero. Obviously, the brightest intensity occurs when 00 is loaded into the counter.

The connection diagram in Fig. 9–38 for the CD40102B is identical to that in

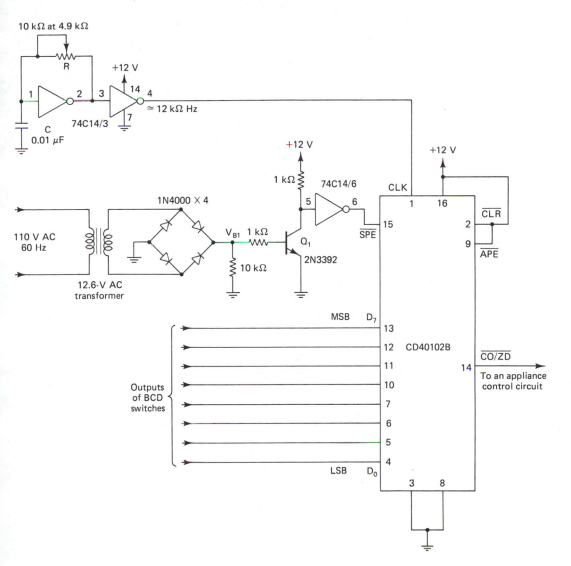

Figure 9–38. CD40102B configured for light-intensity control.

Fig. 9–34 except for the $\overline{\text{APE}}$ and $\overline{\text{SPE}}$ inputs. For more information on the pin functions of a CD40102B, the reader may refer to Section 9–5.

Sync signal circuitry. Next, let us consider the *mode* of operation that should be used to load a BCD number into the counter. Since we wish to maintain constant intensity, we need to load the same BCD number at a specific time during the AC cycle and thereafter keep on reloading it as long as desired. However, the maximum time allowed for the counter to decrement any number between 00 and 99 to zero is 8.33 ms. Therefore, we need to use a synchronous mode to jam the BCD number into the counter every 8.33 ms.

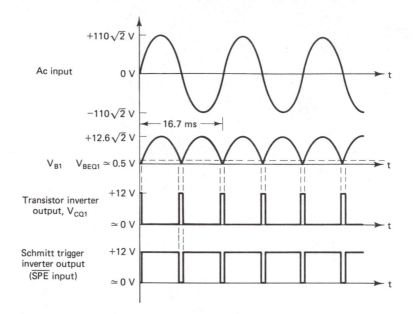

Figure 9–39. Sync signal ($\overline{SPE}$) and the associated waveforms for the circuit of Figure 9–38.

The synchronous preset enable ($\overline{SPE}$) input signal may be derived from 110-V AC, 60-Hz line voltage. To do this, we will use a step-down 12.6-V AC transformer, a full-wave bridge rectifier, and inverters. A step-down audio transformer with 12.6 V AC and ≤200 mA can be used because it is readily available. Also, 1N4000 rectifiers may be used for the full-wave bridge. In addition, transistor and Schmitt trigger inverters may also be used to obtain the proper polarity and amplitude signal for the $\overline{SPE}$ input. The sync signal and the associated waveforms are shown in Fig. 9–39.

Control circuit. We want to use the output $\overline{CO/ZD}$ of the counter to vary the voltage applied to a light bulb so that it will result in changing the light intensity. A control circuit will consist of a voltage-to-current converter, an optically isolated triac driver, and an extra triac. Obviously, it will be similar to that used for the appliance timer.

The operation of the control circuit is as follows. The output $\overline{CO/ZD}$ goes LOW in ≤8.33 ms, based on a BCD number (0 to 99) loaded into the counter. When $\overline{CO/ZD}$ goes LOW, an optically isolated triac driver turns on and triggers the external triac. The end result of this chain action is that the light bulb turns on. However, the intensity of light bulb depends on the time the $\overline{CO/ZD}$ output goes LOW during the 0 to 8.33 ms. In fact, the smaller the BCD number, the brighter the light, or the larger the number, the dimmer the light.

The schematic diagram of a control circuit and the resulting waveforms are shown in Fig. 9–40. Note that in this figure we have used an MOC3030 zero-voltage-crossing optically isolated triac driver instead of an MOC3010.

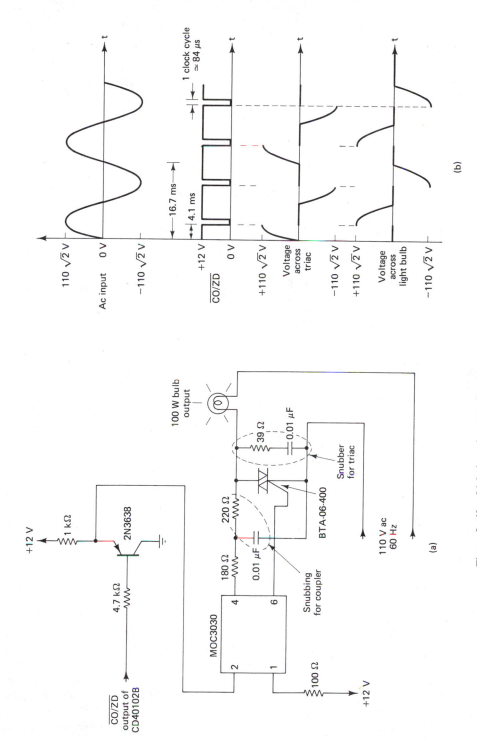

Figure 9–40. Light intensity controller: (a) control circuit, and (b) its waveform.

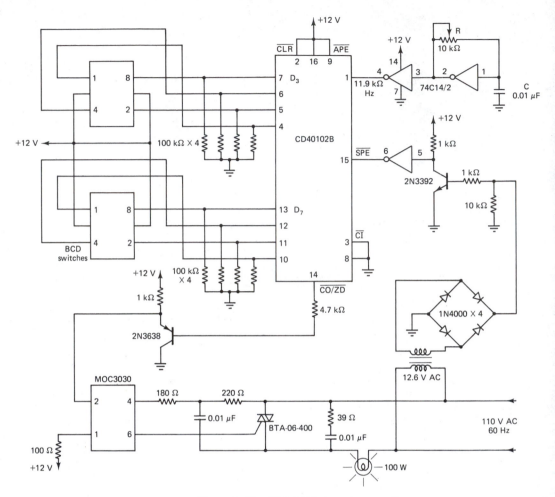

Figure 9–41. Digital light-intensity controller.

Light bulb. The final block in the light-intensity controller of Fig. 9–37 is a light bulb. The output of a control block drives the light bulb.

Schematic Diagram

The complete schematic diagram of the light intensity controller is shown in Fig. 9–41. The overall circuit operation is as follows. A CD40102B is loaded every 8.3 ms and is decremented on each LOW-to-HIGH transition of the clock. However, when the counter is decremented to zero its output changes from HIGH to LOW. This action triggers ON the triac driver MOC3030 and in turn the BTA-06–400 triac, which causes the light to come on. Again the counter is loaded when the $\overline{SPE}$ goes LOW and the cycle repeats. Thus the intensity of light depends on the value of an input BCD number.

The light-intensity controller is an open-loop control system. Its input may be calibrated in terms of a light intensity.

EXAMPLE 9–6

For the circuit of Fig. 9–41, draw waveforms for the output $\overline{\text{CO/ZD}}$, the voltage across a triac, and the voltage across a light bulb if the BCD switches are set to 49.

SOLUTION Since the frequency of the input clock is 11.9 kHz the time it will take the counter to reduce 49 to zero is

$$(49)\frac{1}{11.9 \text{ kHz}} \simeq 4.1\text{ms}$$

Thus the output of the counter will go logic LOW every 4.1 ms and will remain low for 1/11.9 kHz = 84.0 μs. Moreover, when the output $\overline{\text{CO/ZD}}$ of the counter goes from logic HIGH to LOW, the triac turns ON and remains on until the polarity of the ac line voltage changes. When the triac is ON, the voltage across it is almost zero and the voltage across the light bulb is equal to the line voltage. Remember that at any given time the sum of the voltages across the triac and the light bulb must be equal to the ac line voltage. The resulting voltage waveforms are shown in Fig. 9–40b.

9–7 ANALOG VERSUS DIGITAL CONTROL SYSTEMS

Chapter 8 presented analog control systems and Chapter 9, digital control systems. Having studied these two types of control systems, you may be wondering how one should go about selecting a type of system for a given application? Although the answer is not that simple and straightforward, here are some important factors to think about.

1. Assuming that for a given system the performance parameters and constraints are clearly defined, set up priorities between the accuracy and the speed of response.
2. Check what makes the system more economical—analog or digital techniques—without sacrificing the stability of the system.
3. Which techniques would result in a simple, yet easily maintainable system?
4. Before the system is actually fabricated (if possible), analyze system performance using the appropriate techniques.

Generally, digital control systems are easy to design because of the nature of digital techniques. In addition, a variety of digital control components and special-purpose ICs are available in the marketplace. Most digital circuits, including microprocessors, memories, and peripheral ICs, are predesigned packages; these ICs save the user a lot of design and testing time. However, sometimes because of the system contraints, it may not be possible to use digital techniques. At other times, it may be advantageous to use both analog and digital techniques for a given system.

SUMMARY

1. In digital control systems digital techniques are used to regulate energy inputs that control the desired outputs. The controlled output is generally used to identify the type of control system. For example, in a temperature control system, temperature is the controlled output.

2. In a closed-loop temperature control system, the actual temperature is compared with the desired temperature. This action generates an error signal that drives a controller. The controller, in turn, drives the actual temperature toward its desired value. When the actual temperature is the same as the desired temperature, the error signal is almost zero and the controller is disabled.

3. The use of digital techniques in speed control systems make these systems more accurate, reliable, and simpler.

4. In position control systems or servomechanisms, the heart of the system is an electromechanical device called a stepper motor. The stepper motor is used in applications such as an $x-y$ plotter, microcomputer printers, engineering and litho cameras, and even in robots. A stepper-controlled system consists of a stepper, a digital translator, and an electronic power drive.

5. A coin-operated car wash, a vacuum cleaner, a washer, a dryer, and video games are some of the examples in which a digital timer is used. The basic building blocks of a typical timer are the counters and divide-by-N networks.

6. Accuracy, small size, and ease of operation are some of the advantages of a digital light-intensity controller.

QUESTIONS

9-1. Explain the major difference between a digital control system and an analog control system.

9-2. Why is a closed-loop temperature control system preferred over an open-loop control system? Give at least one example where this is true.

9-3. What modifications are necessary if the circuit in Fig. 9-13 is used to monitor temperatures between 100° and 150°F? Explain.

9-4. Is it possible to replace DAC (1408) and the current-to-voltage converter (A_1) by a single IC to achieve the same function? What are the advantages and disadvantages of doing so? Refer to Fig. 9-23a.

9-5. Describe the operation of the power drive circuit in Fig. 9-23a during clockwise and counterclockwise rotation of the motor.

9-6. What are the different types of electronic power drives that are used for stepper-operated loads? Which is most commonly used, and why?

9-7. In Fig. 9-31, what is the function of the CD4098 monostable multivibrator?

9-8. Why is it essential to have "hold" and "emergency disable" features for stepper-motor-controlled systems?

9-9. What modifications are necessary in the circuit of Fig. 9-36 if the maximum time set is 20 min?

9-10. In the circuit of Fig. 9-36, the load is turned on as soon as the START switch SW$_1$ is depressed. What modifications are necessary in this circuit if the load is to be turned on 10 min after the START switch is depressed?

9–11. In the circuit of Fig. 9–41, why is it necessary to have a 120-Hz sync signal?

9–12. Explain the significance of an 11.9-kHz clock. Refer to Fig. 9–41.

PROBLEMS

9–1. Calculate the output voltage of LM334 at 80°F. Assume that the output voltage of LM334 changes at the rate of 10 mV/K. Refer to Fig. 9–4.

9–2. Calculate the gain of the sense amplifier A_3 if the output of A_1 is 2 V, $R_{sense} = 0.1\ \Omega$, and a given dc motor draws 0.1 A at its maximum speed. Refer to Fig. 9–23a.

9–3. In the circuit of Fig. 9–23a, determine the output frequency of the VFC LM331 if $V_e = 1$ V, $R_1 = 82$ kΩ, $C_1 = 0.01$ μF, $R_{ref} = 15$ kΩ, and $R_2 = 100$ kΩ.

9–4. In the circuit of Fig. 9–31, if the clock rate is 1000 cycles/s, how many steps will the motor advance in the full-step and double-step modes?

9–5. In Fig. 9–31, assume that a lead screw is used with the stepper motor to drive a load. The step angle of the motor is 1.8°, and the motor speed is 1000 steps/s. How much linear distance (in inches) will the load move in 30 s if the lead screw pitch is 5 threads/in.?

9–6. For the circuit of Fig. 9–36, determine the value of R and C if the clock rate is 0.3 cycle/s.

9–7. For the appliance timer of Fig. 9–36, design a buzzer circuit as an audio indicator, which is triggered when the appliance turns off. Also, the buzzer should turn off in ≤3 s. Draw the schematic diagram of the buzzer circuit.

9–8. In the digital light-intensity controller of Fig. 9–41, it is recommended that $\overline{APE}$ input be used instead of $\overline{SPE}$ to load the BCD input number into the counter. Design the necessary circuit to accomplish this and draw its circuit diagram.

LAB EXPERIMENT 9–1
STEPPER MOTOR TRANSLATOR/DRIVER
AND DRIVE CIRCUIT

Objectives. To learn the operation of a stepper translator and stepper drive circuit. After completing this experiment, you should be able to:

1. Explain the operation of a stepper motor translator/driver.
2. Explain the relationship between drive signals.
3. Explain the difference between full-step and double-step operating modes.
4. Discuss the difference between unipolar and chopped (switching) drive circuits.
5. Select and design a proper drive circuit for a given stepper motor and a translator.

Equipment

Dual-trace oscilloscope

+5-V power supply

+12-V power supply

Materials

UCN4202A stepper motor translator/driver

12-V, 330-mA, or equivalent stepper motor

74C14 hex Schmitt trigger inverters

Two 1-MΩ potentiometers

100-kΩ resistor

Six 1-kΩ resistors

Four 10-kΩ resistors

3.3-µF, 16-V capacitor

1-µF, 16-V capacitor

0.05-µF, 16-V capacitor

10-Ω, 10-W resistor

Procedure

1. Connect the circuit as shown in Fig. LE9–1.1. Use three pieces of wire as switches at pins 1, 9, and 12 of the 4202A. Make sure that the switch positions are as indicated in Fig. LE9–1.1.

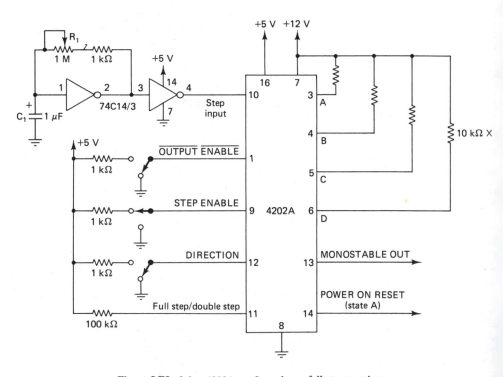

Figure LE9–1.1. 4202A configured as a full-step translator.

Digital Control Systems   Chap. 9

2. Connect one channel of an oscilloscope to pin 4 of the 74C14 and adjust the 1-MΩ potentiometer until the signal frequency at this pin is 100 Hz. Monitor output A at pin 3 of the 4202A using the other channel of the scope. Use the proper triggering mode. Draw the STEP INPUT and output A waveforms in Fig. LE9–1.4a.

3. Monitor simultaneously outputs A and B (at pins 3 and 4) of the 4202A using the oscilloscope. Draw the output B waveform in Fig. LE9–1.4a.

4. Repeat step 3 for outputs C and D.

5. Monitor the STEP INPUT (pin 10) and MONOSTABLE OUT (pin 13) waveforms using both channels of the oscilloscope and draw the MONOSTABLE OUT waveform in the space provided in Fig. LE9–1.4a.

6. Check the effect of OUTPUT ENABLE input on the MONOSTABLE OUT by connecting the OUTPUT ENABLE (pin 1) to ground and then to the +5-V through 1-kΩ resistor; enter the results in Table LE9–1.1.

7. Repeat step 6 for the STEP ENABLE (pin 9) input.

8. Simultaneously monitor the STEP INPUT and the POWER ON RESET output using both channels of the oscilloscope. Then draw the POWER ON RESET waveform in Fig. LE9–1.4a. Observe the POWER ON RESET output by turning the power (+5 V) to the 4202A (pin 16) on and off.

9. Connect the DIRECTION input (pin 12) to the +5-V supply through 1-kΩ resistor and monitor the STEP INPUT and output A waveforms simultaneously on the oscilloscope using both channels. Draw these waveforms in Fig. LE9–1.4b.

10. Change the channel connected to STEP INPUT over to output B. Observe and compare the output B waveform to output A and draw it in Fig. LE9–1.4b.

11. Monitor outputs B and C and C and D respectively; then draw the output waveforms for C and D in Fig. LE9–1.4b.

12. Repeat step 5 except draw the MONOSTABLE OUT waveform in Fig. LE9–1.4b.

13. With the oscilloscope connected to outputs A and B (pins 3 and 4), alternately connect OUTPUT ENABLE (pin 1) of the 4202A to ground and the +5-V supply through the 1-kΩ resistor for few times. Observe the effect of the OUTPUT ENABLE input on outputs A and B on the oscilloscope and record the result in Table LE9–1.1.

14. Repeat step 13 for outputs C and D.

15. With scope connected to outputs C and D, alternately connect STEP ENABLE (pin 9) to ground and +5 V a few times and observe the outputs C and D simultaneously. Record changes in outputs C and D in Table LE9–1.1.

16. Repeat step 15 for outputs A and B.

17. Next connect a 0.05-μF capacitor between pins 11 and 8 of the 4202A. With OUTPUT ENABLE LOW, DIRECTION INPUT LOW, and STEP ENABLE HIGH, monitor the STEP INPUT and output A waveforms on the oscilloscope. Draw these waveforms in Fig. LE9–1.5a.

18. Monitor output waveforms A and B, B and C, and C and D two at a time using the oscilloscope and draw them in Fig. LE9–1.5a.

19. Repeat steps 17 and 18 except that DIRECTION INPUT is pulled HIGH through 1-kΩ resistor. Draw waveforms in Fig. LE9–1.5b.

20. First remove the 0.05-μF capacitor from pins 11 and 8. Then construct the oscillator circuit using part of the remaining 74C14 inverters as shown in Fig. LE9–1.2.

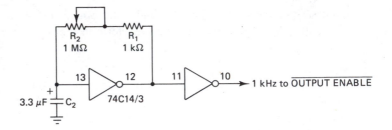

Figure LE9–1.2. Chop oscillator.

Connect one channel of the oscilloscope to the output pin 10 and adjust the 1-MΩ potentiometer until the output frequency is 1 kHz.

21. Connect pin 10 of 74C14 to the $\overline{\text{OUTPUT ENABLE}}$ of 4202A. Connect the other channel of the oscilloscope to output A (pin 3). Draw the output A waveform in Fig. LE9–1.4c.

22. Monitor outputs B, C, and D and draw their waveforms in Fig. LE9–1.4c.

23. Disconnect $\overline{\text{OUTPUT ENABLE}}$ from the chop oscillator and reconnect it to ground.

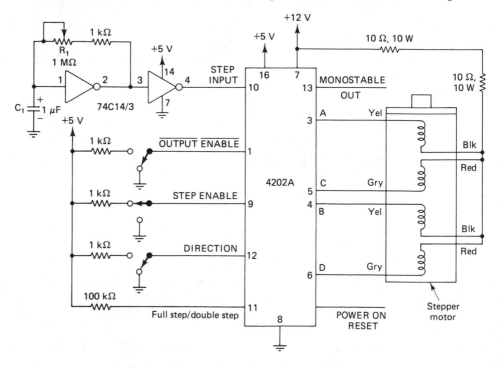

Figure LE9–1.3. Unipolar (L/R) stepper motor drive circuit.

Remove the four 10-kΩ resistors from outputs A, B, C, and D and connect the stepper motor instead, as shown in Fig. LE9−1.3. The stepper should be running now; if it does not, check the connection diagram carefully.

24. Verify operation of the DIRECTION CONTROL (DC) input as follows. When the DC input is held LOW, the direction of stepper shaft rotation is clockwise; when it is held HIGH, the direction of shaft rotation is counterclockwise.

25. First connect the $\overline{\text{OUTPUT ENABLE}}$ input to +5 V through the 1-kΩ resistor and try to turn the shaft of the stepper by hand. Enter the result in Table LE9−1.1. Next ground the $\overline{\text{OUTPUT ENABLE}}$ and observe the stepper shaft. Enter your observations in Table LE9−1.1.

26. Ground the STEP ENABLE input and try to turn the stepper shaft again by hand. Enter the result in Table LE9−1.1. Pull the STEP ENABLE input HIGH through the 1-kΩ resistor and again observe the motor shaft. Enter the observation in Table LE9−1.1.

27. To obtain minimum and maximum speed of rotation of the motor shaft, monitor the STEP INPUT on the oscilloscope. Then change the frequency of the STEP INPUT slowly by varying the 1-MΩ pot (R_1) and simultaneously, observe the motor shaft. Enter minimum and maximum speeds (steps/s) of the stepper motor in Table LE9−1.2.

28. Reconnect a 0.05-µF capacitor between pins 11 and 8 and repeat step 27.

29. Using the oscilloscope, readjust the oscillator in Fig. LE9−1.2 to 1 Hz. Disconnect the $\overline{\text{OUTPUT ENABLE}}$ pin from ground and reconnect it to oscillator output pin 10. Also readjust the STEP INPUT to 100 Hz. Observe the stepper shaft and write down your observations under Table LE9−1.2.

30. Modify the circuit in Fig. LE9−1.3 so that it will cause the motor shaft to move 100 steps clockwise in 1 s, then stop for 1 s and then move again counterclockwise 100 steps in 1 s and repeat the sequence. Include a complete circuit diagram in your report and verify the operation of the circuit with your instructor.

Test data

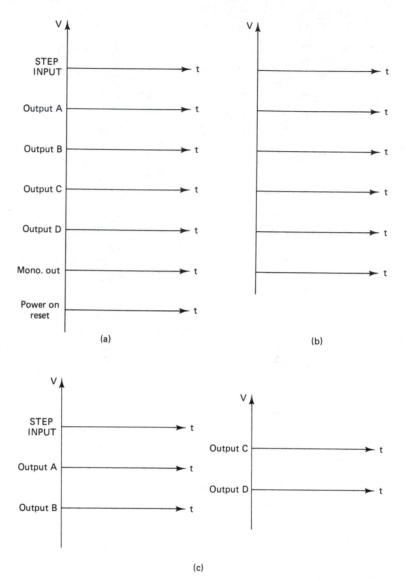

(a)

(b)

(c)

Figure LE9–1.4. Full-step waveforms: (a) direction control pin 12 at logic LOW (0 V); (b) direction control pin 12 at logic HIGH (+5 V); (c) external "CHOP" signal connected to $\overline{\text{OUTPUT ENABLE}}$ (pin 10).

Digital Control Systems   Chap. 9

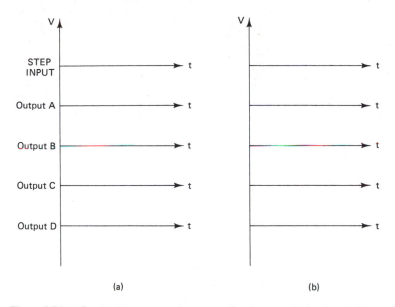

Figure LE9–1.5. Double-step waveforms: (a) direction control pin 12 at logic LOW (0 V); (b) direction control pin 12 at logic HIGH (+5 V).

TABLE LE9–1.1

Function	Logic level	Monostable out	A	B	C	D	Stepper motor *is*
				Outputs			
OUTPUT	HIGH						
ENABLE	LOW						
STEP	HIGH						
ENABLE							
	LOW						

TABLE LE9–1.2

Operating mode	Motor speed (steps/s)	
	Minimum	Maximum
Full step		
Double step		

Comments:

Evaluations. The responses to the following questions and/or conclusions must be supported whenever possible by the applicable theory.

1. In full-step operating mode, what is the time relationship between the STEP INPUT signal and outputs *A*, *B*, *C*, and *D*?
2. When DIRECTION INPUT is grounded, in what sequence do outputs *A*, *B*, *C*, and *D* change?
3. What is the sequence in which outputs *A*, *B*, *C*, and *D* change when the DIRECTION INPUT is pulled HIGH?
4. What is the time relationship between STEP INPUT and outputs *A*, *B*, *C*, and *D* in the double-step operating mode?
5. In the full-step operating mode, what is the time relationship between STEP INPUT and MONOSTABLE OUT?
6. In the double-step operating mode, what determines the on-time of outputs *B* and *D*?
7. Explain the effect of $\overline{\text{OUTPUT ENABLE}}$ and STEP ENABLE inputs on outputs *A*, *B*, *C*, and *D* and in turn on the use of a stepper motor.
8. Compare and contrast the double-step operating mode to full-step operating mode.
9. Explain the significance of "chopping" output signals *A*, *B*, *C*, and *D* (refer to steps 21 and 22 of the Procedure). When can this technique be used?

LAB EXPERIMENT 9–2

DIGITAL LIGHT-INTENSITY CONTROLLER

Objectives. To learn what the light-intensity controller is and how it works. At the end of this experiment, you should be able to:

1. Apply the concepts learned in this experiment to other control applications.
2. Explain the need for a light-intensity controller.
3. Discuss the operation of a given light-intensity controller.
4. Design a light-intensity controller for a desired application.

Equipment

Dual-trace oscilloscope
+5-V power supply

Materials

CD40102B synchronous down counter
74C14 hex Schmitt trigger inverters
MOC 3030 optically isolated triac driver
BTA-06−400 or equivalent triac
MCT6 dual phototransistor optoisolator
2N4355 or equivalent *pnp* transistor
2N3392 or equivalent *npn* transistor
Two 1N4003 diodes
12.6-V center-tap transformer
Eight 100-kΩ resistors
Two 10-kΩ resistors
Two 4.7-kΩ resistors
Two 1-kΩ resistors
Two 470-Ω resistors
330-Ω resistor
Three 220-Ω resistors
39-Ω resistor
0.047-μF capacitor
Two 0.01μF capacitors
Two BCD switches
40-W light bulb

Procedure

1. Design an *RC* oscillator using a 74C14 to have a maximum frequency of oscillation
   equal to 15 kHz. The connection diagram for such an oscillator is shown in Fig.
   LE9−2.1.

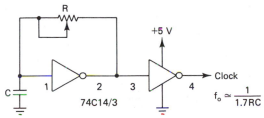

Figure LE9−2.1. *RC* oscillator using 74C14.

2. Using the designed values of R and C, connect the circuit as shown in Fig. LE9–2.1. To set oscillator frequency to approximately 11.9 kHz, adjust the pot R; use an oscilloscope to accomplish this.

3. Connect the circuit as shown in Fig. LE9–2.2 and apply dc as well as ac voltages. Use the oscillator output in Fig. LE9–2.1 as a clock.

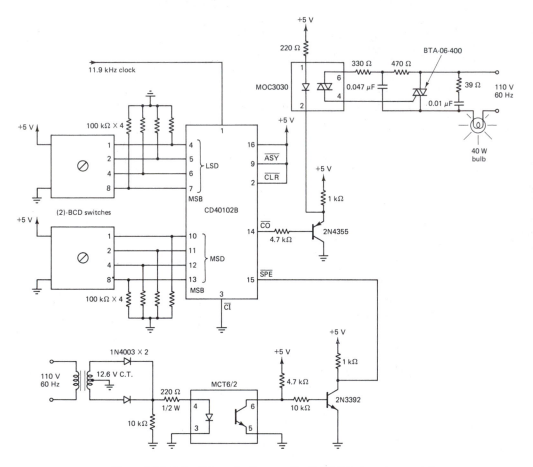

Figure LE9–2.2. Schematic diagram of a digital light-intensity controller.

4. Set the input BCD switches to 10_{BCD} and using both channels of an oscilloscope, monitor the voltage waveforms at pin 14 and 15 of the 40102B counter. Draw these waveforms in Fig. LE9–2.3a using the same scale. Observe the brightness of the light bulb.

5. Change the input switches to the following BCD numbers, one at a time: 11, 12, 13, 14, 15, 16, 17, 18, and 19. Observe the brightness of the light at each setting.

6. Repeat step 5 for the following BCD inputs: 00, 20, 30, 40, 50, 60, 70, 80, and 90.

7. Repeat step 4 for BCD input 99_{BCD}. Draw the waveforms in Fig. LE9–2.3b.

Test data

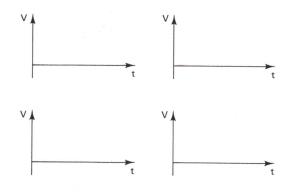

Figure LE9–2.3.   (a) Step 4 waveforms; (b) step 7 waveforms.

Evaluation.   The responses to the following questions and/or conclusions must be supported whenever possible by the applicable theory.

1. What is the function of MOC 3030 in Fig. LE9–2.2?
2. Why is the MCT6 needed in Fig. LE9–2.2?
3. Can the inverters, such as 74C14, be used instead of 2N4355 and 2N3392 in Fig. LE9–2.2? Explain.
4. What values of the BCD inputs result in the lowest (bulb off) and highest brightness of the bulb? Explain.
5. What maximum-wattage bulb can be used safely with BTA-06–400, a 6-A triac?
6. Explain briefly another application of the circuit in Fig. LE9–2.2.
7. How often is the input BCD number loaded into the down counter?
8. Explain why the frequency of the clock signal must be approximately equal to 11.9 kHz for the circuit in Fig. LE9–2.2 to work properly.
9. In how many different steps can you change the intensity of a light bulb with BCD inputs? How can the number of steps be increased?
10. What changes are required in the circuit of Fig. LE9–2.2 if the frequency of AC is 50 Hz?
11. Can an optically isolated triac driver such as MOC3010 be used instead of MOC3030 in the circuit of Fig. LE9–2.2? Explain why or why not.

chapter 10

MICROPROCESSOR-BASED CONTROL SYSTEMS

10–1 INTRODUCTION

In Chapter 8 we studied linear control systems. In linear control systems analog techniques are used to regulate an energy input to control a desired output. The characteristic of linear control systems is that they obey the superposition theorem. On the other hand, Chapter 9 presented digital control systems in which digital techniques are used to control the desired output. Most of the digital systems are less complex, easier to maintain, and cheaper than the equivalent linear systems. Since the advent of microprocessors in the early 1970s, the design of control systems has been made easier and more economical. In addition, most digital control systems can be converted very easily into microprocessor-controlled systems. In this chapter we examine microprocessor-controlled systems and evaluate their performance. To verify how easy it is to convert digital control systems to microprocessor-controlled systems, we will discuss some of the same systems that we studied in Chapter 9. The systems that are discussed here include temperature control, motor speed control, appliance timer, and light-intensity control.

10–2 MICROPROCESSOR-CONTROLLED HOME HEATING AND COOLING SYSTEM

A conventional home heating and cooling system using a thermostat is basically an analog control system. Because of its nature, an analog control system is less accurate and uneconomical than its equivalent digital temperature control system. Generally, in a conventional home heating and cooling system the actual temperature in a given room is never the same as its set value and may differ from its set input by as much as 5°F. A microprocessor-controlled home heating and cooling system, on the other hand, is more accurate in that the actual temperature can be made equal to the set temperature. Generally, a microprocessor-controlled home heating and cooling system is more efficient and economical than a conventional home heating system. A microprocessor-controlled system can be made more economical by turning it on and off at predetermined times. For example, in winter, during weekdays from 8 A.M. to 5 P.M. when no one is home, the temperature in the house may be maintained at 50°F. However, from 5 P.M. to 11 P.M., when the occupants are at home, the temperature may be programmed to be at 65°F. Between 11 P.M. and 6 A.M. it may again be programmed to be at 55°F. From 6 A.M. to 8 A.M. it may again be maintained at 65°F. There are commercial units on the market designed to operate this way. In microprocessor-controlled systems the actual as well as the set temperature may be displayed using LEDs instead of using thermostats as is done in conventional home heating systems. The advantage of display is that the displayed temperature is much more convenient to read than a thermostat.

In this section we study a microprocessor-based home heating and cooling system and examine the concepts and principles used in its construction. This system is used for heating as well as for cooling. The home heating and cooling system presented here is a prototype of an actual home heating and cooling system in which the heater is simulated by a light bulb and the air conditioner is simulated by a dc motor. However, the underlying principles can be applied very easily to a practical home heating system.

Block Diagram

The block diagram of a microprocessor-controlled home heating and cooling system is shown in Fig. 10–1. It consists of the following blocks: temperature sensor, analog-to-digital converter, microcomputer, control block, and heater and air conditioner.

System Description

Before we design the circuit for the home heating and cooling system, it is important to understand the function of each block in Fig. 10–1. The function of the temperature

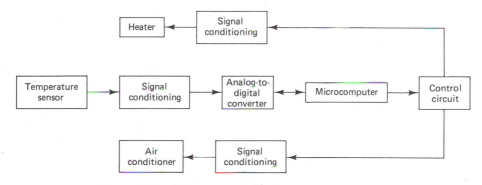

Figure 10–1. Microprocessor-controlled home heating system.

sensor block is to sense the temperature in a given room. The signal generated by the temperature sensor may be either a current or a voltage that is directly proportional to the variation in temperature. The output signal of the temperature sensor is modulated such that it is suitable for the analog-to-digital block. This function is accomplished by the block labeled "signal conditioning." The characteristics of the signal conditioning block are high input impedance and low output impedance. It also provides adequate amplification for the signal. The gain of the signal conditioning block is selected so that its output signal at the lowest and the highest temperatures is within the input range of the analog-to-digital converter (ADC).

The function of the ADC is to change an analog output signal of the signal conditioning block into an equivalent digital signal. The ADC is necessary here because microprocessors process only digital (binary) information. The function of the microcomputer block in Fig. 10–1 is to process the binary information received and to turn on the appropriate load (heater or air conditioner) depending on the difference between the actual and set temperatures. This is done through a program stored in a microcomputer memory. This program is executed continuously and allows the user to redefine the upper and lower temperature limits as needed. The advantages inherent in the use of microcomputer are flexibility, reliability, and accuracy. Thus the microcomputer block in Fig. 10–1 is an information processor as well as a controller block.

The choice of the microprocessor for this application is not critical; hence any microprocessor, such as a 6808, 8085, 6502, or Z80, may readily be used. Although a dedicated stand-alone unit may be designed for a home heating system, a commercial microcomputer trainer may very easily be used to verify system operation and prove the basic operating principles.

In Fig. 10–1 the output of the microcomputer is applied to a control block, which in turn controls the loads: heater and air conditioner. The output signals of a control block are not capable of driving the loads directly and hence require signal conditioning. This is done by signal conditioning blocks. There are two signal conditioning blocks, one for each load. These blocks also provide the interfacing and level translation necessary to isolate microcomputer outputs from the high power loads. Finally, the outputs of the signal conditioning blocks are applied as inputs to the heater and air conditioner blocks which control their operation. Based on the electrical specifications of the heater and the air conditioner, corresponding signal conditioning blocks are designed.

System Specifications

We will design a microprocessor-based home heating system that will be able to maintain the temperature at a desired value or within a certain range. In addition, the user should be able to select upper and lower temperature limits as desired. When the actual room temperature falls below a set lower limit, the heater comes on. The heater remains on until the temperature equals or exceeds the set lower limit. On the other hand, if the actual room temperature exceeds the set upper limit, the air conditioner comes on. It remains on until the room temperature drops below the set upper limit. When the room temperature is within the set limits, both the heater and air conditioner remain off. To verify the concepts and prove the validity of the microprocessor-based home heating system, a prototype may be built in which a heater is simulated by a light bulb and an air conditioner by a dc motor. The current room temperature will be displayed continuously and updated using seven-segment LEDs. In addition to its flexibility, the system should be stand-alone and cost-effective.

System Design

The block diagram of the home heating and cooling system shown in Fig. 10–1 is used to obtain its schematic diagram. First, each block is considered individually and an appropriate circuit is designed to satisfy its functional requirements. When all the blocks are designed, their schematic diagrams are breadboarded and tested one at a time against their desired performances. Once the testing of individual blocks is finished, all the blocks are connected together according to the block diagram of Fig. 10–1. At this point, the problems that arise due to mismatching or improper interfacing between the blocks may be solved. Finally, the program is written and tested with the designed system. The software and/or the hardware is modified until the system performs satisfactorily. In the following sections we discuss hardware and software considerations for the home heating system so that the reader may have clear understanding of both. Often in microprocessor-based systems, a trade-off between hardware and software is sought to obtain the most efficient and economical system. Now let us examine the hardware and software of the home heating system.

System Hardware

Microcomputer selection. Before we start the design of hardware circuitry for a home heating and cooling system, it is important to know what processor or microcomputer we are to use. The type of microcomputer and its electrical specifications, together with its associated features, will dictate the extent of hardware and software design. For experimental purposes it may be advantageous to use an existing microcomputer trainer, which should save a lot of design time and also help avoid frustration. The major thrust of this chapter is to design the necessary interface circuitry for a certain microcomputer trainer in controlling the operation of a desired system. Once the interface circuitry is designed, it may very easily be tested with the chosen trainer. If the system works satisfactory, a stand-alone system may be designed, if needed. In short, the use of an existing microcomputer trainer should be handy and prove to be an useful tool in testing the designed interface circuitry.

In this chapter we use the Heath Kit ET3400A microprocessor trainer. The electrical specifications and features of this trainer are as follows:

1. Uses 6808 microprocessor
2. 512 bytes of RAM
3. 1 kilobyte of ROM
4. 17-key hexadecimal keyboard
5. Six digits of hexadecimal display
6. INCH (inputs a character), ENCODE, OUTCH (outputs a character), OUTHEX, OUTBYT,

Figure 10-2. ET3400A microprocessor trainer. (Courtesy of Heath Company; St. Joseph, Mich.)

OUTST1, DSPLAY, IHB (output two hex characters), and REDIS subroutines in ROM

7. All data, address, and control lines brought up on the front panel for easy access

In short, ET3400A is a low-cost microprocessor-based system that can be used as a design aid for developing special interfaces. The top view of ET3400A is shown in Fig. 10-2.

Temperature sensor. First, we will design the temperature sensor circuitry. In fact, we can use the circuit in Fig. 9-4. For more information the reader may refer to Section 9-2.

Level translator/signal conditioning stage. There are a variety of ways that we can condition the output signal of the sensor to make it suitable for the ADC0804. The obvious choice are the voltage follower and the differential amplifier circuits of Fig. 9-5. For more information on the circuit, the reader may refer to Section 9-2.

Analog-to-digital converter. Next, we consider an analog-to-digital converter so that an input analog signal is converted into a digital signal suitable for microprocessor to process. We will use an 0804 ADC, which was discussed in Chapter 9. For the electrical specifications and pin functions of the ADC0804, the reader may refer to Section 9-2 (Fig. 9-6).

Microprocessor interface. In this section we examine the use of the ADC0804 with a 6800/6808 microprocessor to control a home heating system. Note that the control bus for the 6800/6808 microprocessor does not use the $\overline{RD}$ and $\overline{WR}$ strobe signals; instead, it employs a single R/$\overline{W}$ line. In addition, in the 6800 system all I/O devices are memory mapped. In the ET3400A microcomputer trainer no other devices are addressed at hex addresses 0200 to BFFF and C200 to FBFF.

Thus we have to generate $\overline{RD}$ and $\overline{WR}$ signals for the ADC0804 and READ ENABLE ($\overline{RE}$) signal for the ET3400A trainer using a decoder. Although there are various ways to configure the decoders to generate these signals, we will use the most commonly available 74LS138 three-to-eight-line decoder. Also, we will assign hex address 8000 to the converter. The pin diagram and the function table for the 74LS138 are shown in Fig. 10–3.

We can decode hex address 8000 assigned to ADC0804 using an ET3400A trainer, a 74LS138 decoder, and a ADC0804 converter as shown in Fig. 10–4. Note that we have used the upper four most significant address lines, A_{15}, A_{14}, A_{13}, A_{12} and $\overline{VMA\phi2}$ signal inputs to the 74LS138 decoder. Specifically, the A_{14}, A_{13}, and A_{12} address lines are the "select" inputs and A_{15} and $\overline{VMA\phi2}$ are the enable signals, since the status of VMA indicates a valid memory address and ϕ_2 controls all data and address activity. The $\overline{VMA\phi2}$ signal is necessary for proper operation of the interface circuitry of Fig.

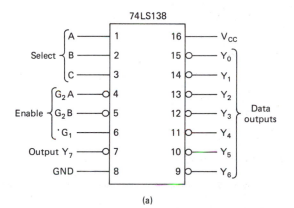

(a)

Inputs					Outputs							
Enable		Select										
G_1	G_2	C	B	A	Y_0	Y_1	Y_2	Y_3	Y_4	Y_5	Y_6	Y_7
X	H	X	X	X	H	H	H	H	H	H	H	H
L	X	X	X	X	H	H	H	H	H	H	H	H
H	L	L	L	L	L	H	H	H	H	H	H	H
H	L	L	L	H	H	L	H	H	H	H	H	H
H	L	L	H	L	H	H	L	H	H	H	H	H
H	L	L	H	H	H	H	H	L	H	H	H	H
H	L	H	L	L	H	H	H	H	L	H	H	H
H	L	H	L	H	H	H	H	H	H	L	H	H
H	L	H	H	L	H	H	H	H	H	H	L	H
H	L	H	H	H	H	H	H	H	H	H	H	L

(b)

Figure 10–3. 74LS138: (a) pin diagram; (b) function table, $G_2 = G_2A + G_2B$. H, high level; L, low level; X, irrelevant. (Courtesy of Texas Instruments, Inc.)

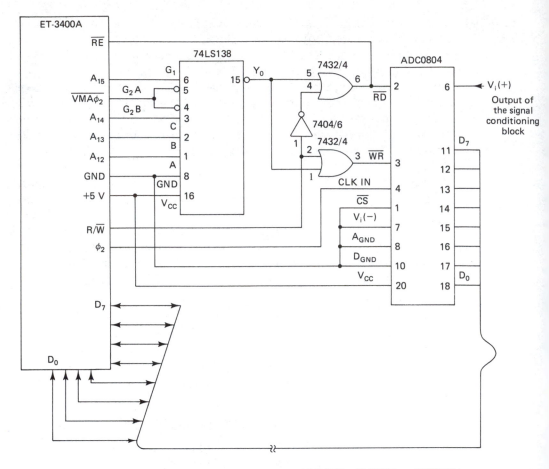

Figure 10–4. Connection diagram of 74LS138, ADC0804, and ET3400A.

10–4. According to the function table of the 74LS138 decoder, when A_{15} is HIGH and $\overline{\text{VMA}\phi2}$, A_{14}, A_{13}, and A_{12} are LOW, output Y_o is LOW. Y_o is ORed with the R/$\overline{\text{W}}$ signal of the 6808 microprocessor to produce $\overline{\text{RD}}$ and $\overline{\text{WR}}$ signals for the ADC0804 converter and an $\overline{\text{RE}}$ signal for the ET3400A trainer. When LOW, $\overline{\text{RE}}$, the read enable signal, activates the tri-state data bus drivers, and data on the data bus are read by the processor. Note that the ϕ_2 clock of the 6808 is connected to the CLK IN, pin 4, of the converter and the CHIP SELECT ($\overline{\text{CS}}$) of the converter is grounded. The remaining pin connections of the converter are self-explanatory. Thus, when A_{15} is logic 1, A_{14}, A_{13}, A_{12}, and $\overline{\text{VMA}\phi2}$ are logic 0, and R/$\overline{\text{W}}$ is logic 0; output Y_o of the 74LS138 is logic 0, which in turn causes the ADC0804 converter's READ ($\overline{\text{RD}}$) input to be logic 1 and WRITE ($\overline{\text{WR}}$) input to be logic 0. This starts the analog-to-digital conversion process of the ADC0804 in response to its input V_i at pin 6. Remember that V_i is the output of the signal conditioning block, which represents the actual temperature in the room. When A_{15} is logic 1, A_{14}, A_{13}, A_{12}, $\overline{\text{VMA}\phi2}$ are logic 0, and R/$\overline{\text{W}}$ is logic 1, $\overline{\text{RE}}$ and $\overline{\text{RD}}$, pin 2 of the converter, are logic 0, and pin 3, $\overline{\text{WR}}$, is logic 1; the output data of the converter are put on the data bus and read by the microprocessor with the help of the ϕ_2 clock. In other words, each time address 1000 XXXX XXXX XXXX$_2$ is on the address bus, $\overline{\text{VMA}\phi2}$ is LOW, and $R/\overline{W}$ is HIGH, we should be able to read the output data of the converter. The type of decoding used in Fig. 10–4 is referred to as partial decoding because all 16 address lines are not used to decode the hex address 8000 assigned to the converter.

Control circuitry. Based on the difference between the set temperature and the actual temperature in a given room, either a heater or an air conditioner is turned ON. In our discussion here, for simplicity we will simulate the operation of the heater and air conditioner by using a light bulb and a dc motor, respectively. Recall that in a 6800 system, I/O devices are treated as memory addresses. Since both the heater and air conditioner cannot be on at the same time, we will assign them hex address 9000. Obviously, we have to decode address 9000. The simplest way to accomplish this is to use the 74LS138 connection diagram in Fig. 10–4 but to use the output Y_1. In other words, Y_1 is logic 0 if A_{15} and A_{12} are logic 1 and A_{14}, A_{13}, and $\overline{\text{VMA}\phi2}$ are all logic 0 (see the function table of 74LS138 in Fig. 10–3b). Thus each time A_{15}, A_{14}, A_{13}, A_{12}, A_{11}, A_{10}, A_9, A_8, A_7, A_6, A_5, A_4, A_3, A_2, A_1, $A_0 = 1001$ XXXX XXXX XXXX$_2$ and $\overline{\text{VMA}\phi2}$ is logic 0, the output Y_1 of the decoder 74LS138 is logic 0. Output Y_1 must be part of the control circuit which determines the operation of the heater and air conditioner as a function of temperature. There are a variety of ways that we can use Y_1 to control operation of the heater and air conditioner. One way is to use a 4-bit bistable latch 74LS75. The connection diagram and the function table for each latch of the 74LS75 are shown in Fig. 10–5. From the function table it is obvious that the information present at a data D input transferred to the Q output when the enable input G is HIGH and the Q output follows the data input as long as the enable remains HIGH. When the enable goes LOW, the information that was present at the data D input at the time transition occurred is retained at the Q output until the enable is allowed to go high again.

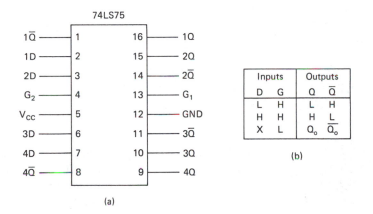

Figure 10–5. 74LS75 4-bit bistable latch: (a) pin diagram; (b) function table, H, HIGH level; L, LOW level; X, irrelevant; Q_0, level of Q before the HIGH-to-LOW transition of G; G_1, enable for latches 1 and 2; G_2, enable for latches 3 and 4. (Courtesy of Texas Instruments, Inc.)

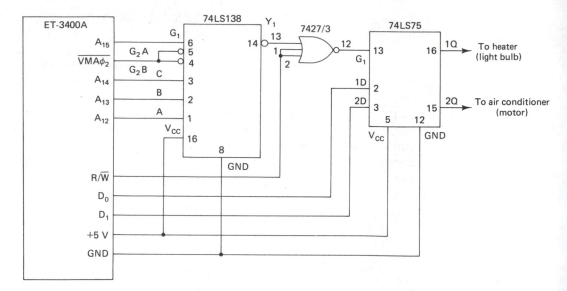

Figure 10–6. 74LS75 and associated circuitry to control the loads.

To control operation of the heater and the air conditioner, the 74LS75 latch may be configured as shown in Fig. 10–6. In this figure data inputs 1D and 2D of the 74LS75 latch are data lines D_0 and D_1 of the data bus, which control the operation of the heater and air conditioner, respectively. For example, if the temperature in a room is less than the desired value, we may use data line D_0 to turn the heater ON, or if the temperature in the room is higher than its desired setting, we may use data line D_1 to turn the air conditioner ON. To set D_0 and D_1 to the desired level, we need to utilize the R/$\overline{\text{W}}$ signal of the ET3400A. As shown in Fig. 10–6, the R/$\overline{\text{W}}$ signal is NORed with Y_1 output of the 74LS138 decoder, and the output of the NOR gate is used to enable the 74LS75 latch.

Next, we consider the signal conditioning circuit for the loads: heater and air conditioner.

Signal conditioning circuits for the loads. Outputs 1Q and 2Q of the 74LS75 latch are used to activate the loads: heater and air conditioner. However, these are digital signals and hence require conditioning so that they can drive the loads. The type of signal conditioning needed depends on the electrical specifications of the loads. Therefore, to drive an ac-operated heater (light bulb), we will use output 1Q of the 74LS75 latch and apply it through an optically isolated triac driver MOC3010, which in turn controls operation of the heater (light bulb). The function of the MOC3010 is to isolate and provide interfacing between digital and analog signals. It also provides the power necessary to drive the load. The pin diagram and the internal structure of MOC3010 are shown in Fig. 9–10.

If input 1D of 74LS75 is HIGH and enable G_1 is also HIGH, output 1Q goes high. When this happens, the input diode of the MOC3010 is turned ON, which in turn turns the heater ON (Fig. 10–7). Output 2Q of 74LS75 is used to control operation of the air

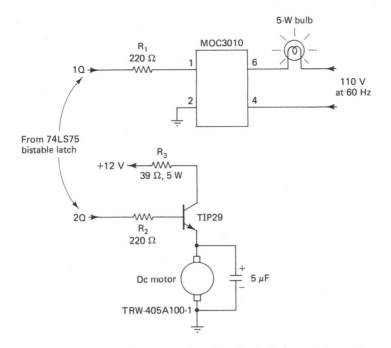

Figure 10–7. Signal conditioning circuit to drive loads—heater and air conditioner.

conditioner. In Fig. 10–7, the air conditioner is simulated by a dc motor. If input 2D is HIGH and enable input G_1 is HIGH, output 2Q of the 74LS75 goes HIGH, which turns the motor ON. To condition the output 2Q so that it can drive the motor, a transistor current amplifier is used. Specifically, a dc motor is connected in series with the emitter of a TIP29 transistor. The electrical specifications, mainly the voltage and current ratings of a dc motor, dictate the selection of a transistor. In Fig. 10–7, TRW–405A100–1, a 12-V, 300-mA, 4800-rpm motor, is used. In addition, transistor TIP29 is rated for $I_{C(\text{max})} = 1.0$ A and $V_{CE(\text{max})} = 40$ V, and hence should be adequate to handle the operation of the motor. A 5-μF capacitor is used across the motor to reduce the inductive spikes so that trouble-free operation may be ensured. This completes the design and functional details of the home heating and cooling system hardware. The complete schematic diagram of the system hardware is shown in Fig. 10–8.

So far we have studied the system hardware and circuit operation of a home heating and cooling system. The next step is to examine the system software so that satisfactory and economical system operation can be achieved.

System Software

First let us reexamine the functions that we want to accomplish through software. This task should be easier because we already know what the system hardware does. Recall that we want the home heating and cooling system to be flexible enough so that the user can select upper and lower temperature limits as needed. These limits obviously can be set in software and will be entered before the program is executed. Thus when

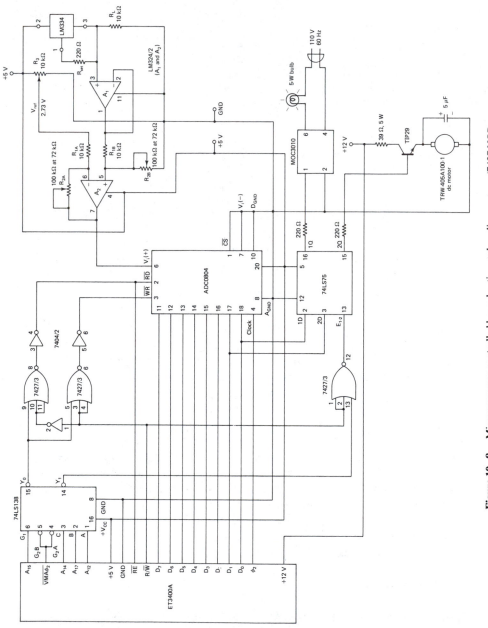

Figure 10–8. Microprocessor-controlled home heating and cooling system. (7427 NOR gates and 7404 INTERVERTS are used as OR gates.)

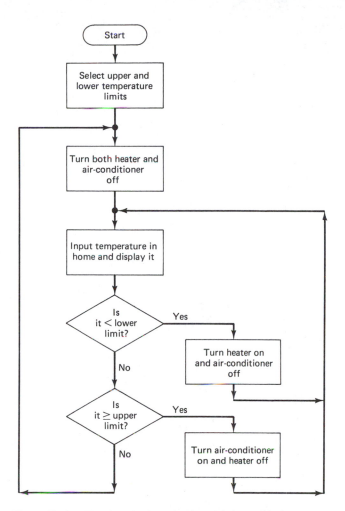

Figure 10-9. Flowchart for home heating and air-conditioning system.

we write software, we will set aside memory locations for each of the upper and lower temperature limits. Any time the temperature in a home goes below the set lower limit, the heater should turn ON. The heater should remain ON until the actual temperature in the home exceeds the set lower limit. On the other hand, whenever the temperature in the home exceeds the set upper temperature limit, the air conditioner should turn ON. The air conditioner should remain ON until the actual temperature in the home drops below the set upper temperature limit. However, if the actual temperature is within the set lower and upper limits, both the heater and air conditioner should remain OFF. In addition, we would like to display the current temperature in the home. Before writing the program, let us take a look at the flowchart. Figure 10-9 shows the flowchart for a home heating and cooling system. It should be fairly straightforward to write a program from the flowchart. We will start the program at memory location 0000_H because ET3400A RAM resides at 0000_H through $00FF_H$. We will also assume that 65°F is the

Location		Mnemonic	Op-code		Comments
0000		CLRA	4F		
0001		STAA	B7		Turn both
0002		90	90		devices off.
0003		00	00		
0004	START	STAA	B7		
0005		80	80		Start A/D.
0006		00	00		
0007		LDAB	C6		
0008		0C	0C		
0009	LOOP1	DECB	5A		100 μs plus
000A		BNE	26		delay.
000B		LOOP1	FD		
000C		LDAA	B6		
000D		80	80		Get data
000E		00	00		from A/D.
000F		LSRA	44		
0010		ADDA#	8B		
0011		00	00		Convert to
0012		DAA	19		BCD format
0013		ADD#	8B		add 32°
0014		32	32		offset.
0015		DAA	19		
0016		JSR	BD		Output
0017		OUT	FE		temperature.
0018		BYTE*	20		
0019		CMPA#	81		Check if temp.
001A		65	65		< 65°F. If not
001B		BCC	24		go compare if
001C		ABOVE	07		≥ 70°F.
001D		LDAA#	86		
001E		01	01		If yes turn
001F		STAA	B7		heater on and
0020		90	90		turn off air-
0021		00	00		conditioner.
0022		BRA	20		Goto time
0023		AGAIN	0F		delay.
0024	ABOVE	CMPA#	81		Check if
0025		70	70		temp ≥ 70°F.
0026		BCS	25		If not go turn
0027		OFF	07		both devices
0028		LDAA#	86		off.
0029		02	02		Turn on air-
002A		STAA	B7		conditioner
002B		90	90		and turn off
002C		00	00		heater.
002D		BRA	20		Goto time
002E		AGAIN	04		delay.
002F	OFF	CLRA	4F		
0030		STAA	B7		Turn both air-
0031		90	90		conditioner
0032		00	00		and heater off.
0033	AGAIN	JSR	BD		
0034		RE	FC		Relocate
0035		DIS*	BC		display.
0036		LDX#	CE		
0037		FF	FF		Time delay for
0038		FF	FF		spacing temp.
0039	LOOP2	DEX	09		sampling.
003A		BNE	26		
003B		LOOP2	FD		
003C		BRA	20		Go sampling
003D		START	C6		temperature.

Figure 10–10. Program for home heating and cooling system. (Monitor programs*, courtesy of Heath Company; St. Joseph, Mich.)

lower temperature limit and 70°F is the upper temperature limit. The program is shown in Fig. 10–10. In the program, first, both devices, the heater and the air conditioner, are turned OFF. Then we enable the ADC and allow it sufficient conversion time (>100 μs) to convert to a digital signal the analog input that represents the temperature in the home. Then we input the digital signal into the microprocessor. There digital data are converted to BCD and changed into degrees Fahrenheit. After that, the data are displayed on the LEDs using the monitor routine called OUTBYT. Next, the home temperature is compared to the lower-temperature limit of 65°F. If it is not less than 65°F, it is checked to see if it is $\geq$70°F; if not, both the heater and the air conditioner are turned OFF. On the other hand, if the home temperature is $<$65°F, the heater is turned ON by "setting" bit D_0 of the data bus. If the temperature in the home is $\geq$70°F, however, the air conditioner is turned ON by "setting" bit D_1 of the data bus (Fig. 10–8). This procedure is repeated periodically where the sampling time is determined by the time delay. Also, to help read displayed temperature, a monitor program REDIS is used. Note that the user can choose upper and lower temperature limits and simply key them in memory locations 001A and 0025. Once this is done, the next thing to do is to execute the program.

The basic concepts illustrated above can very easily be extended to a more sophisticated home heating and cooling system in which temperatures can be changed automatically to the desired values at predetermined times.

This concludes the discussion of the program of Fig. 10–10 for a home heating and cooling system. Note that there are a number of ways that a software program can be written to support the hardware of Fig. 10–8. In fact, the reader is encouraged to try some different ways when working on Lab Experiment 10–1.

10–3 MICROPROCESSOR-BASED DC MOTOR SPEED CONTROL SYSTEM

In Chapter 9 we studied a digital dc motor speed control system. In this section we see how we can integrate microprocessor trainer ET3400A to control the speed of a dc motor. Obviously, the objective is to use ET3400A with the digital dc motor speed control system of Fig. 9–23a with minimum modifications. Let us begin with the block diagram of a microprocessor-based dc motor speed control system.

Block Diagram

The block diagram of a microprocessor-based dc motor speed control system is shown in Fig. 10–11. It consists of a microprocessor trainer, a DAC, a controller, a controlled process, and a measuring device blocks. Each block has to be examined closely to ensure necessary modifications in the circuit of Fig. 9–23a.

System Description

The function of the microprocessor block is to supply input binary data to the DAC, to supply address and control lines for address decoding, and to store the program that

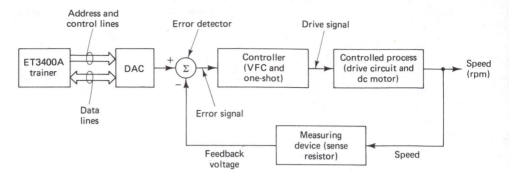

Figure 10–11. Block diagram of a microprocessor-controlled dc motor speed control system.

controls and displays the motor speed and direction. The function of the DAC is to convert the digital input data that were supplied by the trainer into an appropriate analog output signal. This analog output voltage represents the *desired* speed and is compared with the feedback signal, which represents the actual speed of a dc motor. The difference between the desired speed voltage and the actual speed voltage is referred to as the *error voltage*. The error voltage is transformed into an appropriate drive signal by the controller block (Fig. 10–11). The controller block consists of a voltage-to-frequency and pulse-width-modulation circuit. The output of a controller block is called the *drive signal*, which is an input to the controlled process block. The drive signal is acted on by the controlled process block to produce a desired speed. The controlled process block consists of a power-drive circuit and a dc motor. The output of this block is an actual speed. An actual motor speed is sensed by a measuring device and converted into a proportional voltage. This voltage is then compared to the input voltage, which is proportional to the desired speed setting (Fig. 10–11). The type of feedback used in Fig. 10–11 is negative, which helps to drive an actual speed toward the desired speed.

System Specifications

We wish to design a microprocessor-based dc motor speed control system in which the user is to select a desired motor speed. The speed is to be programmed in hex and can be displayed as LO, MEDIUM, or HIGH, based on the number programmed in. If the programmed number is between 01_H and 54_H, the corresponding speed will be displayed as LO. If the programmed number is between 55_H and $A9_H$, the corresponding speed will be displayed as MEDIUM, and if the programmed number is between AA_H and FF_H, the speed will be displayed as HIGH. This is because the larger the input number, the higher the speed. Note that the motor shaft rotation can also be programmed as either clockwise or counterclockwise. In addition, in software the speed of the motor may be increased or decreased at regular intervals. Since we use negative feedback, it is possible to maintain the speed of the motor at a certain value as long as desired. The performance and the speed accuracy of the motor depend on the design of the pulse-width-modulation circuitry, power drive circuitry, and electrical specifications of the motor.

System Design

As any other microprocessor-based system, the dc motor speed control system involves hardware and software design. As far as hardware design is concerned, we need to make necessary modifications in the circuit of Fig. 9–23a so that we can use a microprocessor to control it. First, let us consider the hardware design.

System Hardware

To use an ET3400A trainer with the hardware in Fig. 9–23a, we need to design interface circuitry, which will include an address decoder and a motor direction control circuitry.

Address decoder and latches. Recall that in Fig. 9–23a the binary switch positions determine the speed of the motor. In other words, the motor speed is changed by changing the switch positions. When we use the ET3400A trainer the binary input switches will be replaced by the data bus. However, to write the data on the data bus into the peripheral device, they must be assigned an address. The peripheral device here is a DAC1408 (Fig. 9–23a). Since the DAC1408 does not have a provision to be addressed directly by the microprocessor, we need to use a separate address decoder circuit. As in the preceding section, we will use a 74LS138 three-to-eight-line decoder as an address decoder. In fact, we may assign the same address 8000_H to the DAC, so that we can use the 74LS138 decoder circuit of Fig. 10–4. Also, we will use an 8-bit latch 74LS373 since the data bus cannot be connected directly to the DAC1408. The 8-bit latch will store the binary data that are used to determine the speed of the motor until new data are to be written. Let us briefly examine the important electrical features of 74LS373 octal D-type latches. These include:

Eight latches in a single 20-pin package

Full parallel access for loading

Buffered control inputs

Three-state-bus driving outputs

Enable input having hysteresis to improve noise rejection

The 74LS373 has eight transparent D-type latches. When the enable (G) input is HIGH, the Q outputs follow the data (D) inputs. If the enable input is taken LOW, the outputs are latched at the level of data that was set up. The pin configuration and the function table of the 74LS373 are shown in Fig. 10–12.

A buffered output enable control ($\overline{OC}$) input can be used to place the outputs in either a normal HIGH or LOW logic level or a high-impedance state. In the high-impedance state the outputs neither load nor drive the bus lines. In addition, the old data can be retained or new data can be entered when the outputs are off. We will connect the data bus of a ET3400A to the data inputs of the 74LS373 latches. The outputs of these latches will be inputs to the DAC1408. Also, we will connect the $\overline{OC}$ input to ground so that the outputs of the latches can be placed in either a HIGH or a LOW logic level. Finally, the G input for the 74LS373 can be derived

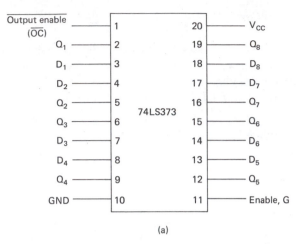

(a)

Output control	Enable G	Input D	Output Q
L	H	H	H
L	H	L	L
L	L	X	Q_o
H	X	X	Z

(b)

Figure 10–12. 74LS373 octal D-type transparent latches: (a) pin configuration; (b) function table. H, HIGH level; L, LOW level; X, irrelevant; Z, high impedance (off). (Courtesy of Texas Instruments, Inc.)

by using the Y_o output of the 74LS138 and the R/$\overline{\text{W}}$ signal of the ET3400A. Specifically, we can NOR the Y_o and R/$\overline{\text{W}}$ signals and connect the output of the NOR gate to the G input. The complete diagram of the address decoder and latches is shown in Fig. 10–13. In this figure when Y_o and R/$\overline{\text{W}}$ are LOW, the G input will be HIGH and the latches will be enabled, which in turn will latch the data on the bus into the 74LS373. However, the only time the Y_o and R/$\overline{\text{W}}$ go LOW is when the processor writes to address 8000_H. Also, the data latched into the latches appear at the outputs and remain there until a new number is written by another write cycle.

Direction control circuitry. Since we are using a microprocessor, it should be easier to control the desired motor speed direction through software. However, to accomplish this in software, we need to make hardware modifications in the circuit of Fig. 9–23a. Specifically, we will replace the CW/CCW direction control switch with a D latch such as 74LS75. However, the latch then needs to be assigned a separate address. Therefore, let us assign address 9000_H to the latch so that we can still use the 74LS138 address decoder. For more information on 74LS75 latches, the reader may refer to Fig. 10–5. The complete connection diagram for the direction control circuitry is shown in Fig. 10–14.

In Fig. 10–14, address 9000_H is partially decoded with the use of address lines

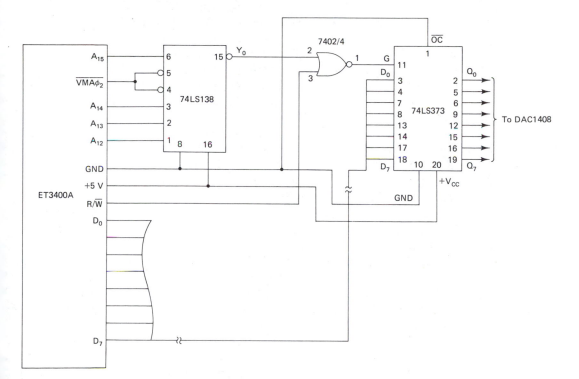

Figure 10–13. Connection diagram of the address decoder to decode address 8000_H and octal D-type latches for the 8-bit data input to the DAC-1408.

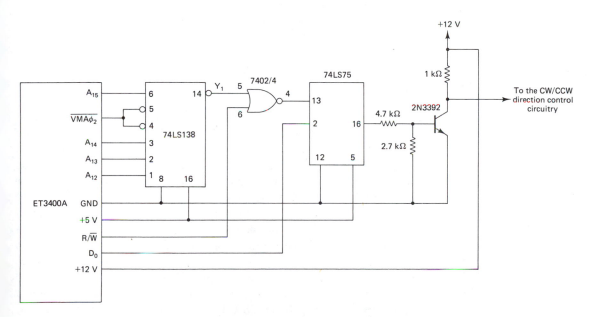

Figure 10–14. Direction control circuitry.

A_{15}, A_{14}, A_{13}, A_{12}, and $\overline{\text{VMA}\phi 2}$. When the address 9000_H is on the address bus, the output Y_1 of the 74LS138 decoder goes low. This output is then NORed with the R/$\overline{\text{W}}$ signal of the microprocessor using 7402 quad two-input NOR gates. The output of the NOR gate, in turn, drives the enable input of the 74LS75 4-bit latch. The data line D_0 is the data input for the latch. When the enable input of the latch is HIGH, the data at the D_0 input is latched into it. The output of the latch is then transformed into a 12-V signal using a transistor inverter. The level translation is required because the direction control signal is to drive the CMOS 74C14 gates. A desired direction can be programmed by setting or resetting the D_0 line of the data bus. Let us assume that when D_0 is logic 1, the direction of rotation of the motor shaft is clockwise, and when it is logic 0, the direction of rotation of the motor shaft is counterclockwise.

Thus the interfacing circuit, including the replacements for the binary switch for digital inputs and the CW/CCW direction switch, consists of a 74LS138 decoder, 74LS373 octal latches, 74LS75 D latches, a transistor inverter, and 74LS02 NOR gates (see Figs. 9–23a and 10–14).

Next let us briefly review the remaining dc motor speed control circuit of Fig. 9–23a. As mentioned earlier, we want to modify the dc motor speed control circuit of Fig. 9–23a so that we can use the ET3400A trainer to operate it. So far we have replaced the eight-position bit switch by the address decoder and the octal D-type latches so that the processor can write a desired binary number into the DAC1408 (Fig. 10–13). Also, we replaced the CW/CCW switch by the 74LS75 latch, as shown in Fig. 10–14. However, close examination of the circuit in Fig. 9–23a reveals that we do not need to make further modifications in the circuit. Specifically, the digital-to-analog converter, the error detector (differential amplifier), the controller block (VFC and one-shot), and the controlled process block (motor power drive and motor) discussed in Section 9–3 can be used as shown in Fig. 9–23a without any changes. Thus the schematic diagram of a microprocessor-based dc motor speed control system is shown in Fig. 10–15. This completes the hardware design of the system. Next, let us study the software requirements of the system.

System Software

Recall that in the microprocessor-based dc motor speed control system, the user is to select and program a desired motor speed in hex. Moreover, the speed is to be displayed as LO, MEDIUM, or HIGH, based on the number programmed in. Also, the motor shaft rotation can be programmed as either clockwise or counterclockwise. To satisfy the requirements, let us consider the flowchart first. Figure 10–16 shows the flowchart for a microprocessor-based dc motor speed control system.

We begin the program by loading a desired motor speed and a direction of rotation into the processor. The hex numbers are then stored into a DAC1408 and a 74LS75 latch residing at locations 8000_H and 9000_H, respectively. Next, the motor direction stored into the latch is checked and displayed accordingly as either C for clockwise, and CC for counterclockwise. A time-delay loop is used to hold the direction displayed stationary before the program jumps to check the speed of a motor. Next, the number representing a motor speed is compared against the LO and MEDIUM speed limits of 54_H and $A9_H$, respectively. Specifically, if the number loaded is $\leq 54_H$, it is displayed as

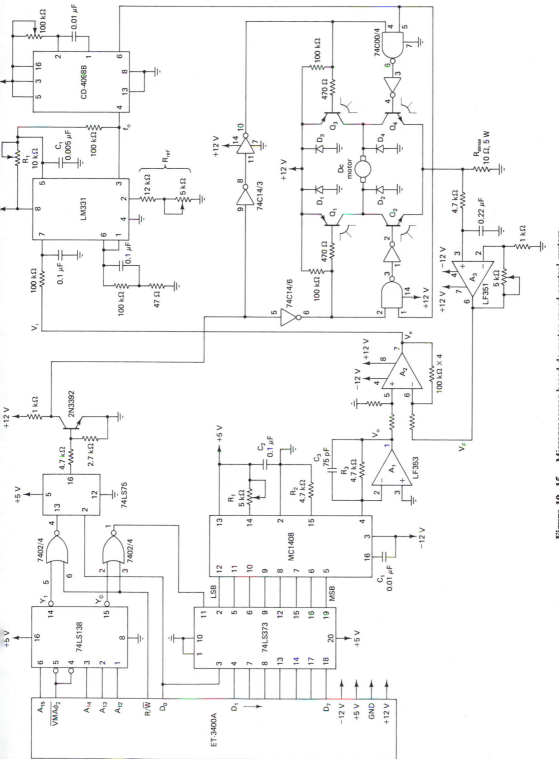

Figure 10–15. Microprocessor-based dc motor speed control system.

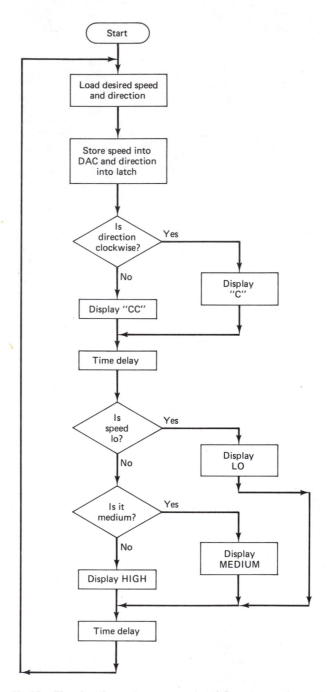

Figure 10–16. Flowchart for a microprocessor-based dc motor speed control system.

LO. On the other hand, if the number is $\leq A9_H$, it is displayed as MEDIUM, otherwise displayed as HIGH. Next, a time-delay loop is used to display the desired speed before the program jumps to the beginning to repeat the cycle. Thus the program will display the direction and the motor speed alternately. The user may break the cycle to enter a new speed and a direction of rotation for a motor.

Once we have a flowchart, it should be fairly straightforward to write a program. Therefore, a typical program using the flowchart of Fig. 10–16 is written as shown in Fig. 10–17. The program uses output string of characters (OUTSTR) and reset display (REDIS) monitor subroutines to show the direction of rotation and the speed of the motor. Also, memory locations 0070_H and 0071_H are reserved for the direction and motor speed, respectively. Before the program is executed, the desired direction of rotation and the speed must be entered into these memory locations.

The program of Fig. 10–17 strictly follows the flowchart in Fig. 10–16. Therefore, it is fairly easy to follow and should need no further explanation.

Now let us briefly evaluate the microprocessor-based speed control system of Fig. 10–15 against its design specifications. We made necessary modifications in the system of Fig. 9–23a so that it can be used with ET3400A. The system is flexible enough so that the user can select any motor speed from 01_H to FF_H and choose a desired direction of rotation as well. In addition, the system displays the chosen direction and current speed of the motor. The system is fairly versatile in that different-size dc motors may be used in it provided that the VFC, the one-shot, and the speed-to-voltage converter circuits are recalibrated for each of these motors based on their electrical specifications. Even the software may be modified easily so that a motor moves clockwise for a certain duration and then reverses its direction, or a motor speed may change at a certain rate to reach a predetermined value. Thus the microprocessor-based dc motor speed control system of Fig. 10–15 does satisfy all the design specifications.

10–4 MICROPROCESSOR-CONTROLLED APPLIANCE TIMER

Next, we consider the appliance timer circuit that was presented in Chapter 9 (see Fig. 9–36). Our objective here is to control timer operation using a ET3400A microprocessor trainer without redesigning the entire circuit or making any drastic changes in the circuit of Fig. 9–36. Obviously, to accomplish this, we need to reexamine the circuit with reference to the capabilities of the ET3400A. To facilitate these modifications, let us start with the block diagram of a microprocessor-controlled appliance timer.

Block Diagram

The block diagram of a microprocessor-controlled appliance timer is shown in Fig. 10–18. It consists of a microprocessor trainer, a counter, a signal conditioning circuit, and an appliance.

System Description

In the block diagram of Fig. 10–18, the function of the microprocessor trainer is to provide data inputs and a clock to the counter. Recall that ET3400A uses a 6808 micropro-

Address		Mnemonic	Op-code		Comments
0000	START	LDAA	96		Load a desired speed
0001		SPEED	70		in accumulator A.
0002		STAA	B7		Store the desired
0003		80	80		speed into the
0004		00	00		DAC1408.
0005		LDAB	D6		Load a desired
0006		DIRECTION	71		direction in
					accumulator B.
0007		STAB	F7		Store the desired
0008		90	90		direction into the
0009		00	00		latch 74LS75.
000A		CMPB	C1		Is direction
000B		#00	00		clockwise?
000C		BEQ	27		If yes branch to
000D		CW	0C		display C.
000E		JSR	BD		
000F		FE	FE		
0010		52*	52		
0011		00	00		
0012		00	00		
0013		C	0D		If not display CC.
0014		C	0D		
0015		00	00		
0016		00	00		
0017		80	80		
0018		BRA	20		
0019		DELAY1	0A		
001A	CW	JSR	BD		
001B		FE	FE		
001C		52	52		
001D		00	00		
001E		00	00		
001F		00	00		Display C.
0020		C	0D		
0021		00	00		
0022		00	00		
0023		80	80		
0024	DELAY1	LDX	CE		
0025		TIME	FF		
0026		DELAY	FF		Time delay for
0027	WAIT	DEX	09		direction displayed.
0028		BNE	26		
0029		WAIT	FD		
002A		JSR	BD		
002B		RE	FC		Relocate display.
002C		DIS*	BC		
002D		CMPA	81		Is speed LO?
002E		#LO	54		
002F		BCC	24		If not go to check
0030		MEDIUM	0C		if it is MEDIUM.
0031		JSR	BD		
0032		FE	FE		
0033		52	52		
0034		00	00		
0035		00	00		
0036		L	0E		If yes, display LO.
0037		0	1D		
0038		00	00		
0039		00	00		
003A		80	80		
003B		BRA	20		
003C		DELAY2	1B		

Figure 10–17. Program for a microprocessor-based speed control system. (Monitor programs*, courtesy of Heath Company; St. Joseph, Mich.)

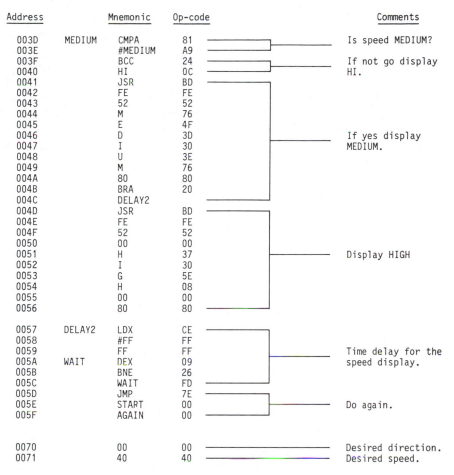

Address		Mnemonic	Op-code		Comments
003D	MEDIUM	CMPA	81		Is speed MEDIUM?
003E		#MEDIUM	A9		
003F		BCC	24		If not go display
0040		HI	0C		HI.
0041		JSR	BD		
0042		FE	FE		
0043		52	52		
0044		M	76		
0045		E	4F		
0046		D	3D		If yes display
0047		I	30		MEDIUM.
0048		U	3E		
0049		M	76		
004A		80	80		
004B		BRA	20		
004C		DELAY2			
004D		JSR	BD		
004E		FE	FE		
004F		52	52		
0050		00	00		
0051		H	37		Display HIGH
0052		I	30		
0053		G	5E		
0054		H	08		
0055		00	00		
0056		80	80		
0057	DELAY2	LDX	CE		
0058		#FF	FF		
0059		FF	FF		Time delay for the
005A	WAIT	DEX	09		speed display.
005B		BNE	26		
005C		WAIT	FD		
005D		JMP	7E		
005E		START	00		Do again.
005F		AGAIN	00		
0070		00	00		Desired direction.
0071		40	40		Desired speed.

Figure 10–17. (*continued*)

cessor, which treats the input/output device as a memory location. Hence a hex address of 8000 may be assigned to the counter. Obviously, a decoder circuit must be used to decode the address 8000_H.

Normally, an appliance is OFF. However, when it is turned ON, the on-time is determined by the number latched into the down counter and the counter's clock frequency. In other words, for a given input range, that is, 01_H to FF_H, the appliance on-

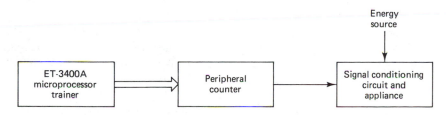

Figure 10–18. Block diagram of microprocessor-controlled appliance timer.

time may be increased or decreased by selecting an appropriate clock frequency for the counter. Since the microprocessor trainer is used, we should be able to display the desired time an appliance is ON.

The counter in the block diagram of Fig. 10–18 is a presettable down counter CD40102B. Initially, the output of the counter is disabled and the appliance is turned OFF. However, as soon as a desired number is loaded into the counter, the output of the counter is enabled and the appliance is turned ON. The appliance remains ON until the content of the counter is decremented to zero. However, the appliance is turned OFF when the content of the counter is zero. The process may be repeated by reloading the counter.

The final block in Fig. 10–18 is a signal conditioning and appliance block. The type of signal conditioning needed is determined by the type and size of appliance (load). In addition, the output of the counter is not capable of driving the load directly; hence a signal conditioning circuit is needed to drive it.

System Specifications

We will design a microprocessor-controlled appliance timer system in which the user is to select the desired on-time for an appliance. The time will be programmed in hex and may be changed if desired before the program is executed. Also, this on-time will be displayed in BCD on the LED display of the ET3400A trainer. The displayed time will be updated at regular intervals until it is zeroed out. More important, the on-time will vary from 2 to 198 s.

System Design

The microprocessor-controlled appliance timer involves system hardware and software design. In fact, the system will be designed according to its specifications, which are listed above. Some of these specifications may easily be achieved in software, whereas others may be accomplished through hardware. To gain a clear understanding of system hardware and software capabilities, let us consider each separately.

System Hardware

The system hardware for a microprocessor-controlled appliance timer will be designed by using the block diagram of Fig. 10–18. As mentioned earlier, our main objective here is to use the appliance timer circuit of Fig. 9–36 with the microprocessor ET3400A trainer. This means that we need to make necessary modifications in the circuit of Fig. 9–36 so that it can be used with the ET3400A trainer. With reference to the block diagram of Fig. 10–18, the only additional circuitry that we need to design is an interface between the ET3400A and a counter. Specifically, we need to design an address decoder and obtain a clock signal and data inputs for the down counter CD40102B. Let us examine each of these circuits in detail.

Address decoder. Since 6800 family microprocessors treat the peripheral devices as memory locations, we will assign the address 8000_H to the down counter

CD40102B. In fact, we will use a similar decoding scheme as that used for the microprocessor-controlled home heating system. Specifically, we will use a 74LS138 three-to-eight-line decoder, and the A_{15}, A_{14}, A_{13}, A_{12}, and $\overline{VMA\phi2}$ signals of the ET3400A to decode the address 8000_H, as shown in Fig. 10–19. When A_{15}, A_{14}, A_{13}, and A_{12} are logic 1, 0, 0, and 0, respectively and $\overline{VMA\phi2}$ is LOW, the output Y_o of the decoder 74LS138 is selected (logic LOW). This output Y_o is ORed with the READ–WRITE ($R/\overline{W}$) signal of the processor because the $R/\overline{W}$ signal is necessary to write (jam) the desired time into the presettable down counter. The output of the OR gate, which is formed by using 7402 two-input quad NOR gates, will be connected to the ASYNCHRONOUS PRESET ENABLE ($\overline{APE}$) of a presettable down counter. When we do a write operation to the counter at hex address 8000, the $\overline{APE}$ input will be momentarily pulled LOW, which will jam the BCD number into the down counter.

Clock input. Recall that in normal operation, the counter is decremented by one count on each positive transition of the clock. In other words, for a given count the frequency of the clock determines the time it takes for the counter to reach the zero count. The ET3400A trainer has two clock signals, 60 Hz and 1 Hz, which are readily available on the front panel.

However, the requirement is to choose an appliance on-time from 2 to 198 s. Therefore, we will use a 1-Hz clock and convert it into a 2-s clock using a 74LS76 *JK* flip-flop, as shown in Fig. 10–20.

In addition, when the counter reaches zero or the output $\overline{CO/ZD}$ is zero, counting should be stopped; that is, the clock input should be disabled. This can be accomplished by ANDing the 1-Hz clock input and the $\overline{CO/ZD}$ output as show in Fig. 10–20.

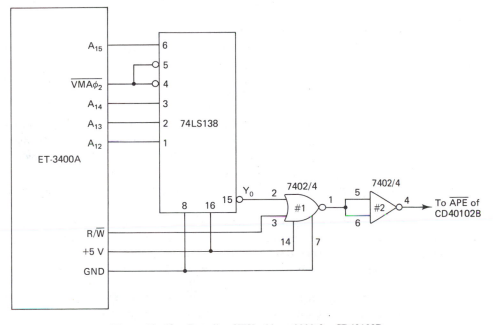

Figure 10–19. Decoding HEX address 8000 for CD40102B.

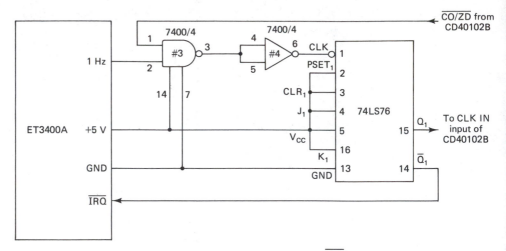

Figure 10–20. CLK IN signal for CD40102B and $\overline{\text{IRQ}}$ for ET3400A.

Recall that the system requirement is to display the timer on-time and also to update it. A display routine may be used to display the on-time. However, to update the displayed time periodically, we shall use the hardware-interrupt capability of the microprocessor. To accomplish this, we will use the $\overline{\text{Q1}}$ output of the *JK* flip-flop and apply it to the interrupt request $(\overline{\text{IRQ}})$ input of the ET3400A trainer (Fig. 10–20). On each HIGH-to-LOW transition of the $\overline{\text{Q1}}$ output, the microprocessor will be interrupted, which will serve the interrupt routine to display the updated time. This process will continue until the count reaches zero. At this time a new number may be jammed into the counter or the previous program may be executed again.

Two 4-bit BCD words. The data inputs, two 4-bit BCD words for the down counter, will be provided by the data bus of the ET3400A. However, note that the down counter in Fig. 9–36 operates on a +12-V supply and all the input and output signals of the ET3400A are either 0 or 5 V. This means that we need a level-translation stage between the counter and the ET3400A.

Level-translation stage. To make ET3400A signals compatible to the down counter signals, we need a level-translation stage. These signals include 8-bit data inputs, a CLK IN, and an $\overline{\text{APE}}$ signal. Specifically, level translation involves changing 5-V signals into 12-V signals. This can be done by the use of noninverting open-collector buffers such as 7407 or 7417 (Fig. 10–21). In this figure 2.2-kΩ collector resistors are used to limit the output currents.

Also, the output $\overline{\text{CO/ZD}}$ of the down counter CD40102B is either 0 or 12 V, based on the content of the counter. However, the output is an input to the NAND gate 3 (Fig. 10–20). Therefore, it must be converted to ≃ 5 V. This is accomplished by the voltage-divider network shown in Fig. 10–21.

Signal conditioning stage. The final block in Fig. 10–18 is to condition the output signal of the down counter to turn an appliance on or off. If the content of the counter is nonzero, its output $\overline{\text{CO/ZD}}$ is HIGH and the appliance should be ON. On

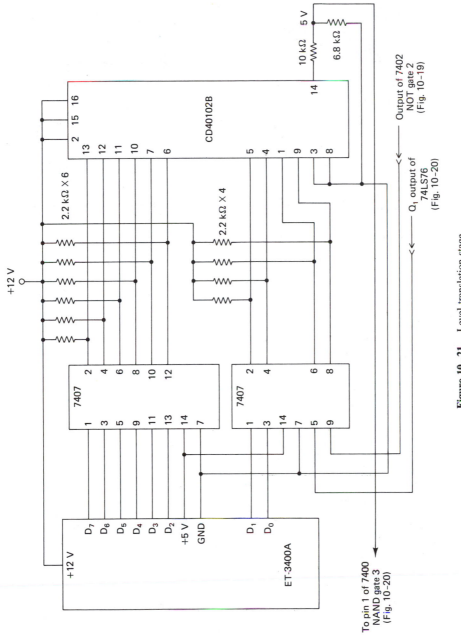

Figure 10–21. Level translation stage.

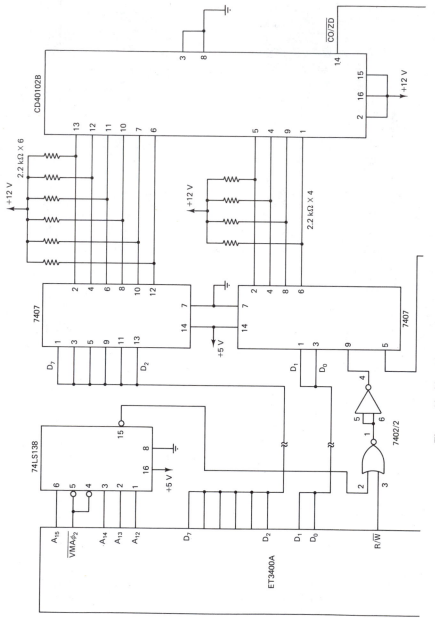

Figure 10–22. Microprocessor-controlled appliance timer.

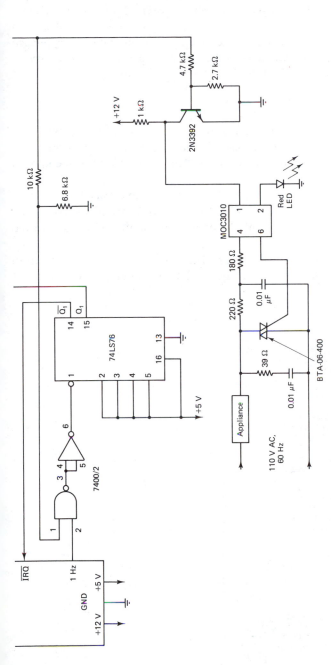

Figure 10–22. (*continued*)

the other hand, if the content of the counter is zero, its output is zero and the appliance should be OFF. Obviously, we will use identical signal conditioning circuitry to that of Fig. 9–11. Namely, we will use a driver stage, an optically isolated triac driver, snubber circuits, and a high-current switching triac. This completes our discussion of system design. The complete schematic diagram of the microprocessor-controlled appliance timer is shown in Fig. 10–22. Next we consider the system software necessary to support the schematic diagram of Fig. 10–22.

System Software

Before we draw the flowchart, we need to consider the system functions that we wish to accomplish in software. We wish to display the appliance on-time and update it continuously until decremented to zero. The flowchart that will enable us to accomplish this task is shown in Fig. 10–23.

There are a variety of ways to write a software program to satisfy the flowchart of Fig. 10–23. One of the ways is shown in Fig. 10–24. The program is fairly simple and self-explanatory. Hence the reader should be able to follow it with the help of the flowchart of Fig. 10–23.

The memory location 0005_{16} contains an appliance on-time which can vary from 01_{16} to 63_{16} or 01_{10} to 99_{10}. However, remember that actual on-time is a function of

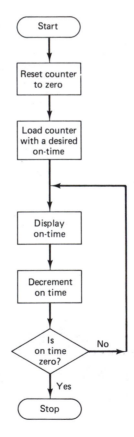

Figure 10–23. Flowchart for a microprocessor-controlled appliance timer program.

Address		Mnemonic	Op-code	Comments
0000		CLRB	5F	
0001		STAB$	F7	Clear down counter.
0002		"DOWN	80	
0003		COUNTER"	00	
0004		LDAB#	C6	
0005		"ON TIME"	0F	Load desired time in
0006		ABA	1B	accumulator B and
0007		ADDA#	8B	
0008		00	00	convert it to BCD.
0009		DAA	19	
000A		STAA$	B7	
000B		"DOWN	80	Load desired time in
000C		COUNTER"	00	the down counter.
000D	REPEAT	JSR	BD	
000E		FE	FE	Display on time.
000F		20*	20	
0010		WAI	3E	Wait for interrupt.
0011		DECB	5A	Decrement on time which is in hex.
0012		CMPB#	C1	Is on time in hex
0013		00	00	less than zero?
0014		BLT	2D	If yes then stop
0015		STOP	17	otherwise continue.
0016		CLRA	4F	
0017		ABA	17	
0018		ADD#	8B	Convert on time
0019		00	00	into BCD.
001A		DAA	19	
001B		JSR	BD	
001C		FC	FC	Relocate display.
001D		BC*	BC	
001E		LDX#	CE	
001F		"TIME	FF	
0020		DELAY"	FF	Time delay for
0021	BACK	DEX	09	spacing time
0022		BNE	26	samples.
0023		BACK	FD	
0024		BRA	20	
0025		REPEAT	E7	Go sample time.
0026	STOP	WAI	3E	Stop processing.
00F7		RTI	3B	Return from interrupt.

Figure 10–24. Program for a microprocessor-controlled appliance timer. (Monitor programs*, courtesy of Heath Company; St. Joseph, Mich.)

clock frequency as well the number stored in location 0005_{16}. In the circuit of Fig. 10–22, the clock frequency is 0.5 Hz; hence the on-time can be varied from 2 to 198 s, corresponding to 01_{16} to 63_{16}, respectively. Obviously, longer on-times may be obtained with lower clock frequencies.

10–5 MICROPROCESSOR-BASED LIGHT-INTENSITY CONTROLLER

In this section we examine how we can use a microprocessor trainer to control the intensity of light for a known duration. As mentioned earlier, often there is a need to control the intensity of a light source at a given lux value for a certain duration, especially in film-exposing applications. This task should be much easier with a microprocessor, simply because of the stored-program provision. Recall that we studied the digital light-intensity controller in Section 9–6. However, our main objective here is to use the digital light-intensity controller of Fig. 9–41 with an ET3400A microprocessor trainer, and that, too, with the fewest modifications. Let us begin with the block diagram of a microprocessor-based light-intensity controller.

Block Diagram

The block diagram of a microprocessor-based light-intensity controller is shown in Fig. 10–25. It consists of a microprocessor trainer, a down counter, a signal conditioning circuit, and a load.

System Description

In the block diagram of Fig. 10–25, the function of the microprocessor trainer is to provide data inputs and a clock for the down counter. It also provides address lines and control signals to decode the address assigned to the down counter. More important, we can use the monitor routines of the trainer to display the intensity of a light source. As in Section 10–4, let us assign address 8000_H to the down counter. Obviously, a decoder circuit must be used to decode this address. Naturally, the choice is the same circuit as that used with the microprocessor-controlled appliance timer.

Normally, the light source to be controlled is off. However, based on a desired light intensity, a corresponding number is jammed into the down counter. When the down counter is decremented to zero, its output goes LOW, which turns the light source ON. Recall that the intensity of light varies depending on the time the light source is ON during the 60-Hz ac cycle. In fact, the longer the light source is ON, the brighter the intensity of light, or the shorter the light source is ON, the dimmer the light. In other words, ideally, hex number FF should result in a dim light and 00_H in a bright light. In addition, a certain light intensity may be maintained as long as needed. This is accomplished by jamming the same hex number into the down counter.

The function of the down counter block in Fig. 10–25 is to process the hex number that is loaded into it. The size of the number and the clock input determine

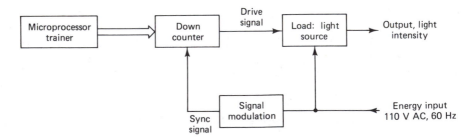

Figure 10–25. Block diagram of a microprocessor-based light intensity controller.

the time it takes the counter to decrement its content to zero. The output of the counter is processed such that when it is LOW (logic 0), it turns the load ON, and when HIGH (logic 1), it turns the load OFF.

The block labeled "load" represents the light source, which is controlled by the output of the counter. The load is operated off a 110-V, 60-Hz energy input.

The final block in Fig. 10–25 is a signal modulation. The function of this block is to provide a signal that is in synchronization with the 60 Hz so that a desired hex number is loaded into the counter at fixed but regular intervals. To maintain the intensity of light at a certain level, it is essential to load a specific number into the counter at regular intervals.

System Specifications

We wish to design a microprocessor-based light-intensity controller in which the user is to select a desired light intensity and its duration. The desired intensity is programmed in hex and ideally may be changed through 256 different levels.

However, to obtain 256 different intensity levels, we need to use CD40103B, an 8-bit presettable down counter instead of CD40102B, a two 4-bit BCD counter. Nevertheless, the pin diagram for both counters is the same. In addition, the intensity will be displayed as DIM, MEDIUM or BRIGHT on the LED display of the ET3400A trainer.

System Design

As noted in previous sections, a microprocessor-controlled system involves both hardware and software design. We first consider the hardware design.

System Hardware

To design a microprocessor-based light-intensity controller, we will use the block diagram of Fig. 10–25. First we need to design an interface circuitry between ET3400A and a down counter, that is, an address decoder and a clock signal for the down counter. Let us begin with the address decoder.

Address decoder. If we assign the address 8000_H to the down counter and chose to use a 74LS138 three-to-eight-line decoder and 74LS373 latches, the decoding circuit should be similar to that shown in Fig. 10–13, except that the Q outputs of the 74LS373 will be connected to a down counter and not to a DAC. For more information on 74LS138 and 74LS373, the reader may refer to Sections 10–2 and 10–3, respectively. The decoder diagram, including 74LS373 latches, is shown in Fig. 10–26. When a write operation is performed to address 8000_H, the data on the bus will be latched into the latches and will appear at the Q outputs.

Clock input. The clock needed for CD40103B can be derived easily from the ϕ_2 signal of the 6808. Recall that in Fig. 9–41, since a 120-Hz signal was applied to

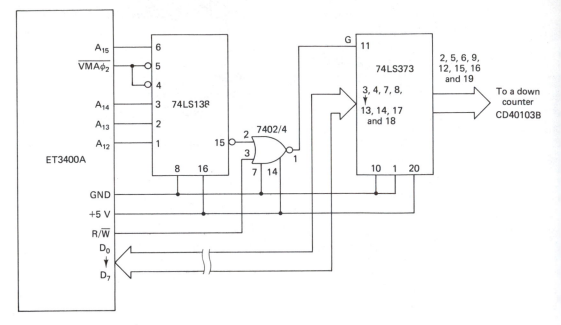

Figure 10-26. Address decoder for a microprocessor-based light-intensity controller.

the $\overline{\text{SPE}}$ input, the counter was loaded every 8.3 ms. Therefore, the frequency of the clock needed to decrement a maximum count of 255_{10} to zero in ≤ 8.3 ms is

$$\text{clock frequency} = \frac{255}{8.3 \text{ ms}} \qquad (10-1)$$

$$= 30.6 \text{ kHz}$$

However, the frequency of ϕ_2 is 1 MHz; therefore, we need to use a divide-by circuit to obtain a clock of approximately 31 kHz from ϕ_2. Remember that a decade counter such as 74LS90 can be used as a divide-by-2, by-5, or by-10 network.

The pin diagram and the bi-quinary and reset/count function tables for 74LS90 are shown in Fig. 10-27. If we configure 74LS90 as a divided-by-5 counter and connect such two counters in series, we can obtain a clock frequency of 40 kHz from ϕ_2 since 1 MHz/25 = 40 kHz (Fig. 10-28). Thus with a 40-kHz clock, a count of 255_{10} will take only 6.38 ms to decrement to zero. Obviously, a clock frequency of <30.6 kHz will not be satisfactory because with a maximum count of 255_{10}, the counter will not decrement to zero before 8.3 ms.

So far we have designed the address decoder and the clock circuits for the light-intensity controller so that the circuit of Fig. 9-41 can be used with an ET3400 trainer. Specifically, in Fig. 9-41, we will replace the eight-position bit switch and the 74C14 clock source by the address decoder of Fig. 10-26 and the clock of Fig. 10-28, respectively. In addition, there are two more circuits that we need to reexamine: the SYNCHRO-NOUS PRESET ENABLE ($\overline{\text{SPE}}$) signal circuit and the output ($\overline{\text{CO/ZD}}$) signal conditioning circuit. These circuits need minor modifications namely in the component values because of the change in supply voltage from +12 V to +5 V. The new component values are

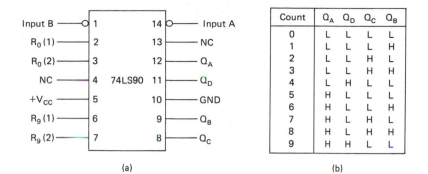

Count	Q_A	Q_D	Q_C	Q_B
0	L	L	L	L
1	L	L	L	H
2	L	L	H	L
3	L	L	H	H
4	L	H	L	L
5	H	L	L	L
6	H	L	L	H
7	H	L	H	L
8	H	L	H	H
9	H	H	L	L

(a)   (b)

Reset Inputs				Output			
$R_0(1)$	$R_0(2)$	$R_9(1)$	$R_9(2)$	Q_D	Q_C	Q_B	Q_A
H	H	L	X	L	L	L	L
H	H	X	L	L	L	L	L
X	X	H	H	H	L	L	H
X	L	X	L	Count			
L	X	L	X	Count			
L	X	X	L	Count			
X	L	L	X	Count			

(c)

Figure 10–27. 74LS90: (a) pin diagram; (b) bi-quinary function table (output Q_D is connected to input A); (c) reset/count function table. (Courtesy of Texas Instruments, Inc.)

selected such that the above two circuits perform satisfactorily. For the design information on the sync circuit and the clock reader may refer to Section 9–6. The complete schematic diagram of a microprocessor-based light-intensity controller is shown in Fig. 10–29. This completes our discussion on the hardware design; let us study the system software next.

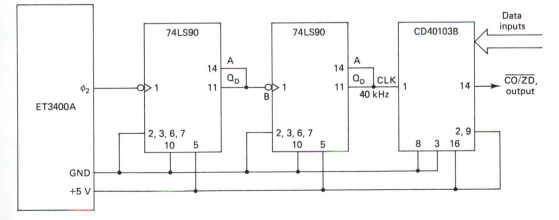

Figure 10–28. 74LS90 decade counters used to obtain 40-kHz clock from ϕ_2.

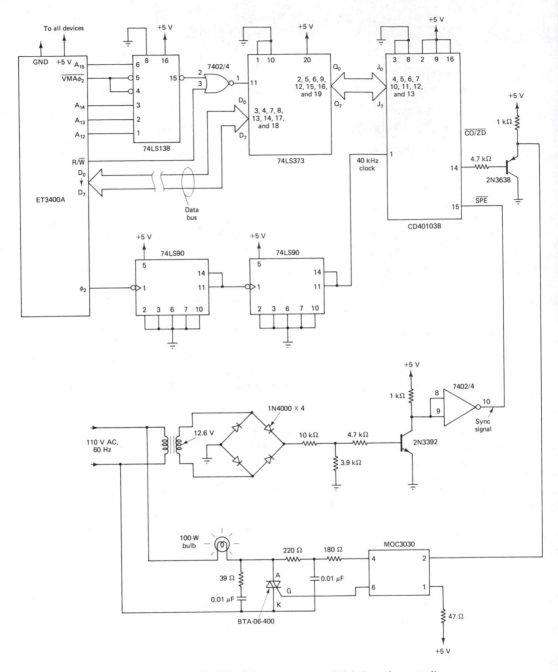

Figure 10–29. Microprocessor-based light-intensity controller.

System Software

Besides loading the desired intensity count into the counter, we also need to display it. In fact, we wish to display intensity as DIM, MEDIUM, or BRIGHT, based on the number that is loaded into the counter. Namely, if the count is $< 55_H$, the intensity will be displayed as BRIGHT, if between 55_H, and $A9_H$, it will be displayed as MEDIUM, and if $\geq AA_H$, it will be displayed as DIM.

Let us begin with the flowchart. The flowchart that will enable us to accomplish the task described above is shown in Fig. 10–30. The program may be written simply by following the flowchart of Fig. 10–30. One of the possible programs is shown in Fig. 10–31. Note that monitor subroutine OUTSTO is used. It outputs messages by displaying up to six characters, one word at a time. One of the requirement of the routine, however, is that DECIMAL POINT (DP) must be lit to indicate the end of the string to exit OUTSTO. DP is placed in the seventh display position to fulfill the requirement without actually being displayed. In addition, we are using hex code 76 for M, which looks like n.

In order to display light intensity as DIM, MEDIUM, or BRIGHT we have divided the $00_H \rightarrow FF_H$ range into three groups. $00_H \rightarrow 54_H$, $55_H \rightarrow A9_H$, and $AA_H \rightarrow FF_H$. These subgroups represent BRIGHT, MEDIUM, and DIM intensity levels, respectively.

The program in Fig. 10–31 is straightforward and easy to understand. First a hex number is chosen based on the desired intensity and loaded into the counter. The

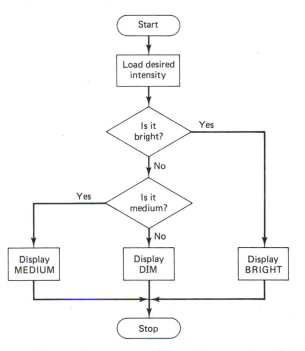

Figure 10–30. Flowchart for a program used in a microprocessor-based light-intensity controller.

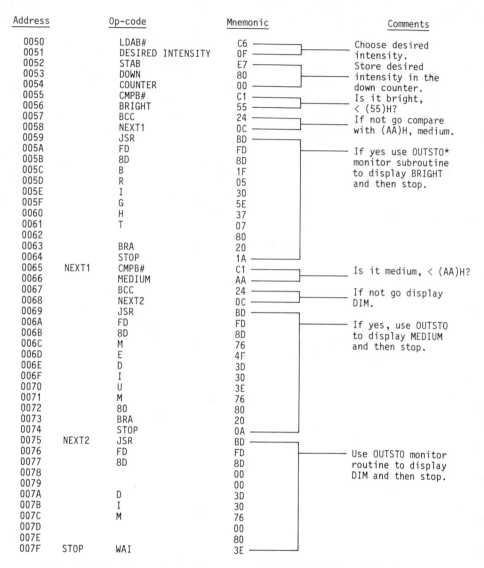

Address	Op-code		Mnemonic	Comments
0050		LDAB#	C6	Choose desired intensity.
0051		DESIRED INTENSITY	OF	
0052		STAB	E7	Store desired intensity in the down counter.
0053		DOWN	80	
0054		COUNTER	00	
0055		CMPB#	C1	Is it bright, < (55)H?
0056		BRIGHT	55	
0057		BCC	24	If not go compare with (AA)H, medium.
0058		NEXT1	0C	
0059		JSR	BD	If yes use OUTSTO* monitor subroutine to display BRIGHT and then stop.
005A		FD	FD	
005B		8D	8D	
005C		B	1F	
005D		R	05	
005E		I	30	
005F		G	5E	
0060		H	37	
0061		T	07	
0062			80	
0063		BRA	20	
0064		STOP	1A	
0065	NEXT1	CMPB#	C1	Is it medium, < (AA)H?
0066		MEDIUM	AA	
0067		BCC	24	If not go display DIM.
0068		NEXT2	0C	
0069		JSR	BD	If yes, use OUTSTO to display MEDIUM and then stop.
006A		FD	FD	
006B		8D	8D	
006C		M	76	
006D		E	4F	
006E		D	3D	
006F		I	30	
0070		U	3E	
0071		M	76	
0072		80	80	
0073		BRA	20	
0074		STOP	0A	
0075	NEXT2	JSR	BD	Use OUTSTO monitor routine to display DIM and then stop.
0076		FD	FD	
0077		8D	8D	
0078			00	
0079			00	
007A		D	3D	
007B		I	30	
007C		M	76	
007D			00	
007E			80	
007F	STOP	WAI	3E	

Figure 10–31. Program for the microprocessor-based light-intensity controller. (OUTSTO monitor routine courtesy of Heath Company; St. Joseph, Mich.)

hex number is then compared with the predetermined ranges and displayed accordingly using subroutine OUTSTO.

The program in Fig. 10–31 may be rewritten in a number of different ways to include some extra functions. For example, we may increase or decrease the intensity of light in a certain order for a specific time. The reader is encouraged to try a few different variations and write programs to accomplish these.

SUMMARY

1. A microprocessor-controlled system involves both hardware and software design. Obviously, both hardware and software designs of a system are functions of the electrical specifications and the type of microprocessor used. In addition, an interface circuitry between the processor and the system is an important part of a microprocessor-based system.

2. A microprocessor-controlled home heating and cooling system is more efficient and economical than a conventional home heating system because it can be programmed to come ON and turn OFF at predetermined times. Moreover, it is more accurate and reliable since it uses ICs and digital techniques to control the temperature. Above all, the current temperature is displayed and updated continuously.

3. In a microprocessor-based dc motor speed control system the user can select a desired motor speed and the direction of rotation as needed. In addition, the chosen speed may be maintained at a fixed value for a predetermined time. Also, for convenience the chosen speed and the direction are displayed continuously.

4. In a microprocessor-based appliance timer a given appliance may be turned on for a predetermined time. A different on-time may be programmed for different appliances or a fixed on-time may be used repeatedly for a given appliance. Also, the time selected is displayed and updated continuously until the appliance is turned off.

5. A microprocessor-based light-intensity controller provides the user with the flexibility of selecting various levels of intensity. A specific intensity may be programmed and can also be maintained as long as desired based on the application needs. The intensity programmed is also displayed as either DIM, MEDIUM, or BRIGHT.

QUESTIONS

10-1. List three important considerations in selecting a microprocessor for a microprocessor-based control system.

10-2. What are the important advantages of a microprocessor-based temperature control system over a conventional temperature control system?

10-3. In Fig. 10-8, explain the use of 74LS75 4-bit bistable latches.

10-4. Explain the use of 74LS373 latches in the circuit of Fig. 10-13.

10-5. Briefly explain the operation of the pulse-width-modulation (PWM) circuit used in Fig. 10-15 to drive the motor. List two advantages of the PWM technique.

10-6. What is the function of the 74LS76 and 7407s in the circuits of Figs. 10-20 and 10-21, respectively?

10-7. What determines the maximum on-time of the appliance in the circuit of Fig. 10-22?

10-8. Compare and contrast the appliance timer of Fig. 10-22 with a commercial timer. Use a specific example to support your explanation.

10-9. List and briefly explain at least one practical application of a light-intensity controller.

10-10. In the circuit of Fig. 10-29, why is it necessary to use a sync signal?

10–11. Can we use a clock frequency of 20 kHz for the CD40103B down counter in the circuit of Fig. 10–29? Why or why not?

PROBLEMS

10–1. In the microprocessor-controlled home heating and cooling system of Fig. 10–8, octal *D*-type latches such as 74LS373 are to be used to interface the data bus to the analog-to-digital converter. Modify the system as necessary to implement this change, and draw the complete circuit diagram.

10–2. Write a program for the microprocessor-controlled home heating and cooling system of Fig. 10–8 so that the temperature can be displayed in degrees Celsius.

10–3. In the dc motor speed control system of Fig. 10–15, it is decided to use the pulse-width-modulated signal to drive both Q_1–Q_4 and Q_2–Q_3 transistor pairs. Modify the circuit as required, and redraw it.

10–4. Write a program for the dc motor speed control system of Fig. 10–15 so that the motor alternately moves clockwise and counterclockwise for 30 s at a speed of 100 rpm. If necessary, you may also modify the hardware in the circuit of Fig. 10–15 to accommodate the change.

10–5. The microprocessor-controlled appliance timer in Fig. 10–22 is converted over to a +5-V supply. Make necessary modifications in the circuit and then redraw the entire circuit.

10–6. Make necessary modifications in the circuit of Fig. 10–22 so that the appliance on-time may be varied from 25 s to over 40 min. Draw the complete circuit.

10–7. If the circuit of Fig. 10–22 is connected and the power is initially applied to it, the chances are that the appliance will be on. This is because the content of the down counter may not be zero. Modify the timer circuit such that the appliance does not turn on prematurely.

10–8. Write a program for the circuit of Fig. 10–22 so that the programmed appliance on-time and the updated on-time are displayed essentially in that order until the appliance is turned off.

10–9. Modify the microprocessor-based light-intensity controller of Fig. 10–29 so that the user may select the level of intensity and the duration of intensity as well. You may make changes in hardware, in software, or in both to satisfy this requirement.

LAB EXPERIMENT 10–1

MICROPROCESSOR-BASED TEMPERATURE CONTROL SYSTEM

Objective. To breadboard a microprocessor-based temperature control system and write a program to verify its operation. After completing this experiment, you should be able to:

1. Explain the operation of the system.
2. Design an address decoder using a 74LS138 three-to-eight-line decoder.

3. Design an analog-to-digital converter for a given range of analog voltages.
4. Design a circuit to interface high-voltage analog and low-voltage digital signals.
5. Write a program for a given hardware configuration.

Equipment

ET3400 or 3400A microprocessor trainer
Digital multimeter
Oscilloscope

Materials

74LS138 three-to-eight-line decoder
7404 hex inverters
7427 triple three-input NOR gates
ADC0804 ADC or equivalent
74LS75 4-bit bistable latches
LM324 op amp
LM334 adjustable current source
MOC3010 optically isolated triac driver
5-W light bulb
TRW-405A100-1 dc motor or equivalent
TIP29 *npn* power transistor
Two 100-kΩ pots
10-kΩ pot
Three 10-kΩ resistors
Three 220-Ω resistors
39-Ω, 5-W resistor
5-μF, 25-V capacitor

Procedure

1. Construct the circuit of Fig. LE10–1.1 using the following steps.
   (a) Connect the temperature transducer, the voltage follower, and the differential amplifier circuits. Make sure that the 100-kΩ potentiometers are adjusted to 72 kΩ before connecting them with the differential amplifier. Also, adjust the temperature reference 10-kΩ potentiometer to 2.73 V. Measure the output voltage of the differential amplifier and record it in Table LE10–1.
   (b) Connect the dc motor circuit as shown in Fig. LE10–1.1 except the 220-Ω base resistor. Now connect one end of the 220-Ω resistor to the base of TIP transistor. Check the operation of the motor by alternately connecting the other end of the resistor to ground and +5-V supply. Make appropriate corrections until the circuit works.

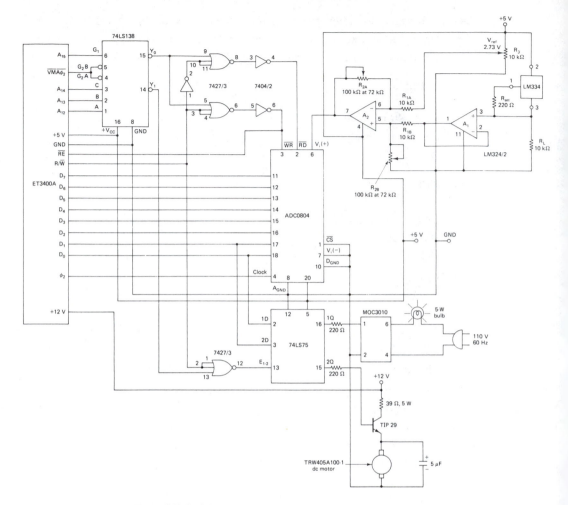

Figure LE10–1.1 Microprocessor-controlled home heating and cooling system.

(c) Connect the light-bulb circuit, and check its operation by alternately connecting the 220-Ω resistor to ground and to the +5-V supply. Also, make necessary changes until the circuit works satsifactorily.

(d) Connect the remaining circuit of Fig. LE10–1.1 Check to make sure that all parts of the circuit work properly. If not, isolate the problem and correct it.

2. Write a program so that the temperature is maintained between 68° and 72°F. Also, make sure that the current temperature is displayed on the trainer LEDs in degrees Fahrenheit. Remember that the address of the ADC is 8000_H and that of the latches is 9000_H.

3. Input the program in step 2 into the trainer and execute it to ensure that it satisfies the requirements in step 2. Make sure that the displayed temperature is approximately equal to the actual temperature in the room.

4. Modify the program in step 2 so that the temperature can be displayed in degrees Celsius instead of degrees Fahrenheit.

5. Repeat step 3.

6. Complete Table LE10−1.

7. Submit both programs, including flowcharts, with your lab report. Also, include any hardware modifications that may have been made in the circuit of Fig. LE10−1.

Test data

TABLE LE10−1

Actual temperature	Output voltage of the differential amplifier (V)	Temperature displayed	Difference in temperature
_____ °F			
_____ °C			

Evaluation. The responses to the following questions and/or conclusions must be supported whenever possible by the applicable theory.

1. In Fig. LE10−1.1, why is it necessary to adjust the reference voltage V_{ref} to 2.73 V?

2. What is the function of MOC3010 in Fig. LE10−1.1?

3. Why is it necessary to use 74LS75? Refer to Fig. LE10−1.1.

4. Is there any difference between the actual temperature and the temperature displayed? Refer to Table LE10−1. Give reasons if temperatures differ by more than 5°.

5. Comment on the overall performance of the microprocessor-based temperature control system. Make specific comments about the hardware and software portions of the system.

LAB EXPERIMENT 10−2

MICROPROCESSOR-BASED LIGHT-INTENSITY CONTROLLER

Objective. To breadboard a microprocessor-based light-intensity controller and write a program to verify its operation. At the end of this experiment, you should be able to:

1. Design an address decoder using a 74LS138 decoder and 74LS373 latches.

2. Design a divide-by-N network using 74LS90 counters.

3. Design an interface circuit to isolate analog and digital voltages.
4. Design a circuit to obtain a pulse waveform using a line voltage of 110 V at 60 Hz.
5. Write a program for the given hardware.
6. Interface given hardware with an ET3400A trainer.

Equipment

ET3400 or 3400A microprocessor trainer
Digital multimeter
Oscilloscope

Materials

74LS138 three-to-eight-line decoder
74LS373 octal *D*-type latches
CD40103B down counter
7402 NOR gates
Two 74LS90 decade counters
MOC3030 optically isolated triac driver
2N3638 *pnp* transistor
2N3392 *npn* transistor
12.6-V AC $\leq$ 500-mA transformer
BTA-06–400 or equivalent triac
100-W bulb
Four 1N4000 diodes
10-kΩ resistor
Two 4.7-kΩ resistors
3.9-kΩ resistor
Two 1-kΩ resistors
220-Ω resistor
180-Ω resistor
47-Ω resistor
39-Ω resistor
Two -0.01-μF, 250-V capacitors

Procedure

1. Construct the circuit of Fig. LE10–2.1 using the following steps.
   (a) Connect the sync signal circuit for a down counter CD40103B. However, do not connect the output of the inverter formed by 7402 to the $\overline{\text{SPE}}$, pin 15 of the CD4010B.

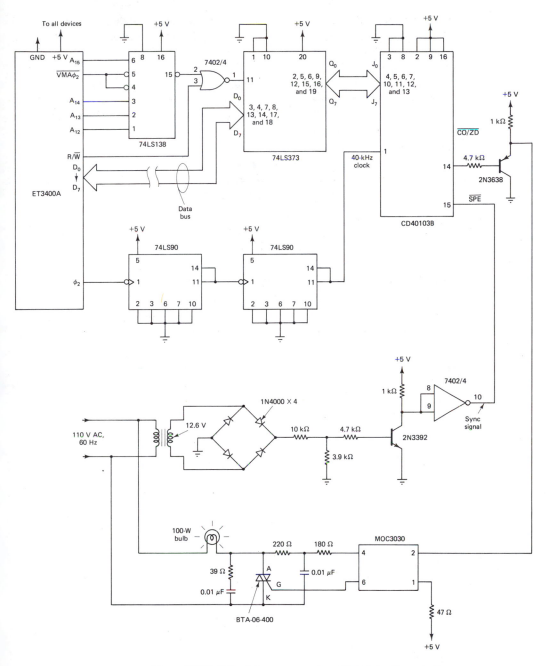

Figure LE10–2.1. Microprocessor-based light-intensity controller.

(b) Using an oscilloscope, observe the voltage waveform across the secondary of a 12.6-V transformer, at the collector of a 2N3392, and at pin 10 of a 7402 inverter. Draw these waveforms in Fig. LE10–2.2a.

(c) Connect the load circuit (a light bulb) of the light-intensity controller except that the 4.7-kΩ resistor is not connected to the $\overline{CO/ZD}$ output of the CD40103B. Now connect the 4.7-kΩ resistor alternately to ground and to +5 V to check the operation of the light bulb. Make necessary changes in the load circuit until it works properly.

(c) Connect the divide-by-25 circuit as shown in Fig. LE10–2.1. Using an oscilloscope, measure ϕ_2 and the output waveform at pin 11 of the second decade counter. Draw the observed waveforms in Fig. LE10–2.2b.

(e) Connect the remaining circuit in Fig. LE10–2.1. Make sure that all parts of the circuit work satisfactorily.

2. Write a program so that the light intensity can be varied from dim to bright. Also, display the intensity as DIM, MEDIUM, or BRIGHT. Remember that 8000_H is the address of the down counter.

3. Input the program in step 2 into the trainer and execute it. Make sure that the program satisfies the requirements in step 2.

4. Using the program in step 2, determine the binary number at which the light-bulb is off. Also, determine the number at which the bulb is maximum bright. Record these values in Table LE10–2.

5. Write another program so that the intensity of a light bulb can be increased from dim to bright and then from bright to dim at a certain rate. Input this program into the trainer and verify its validity.

6. Submit the programs, including flowcharts, with your report.

Test data

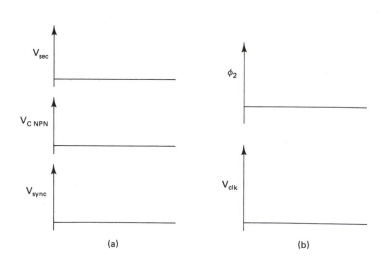

(a) (b)

Figure LE10–2.2

TABLE LE10-2

Light-bulb status	Input binary number
Bulb off	
Bulb maximum bright	

Evaluation. The responses to the following questions and/or conclusions must be supported whenever possible by the applicable theory.

1. Why is it necessary to use a divide-by-25 network? Refer to Fig. LE10–2.1.
2. What is the function of the sync circuit in Fig. LE10–2.1?
3. In Fig. LE10–2.1 can we use $\overline{APE}$ input instead of $\overline{SPE}$ to load a desired intensity number into the counter? Why or why not?
4. How many different intensity levels are there for the light bulb in Fig. LE10–2.1? Explain your answer briefly.
5. An input binary number is loaded into the down counter approximately every 8.3 ms (120 Hz) by the $\overline{SPE}$ input. Why? Explain.
6. What is the significance of the 40-kHz clock used for the down counter? What happens if we use a 20- or 60-kHz clock? Explain.

STEP RESPONSE OF A FIRST-ORDER SYSTEM

Any first-order closed-loop system, represented by the block diagram in Fig. A–1a, can also be represented by the following first-order differential equation:

$$\tau_s \dot{y} + y(t) = K_s x(t) \tag{A–1}$$

where $K_s = G_o/(1 + G_o K_f)$ = system constant
$\tau_s = \tau/(1 + G_o K_f)$ = system time constant
G_o = open-loop gain
$G_o K_f$ = loop gain

The transient response analysis must include a possibility of initial conditions that exist at the time of application of step input; accordingly, as shown in Fig. A–1b, the step input of magnitude E is applied at $t = t_1$ when the output is $y(0)$, which is due to the input $x(0)$.

The input may be described by

$$x(t) = x(0) + Eu(t - t_1)$$

and its Laplace transform by

$$\mathcal{L}[x(t)] = X(s) = \frac{x(0) + Ee^{-t_1 s}}{s}$$

Taking Laplace transform of Eq. (A–1) and substituting the expression for $X(s)$ above, the following result, which includes the partial-fraction expansion, is obtained after some algebraic manipulations:

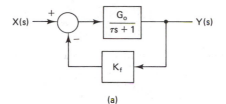

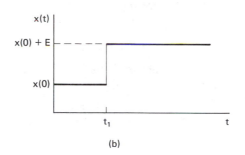

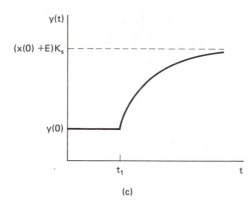

Figure A–1. First-order system transient response: (a) system block diagram; (b) step input; (c) response.

$$Y(s) = \frac{K_s x(0)/\tau_s}{s(s + 1/\tau_s)} + \frac{y(0)}{s + 1/\tau_s} + \frac{(K_s E/\tau_s)e^{-st_1}}{s(s + 1/\tau_s)} \qquad \text{(A–2)}$$

The first term in Eq. (A–2) is expanded by partial fractions as

$$\frac{K_s x(0)/\tau_s}{s(s + 1/\tau_s)} = \frac{A}{s} + \frac{B}{s + 1/\tau_s} \qquad \text{(A–3)}$$

By the Heaviside method the constants A and B are evaluated to be

$$A = K_s x(0) \qquad \text{and} \qquad B = -K_s x(0)$$

Using these constants in Eq. (A–3) and taking the inverse Laplace transform, the following result is obtained:

$$\mathscr{L}^{-1}\left[\frac{K_s x(0)/\tau_s}{s(s + 1/\tau_s)}\right] = K_s x(0) - K_s x(0)e^{-t/\tau_s} \tag{A-4}$$

The inverse Laplace transform of the second term in Eq. (A–2) is readily obtained as

$$\mathscr{L}^{-1}\left[\frac{y(0)}{s + 1/\tau_s}\right] = y(0)e^{-t/\tau_s} \tag{A-5}$$

To obtain the inverse Laplace transform of the last term in Eq. (A–2), the following theorem must be used:

$$\mathscr{L}^{-1}[F(s)e^{-st_1}] = f(t - t_1)u(t - t_1) \tag{A-6}$$

The expression for $F(s)$ is obtained from the last term in Eq. (A–2) and expanded by partial fractions to yield the following:

$$F(s) = \frac{K_s E/\tau_s}{s(s + 1/\tau_s)} = \frac{K_s E}{s} - \frac{K_s E}{s + 1/\tau_s}$$

and

$$\mathscr{L}^{-1}[F(s)] = f(t) = K_s E - K_s E e^{-t/\tau_s}$$

When the $f(t)$ above is shifted by t_1, the result is

$$f(t - t_1) = K_s E - K_s E e^{-(t-t_1)/\tau_s}$$

Using the expression for $f(t - t_1)$ in Eq. (A–6), the inverse Laplace transform of the last term in Eq. (A–2) follows:

$$\mathscr{L}^{-1}\left[\frac{(K_s E/s)e^{-st_1}}{s(s + 1/\tau_s)}\right] = (K_s E - K_s E e^{-(t-t_1)/\tau_s})[u(t - t_1)] \tag{A-7}$$

Using Eqs. (A–4), (A–5), and (A–7) in the inverse Laplace transform of Eq. (A–2), the following restult is obtained:

$$\mathscr{L}^{-1}[Y(s)] = y(t) = K_s x(0)(1 - e^{-t/\tau_s}) + y(0)e^{-t/\tau_s} + K_s E(1 - e^{-(t-t_1)/\tau_s})[u(t - t_1)]$$

The equation above represents the response of the first-order closed-loop system to the step input, which occurs at $t = t_1$. Without loss of generality we let $t_1 = 0$, as its value was arbitrary from the start. When this is done, the response above equation reduces to the following simpler form:

$$y(t) = [x(0) + E]K_s - \{[(x(0) + E]K_s - y(0)\}e^{-t/\tau_s} \tag{A-8}$$

Letting $y_f = [x(0) + E]K_s$, the steady state or the final value of y, and $y_i = y(0)$, the initial value of y, Eq. (A–8) transforms to the recognizable standard form

$$y(t) = y_f - (y_f - y_i)e^{-t/\tau_s} \tag{A-9}$$

The graphical form of the response associated with Eq. (A–9) is shown in Fig. A–1c.

HYDRAULIC SYSTEM

appendix B

The analysis presented here is based on contribution to hydrodynamics by Bernoulli and Torricelli. Their efforts made possible the interrelationship of pressure, velocity, and elevation which is associated with a flowing liquid. Our objective here is to determine the relationship between the velocity of liquid exiting the tank and the liquid head, and the conditions under which this relationship is linear. The result will play an important role in the liquid-level control system described in Chapter 8.

The analysis begins with the tank shown in Fig. B-1. Its cross-sectional area is A_1 and the cross-sectional area of the opening through which the liquid flows from the

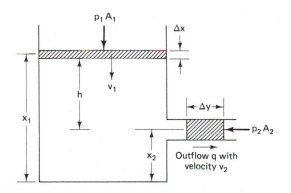

Figure B-1. Liquid flow from a tank under the action of the head.

tank is A_2. The pressure above the liquid in the tank is p_1 and the pressure that opposes the flow of liquid through the opening is p_2. In time Δt a volume of liquid in the tank (shown by the shaded area) advances a distance Δx with an average velocity $v_1 = \Delta x/\Delta t$. The work done on the liquid system by the pressure p_1 is $p_1 A_1 \Delta x$. Simultaneously, a volume of liquid advances a distance Δy through the opening. The work done by the liquid system against the pressure p_2 is $p_2 A_2 \Delta y$. Assuming the liquid to be incompressible, the two displacement volumes must be equal. If the mass of each displacement volume is m and if the density of liquid is ρ, it follows that the net work done on the liquid system may be expressed as

$$W_n = p_1 A_1 \Delta x - p_2 A_2 \Delta y = (p_1 - p_2)\frac{m}{\rho} \tag{B-1}$$

As the total energy of any system must remain constant, the net work done on the liquid system must represent the system's kinetic energy change—its potential energy change plus any losses. This may be expressed as

$$W_n = \Delta(KE) + \Delta(PE) + F_v \tag{B-2}$$

where

$$\Delta(KE) = \tfrac{1}{2}m(v_1^2 - v_2^2)$$

$$\Delta(PE) = mg(x_1 - x_2)$$

The term F_v in Eq. (B-2) is associated with viscous friction losses. It depends on the type of liquid and its velocity. Viscosity may be regarded as friction that is internal to the liquid. Viscous liquid may be modeled as a stack of thin sheets or disks. An attempt to slide one sheet past another results in opposing (viscous) frictional force; hence an applied force is needed to overcome this friction. Most liquids exhibit viscosity; however, for a nonviscous liquid the motion could be characterized by layers that move uniformly with same velocity. This type of motion through the tank outlet pipe is shown in Fig. B-2a.

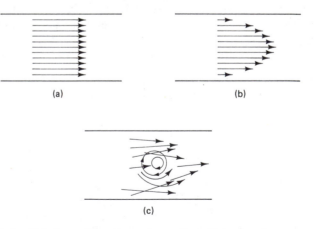

(a)   (b)

(c)

Figure B-2. Velocity profiles: (a) nonviscous flow; (b) laminar flow; (c) turbulent flow.

The flow of a viscous liquid can be either laminar or turbulent. *Laminar flow* is characterized by thin sheets of liquid that move by sliding over each other. The sheet closest to the stationary wall (such as the wall of a pipe through which the liquid flows) has zero velocity, and the sheet that moves midway between walls has maximum velocity. This type of parabolic velocity profile is shown in Fig. B–2b. *Turbulent flow* exhibits irregular and complicated motion. This type of flow is illustrated in Fig. B–2c.

Associated with fluid dynamics are several important universal constants. They are dimensionless and are formed as ratios of forces that act to produce a specified fluid motion. *Weber's number* is associated with surface tension when it is the dominant force; the *Cauchy* or *Mach number* is related to elastic forces; and the *Froude number* applies to cases where gravitational forces are dominant. The *Reynolds number*, defined as

$$N_r = \frac{v\rho D}{\eta}$$

where V is the velocity, ρ the density, D the pipe diameter and η the viscosity, applies to the flow of viscous fluids. On the basis of experiments, the numerical value of the Reynolds number can be used to predict whether the given viscous liquid flow is laminar or turbulent. Values below 2000 generally describe laminar flow, and those above 3000 characterize turbulent flow. In the present analysis, laminar flow is assumed; furthermore, frictional losses in the short section of outlet pipe in Fig. B–1 are negligible; hence F_v is negligible in relation to the remaining terms in Eq. (B–2). Combining Eqs. (B–1) and (B–2) yields

$$(p_1 - p_2)\frac{m}{\rho} = \tfrac{1}{2}m(v_1^2 - v_2^2) + mg(x_1 - x_2) \tag{B–3}$$

In Fig. B–1 the tank is open to the atmosphere; hence $p_1 = p_2 = p_a$. The liquid head is defined as $h = x_1 - x_2$ and the velocity v_1 is assumed to be much less than v_2. Under these conditions Eq. (B–3) reduces to $\tfrac{1}{2}mv_2^2 = mgh$. Solving for the outlet liquid velocity gives us

$$v_2 = (2gh)^{1/2} \qquad \text{ft/s} \tag{B–4}$$

and the flow through the outlet pipe is

$$q = A_2 v_2 = A_2(2gh)^{1/2} \qquad \text{ft}^3/\text{s} \tag{B–5}$$

According to Eq. (B–5), the variation of flow versus the head h is parabolic, as shown in Fig. B–3. Expanding Eq. (B–5) by Taylor's series in the neighborhood of $h = a$, the following results:

$$q(h) = q(a)\left[1 + \frac{1}{2a}(h - a) - \frac{1}{(2a)^2}(h - a)^2 + \frac{1}{(2a)^3}(h - a)^3 - \cdots\right]$$

where $q(a) = A_2(2ga)^{1/2}$. If $(h - a) \ll 1$, the higher-order terms may be neglected and the expansion may be approximated by

$$q(h) = \frac{q(a)}{2a}h + \frac{q(a)}{2} \qquad \text{ft}^3/\text{s} \tag{B–6}$$

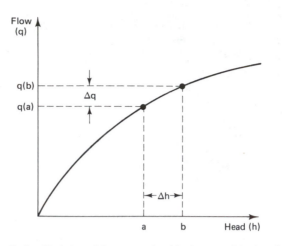

Figure B–3. Variation of flow versus head in the case of laminar flow.

The expression (B–6) shows a linear variation between the flow and the head in the neighborhood $h = a$. If the operating point shifts from $h = a$ to $h = b$, the corresponding increment in flow is expressed as

$$\Delta q = q(h + \Delta h) - q(h) = \frac{q(a)}{2a}(h + \Delta h) + \frac{q(a)}{2} - \left[\frac{q(a)h}{2a} + \frac{1}{2}q(a)\right]$$

hence

$$\Delta q = \frac{q(a)}{2a} \Delta h \qquad\qquad\qquad (B-7)$$

The differential rate of change of flow with respect to h at $h = a$, which is the slope of the line tangent to the curve in Fig. B–3 at $h = a$, may be obtained from Eq. (B–7) in the limit as $\Delta h \to 0$ in the ratio $\Delta q/\Delta h$ or from Eq. (B–6) through differentiation as $2h/q$. The hydraulic resistance R_h is defined as the reciprocal of this slope; hence

$$R_h = \frac{2h}{q} \qquad \text{s/ft}^2 \qquad \text{(dynamic)} \qquad (B-8)$$

This is the resistance of the outlet valve and pipe to the flow of liquid. If the flow is nonlinear as shown in Fig. B–3, the resistance depends on the flow rate being low for low flow rates, and increases with an increase in the flow rate. The static resistance is defined as the slope of the line drawn from the origin to the operating point; hence

$$R_h = \frac{h}{q} \qquad \text{s/ft}^2 \qquad \text{static} \qquad (B-9)$$

In cases where the flow is laminar and the hydraulic friction is negligible, R_h may be related to the cross-sectional area A_2 of the outlet pipe and the head h by substituting Eq. (B–5) in Eq. (B–9) as follows:

$$R_h = A_2^{-1} \frac{h^{1/2}}{2g} \qquad \text{s/ft}^2 \qquad (B-10)$$

DATA SPECS

CARBORUNDUM THERMISTORS

ROD THERMISTORS

Rod thermistors are manufactured by extruding and firing a suitable ceramic mixture. For a given type of thermistor, low-, medium-, or high-sensitivity, a number of mixtures are available: in general, a mixture yielding a higher resistivity product also yields a higher beta or temperature sensitivity.

Standard rod thermistors are shown in the table below. They have a resistance tolerance of ±10% at 25°C, and they are neither coated nor marked. Special rod thermistors may be ordered with intermediate resistance values, other resistance tolerances, other calibration temperatures, coating, marking, special lead preparation, etc. Parts without wire leads are also available.

Resistance Ohms ±10% @ 25°C	Dash Number	MEDIUM TEMPERATURE COEFFICIENT Type "F" Rods Beta ±10%			HIGH TEMPERATURE COEFFICIENT Type "H" Rods Beta ±5%	
		Style 0325F5	Styles 0540F4- 0550F5-	Style 1099F5-	Style 0325H5-	Styles 0550H4- 0550H5-
2.2	-402	—	—	1300	—	—
3.3	-403	—	—	1350	—	—
4.7	-404	—	1300	1400	—	—
6.8	-406	—	1350	1450	—	—
10.0	-410	1300	1400	1500	—	—
15.0	-411	1350	1450	1550	—	—
22.0	-412	1400	1500	1600	—	—
33.0	-413	1450	1550	1600	—	—
47.0	-414	1500	1600	1650	—	—
68.0	-416	1550	1600	1650	—	—
100	-420	1600	1650	1700	—	—
150	-421	1600	1650	1700	—	—
220	-422	1650	1700	1750	—	—
330	-423	1650	1700	1750	—	—
470	-424	1700	1750	1800	—	—
680	-426	1700	1750	1800	—	—
1,000	-430	1750	1800	1850	—	3000
1,500	-431	1750	1800	1850	—	3100
2,200	-432	1800	1850	1900	3000	3150
3,300	-433	1800	1850	1900	3100	3200
4,700	-434	1850	1900	1950	3150	3300
6,800	-436	1850	1900	1950	3200	3400
10,000	-440	1900	1950	2000	3300	3500
15,000	-441	1900	1950	2000	3400	3600
22,000	-442	1950	2000	2050	3500	3700
33,000	-443	1950	2000	2050	3600	3800
47,000	-444	2000	2050	2100	3700	3900
68,000	-446	2000	2050	2100	3800	4000
100,000	-450	2050	2100	2150	3900	4100
150,000	-451	2050	2100	2150	4000	4150
220,000	-452	2100	2150	2200	4100	4200
330,000	-453	2100	2150	2200	4150	4250
470,000	-454	2150	2200	2200	4200	4300
680,000	-456	2150	2200	—	4250	4350
1,000,000	-460	2200	2200	—	4300	4400
1,500,000	-461	2200	—	—	4350	—
2,200,000	-462	2200	—	—	4400	—

Order by the style number followed by the dash number. For example, part number 0540F4-404 is a thermistor of style 0540F4-, of 4.7 ±10% ohms resistance, and with a beta of 1300 ±10% K.

FIGURE 1

FIGURE 2

Style	Figure	Dimensions (inches)			AWG	Dissipation Constant (mw/°C)	Time Constant (seconds)
		A	B	C			
0325F5- 0325H5-	1	0.12	0.30	1.00	24	6.0	10
0540F4- 0550H4-	2 2	0.20 0.20	0.50 0.63	1.25 1.25	20 20	9.5 11.0	22 25
0550F5- 0550H5-	1	0.22	0.63	1.38	20	11.0	25
1099F5-	1	0.38	1.13	1.38	20	24.0	90

The Carborundum Company, Electric Products Division, Post Office Box 339, Niagara Falls, N.Y. 14302 Tel: (716) 278-2521 Telex: 91-555 (CARBO GLO NGF)

CARBORUNDUM

Courtesy of Sohio Engineered Materials Co./Carborundum.

WIRE GAGES — Iso-Elastic Paper/Bakelite Carrier

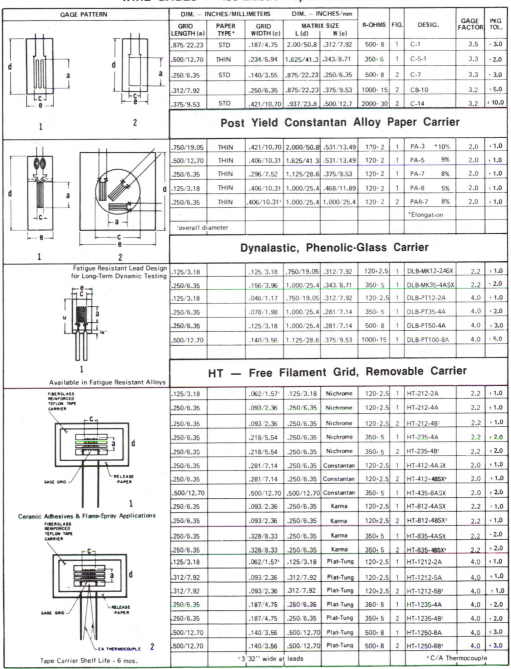

GAGE PATTERN	DIM. — INCHES/MILLIMETERS			DIM. — INCHES/mm		R-OHMS	FIG.	DESIG.	GAGE FACTOR	PKG TOL.
	GRID LENGTH (a)	PAPER TYPE*	GRID WIDTH (c)	MATRIX SIZE L (d)	W (e)					
	.875/22.23	STD	.187/4.75	2.00/50.8	.312/7.92	500·8	1	C-1	3.5	·3.0
	.500/12.70	THIN	.234/5.94	1.625/41.3	.343/8.71	350·6	1	C-5-1	3.3	·2.0
	.250/6.35	STD	.140/3.55	.875/22.23	.250/6.35	500·8	2	C-7	3.3	·3.0
	.312/7.92		.250/6.35	.875/22.23	.375/9.53	1000·15	2	CB-10	3.2	·5.0
	.375/9.53	STD	.421/10.70	.937/23.8	.500/12.7	2000·30	2	C-14	3.2	± 10.0

Figures 1, 2

Post Yield Constantan Alloy Paper Carrier

GRID LENGTH (a)	PAPER TYPE	GRID WIDTH (c)	L (d)	W (e)	R-OHMS	FIG.	DESIG.	GAGE FACTOR	PKG TOL.
.750/19.05	THIN	.421/10.70	2.000/50.8	.531/13.49	120·2	1	PA-3 *10%	2.0	·1.0
.500/12.70	THIN	.406/10.31	1.625/41.3	.531/13.49	120·2	1	PA-5 9%	2.0	±1.0
.250/6.35	THIN	.296/7.52	1.125/28.6	.375/9.53	120·2	1	PA-7 8%	2.0	·1.0
.125/3.18	THIN	.406/10.31	1.000/25.4	.468/11.89	120·2	1	PA-8 5%	2.0	±1.0
.250/6.35	THIN	.406/10.31'	1.000/25.4	1.000/25.4	120·2	2	PAR-7 8%	2.0	·1.0
							*Elongation		

'overall diameter

Figures 1, 2

Dynalastic, Phenolic-Glass Carrier

Fatigue Resistant Lead Design for Long-Term Dynamic Testing

GRID LENGTH (a)		GRID WIDTH (c)	L (d)	W (e)	R-OHMS	FIG.	DESIG.	GAGE FACTOR	PKG TOL.
.125/3.18		.125/3.18	.750/19.05	.312/7.92	120·2.5	1	DLB-MK12-2ASX	2.2	·1.0
.250/6.35		.156/3.96	1.000/25.4	.343/8.71	350·5	1	DLB-MK35-4ASX	2.2	·2.0
.125/3.18		.046/1.17	.750/19.05	.312/7.92	120·2.5	1	DLB-PT12-2A	4.0	·1.0
.250/6.35		.078/1.98	1.000/25.4	.281/7.14	350·5	1	DLB-PT35-4A	4.0	·2.0
.250/6.35		.125/3.18	1.000/25.4	.281/7.14	500·8	1	DLB-PT50-4A	4.0	·3.0
.500/12.70		.140/3.56	1.125/28.6	.375/9.53	1000·15	1	DLB-PT100-8A	4.0	·5.0

Figure 1 — Available in Fatigue Resistant Alloys

HT — Free Filament Grid, Removable Carrier

FIBERGLASS REINFORCED TEFLON TAPE CARRIER

GRID LENGTH (a)		GRID WIDTH (c)	L (d)	W (e)	R-OHMS	FIG.	DESIG.	GAGE FACTOR	PKG TOL.
.125/3.18		.062/1.57'	.125/3.18	Nichrome	120·2.5	1	HT-212-2A	2.2	·1.0
.250/6.35		.093/2.36	.250/6.35	Nichrome	120·2.5	1	HT-212-4A	2.2	·1.0
.250/6.35		.093/2.36	.250/6.35	Nichrome	120·2.5	2	HT-212-4B'	2.2	·1.0
.250/6.35		.218/5.54	.250/6.35	Nichrome	350·5	1	HT-235-4A	2.2	·2.0
.250/6.35		.218/5.54	.250/6.35	Nichrome	350·5	2	HT-235-4B'	2.2	±2.0
.250/6.35		.281/7.14	.250/6.35	Constantan	120·2.5	1	HT-412-4ASX	2.0	·1.0
.250/6.35		.281/7.14	.250/6.35	Constantan	120·2.5	2	HT-412-4BSX'	2.0	·1.0
.500/12.70		.500/12.70	.500/12.70	Constantan	350·5	1	HT-435-8ASX	2.0	±2.0
.250/6.35		.093/2.36	.250/6.35	Karma	120·2.5	1	HT-812-4ASX	2.2	·1.0
.250/6.35		.093/2.36	.250/6.35	Karma	120±2.5	2	HT-812-4BSX'	2.2	·1.0
.250/6.35		.328/8.33	.250/6.35	Karma	350±5	1	HT-835-4ASX	2.2	·2.0
.250/6.35		.328/8.33	.250/6.35	Karma	350·5	2	HT-835-4BSX'	2.2	±2.0
.125/3.18		.062/1.57'	.125/3.18	Plat-Tung	120·2.5	1	HT-1212-2A	4.0	·1.0
.312/7.92		.093/2.36	.312/7.92	Plat-Tung	120·2.5	1	HT-1212-5A	4.0	·1.0
.312/7.92		.093/2.36	.312/7.92	Plat-Tung	120·2.5	2	HT-1212-5B'	4.0	·1.0
.250/6.35		.187/4.75	.260/6.36	Plat-Tung	350·5	1	HT-1235-4A	4.0	·2.0
.250/6.35		.187/4.75	.250/6.35	Plat-Tung	350·5	2	HT-1235-4B'	4.0	·2.0
.500/12.70		.140/3.56	.500/12.70	Plat-Tung	500·8	1	HT-1250-8A	4.0	·3.0
.500/12.70		.140/3.56	.500/12.70	Plat-Tung	500·8	2	HT-1250-8B'	4.0	·3.0

GAGE GRID — RELEASE PAPER

Figure 1 — Ceramic Adhesives & Flame-Spray Applications

FIBERGLASS REINFORCED TEFLON TAPE CARRIER

GAGE GRID — RELEASE PAPER — CA THERMOCOUPLE

Figure 2

Tape Carrier Shelf Life - 6 mos.

'3/32" wide at leads   'C/A Thermocouple

*STD paper use Duco. THIN paper use SR-4.

Courtesy of BLH Electronics, Canton, MA.

GENERAL PURPOSE GAGES

**DWG.	GAGE TYPE	NOMINAL RESISTANCE OHMS	NOMINAL GF	ACTIVE GAGE LENGTH('')
A	SPB1-06-12	120	+110	.06
	SPB1-12-12	120	+115	.12
	SPB1-20-35	350	+118	.20
B	SPB2-06-12	120	+110	.06
	SPB2-12-12	120	+115	.12
	SPB2-20-35	350	+118	.20
C	SPB3-06-12	120	+110	.06
	SPB3-12-12	120	+115	.12
	SPB3-20-35	350	+118	.20
D	SP4-17-35	350	+118	.17
E	SP5-06-12	120	+110	.06
	SP5-10-12	120	+115	.10
	SP5-17-35	350	+118	.17

All gages with B's included in the nomenclature are Phenolic Glass backed and are rated for service to 500F. Other types are unbacked (free handling) and are rated for 700F service.

The nominal non-linearity constant (C) for General Purpose Gages is 3600.

**See page 32 for gage drawings

Semiconductor gages listed in this section are for applications where self-temperature compensation is not required and where optimum linearity is advantageous. They are particularly useful for general transducer work and for gaging where dummy, half- or full-bridge installations will be used.

Temperature compensation of bridges is most successful with units of the type listed here and transducers with specifications equivalent to those usually associated with wire and foil type transducers are readily obtainable.

Although no semiconductor strain gage is perfectly linear, the general purpose type of semiconductor gage exhibits the best combination of linearity, zero shift and strain sensitivity versus temperature. Figures 1 and 2 shows typical strain sensitivity and apparent strain.

All semiconductor gages are supplied with tables of unit resistance change versus strain level and temperature. By using techniques described in the BLH Semiconductor Strain Gage Handbook (which can be obtained from BLH), data from strain measurements can be readily reduced to provide accurate results.

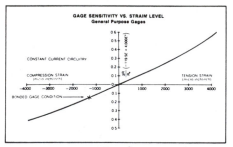

Figure 1

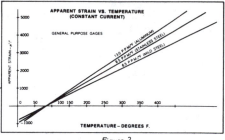

Figure 2

ZERO SHIFT COMPENSATED GAGES

**DWG.	GAGE TYPE	NOMINAL RESISTANCE OHMS	NOMINAL GF	ACTIVE GAGE LENGTH('')
A	SNB1-06-12S6	120	−103	.06
	SNB1-06-12S9	120	−106	.06
	SNB1-06-12S13	120	−110	.06
	SNB1-16-35S6	350	−103	.16
	SNB1-16-35S9	350	−106	.16
	SNB1-16-35S13	350	−110	.16
B	SNB2-06-12S6	120	−103	.06
	SNB2-06-12S9	120	−106	.06
	SNB2-06-12S13	120	−110	.06
	SNB2-16-35S6	350	−103	.16
	SNB2-16-35S9	350	−106	.16
	SNB2-16-35S13	350	−110	.16
C	SNB3-06-12S6	120	−103	.06
	SNB3-06-12S9	120	−106	.06
	SNB3-06-12S13	120	−110	.06
	SNB3-16-35S6	350	−103	.16
	SNB3-16-35S9	350	−106	.16
	SNB3-16-35S13	350	−110	.16

The nominal non-linearity constant (C) for above gages is:

G.F.	−103	−106	−110
C_2	13,000	13,600	14,500

**See page 32 for gage drawings

Semiconductor self-temperature compensated gages are unique in that they are designed and manufactured to match the expansion coefficient of typical structural materials. Their properties are such that the increase in resistivity with temperature is cancelled by the decrease in resistance caused by expansion of the material to which they are bonded. Section III in the Semiconductor Strain Gage Handbook (which can be obtained from BLH) describes this in detail.

The S number which follows each gage type describes the specimen material for which the gages are matched. S6 gages, then, are formulated for use on mild steel and other materials having expansion coefficients of approximately 6-1/2 PPM/°F. Accordingly, S9 and S13 are for common stainless steels and aluminum respectively. (continued next page)

Courtesy of BLH Electronics, Canton, MA.

The temperature compensation span for self-compensated semiconductor gages is from 50 to 150F when installed with a room temperature curing cement. Because of the non-linearity of semiconductor gages and properties of various adhesives, however, it is possible to shift the temperature span over which compensation is achieved by using proper gage and elevated temperature curve.

Similar results can be obtained with semiconductor elements which have S numbers other than those particularly suited to the specimen material. Complete instructions are included in each package of gages showing the entire range and possibilities as well as the particular method of data reduction for each type of installation.

Selected material N-type semiconductor gages represent the most economical type of temperature-compensated semiconductor strain sensors because only one bridge arm is involved and no matching of components or use of additional elements is required.

Installation is simple and straightforward - similar to that of standard type semiconductor strain gages. These units are specifically designed for use with constant voltage Wheatstone bridge circuits, the instrumentation universally used for strain gage work. In addition, their non-linearity is opposite to that of the Wheatstone bridge and good linearity compensation generally results from their use.

Where temperature varies, N-type gages usually represent the lowest and most accurate method of temperature compensation.

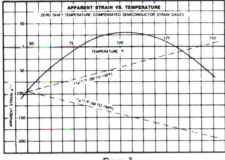

Figure 3

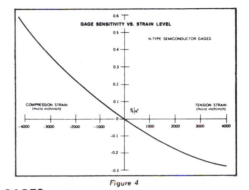

Figure 4

SHORT GAGES

**DWG.	GAGE TYPE	NOMINAL RESISTANCE OHMS	NOMINAL GF	ACTIVE GAGE LENGTH INCHES
A	SPB1-03-12	120	+110	.03
	SPB1-06-12	120	+110	.06
	SPB1-07-35	350	+130	.07
	SPB1-09-100	1,000	+145	.09
	SPB1-12-12	120	+115	.12
	SPB1-12-35	350	+130	.12
B	SPB2-03-12	120	+110	.03
	SPB2-06-12	120	+110	.06
	SPB2-07-35	350	+130	.07
	SPB2-09-100	1,000	+145	.09
	SPB2-12-12	120	+115	.12
	SPB2-12-35	350	+130	.12
C	SPB3-03-12	120	+110	.03
	SPB3-06-12	120	+110	.06
	SPB3-07-35	350	+130	.07
	SPB3-09-100	1,000	+145	.09
	SPB3-12-12	120	+115	.12
	SPB3-12-35	350	+130	.12
E	SP5-03-12	120	+110	.03
	SP5-06-12	120	+110	.06
	SP5-06-35	350	+130	.06
	SP5-09-100	1,000	+145	.09
	SP5-10-12	120	+115	.10
	SP5-11-35	350	+130	.11

**See page 32 for gage drawings

The ability to produce short semiconductor gages lies in the fact that semiconducting materials are used and long filaments or grid shapes are not required.

Higher resistivity material, however, requires a small sacrifice in zero shift versus temperature. Figure 5 shows typical apparent strain curves on tool steel for gages with gage factors of +116, +138, and +148. Linearity improves, however, and the use of high resistivity elements in transducer work is recommended. Handling of the gages is identical to that of other semiconductor sensors.

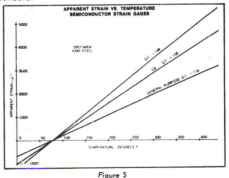

Figure 5

Courtesy of BLH Electronics, Canton, MA.

OEM HALL GENERATORS

FH-500 SERIES

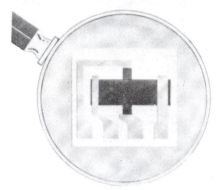

MODELS FH-520 and FH-540
with Flexstrip™ lead

- **HIGH SENSITIVITY**

- **HIGH INPUT IMPEDANCE**

- **GOOD TEMPERATURE CHARACTERISTICS**

- **LONG TERM RELIABILITY**

- **LOW CURRENT REQUIREMENT**

Opens new areas of application for HALL-PAK Generators

- Proximity Switches
- Linear/Angular Transducers
- Brushless dc Motors
- Ignition Systems
- Magnetic Card Readers
- Current Probes
- Magnetic Tape Heads

- Nondestructive Memory Readouts
- Clamp-on Ammeters
- Gaussmeters
- Power Transducers
- Guidance Systems
- Electronic Compasses

GENERAL DESCRIPTION

FH-500 Series Hall generators are miniature solid state Hall effect magnetic field sensing devices. There are four electrical connections to each Hall generator. Two carry the control current (I_c) and two supply the Hall output voltage (V_H). The Hall output voltage is directly proportional to the product of the control current and the magnetic field component normal to the Hall active area. The Series FH-500 lead strip is composed of printed circuit leads encased in DuPont's Kapton and terminating in contacts .075″ on center. This flexible and tough lead strip provides for exceptionally easy handling. The lead strip is available in a variety of configurations. The FH-500 is available in two different sensitivity ranges: 10.0 mV/kG and 12.0 mV/kG. More detailed specifications are listed on the reverse of this sheet.

83045D   LITHO IN U.S.A.   70060

Courtesy F. W. Bell, Inc.

Electrical Specifications

POLARITY: With field direction (**B**) as shown and I_c entering the $I_c(+)$ terminal, the positive Hall voltage will appear at the $V_H(+)$ terminal.

NOTE: Unless otherwise specified, all specifications apply at nominal control current with T 25 C. Heat sinking can enhance performance in several respects. For information concerning operating on a heat sink, such as the pole face of a magnet, contact F.W. Bell, Inc.

	FH-520	FH-540
Input Resistance, R_{in}	20-40 Ω	40-80 Ω
Output Resistance, R_{out}	2.2 R_{in} approx.	2.2 R_{in} approx.
Magnetic Sensitivity γ_B, min. @ I_{cn}	10.0 mV/kG	12.0 mV/kG
Product Sensitivity, γ_{IB}, min.	0.4 V/A•kG	0.8 V/A•kG
Resistive Residual Voltage, V_M @ I_{cn}, B O	5 mV max.	6 mV max.
Nominal Control Current, I_{cn}	25 mA	15mA
Maximum Continuous Control Current, I_{cmos}	50 mA	30 mA
Mean Temperature Coefficient of V_H (-20 C to +80 C), β_T	-0.1%/ C max.	-0.1%/ C max.
Mean Temperature Coefficent of Resistance, α_T (-20 C to +80 C)	0.1%/ C max.	0.1%/ C max.
Temperature Dependence of Resistive Residual Voltage, D_T @ I_{cn} (-20 C to +80 C)	10 μV/ C max.	10 μV/ C max.
Thermal Resistance, Hall Plate to Ambient, $R\theta_{P-A}$	0.8 C/mW	0.8 C/mW
Thermal Resistance, Hall Plate to Encapsulation, $R\theta_{P-E}$	0.04 C/mW	0.04 C/mW
Operating Temperature Range	-55 C to +100 C	-55 C to +100 C
Storage Temperature Range	-55 C to +120 C	-55 C to +120 C

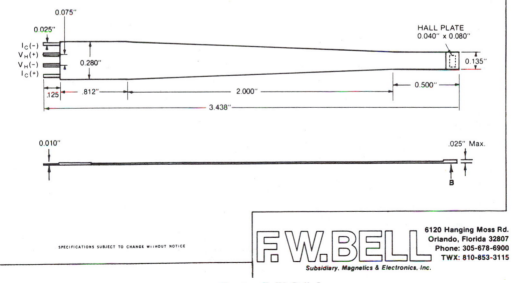

SPECIFICATIONS SUBJECT TO CHANGE WITHOUT NOTICE

F.W.BELL
Subsidiary, Magnetics & Electronics, Inc.

6120 Hanging Moss Rd.
Orlando, Florida 32807
Phone: 305-678-6900
TWX: 810-853-3115

Courtesy F. W. Bell, Inc.

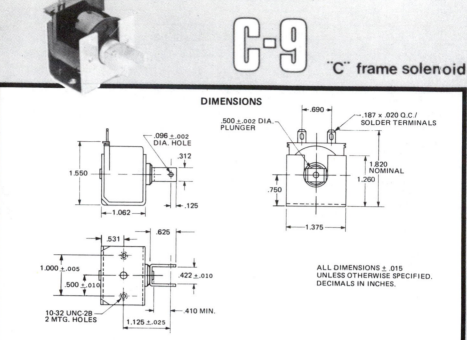

DELTROL controls / DIVISION OF DELTROL CORP

C-9

"C" frame solenoid

DIMENSIONS

.096 ±.002 DIA. HOLE
.312
1.550
1.062
.125

.500 ±.002 DIA. PLUNGER
.690
.187 x .020 Q.C./ SOLDER TERMINALS
1.820 NOMINAL
1.260
.750
1.375

.531
.625
1.000 ±.005
.500 ±.010
.422 ±.010
10-32 UNC-2B 2 MTG. HOLES
1.125 ±.025
.410 MIN.

ALL DIMENSIONS ± .015
UNLESS OTHERWISE SPECIFIED.
DECIMALS IN INCHES.

Standard Power Ratings

Duty Cycle	Nominal AC Power (Seated)	Nominal DC Power
continuous	13.7VA	8.6W
intermittent	52VA	15.5W
pulse	137VA	86W

Standard Coil Ratings AC Coils

Voltage 50/60 Hz	Coil Resistance ± 10% @ 25° C		
	continuous	intermittent	pulse
24	9.3 ohms	3.3 ohms	1.6 ohms
120	240 ohms	84.1 ohms	37.6 ohms
240	960 ohms	342 ohms	143 ohms

Standard Coil Ratings DC Coils

Voltage	Coil Resistance ± 10% @ 25° C		
	continuous	intermittent	pulse
12	16.5 ohms	9.2 ohms	1.6 ohms
24	66.5 ohms	37.6 ohms	6.7 ohms
110	1430 ohms	788 ohms	143 ohms

Specification Data

weight: 5 oz.
coil treatment: plastic encapsulation, class A rating — standard
maximum operating temperature:
class A — 105° C (221° F)
class B — 130° C (266° F)
dielectric strength: 30 volts and under: 500 VRMS. Over 30 volts: 1000 VRMS plus 2x rated voltage for one minute
plunger guide material: plastic
life expectancy: 10^7 cycles under optimum load conditions
terminals: .187" quick-connect/solder lug standard — .250" quick-connects and wire leads optional

24

Courtesy of Deltrol Controls.

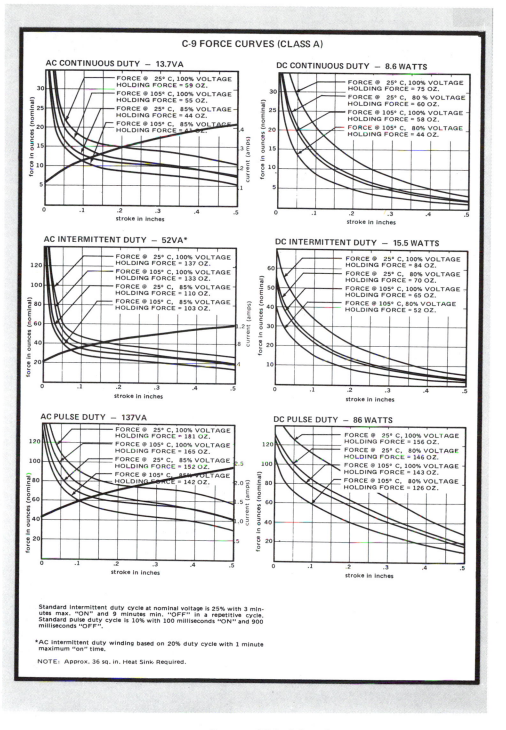

C-9 FORCE CURVES (CLASS A)

AC CONTINUOUS DUTY — 13.7VA

FORCE @ 25° C, 100% VOLTAGE
HOLDING FORCE = 59 OZ.
FORCE @ 105° C, 100% VOLTAGE
HOLDING FORCE = 55 OZ.
FORCE @ 25° C, 85% VOLTAGE
HOLDING FORCE = 44 OZ.
FORCE @ 105° C, 85% VOLTAGE
HOLDING FORCE = 41 OZ.

DC CONTINUOUS DUTY — 8.6 WATTS

FORCE @ 25° C, 100% VOLTAGE
HOLDING FORCE = 75 OZ.
FORCE @ 25° C, 80 % VOLTAGE
HOLDING FORCE = 60 OZ.
FORCE @ 105° C, 100% VOLTAGE
HOLDING FORCE = 58 OZ.
FORCE @ 105° C, 80% VOLTAGE
HOLDING FORCE = 44 OZ.

AC INTERMITTENT DUTY — 52VA*

FORCE @ 25° C, 100% VOLTAGE
HOLDING FORCE = 137 OZ.
FORCE @ 105° C, 100% VOLTAGE
HOLDING FORCE = 133 OZ.
FORCE @ 25° C, 85% VOLTAGE
HOLDING FORCE = 110 OZ.
FORCE @ 105° C, 85% VOLTAGE
HOLDING FORCE = 103 OZ.

DC INTERMITTENT DUTY — 15.5 WATTS

FORCE @ 25° C, 100% VOLTAGE
HOLDING FORCE = 84 OZ.
FORCE @ 25° C, 80% VOLTAGE
HOLDING FORCE = 70 OZ.
FORCE @ 105° C, 100% VOLTAGE
HOLDING FORCE = 65 OZ.
FORCE @ 105° C, 80% VOLTAGE
HOLDING FORCE = 52 OZ.

AC PULSE DUTY — 137VA

FORCE @ 25° C, 100% VOLTAGE
HOLDING FORCE = 181 OZ.
FORCE @ 105° C, 100% VOLTAGE
HOLDING FORCE = 165 OZ.
FORCE @ 25° C, 85% VOLTAGE
HOLDING FORCE = 152 OZ.
FORCE @ 105° C, 85% VOLTAGE
HOLDING FORCE = 142 OZ.

DC PULSE DUTY — 86 WATTS

FORCE @ 25° C, 100% VOLTAGE
HOLDING FORCE = 156 OZ.
FORCE @ 25° C, 80% VOLTAGE
HOLDING FORCE = 146 OZ.
FORCE @ 105° C, 100% VOLTAGE
HOLDING FORCE = 143 OZ.
FORCE @ 105° C, 80% VOLTAGE
HOLDING FORCE = 126 OZ.

Standard intermittent duty cycle at nominal voltage is 25% with 3 minutes max. "ON" and 9 minutes min. "OFF" in a repetitive cycle. Standard pulse duty cycle is 10% with 100 milliseconds "ON" and 900 milliseconds "OFF".

*AC intermittent duty winding based on 20% duty cycle with 1 minute maximum "on" time.

NOTE: Approx. 36 sq. in. Heat Sink Required.

Courtesy of Deltrol Controls.

VERNITECH
Division of VERNITRON CORPORATION

MODEL
VOE-015
MODULAR

THE HOUSE OF CONTROL

OPTICAL INCREMENTAL MODULAR ENCODER

Designed to meet most commercial and military applications, the VOE-015 contains light emitting diodes (LED) and precision chrome on glass code disc for highest reliability and performance. Counts to 1024 pulses per revolution are available. Options include unamplified photosensor outputs or squared outputs TTL or HTL compatible. Units can be furnished with either a single output channel or with a second channel in phase quadrature, and with index marker if required.

For further information please contact our sales or engineering applications departments.

SPECIFICATIONS

MECHANICAL:

Dimensions .... See reverse side

Moment of Inertia ..... 2.0 x 10^{-5} oz-in-Sec^2

Slew Speed .... 5000 RPM

Counting speed .... $\dfrac{3,000,000}{\text{pulses per revolution}}$ RPM
(without index)

Counting speed .... $\dfrac{1,200,000}{\text{pulses per revolution}}$ RPM
(with index)

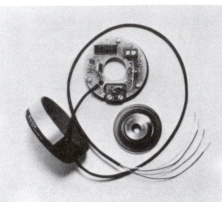

ELECTRICAL:

Output Codes Available ....
1. Single channel with or without index
2. Dual channels in quadrature with or without index

Pulses per Revolution .... Advise requirement

Accuracy pulse to pulse .... ± 1% Standard

Accuracy pulse
to any other pulse .... 1/8 cycle or ± 3′ whichever is greater.*

Illumination Source ...........L.E.D.

Input Power .... +5 ± 5% VDC at 120MA
or
+12 ± 5% VDC at 120MA

Output Levels ....
1. TTL compatible
   Logic 1 — 2.5V min
   Logic 0 — 0.7V max
   Fanout —4 gates
2. HTL compatible
   Logic 1 — 8.5V min
   Logic 0 — 0.7V max
   Fanout — 4 gates

ENVIRONMENTAL:

Operating temperature ...0° to +70°C

Vibration .... 5 to 2 KHZ at 20G

Shock .... 50 G's for 11 MS

*Dependent on shaft accuracy.

Courtesy of Vernitech Division of Vernitron Corp., Deer Park, NY.

VOE-015

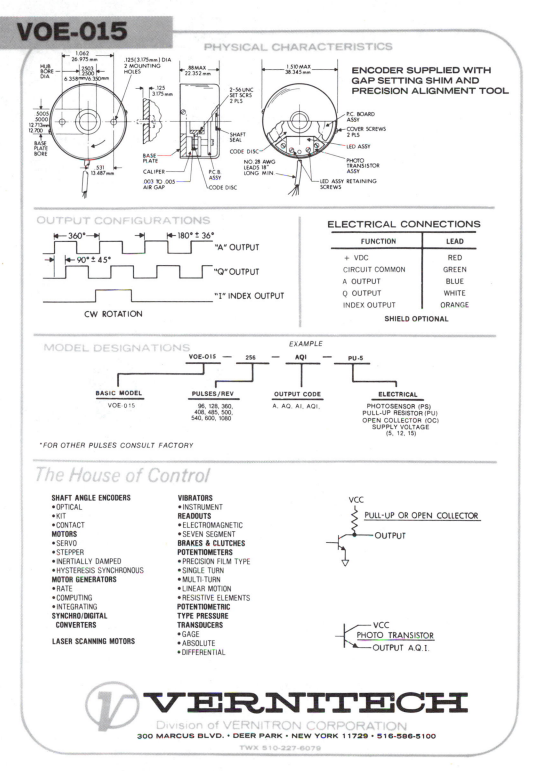

PHYSICAL CHARACTERISTICS

ENCODER SUPPLIED WITH
GAP SETTING SHIM AND
PRECISION ALIGNMENT TOOL

OUTPUT CONFIGURATIONS

"A" OUTPUT

"Q" OUTPUT

"I" INDEX OUTPUT

CW ROTATION

ELECTRICAL CONNECTIONS

FUNCTION	LEAD
+ VDC	RED
CIRCUIT COMMON	GREEN
A OUTPUT	BLUE
Q OUTPUT	WHITE
INDEX OUTPUT	ORANGE

SHIELD OPTIONAL

MODEL DESIGNATIONS

EXAMPLE

VOE-015 — 256 — AQI — PU-5

BASIC MODEL	PULSES/REV	OUTPUT CODE	ELECTRICAL
VOE-015	96, 128, 360, 408, 485, 500, 540, 600, 1080	A, AQ, AI, AQI,	PHOTOSENSOR (PS) PULL-UP RESISTOR (PU) OPEN COLLECTOR (OC) SUPPLY VOLTAGE (5, 12, 15)

FOR OTHER PULSES CONSULT FACTORY

The House of Control

SHAFT ANGLE ENCODERS
- OPTICAL
- KIT
- CONTACT

MOTORS
- SERVO
- STEPPER
- INERTIALLY DAMPED
- HYSTERESIS SYNCHRONOUS

MOTOR GENERATORS
- RATE
- COMPUTING
- INTEGRATING

SYNCHRO/DIGITAL CONVERTERS

LASER SCANNING MOTORS

VIBRATORS
- INSTRUMENT

READOUTS
- ELECTROMAGNETIC
- SEVEN SEGMENT

BRAKES & CLUTCHES

POTENTIOMETERS
- PRECISION FILM TYPE
- SINGLE TURN
- MULTI-TURN
- LINEAR MOTION
- RESISTIVE ELEMENTS

POTENTIOMETRIC TYPE PRESSURE TRANSDUCERS
- GAGE
- ABSOLUTE
- DIFFERENTIAL

VCC

PULL-UP OR OPEN COLLECTOR

OUTPUT

VCC

PHOTO TRANSISTOR

OUTPUT A.Q.I.

VERNITECH
Division of VERNITRON CORPORATION
300 MARCUS BLVD. • DEER PARK • NEW YORK 11729 • 516-586-5100
TWX 510-227-6079

Courtesy of Vernitech Division of Vernitron Corp., Deer Park, NY.

VERNITECH
Division of VERNITRON CORPORATION

THE HOUSE OF CONTROL

MINIATURE HOUSED
OPTICAL INCREMENTAL SHAFT ANGLE ENCODER

Miniature Design for Commercial Applications where size and weight must be minimized. The VOE-05 contains light emitting diodes and precision chrome on glass (or etched metal) code disc for highest reliability and performance. Counts to 256* pulses per revolution are available. Units can be furnished with either a single output channel or with a second channel in phase quadrature, and with index marker if required.

For further information please contact our sales or Applications Engineering Department.

*Can be multiplied with external logic to 1024 pulses per revolution.

SPECIFICATIONS

MECHANICAL:

Dimensions	See reverse side
Torque	.05 oz. in. max.
Moment of Inertia	6.0×10^{-7} oz. in. sec^2
Slew Speed	5000 RPM max
Counting speed (without index)	50 KHZ
Counting speed (with index)	20 KHZ
Radial Load (continuous duty)	4 oz. max.
Axial Load (continuous duty)	4 oz. max.

ELECTRICAL:

Output Codes Available	1. Single channel with or without index 2. Dual channels in quadrature with or without index
Pulses per Revolution	To 256 P.P.R. max.
Accuracy pulse to pulse	± 3° Electrical
Accuracy pulse to any other pulse	± 0.15° mech.
Illumination Source	L.E.D.
Input: Photo Transistor	Vcc 50v max
Photo Transistor power	30 mw max
Led Current	20 ma max (current limiting required)

Output Levels

A and Q Channels into 1K Ohm	1v P.T.P. min
I Channel in to 1K Ohm	0.35v P.T.P min.
Offset; A, Q;	20% of P.T.P. max
Offset; I	15% of P.T.P. max

ENVIRONMENTAL:

Operating temperature	0° to + 60°C
Storage temperature	− 25° to + 85°C
Vibration	5 to 2 KHZ at 20G
Shock	50 G's for 11 MS
Humidity	Non-condensing

Courtesy of Vernitech Division of Vernitron Corp., Deer Park, NY.

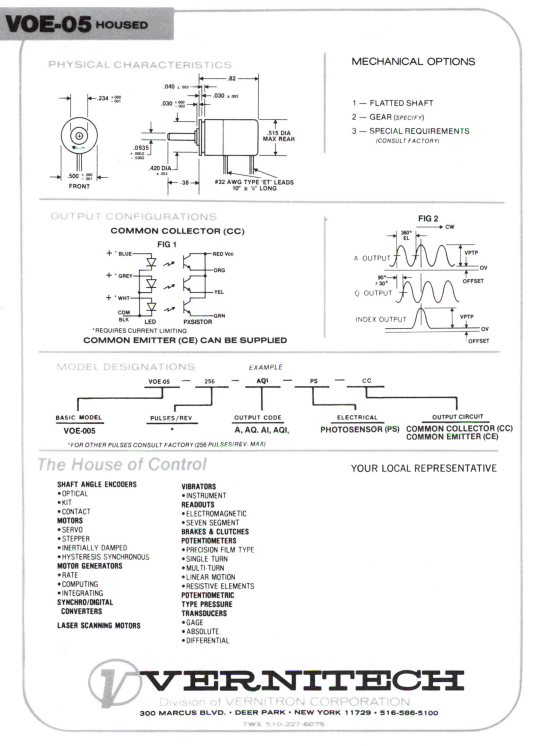

PHYSICAL CHARACTERISTICS

.234 +.000 −.001
.040 ± .003
.82
.030 ± .003
.030 +.000 −.002
.515 DIA MAX REAR
.0935 +.000.0 −.0002
.420 DIA. ± .003
.500 +.000 −.001
FRONT
.38
#32 AWG TYPE 'ET' LEADS 10" ± ½" LONG

MECHANICAL OPTIONS

1 — FLATTED SHAFT
2 — GEAR (SPECIFY)
3 — SPECIAL REQUIREMENTS
(CONSULT FACTORY)

OUTPUT CONFIGURATIONS

COMMON COLLECTOR (CC)

FIG 1

+ * BLUE — RED Vcc
+ * GREY — ORG
+ * WHT — YEL
COM BLK — GRN
LED — PXSISTOR

*REQUIRES CURRENT LIMITING

COMMON EMITTER (CE) CAN BE SUPPLIED

FIG 2

360° EL — CW
A OUTPUT — VPTP / OV
90° ± 30° — OFFSET
Q OUTPUT
INDEX OUTPUT — VPTP / OV / OFFSET

MODEL DESIGNATIONS

EXAMPLE

VOE-05 — 256 — AQI — PS — CC

BASIC MODEL	PULSES/REV	OUTPUT CODE	ELECTRICAL	OUTPUT CIRCUIT
VOE-005	*	A, AQ, AI, AQI,	PHOTOSENSOR (PS)	COMMON COLLECTOR (CC) COMMON EMITTER (CE)

*FOR OTHER PULSES CONSULT FACTORY (256 PULSES/REV. MAX)

The House of Control

SHAFT ANGLE ENCODERS
- OPTICAL
- KIT
- CONTACT

MOTORS
- SERVO
- STEPPER
- INERTIALLY DAMPED
- HYSTERESIS SYNCHRONOUS

MOTOR GENERATORS
- RATE
- COMPUTING
- INTEGRATING

SYNCHRO/DIGITAL CONVERTERS

LASER SCANNING MOTORS

VIBRATORS
- INSTRUMENT

READOUTS
- ELECTROMAGNETIC
- SEVEN SEGMENT

BRAKES & CLUTCHES

POTENTIOMETERS
- PRECISION FILM TYPE
- SINGLE TURN
- MULTI-TURN
- LINEAR MOTION
- RESISTIVE ELEMENTS

POTENTIOMETRIC TYPE PRESSURE TRANSDUCERS
- GAGE
- ABSOLUTE
- DIFFERENTIAL

YOUR LOCAL REPRESENTATIVE

VERNITECH
Division of VERNITRON CORPORATION
300 MARCUS BLVD. • DEER PARK • NEW YORK 11729 • 516-586-5100
TWX 510-227-6079

Courtesy of Vernitech Division of Vernitron Corp., Deer Park, NY.

THE HOUSE OF CONTROL

MULTI-TURN OPTICAL ABSOLUTE ENCODER
CONTINUOUS MECHANICAL ROTATION

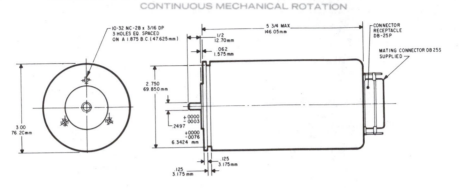

MULTI-TURN CAPABILITY

Counts per turn to 1024 in Binary or Gray code and 1000 counts per turn in BCD. Maximum total counts 524,288 (2^{19}) Binary or Gray and 500,000 counts BCD.

Note: Multi-turn capability is accomplished by internally stacking (2) or (3) Optical Code Discs, interconnected by gearing (Binary, or decimal Ratio). This relationship provides, through electronic circuitry, a continuous Binary or Gray Code (TTL compatible), parallel, non-ambiguous output.

BCD ENCODER
125, 200, 250, 500, 1000 counts per turn

BINARY ENCODER
128, 256, 512 1024 counts per turn

GRAY ENCODER
128, 256, 512, 1024 counts per turn

VERNITECH
Division of VERNITRON CORPORATION
300 MARCUS BLVD., DEER PARK, N.Y. 11729
TELEPHONE (516) 586-5100 TWX 510-227-6079

SPECIFICATIONS

MECHANICAL:

Torque	Unsealed — 0.1 in. oz. Max.
Moment of Inertia	5.5×10^{-4} oz-in-sec^2
Slew Speed	5000 rpm Max.
Counting Speed	30KHz Max.
Radial Load	5 lb. Max.
Axial Load	5 lb. Max.

ELECTRICAL:

Supply Volts	5.0 ±5% VDC
Supply Current	Dependent on Output Code 300 ma to 1000 ma
Output Level	Logic 1 — 2.5 vdc Min. Logic 0 — 0.7 vdc Max. Fanout — 4 gates Other output configurations available on special order
Output Codes	See table
Code Accuracy	± 1/2 least count
Illuminating Source	Light Emitting Diodes
Increasing Count	Clockwise Rotation as Viewed from Shaft End Standard

ENVIRONMENTAL:

Temperature, Operating	0°C to +70°C
Temperature, Storage	−50°C to +85°C
Vibration	5 to 500 hz @ 20g
Shock	50 g's for 11 ms

Courtesy of Vernitech Division of Vernitron Corp., Deer Park, NY.

SINGLE TURN OPTICAL ABSOLUTE ENCODER

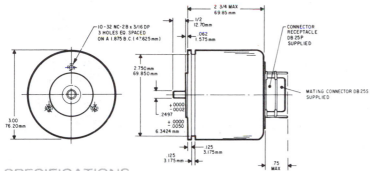

SPECIFICATIONS

MECHANICAL:

Torque	Unsealed — 0.1 in. oz. Max. Sealed — 10 in. oz. Max.
Moment of Inertia	5.5 x 10^{-4} oz-in-sec^2
Slew Speed	5000 rpm Max.
Radial Load	5 lb. Max.
Axial Load	5 lb. Max.

ENVIRONMENTAL:

Temperature, Operating	0°C to +70°C
Temperature, Storage	−50°C to +85°C
Vibration	5 to 500 hz @ 20g
Shock	50 g's for 11 ms

ELECTRICAL:

Supply Volts	5.0 ±5% VDC
Supply Current	Dependent on Output Code 300 ma to 750 ma
Output Level	Logic 1 — 2.5 vdc Min. Logic 0 — 0.7 vdc Max. Fanout — 4 gates Other output configurations available on special order
Output Codes	See table
Code Accuracy	±1/2 least count
Illuminating Source	Light Emitting Diodes
Increasing Count	Clockwise Rotation as Viewed from Shaft End Standard Counter-clockwise on special order

BCD ENCODER
125, 200, 250, 500, 1000 counts per turn

BINARY ENCODER
128, 256, 512 1024 counts per turn

GRAY ENCODER
128, 256, 512, 1024 counts per turn

VERNITECH
Division of VERNITRON CORPORATION
300 MARCUS BLVD. • DEER PARK • NEW YORK 11729 • 516-586-5100
TWX 510-227-6079

Courtesy of Vernitech Division of Vernitron Corp., Deer Park, NY.

VBE
SERIES

THE HOUSE OF CONTROL

CONTACTING   ENCODERS

ABSOLUTE

BINARY CODE

STANDARD FEATURES

1. Output logic DTL — TTL Compatible, 5 VDC
2. Operating Speed — 200 RPM maximum
3. Slew Speed — 1000 RPM
4. Operating Temperature — 0 °C to — 70 °C
5. Life — 5,000,000 Revolutions
6. Increasing Count — CW Rotation
7. Output — Parallel non-ambiguous

OPTIONS

1. Operating Temp — 55 °C to + 105 °C
2. CMos Compatible Logic
3. Other DC Voltages to 28 VDC
4. Connectors — (see drawings)
5. Increasing Count — CCW Rotation

TYPE NUMBER	FULL COUNT	COUNTS PER TURN	TOTAL REV FOR FULL COUNT	WEIGHT (OZ.)	LENGTH "L" (IN.)	MAX. FRICTION TORQUE @ 25 °C (GM-CM)	MOMENT OF INERTIA (GM-CM²)	FIGURE
VBE 11-204	2^8 (256)	256	1	3	1.500	30	5	1
VBE 11-207	2^9 (512)	256	2	4	2.625	30	5	1
VBE 11-210	2^{10} (1,024)	256	4	4	2.625	30	5	1
VBE 11-213	2^{11} (2,048)	256	8	4	2.625	30	5	1
VBE 11-216	2^{12} (4,096)	256	16	4	2.625	30	5	1
VBE 11-219	2^{13} (8,192)	256	32	4	2.625	30	5	1
VBE 11-222	2^{14} (16,384)	256	64	4	2.625	30	5	1
VBE 18-201	2^7 (128)	128	1	4	1.125	45	10	2
VBE 18-204	2^8 (256)	256	1	4	1.125	45	10	2
VBE 18-207	2^9 (512)	256	2	6	2.125	45	10	2
VBE 18-210	2^{10} (1,024)	256	4	6	2.125	45	10	2
VBE 18-213	2^{11} (2,048)	256	8	6	2.125	45	10	2
VBE 18-216	2^{12} (4,096)	256	16	6	2.125	45	10	2
VBE 18-219	2^{13} (8,192)	256	32	6	2.125	45	10	2
VBE 18-222	2^{14} (16,384)	256	64	6	2.125	45	10	2
VBE 18-225	2^{15} (32,768)	256	128	9	3.125	45	10	2
VBE 18-228	2^{16} (65,536)	256	256	9	3.125	45	10	2
VBE 18-231	2^{17} (131,072)	256	512	9	3.125	45	10	2
VBE 18-234	2^{18} (262,144)	256	1,024	9	3.125	45	10	2
VBE 18-237	2^{19} (524,288)	256	2,048	9	3.125	45	10	2
VBE 23-252	2^8 (256)	256	1	8	2.000	75	30	3
VBE 23-254	2^9 (512)	256	2	12	2.750	75	30	3
VBE 23-256	2^{10} (1,024)	256	4	12	2.750	75	30	3
VBE 23-258	2^{11} (2,048)	256	8	12	2.750	75	30	3
VBE 23-260	2^{12} (4,096)	256	16	12	2.750	75	30	3
VBE 23-262	2^{13} (8,192)	256	32	12	2.750	75	30	3
VBE 23-264	2^{14} (16,384)	256	64	12	2.750	75	30	3
VBE 23-266	2^{15} (32,768)	256	128	16	3.750	75	30	3
VBE 23-268	2^{16} (65,536)	256	256	16	3.750	75	30	3
VBE 23-270	2^{17} (131,072)	256	512	16	3.750	75	30	3

Courtesy of Vernitech Division of Vernitron Corp., Deer Park, NY.

VERNITECH
Division of VERNITRON CORPORATION

THE HOUSE OF CONTROL

CONTACTING   ENCODERS

ABSOLUTE

BINARY CODED DECIMAL (BCD)

STANDARD FEATURES

1. Output Logic DTL/TTL Compatible (Size 23)
2. Momentary data storage and inhibit
3. Operating Speed — 200 RPM Maximum
4. Slew Speed — 1000 RPM
5. Operating Temperature — 0 °C to +70 °C
6. Increasing Count — CW Rotation
7. Output — Parallel — Non-Ambiguous
8. Life — 5,000,000 Revolutions
9. 100 and 200 Counts Per Rev. On Size 23

OPTIONS

1. Operating Temperature — −55 °C to +105 °C
2. Various Output Voltage Ratings to 28 VDC
3. Connectors (See Figures)
4. Increasing Count — CCW Rotation

ACTUAL SIZE

TYPE NUMBER	FULL COUNT	COUNTS PER TURN	TOTAL REV FOR FULL COUNT	WEIGHT (OZ.)	LENGTH "L" (IN.)	MAX. FRICTION TORQUE @ 25 °C (GM-CM)	MOMENT OF INERTIA (GM-CM²)	FIGURE
VDE 11-302	3599	100	36	4	1.750	30	5	1
VDE 11-304	999	100	10	4	1.750	30	5	1
VDE 11-305	9999	100	100	6	2.500	30	5	1
VDE 23-301	359	360	1	8	2.000	75	30	3
VDE 23-302	3599	400	9	12	3.000	75	30	3
VDE 23-303	35999	400	90	16	4.125	75	30	3
VDE 23-304	999	400	2.5	12	3.000	75	30	3
VDE 23-305	9999	400	25	12	3.000	75	30	3
VDE 23-306	99999	400	250	16	4.125	75	30	3
VDE 23-307	± 1799	400	9	12	3.000	75	30	3
VDE 23-308	± 17999	400	90	16	4.125	75	30	3
VDE 23-309	± 17959	120	180	16	4.125	75	30	3
VDE 23-310	±179599	400	540	16	4.125	75	30	3
VDE 23-311	± 899	400	4.5	12	3.000	75	30	3
VDE 23-312	± 8999	400	45	12	4.125	75	30	3
VDE 23-313	± 8959	120	90	12	4.125	75	30	3
VDE 23-314	± 89599	400	270	16	4.125	75	30	3
VDE 23-315	± 999	400	5	12	3.000	75	30	3
VDE 23-316	± 9999	400	50	12	4.125	75	30	3

NOTE: The size 11 units require exterior brush selection logic.
± units require exterior up-down counting logic.

Courtesy of Vernitech Division of Vernitron Corp., Deer Park, NY.

THE HOUSE OF CONTROL

single-turn/*infinite* resolution

MODEL 105/106

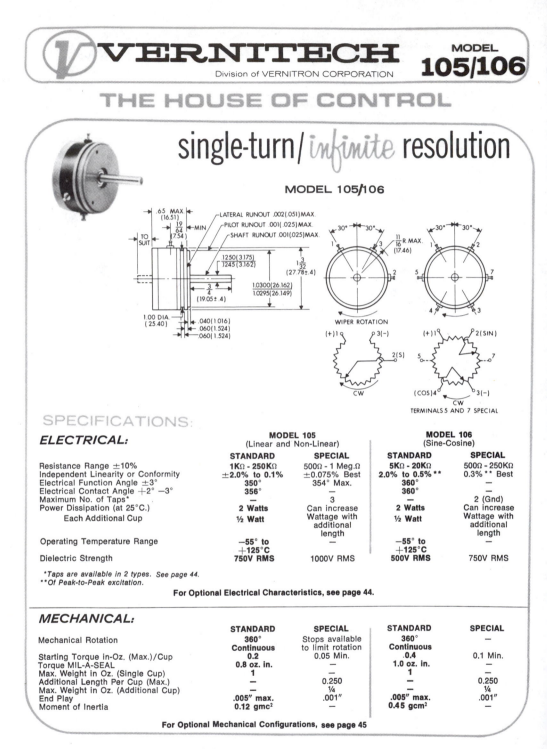

WIPER ROTATION

TERMINALS 5 AND 7 SPECIAL

SPECIFICATIONS:

ELECTRICAL:

	MODEL 105 (Linear and Non-Linear)		MODEL 106 (Sine-Cosine)	
	STANDARD	SPECIAL	STANDARD	SPECIAL
Resistance Range ±10%	1KΩ - 250KΩ	500Ω - 1 Meg.Ω	5KΩ - 20KΩ	500Ω - 250KΩ
Independent Linearity or Conformity	±2.0% to 0.1%	±0.075% Best	2.0% to 0.5% **	0.3% ** Best
Electrical Function Angle ±3°	350°	354° Max.	360°	—
Electrical Contact Angle +2° −3°	356°	—	360°	—
Maximum No. of Taps*	—	3	—	2 (Gnd)
Power Dissipation (at 25°C.)	2 Watts	Can increase	2 Watts	Can increase
Each Additional Cup	½ Watt	Wattage with additional length	½ Watt	Wattage with additional length
Operating Temperature Range	−55° to +125°C	—	−55° to +125°C	—
Dielectric Strength	750V RMS	1000V RMS	500V RMS	750V RMS

*Taps are available in 2 types. See page 44.
**Of Peak-to-Peak excitation.

For Optional Electrical Characteristics, see page 44.

MECHANICAL:

	STANDARD	SPECIAL	STANDARD	SPECIAL
Mechanical Rotation	360° Continuous	Stops available to limit rotation	360° Continuous	—
Starting Torque in-Oz. (Max.)/Cup	0.2	0.05 Min.	.04	0.1 Min.
Torque MIL-A-SEAL	0.8 oz. in.	—	1.0 oz. in.	—
Max. Weight in Oz. (Single Cup)	1	—	1	—
Additional Length Per Cup (Max.)	—	0.250	—	0.250
Max. Weight in Oz. (Additional Cup)	—	¼	—	¼
End Play	.005″ max.	.001″	.005″ max.	.001″
Moment of Inertia	0.12 gmc²	—	0.45 gcm²	—

For Optional Mechanical Configurations, see page 45

Courtesy of Vernitech Division of Vernitron Corp., Deer Park, NY.

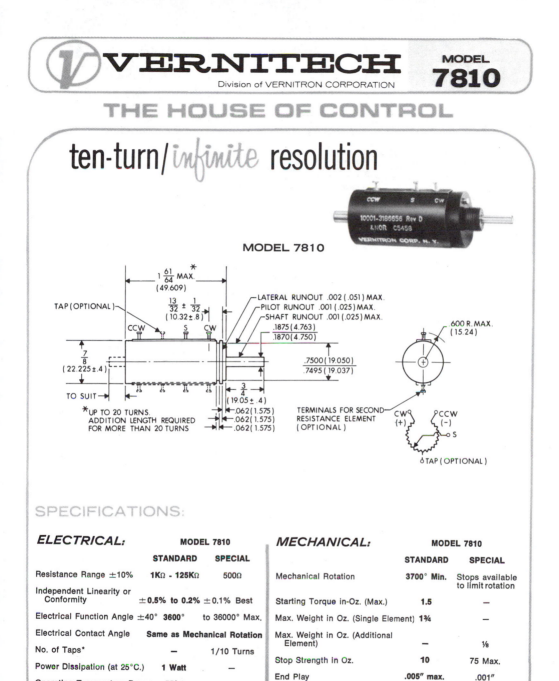

VERNITECH
Division of VERNITRON CORPORATION

THE HOUSE OF CONTROL

ten-turn/*infinite* resolution

MODEL 7810

$1\frac{61}{64}$ MAX.
(49.609)

$\frac{13}{32} \pm \frac{1}{32}$
(10.32±.8)

TAP (OPTIONAL)

CCW   S   CW

LATERAL RUNOUT .002 (.051) MAX.
PILOT RUNOUT .001 (.025) MAX.
SHAFT RUNOUT .001 (.025) MAX.
.1875 (4.763)
.1870 (4.750)

$\frac{7}{8}$
(22.225±.4)

.7500 (19.050)
.7495 (19.037)

.600 R. MAX.
(15.24)

TO SUIT

$\frac{3}{4}$
(19.05 ± .4)

.062 (1.575)
.062 (1.575)
.062 (1.575)

*UP TO 20 TURNS.
ADDITION LENGTH REQUIRED
FOR MORE THAN 20 TURNS

TERMINALS FOR SECOND
RESISTANCE ELEMENT
(OPTIONAL)

CW
(+)

CCW
(-)
S

TAP (OPTIONAL)

SPECIFICATIONS:

ELECTRICAL:

MODEL 7810

	STANDARD	SPECIAL
Resistance Range ±10%	1KΩ - 125KΩ	500Ω
Independent Linearity or Conformity	±0.5% to 0.2%	±0.1% Best
Electrical Function Angle ±40°	3600°	to 36000° Max.
Electrical Contact Angle	Same as Mechanical Rotation	
No. of Taps*	—	1/10 Turns
Power Dissipation (at 25°C.)	1 Watt	—
Operating Temperature Range	−55° to +125°C	—
Dielectric Strength	500V RMS	—

*Taps are available in 2 types. See page 44.
For Optional Electrical Characteristics, see page 44.

MECHANICAL:

MODEL 7810

	STANDARD	SPECIAL
Mechanical Rotation	3700° Min.	Stops available to limit rotation
Starting Torque in-Oz. (Max.)	1.5	—
Max. Weight in Oz. (Single Element)	1¾	—
Max. Weight in Oz. (Additional Element)	—	⅛
Stop Strength in Oz.	10	75 Max.
End Play	.005" max.	.001"
Moment of Inertia	.025 gmc²	

For Optional Mechanical Configurations, see page 45.

Courtesy of Vernitech Division of Vernitron Corp., Deer Park, NY.

MOTOROLA

NPN PHOTOTRANSISTORS AND PN INFRARED EMITTING DIODES

... gallium arsenide LED optically coupled to silicon phototransistors designed for applications requiring electrical isolation, high-current transfer ratios, small package size and low cost; such as interfacing and coupling systems, phase and feedback controls, solid-state relays and general-purpose switching circuits.

- High Isolation Voltage —
  V_{ISO} = 7500 V (Min)

- High Collector Output Current
  @ I_F = 10 mA
  I_C = 5.0 mA (Typ) — 4N25,A,4N26
  2.0 mA (Typ) — 4N27,4N28

- Economical, Compact, Dual-In-Line Package

- Excellent Frequency Response —
  300 kHz (Typ)

- Fast Switching Times @ I_C = 10 mA
  t_{on} = 0.87 μs (Typ) — 4N25,A,4N26
  2.1 μs (Typ) — 4N27,4N28
  t_{off} = 11 μs (Typ) — 4N25,A,4N26
  5.0 μs (Typ) — 4N27,4N28

- 4N25A is UL Recognized
  File Number E54915

**OPTO
COUPLER/ISOLATOR**

TRANSISTOR OUTPUT

*MAXIMUM RATINGS (T_A = 25°C unless otherwise noted).

Rating	Symbol	Value	Unit
INFRARED-EMITTING DIODE MAXIMUM RATINGS			
Reverse Voltage	V_R	3.0	Volts
Forward Current — Continuous	I_F	80	mA
Forward Current — Peak Pulse Width = 300 μs, 2.0% Duty Cycle	I_F	3.0	Amp
Total Power Dissipation @ T_A = 25°C Negligible Power in Transistor Derate above 25°C	P_D	150 2.0	mW mW/°C
PHOTOTRANSISTOR MAXIMUM RATINGS			
Collector-Emitter Voltage	V_{CEO}	30	Volts
Emitter-Collector Voltage	V_{ECO}	7.0	Volts
Collector-Base Voltage	V_{CBO}	70	Volts
Total Device Dissipation @ T_A = 25°C Negligible Power in Diode Derate above 25°C	P_D	150 2.0	mW mW/°C
TOTAL DEVICE RATINGS			
Total Device Dissipation @ T_A = 25°C	P_D	250	mW
Equal Power Dissipation in Each Element Derate above 25°C		3.3	mW/°C
Junction Temperature Range	T_J	-55 to +100	°C
Storage Temperature Range	T_{stg}	-55 to +150	°C
Soldering Temperature (10 s)		260	°C

*Indicates JEDEC Registered Data.

STYLE 1:
PIN 1. ANODE
     2. CATHODE
     3. NC
     4. EMITTER
     5. COLLECTOR
     6. BASE

NOTES:
1. DIMENSIONS A AND B ARE DATUMS.
2. T IS SEATING PLANE.
3. POSITIONAL TOLERANCES FOR LEADS:
   ⊕ ∅ 0.13 (0.005)Ⓜ T AⓂBⓂ
4. DIMENSION L TO CENTER OF LEADS WHEN FORMED PARALLEL.
5. DIMENSIONING AND TOLERANCING PER ANSI Y14.5, 1973.

DIM	MILLIMETERS		INCHES	
	MIN	MAX	MIN	MAX
A	8.13	8.89	0.320	0.350
B	6.10	6.60	0.240	0.260
C	2.92	5.08	0.115	0.200
D	0.41	0.51	0.016	0.020
F	1.02	1.78	0.040	0.070
G	2.54 BSC		0.100 BSC	
J	0.20	0.30	0.008	0.012
K	2.54	3.81	0.100	0.150
L	7.62 BSC		0.300 BSC	
M	0°	15°	0°	15°
N	0.38	2.54	0.015	0.100
P	1.27	2.03	0.050	0.080

CASE 730A-01

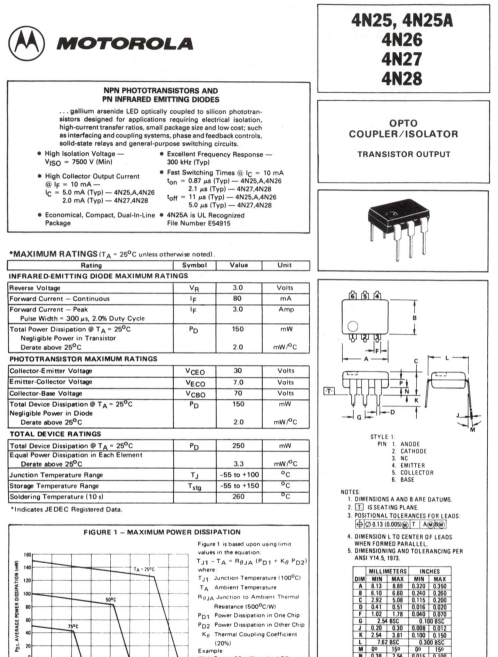

FIGURE 1 — MAXIMUM POWER DISSIPATION

Figure 1 is based upon using limit values in the equation:
$$T_{J1} - T_A = R_{\theta JA} (P_{D1} + K_\theta P_{D2})$$
where:

T_{J1} Junction Temperature (100°C)

T_A Ambient Temperature

$R_{\theta JA}$ Junction to Ambient Thermal Resistance (500°C/W)

P_{D1} Power Dissipation in One Chip

P_{D2} Power Dissipation in Other Chip

K_θ Thermal Coupling Coefficient (20%)

Example:
With P_{D1} = 90 mW in the LED @ T_A = 50°C, the transistor P_D (P_{D2}) must be less than 50 mW.

Courtesy of Motorola, Inc.

4N25, 4N25A, 4N26, 4N27, 4N28

LED CHARACTERISTICS (T_A = 25°C unless otherwise noted)

Characteristic	Symbol	Min	Typ	Max	Unit
*Reverse Leakage Current (V_R = 3.0 V, R_L = 1.0 M ohms)	I_R	–	0.005	100	μA
*Forward Voltage (I_F = 10 mA)	V_F	–	1.2	1.5	Volts
Capacitance (V_R = 0 V, f = 1.0 MHz)	C	–	40	–	pF

PHOTOTRANSISTOR CHARACTERISTICS (T_A = 25°C and I_F = 0 unless otherwise noted)

Characteristic		Symbol	Min	Typ	Max	Unit
*Collector-Emitter Dark Current (V_{CE} = 10 V, Base Open)	4N25, A, 4N26, 4N27	I_{CEO}	–	3.5	50	nA
	4N28		–	–	100	
*Collector-Base Dark Current (V_{CB} = 10 V, Emitter Open)		I_{CBO}	–	–	20	nA
*Collector-Base Breakdown Voltage (I_C = 100 μA, I_E = 0)		$V_{(BR)CBO}$	70	–	–	Volts
*Collector-Emitter Breakdown Voltage (I_C = 1.0 mA, I_B = 0)		$V_{(BR)CEO}$	30	–	–	Volts
*Emitter-Collector Breakdown Voltage (I_E = 100 μA, I_B = 0)		$V_{(BR)ECO}$	7.0	8.0	–	Volts
DC Current Gain (V_{CE} = 5.0 V, I_C = 500 μA)		h_{FE}	–	325	–	–

COUPLED CHARACTERISTICS (T_A = 25°C unless otherwise noted)

Characteristic		Symbol	Min	Typ	Max	Unit
*Collector Output Current (1) (V_{CE} = 10 V, I_F = 10 mA, I_B = 0)	4N25, A, 4N26	I_C	2.0	5.0	–	mA
	4N27, 4N28		1.0	2.0	–	
Isolation Surge Voltage (2, 5)		V_{ISO}				Volts
(60 Hz Peak ac, 5 Seconds)			7500	–	–	
(60 Hz Peak)	*4N25, A		2500	–	–	
	*4N26, 4N27		1500	–	–	
	*4N28		500	–	–	
(60 Hz RMS for 1 Second) (3)	*4N25A		1775	–	–	
Isolation Resistance (2) (V = 500 V)		–	–	10^{11}	–	Ohms
*Collector-Emitter Saturation (I_C = 2.0 mA, I_F = 50 mA)		$V_{CE(sat)}$	–	0.2	0.5	Volts
Isolation Capacitance (2) (V = 0, f = 1.0 MHz)		–	–	0.5	–	pF
Bandwidth (4) (I_C = 2.0 mA, R_L = 100 ohms, Figure 11 (2)		–	–	300	–	kHz

SWITCHING CHARACTERISTICS

			Symbol	Min	Typ	Max	Unit
Delay Time	(I_C = 10 mA, V_{CC} = 10 V	4N25, A, 4N26	t_d	–	0.07	–	μs
		2N27, 4N28		–	0.10	–	
Rise Time	Figures 6 and 8)	4N25, A, 4N26	t_r	–	0.8	–	μs
		4N27, 4N28		–	2.0	–	
Storage Time	(I_C = 10 mA, V_{CC} = 10 V	4N25, A, 4N26	t_s	–	4.0	–	μs
		4N27, 4N28		–	2.0	–	
Fall Time	Figures 7 and 8)	4N25, A, 4N26	t_f	–	8.0	–	μs
		4N27, 4N28		–	8.0	–	

*Indicates JEDEC Registered Data
(1) Pulse Test: Pulse Width = 300 μs, Duty Cycle ≤ 2.0%.
(2) For this test LED pins 1 and 2 are common and phototransistor pins 4, 5, and 6 are common.
(3) RMS Volts, 60 Hz. For this test, pins 1, 2, and 3 are common and pins 4, 5, and 6 are common.
(4) I_F adjusted to yield I_C = 2.0 mA and i_c = 2.0 mA p-p at 10 kHz.
(5) Isolation Surge Voltage, V_{ISO}, is an internal device dielectric breakdown rating.

DC CURRENT TRANSFER CHARACTERISTICS

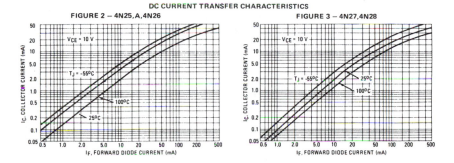

FIGURE 2 — 4N25, A, 4N26   FIGURE 3 — 4N27, 4N28

Courtesy of Motorola, Inc.

TYPICAL ELECTRICAL CHARACTERISTICS

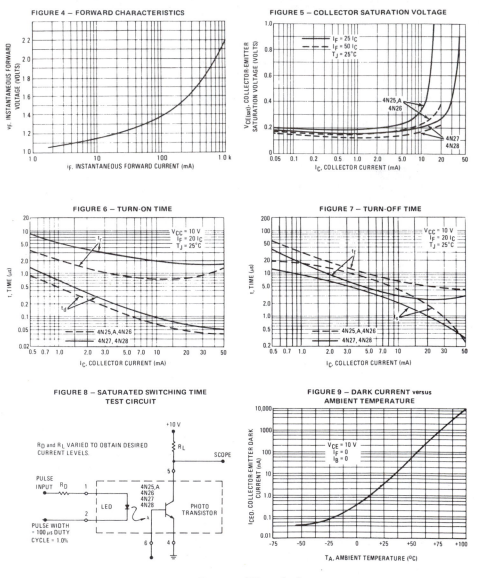

FIGURE 4 – FORWARD CHARACTERISTICS

FIGURE 5 – COLLECTOR SATURATION VOLTAGE

FIGURE 6 – TURN-ON TIME

FIGURE 7 – TURN-OFF TIME

FIGURE 8 – SATURATED SWITCHING TIME TEST CIRCUIT

FIGURE 9 – DARK CURRENT versus AMBIENT TEMPERATURE

Courtesy of Motorola, Inc.

4N25, 4N25A, 4N26, 4N27, 4N28

FIGURE 11 — FREQUENCY RESPONSE TEST CIRCUIT

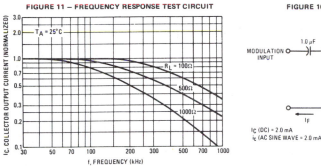

FIGURE 10 — FREQUENCY RESPONSE

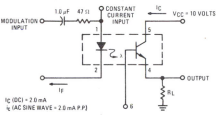

I_C (DC) = 2.0 mA
i_c (AC SINE WAVE = 2.0 mA P.P.)

TYPICAL APPLICATIONS

FIGURE 12 — ISOLATED MTTL TO MOS (P-CHANNEL) LEVEL TRANSLATOR

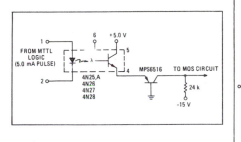

FIGURE 13 — COMPUTER/PERIPHERAL INTERCONNECT

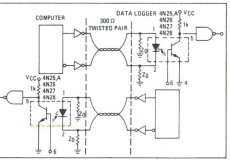

FIGURE 14 — POWER AMPLIFIER

FIGURE 15 — INTERFACE BETWEEN LOGIC AND LOAD

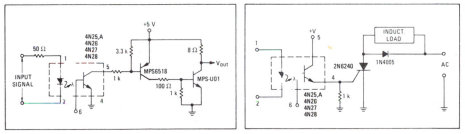

Courtesy of Motorola, Inc.

MOTOROLA

250 V PNP SILICON PHOTO TRIAC DRIVERS

... designed for applications requiring light and infrared LED TRIAC triggering, small size, and low cost.

● Hermetic Package at Economy Prices

● Popular TO-18 Type Package for Easy Handling and Mounting

● High Trigger Sensitivity
H_{FT} = 0.5 mW/cm^2 (Typ-MRD3011)

OPTICALLY TRIGGERED TRIAC DRIVERS

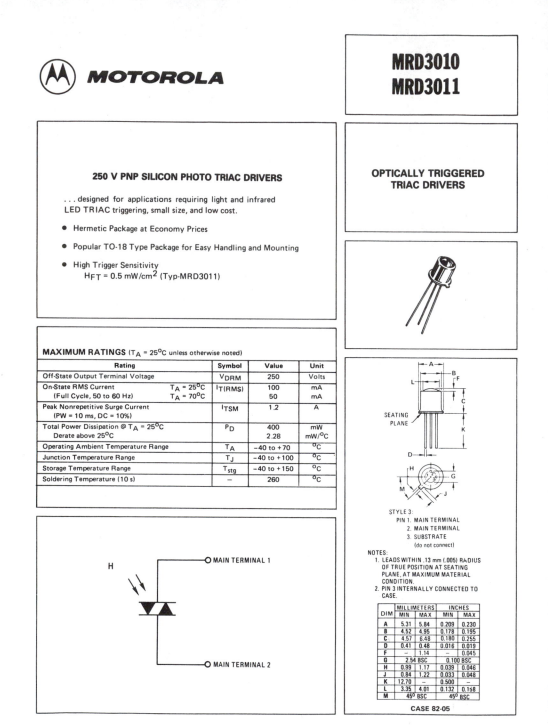

MAXIMUM RATINGS (T_A = 25°C unless otherwise noted)

Rating		Symbol	Value	Unit
Off-State Output Terminal Voltage		V_{DRM}	250	Volts
On-State RMS Current	T_A = 25°C	$I_{T(RMS)}$	100	mA
(Full Cycle, 50 to 60 Hz)	T_A = 70°C		50	mA
Peak Nonrepetitive Surge Current (PW = 10 ms, DC = 10%)		I_{TSM}	1.2	A
Total Power Dissipation @ T_A = 25°C		P_D	400	mW
Derate above 25°C			2.28	mW/°C
Operating Ambient Temperature Range		T_A	−40 to +70	°C
Junction Temperature Range		T_J	−40 to +100	°C
Storage Temperature Range		T_{stg}	−40 to +150	°C
Soldering Temperature (10 s)		−	260	°C

SEATING PLANE

STYLE 3:
PIN 1. MAIN TERMINAL
    2. MAIN TERMINAL
    3. SUBSTRATE
        (do not connect)

NOTES:
1. LEADS WITHIN .13 mm (.005) RADIUS OF TRUE POSITION AT SEATING PLANE, AT MAXIMUM MATERIAL CONDITION.
2. PIN 3 INTERNALLY CONNECTED TO CASE.

DIM	MILLIMETERS		INCHES	
	MIN	MAX	MIN	MAX
A	5.31	5.84	0.209	0.230
B	4.52	4.95	0.178	0.195
C	4.57	6.48	0.180	0.255
D	0.41	0.48	0.016	0.019
F	−	1.14	−	0.045
G	2.54 BSC		0.100 BSC	
H	0.99	1.17	0.039	0.046
J	0.84	1.22	0.033	0.048
K	12.70	−	0.500	−
L	3.35	4.01	0.132	0.158
M	45° BSC		45° BSC	

CASE 82-05

H

○ MAIN TERMINAL 1

○ MAIN TERMINAL 2

Courtesy of Motorola, Inc.

MRD3010, MRD3011

ELECTRICAL CHARACTERISTICS (T_A = 25°C unless otherwise noted)

Characteristic	Symbol	Min	Typ	Max	Unit
DETECTOR CHARACTERISTICS (I_F = 0 unless otherwise noted)					
Peak Blocking Current, Either Direction (Rated V_{DRM}, Note 1)	I_{DRM}	–	10	100	nA
Peak On-State Voltage, Either Direction (I_{TM} = 100 mA Peak)	V_{TM}	–	2.5	3.0	Volts
Critical Rate of Rise of Off-State Voltage, Figure 3	dv/dt	–	2.0	–	V/µs
Critical Rate of Rise of Commutation Voltage, Figure 3 (I_{load} = 15 mA)	dv/dt	–	0.15	–	V/µs
OPTICAL CHARACTERISTICS					
Maximum Irradiance Level Required to Latch Output (Main Terminal Voltage 3.0 V, R_L = 150 Ω) MRD3010 Color Temperature = 2870°K MRD3011	H_{FT}	–	1.0 0.5	5.0 2.0	mW/cm²
Holding Current, Either Direction Initiating Flux Density = 5.0 mW/cm²	I_H	–	100	–	µA

NOTE 1. Test voltage must be applied within dv/dt rating.

FIGURE 1 – ON-STATE CHARACTERISTICS

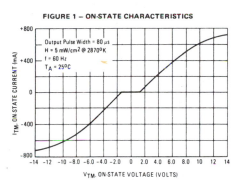

FIGURE 2 – dv/dt TEST CIRCUIT

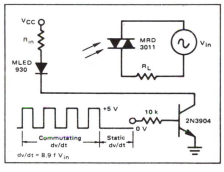

FIGURE 3 – dv/dt versus LOAD RESISTANCE

FIGURE 4 – dv/dt versus TEMPERATURE

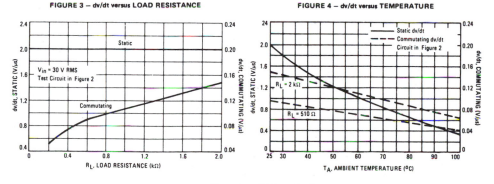

Courtesy of Motorola, Inc.

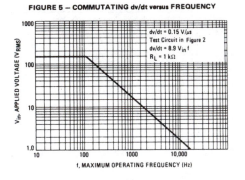

FIGURE 5 — COMMUTATING dv/dt versus FREQUENCY

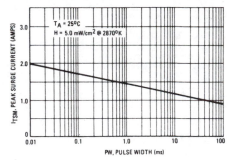

FIGURE 6 — MAXIMUM NONREPETITIVE SURGE CURRENT

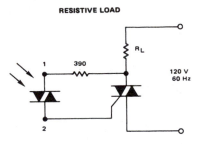

RESISTIVE LOAD

INDUCTIVE LOAD

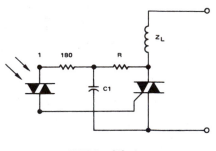

TRIAC $I_{GT} < 15$ mA
R = 2.4 k
C1 = 0.1 µF
TRIAC $I_{GT} > 15$ mA
R = 1.2 kΩ
C1 = 0.2 µF

Courtesy of Motorola, Inc.

MOTOROLA

OPTO SLOTTED COUPLER/INTERRUPTER MODULES

These devices consist of a gallium arsenide infrared emitting diode facing a silicon NPN phototransistor in a molded plastic housing. A slot in the housing between the emitter and the detector provides a means of interrupting the signal. They are widely used in position and motion indicators, end of tape indicators, paper feed controls and arcless switches.

- 1.0 mm Aperture
- Easy PCB Mounting
- Cost Effective
- Industry Standard Configuration
- Uses Long-Lived LPE IRED

OPTO SLOTTED COUPLER

TRANSISTOR OUTPUT

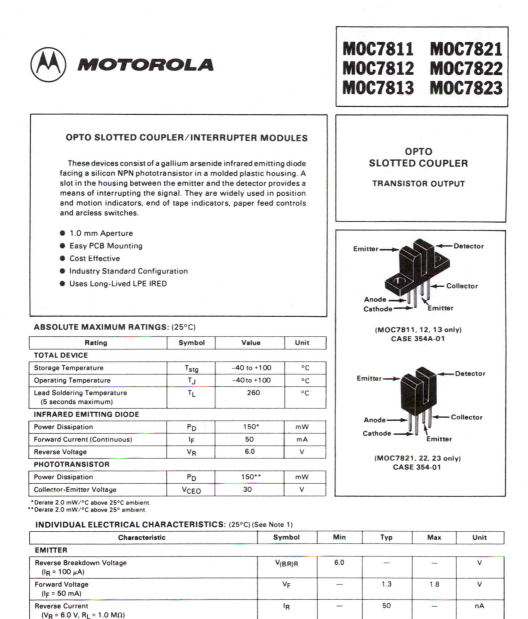

(MOC7811, 12, 13 only)
CASE 354A-01

(MOC7821, 22, 23 only)
CASE 354-01

ABSOLUTE MAXIMUM RATINGS: (25°C)

Rating	Symbol	Value	Unit
TOTAL DEVICE			
Storage Temperature	T_{stg}	–40 to +100	°C
Operating Temperature	T_J	–40 to +100	°C
Lead Soldering Temperature (5 seconds maximum)	T_L	260	°C
INFRARED EMITTING DIODE			
Power Dissipation	P_D	150*	mW
Forward Current (Continuous)	I_F	50	mA
Reverse Voltage	V_R	6.0	V
PHOTOTRANSISTOR			
Power Dissipation	P_D	150**	mW
Collector-Emitter Voltage	V_{CEO}	30	V

*Derate 2.0 mW/°C above 25°C ambient.
**Derate 2.0 mW/°C above 25° ambient.

INDIVIDUAL ELECTRICAL CHARACTERISTICS: (25°C) (See Note 1)

Characteristic	Symbol	Min	Typ	Max	Unit
EMITTER					
Reverse Breakdown Voltage (I_R = 100 µA)	$V_{(BR)R}$	6.0	—	—	V
Forward Voltage (I_F = 50 mA)	V_F	—	1.3	1.8	V
Reverse Current (V_R = 6.0 V, R_L = 1.0 MΩ)	I_R	—	50	—	nA
Capacitance (V = 0, f = 1 MHz)	C_i	—	25	—	pF
DETECTOR					
Breakdown Voltage (I_C = 10 mA, H ≈ 0)	$V_{(BR)CEO}$	30	—	—	V
Collector Dark Current (V_{CE} = 10 V, H ≈ 0)	I_{CEO}	—	—	100	nA

Note 1: Stray irradiation can alter values of characteristics. Adequate shielding should be provided.

Courtesy of Motorola, Inc.

COUPLED ELECTRICAL CHARACTERISTICS: (25°C) (See Note 1)

Characteristics	Symbol	MOC7811/7821			MOC7812/7822			MOC7813/7823			Unit
		Min	Typ	Max	Min	Typ	Max	Min	Typ	Max	
I_F = 5.0 mA, V_{CE} = 5.0 V	$I_{CE(on)}$	0.15	—	—	0.30	—	—	0.60	—	—	mA
I_F = 20 mA, V_{CE} = 5.0 V	$I_{CE(on)}$	1.0	—	—	2.0	—	—	4.0	—	—	mA
I_F = 30 mA, V_{CE} = 5.0 V	$I_{CE(on)}$	1.9	—	—	3.0	—	—	5.5	—	—	mA
I_F = 20 mA, I_C = 1.8 mA	$V_{CE(sat)}$	—	—	—	—	—	0.40	—	—	0.40	V
I_F = 30 mA, I_C = 1.8 mA	$V_{CE(sat)}$	—	—	0.40	—	—	—	—	—	—	V
V_{CC} = 5.0 V, I_F = 30 mA, R_L = 2.5 kΩ	t_{on}	—	12	—	—	12	—	—	12	—	μs
V_{CC} = 5.0 V, I_F = 30 mA, R_L = 2.5 kΩ	t_{off}	—	60	—	—	60	—	—	60	—	μs

Note 1: Stray irradiation can alter values of characteristics. Adequate shielding should be provided.

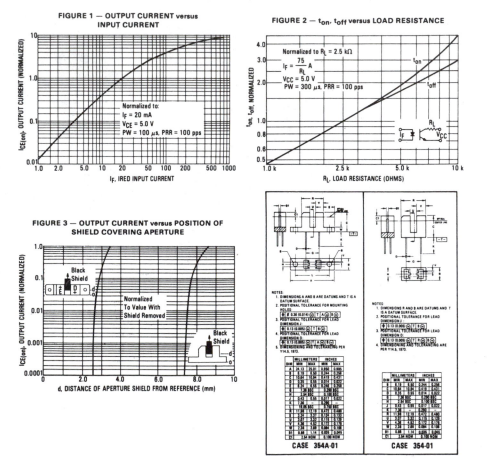

FIGURE 1 — OUTPUT CURRENT versus INPUT CURRENT

FIGURE 2 — t_{on}, t_{off} versus LOAD RESISTANCE

FIGURE 3 — OUTPUT CURRENT versus POSITION OF SHIELD COVERING APERTURE

CASE 354A-01

CASE 354-01

Courtesy of Motorola, Inc.

 **MOTOROLA**

MOC8020
MOC8021

HIGH CTR DARLINGTON COUPLERS

... gallium arsenide LED optically coupled to silicon photodarlington transistors designed for applications requiring electrical isolation, high breakdown voltage, and high current transfer ratios. Provides excellent performance in interfacing and coupling systems, phase and feedback controls, solid state relays, and general purpose switching circuits.

- High Transfer Ratio
  500% — MOC8020   1000% — MOC8021
- High Collector-Emitter Breakdown Voltage—
  $V_{(BR)CEO}$ = 50 Vdc (Min)
- High Isolation Voltage —
  V_{ISO} = 7500 Vac Peak
- UL Recognized, File No. E54915
- Economical Dual-In-Line Package
- Base Not Connected

**OPTO
COUPLER/ISOLATOR**

DARLINGTON OUTPUT

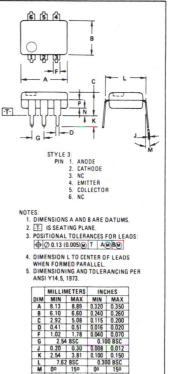

MAXIMUM RATINGS (T_A = 25°C unless otherwise noted.)

Rating	Symbol	Value	Unit
INFRARED-EMITTING DIODE			
Reverse Voltage	V_R	3.0	Volts
Forward Current — Continuous	I_F	50	mA
Forward Current — Peak Pulse Width = 300 μs, 2.0% Duty Cycle	I_F	3.0	Amp
Total Power Dissipation @ T_A = 25°C Negligible Power in Transistor Derate above 25°C	P_D	150 / 2.0	mW / mW/°C
PHOTO DARLINGTON TRANSISTOR			
Collector-Emitter Voltage	V_{CEO}	50	Volts
Emitter-Collector Voltage	V_{ECO}	5.0	Volts
Collector Current — Continuous	I_C	150	mA
Total Power Dissipation @ T_A = 25°C Negligible Power in Diode Derate above 25°C	P_D	150 / 2.0	mW / mW/°C
TOTAL DEVICE			
Total Device Dissipation @ T_A = 25°C Equal Power Dissipation in Each Element Derate above 25°C	P_D	250 / 3.3	mW / mW/°C
Operating Junction Temperature Range	T_J	-55 to +100	°C
Storage Temperature Range	T_{stg}	-55 to +150	°C
Soldering Temperature (10 s)	—	260	°C

STYLE 3:
PIN 1. ANODE
2. CATHODE
3. NC
4. EMITTER
5. COLLECTOR
6. NC

NOTES:
1. DIMENSIONS A AND B ARE DATUMS.
2. ⊤ IS SEATING PLANE.
3. POSITIONAL TOLERANCES FOR LEADS:
   ⊕ ⌀ 0.13 (0.005) Ⓜ T A Ⓜ B Ⓜ
4. DIMENSION L TO CENTER OF LEADS WHEN FORMED PARALLEL.
5. DIMENSIONING AND TOLERANCING PER ANSI Y14.5, 1973.

DIM	MILLIMETERS		INCHES	
	MIN	MAX	MIN	MAX
A	8.13	8.89	0.320	0.350
B	6.10	6.60	0.240	0.260
C	2.92	5.08	0.115	0.200
D	0.41	0.51	0.016	0.020
F	1.02	1.78	0.040	0.070
G	2.54 BSC		0.100 BSC	
J	0.20	0.30	0.008	0.012
K	2.54	3.81	0.100	0.150
L	7.62 BSC		0.300 BSC	
M	0°	15°	0°	15°
N	0.38	2.54	0.015	0.100
P	1.27	2.03	0.050	0.080

CASE 730A-01

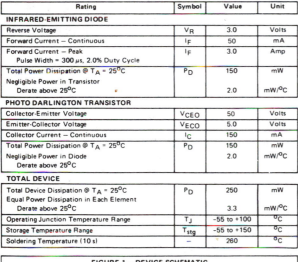

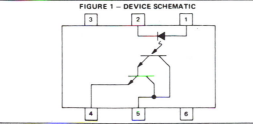

FIGURE 1 — DEVICE SCHEMATIC

Courtesy of Motorola, Inc.

9400, 9401, 9402

Absolute Maximum Ratings

Storage Temperature .... -65°C to +150°C
Operating Temperature
  J Package .... 0°C to 70°C
  L Package .... -40°C to +85°C
$V_{DD} - V_{SS}$ .... 18 V

I_{IN} .... 10 mA
V_{OUT} Max — V_{OUT} Common .... 25 V
$V_{REF} - V_{SS}$ .... -1.5 V
Package Dissipation .... 500 mW
Lead Temperature (Soldering, 10 sec) .... 300°C

Electrical Characteristics, V/F Mode

Unless otherwise specified, V_{DD} = +5V, V_{SS} = —5V, V_{GND} = 0, V_{REF} = —5V, R_{BIAS} = 100KΩ, Full Scale = 10KHz. T_A = 25°C unless Full Temp. Range is specified ·—40°C to +85°C for **L** package, 0°C to 70°C for J package·.

VOLTAGE-TO-FREQUENCY		TSC9401			TSC9400			TSC9402				
Parameter	Definition	Min.	Typ.	Max.	Min.	Typ.	Max.	Min.	Typ.	Max.	Units	Notes
Accuracy Linearity 10KHz	Output Deviation from Straight Line between Normalized Zero and Full Scale Input		0.004	**0.01**		0.01	**0.05**		0.05	**0.25**	% Full Scale	
Linearity 100KHz			0.04	0.08		0.1	0.25		0.25	0.50	% Full Scale	
Gain Temperature Drift	Variation in Gain ·A· due to Temperature Change		±25	±40		±25	±40		±50	±100	ppm/°C Full Scale	1
Gain Variance	Variation from Exact A Compensate by Trimming R_{IN}. V_{REF}. or C_{REF}		±10			±10			±10		% of Nominal	
Zero Offset	Correction at Zero Adjust for Zero Output When Input is Zero		±10	±50		±10	±50		±20	±100	mV	2
Zero Temperature Drift	Variation in Zero Offset Due to Temperature Change		±25	±50		±25	±50		±50	±100	μV/°C	1
Analog Inputs I_{IN} Full Scale	Full Scale Analog Input Current to Achieve Specified Accuracy		10			10			10		μA	
I_{IN} Overrange	Overrange Current			50			50			50	μA	
Response Time	Settling Time to 0.01% Full Scale		2			2			2		Cycles	
Digital Outputs V_{SAT} @ I_{OL}=10μA	Logical "0" Output Voltage			0.4			0.4			0.4	V	3
V_{OUT} Max. — V_{OUT} Common	Voltage Range between Output and Common			18.0			18.0			18.0	V	4
Pulse Frequency Output Width			3.0			3.0			3.0		μsec	
Supply Current I_{DD} Quiescent (L Package) (J Package)	Current Required from Positive Supply During Operation		2.0 2.0	4.0 6.0		2.0 2.0	4.0 6.0		3.0	10.0	mA mA	9
I_{SS} Quiescent (L Package) (J Package)	Current Required from Negative Supply During Operation		-1.5 -1.5	-4.0 -6.0		-1.5 -1.5	-4.0 -6.0		-3.0	-10.0	mA mA	10
V_{DD} Supply	Operating Range of Positive Supply	4.0		7.5	4.0		7.5	4.0		7.5	V	
V_{SS} Supply	Operating Range of Negative Supply	-4.0		-7.5	-4.0		-7.5	-4.0		-7.5	V	
Reference Voltage $V_{REF} - V_{SS}$	Range of Voltage Reference Input	-1.0			-1.0			-1.0			V	

NOTES:
1. Full temperature range.
2. I_{IN} = 0.
3. Full temperature range, I_{OUT} = 10mA.
4. I_{OUT} = 10μA.
5. 10Hz to 100KHz.
6. 5μs min. positive pulse width and 0.5μs min. negative pulse width.

7. $T_r = t_f$ = 20ns.
8. R_L ≥ 2KΩ.
9. Full temperature range, V_{IN} = -0.1V.
10. V_{IN} = -0.1V.
11. I_{IN} connects the summing junction of an operational amplifier. Voltage sources cannot be attached directly but must be buffered by external resistors.

⋇ TELEDYNE SEMICONDUCTOR

Courtesy of Teledyne Semiconductor, Mountain View, CA.

V/F Circuit Description

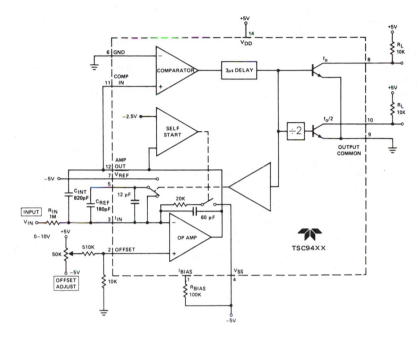

Figure 1. 10Hz to 10KHz V/F Converter

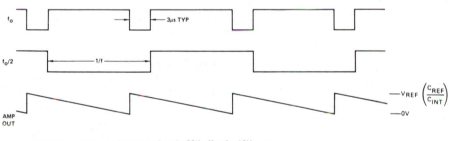

1. To adjust f_{min}, set $V_{IN} = 10mV$ and adjust the 50K offset for 10Hz out.

2. To adjust f_{max}, set $V_{IN} = 10V$ and adjust R_{IN} or V_{REF} for 10KHz out.

3. To increase $f_{OUT}MAX$ to 100KHz change C_{REF} to 27pF and C_{INT} to 75pF.

4. For high performance applications use high stability components for R_{IN}, C_{REF}, V_{REF} (metal film resistors and glass film capacitors). Also separate the output ground (Pin 9) from the input ground (Pin 6).

Figure 2. Output Waveforms

☀ TELEDYNE SEMICONDUCTOR

Courtesy of Teledyne Semiconductor, Mountain View, CA.

9400, 9401, 9402

V/F Circuit Description (Contd.)

The Teledyne 9400 V/F Converter operates on the principal of charge balancing. The input voltage (V_{IN}) is converted to a current (I_{IN}) by the input resistor. This current is then converted to a charge by the integrating capacitor and shows up as a linearly decreasing voltage at the output of the op amp. The zero crossing of the output is sensed by the comparator causing the reference voltage to be applied to the reference capacitor for a time period long enough to virtually charge the capacitor to the reference voltage. This action reduces the charge on the integrating capacitor by a fixed amount ($q = C_{REF} \times V_{REF}$) causing the op amp output to step up a finite amount.

At the end of the charging period, C_{REF} is shorted out dissipating the stored reference charge so that when the output again crosses zero, the system is ready to recycle. In this manner, the continued discharging of the integrating capacitor by the input is balanced out by fixed charges from the reference voltage. As the input voltage is increased, the number of reference pulses required to maintain balance increases causing the output frequency to also increase. Since each charge increment is fixed the increase in frequency with voltage is linear. In addition, the accuracy of the output pulses does not directly effect the linearity of the V/F. It must simply be long enough for full charge transfer to take place.

The 9400 contains a "self-start" circuit to assure that the V/F will always operate properly when power is first applied. In the event that during "Power-on" the op amp output is below the comparator threshold and C_{REF} is already charged, a positive voltage step will not occur. The op amp output will continue to decrease until it crosses the −2.5 volt threshold of the "self-start" comparator. When this happens a resistor is connected to the op amp input causing the output to quickly go positive until the 9400 is once again in its normal operating mode.

The 9400 utilizes both bipolar and MOS transistors on the same substrate, taking advantage of the best features of each. MOS transistors are used at the inputs to reduce offset and bias currents. Bipolar transistors are used in the op amp, for high gain, and on all outputs for excellent current driving capabilities. CMOS logic is used throughout to minimize power consumption.

Pin Functions

Comparator Input — In the V/F mode, this input is connected to the amplifier output (pin 12) and triggers the 3μsec pulse delay when the input voltage passes its threshold. In the F/V mode, the input frequency is applied to the comparator input.

Pulse Freq Out — This output is an open-collector bipolar transistor providing a pulse waveform whose frequency is proportional to the input voltage. This output requires a pull up resistor and interfaces directly with MOS, CMOS and TTL logic.

Freq/2 Out — This output is an open-collector bipolar transistor providing a square wave that is one-half the frequency of the pulse frequency output. This output requires a pull up resistor and interfaces directly with MOS, CMOS, and TTL logic.

Output Common — The emitters of both the freq/2 out and the pulse freq out are connected to this pin. An output level swing from the collector voltage to ground or to the V_{SS} supply may be obtained by connecting to the appropriate point.

R_{BIAS} — Specifications for the 9400 are based on R_{BIAS} = 100K ±10% unless otherwise noted. R_{BIAS} may be varied between the range of $82K \leq R_{BIAS} \leq 120K$.

Amplifier Out — The output stage of the operational amplifier. A negative going ramp signal is available at this pin in the V/F mode. In the F/V mode a voltage proportional to the frequency input is generated.

Zero Adjust — The non-inverting input of the operational amplifier. The low frequency set point is detemined by adjusting the voltage at this pin.

I_{IN} — The inverting input of the operational amplifier and the summing junction when connected in the V/F mode. An input current of 10μA is specified for nominal full scale but an over range current up to 50μA can be used without detrimental effect to the circuit operation.

V_{REF} — A reference voltage from either a precision source or the V_{SS} supply may be applied to this pin. Accuracy will be dependent on the voltage regulation and temperature characteristics of the circuitry.

V_{REF} OUT — The charging current for C_{REF} is derived from the internal circuitry and switched by the break-before-make switch to this pin.

V/F Design Information

Input/Output Relationships — The output frequency is related to the analog input voltage (V_{IN}) by the transfer equation:

$$\text{Frequency Out} = \frac{V_{IN}}{R_{IN}} \times \frac{1}{(V_{REF})(C_{REF})} = f_o$$

External Component Selection

R_{IN} — The value of this component is chosen to give a full scale input current of approximately 10μA.

Example:

$$R_{IN} \cong \frac{V_{IN} \text{ FULL SCALE}}{10\mu A} \qquad R_{IN} \cong \frac{10V}{10\mu A} = 1M\Omega$$

❄ TELEDYNE SEMICONDUCTOR

Courtesy of Teledyne Semiconductor, Mountain View, CA.

V/F Design Information (Contd.)

Note that the value is an approximation, and the exact relationship is defined by the transfer equation. In practice, the value of R_{IN} typically would be trimmed to obtain full scale frequency at V_{IN} FULL SCALE (see adjustment procedure). Metal film resistors with 1% tolerance or better are recommended for high accuracy applications because of their thermal stability and low noise generation.

C_{INT} — Exact value not critical but is related to C_{REF} by the relationship:

$$3C_{REF} \leq C_{INT} \leq 10C_{REF}$$

Improved stability and linearity is obtained when $C_{INT} \geq 4C_{REF}$. Low leakage types are recommended although mica and ceramic devices can be used in applications where their temperature limits are not exceeded. Locate as close as possible to pins 12 and 3.

C_{REF} — Exact value not critical and may be used to trim the full scale frequency (see input/output relation). Glass film or air trimmer capacitors are recommended because of their stability and low leakage. Locate as close as possible to pins 5 and 3.

V_{DD}, V_{SS} — Power supplies of ±5V are recommended. For high accuracy requirements 0.05% line and load regulation and 0.1μF disc decoupling capacitors located near the pins are recommended.

Adjustment Procedure — Figure 1 shows a circuit for trimming the zero location. Full scale may be trimmed by adjusting R_{IN}, V_{REF}, or C_{REF}. Recommended procedure is as follows for a 10KHz full scale frequency.

1. Set V_{IN} to 10mV and trim the zero adjust circuit to obtain a 10Hz output frequency.

2. Set V_{IN} to 10.000V and trim either R_{IN}, V_{REF}, or C_{REF} to obtain a 10KHz output frequency.

If adjustments are performed in this order, there should be no interaction and they should not have to be repeated.

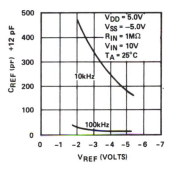

Figure 3. Recommended C_{REF} vs V_{REF}

V/F Single Supply Operation

NOTE:
See Also the TSC7660
Data Sheet;

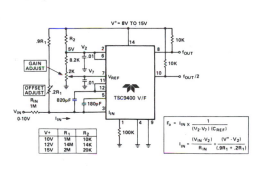

Figure 4. Fixed Voltage—Single Supply Operation

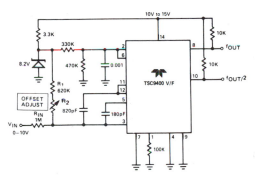

Figure 5. Variable Voltage—Single Supply Operation

9400, 9401, 9402

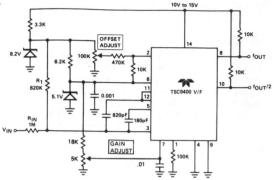

Figure 6. Single Variable Supply Voltage with Offset and Gain Adjust

Electrical Characteristics, F/V Mode

Unless otherwise specified, V_{DD} = +5V, V_{SS} = −5V, V_{GND} = 0, V_{REF} = −5V, R_{BIAS} = 100KΩ, Full Scale = 10KHz. T_A = 25°C unless Full Temp. Range is specified (−40°C to +85°C for **L** package, 0°C to 70°C for J package).

FREQUENCY-TO-VOLTAGE		TSC9401			TSC9400			TSC9402				
Parameter	Definition	Min.	Typ.	Max.	Min.	Typ.	Max.	Min.	Typ.	Max.	Units	Notes
Accuracy Non-Linearity	Deviation from Ideal Transfer Function as a Percentage of Full Scale Voltage		0.01	**0.02**		0.02	**0.05**		0.05	**0.25**	% Full Scale	5
Input Frequency Range	Frequency Range for Specified Non-Linearity	10		100K	10		100K	10		100K	Hz	6
Frequency Inputs Positive Excursion	Voltage Required to Turn Comparator On	0.4		V_{DD}	0.4		V_{DD}	0.4		V_{DD}	V	7
Negative Excursion	Voltage Required to Turn Comparator Off	-0.4		-2V	-0.4		-2V	-0.4		-2V	V	7
Min. Positive Pulse Width	Time between Threshold Crossings		5.0			5.0			5.0		μs	7
Min. Negative Pulse Width	Time between Threshold Crossings		0.5			0.5			0.5		μs	7
Input Impedance		10			10			10			MΩ	
Analog Outputs Output Voltage	Voltage Range of Op Amp Output for Specified Non-Linearity		V_{DD}-1			V_{DD}-1			V_{DD}-1		V	8
Output Loading	Resistive Loading at Output of Op Amp	2K			2K			2K			Ω	
Supply Current I_{DD} Quiescent (L Package) (J Package)	Current Required from Positive Supply During Operation		2.0 2.0	4.0 6.0		2.0 2.0	4.0 6.0		3.0	10.0	mA mA	9
I_{SS} Quiescent (L Package) (J Package)	Current Required from Negative Supply During Operation		-1.5 -1.5	-4.0 -6.0		-1.5 -1.5	-4.0 -6.0		-3.0	-10.0	mA mA	10
V_{DD} Supply	Operating Range of Positive Supply	4.0		7.5	4.0		7.5	4.0		7.5	V	
V_{SS} Supply	Operating Range of Negative Supply	-4.0		-7.5	-4.0		-7.5	-4.0		-7.5	V	
Reference Voltage V_{REF}—V_{SS}	Range of Voltage Reference Input	-1.0			-1.0			-1.0			V	

NOTES:
1. Full temperature range.
2. I_{IN} = 0.
3. Full temperature range, I_{OUT} = 10mA.
4. I_{OUT} = 10μA.
5. 10Hz to 100KHz.
6. 5μs min. positive pulse width and 0.5μs min. negative pulse width.

7. T_r = t_f = 20ns.
8. R_L ≥ 2KΩ.
9. Full temperature range, V_{IN} = -0.1V.
10. V_{IN} = -0.1V.
11. I_{IN} connects the summing junction of an operational amplifier. Voltage sources cannot be attached directly but must be buffered by external resistors.

TELEDYNE SEMICONDUCTOR

Courtesy of Teledyne Semiconductor, Mountain View, CA.

F/V Circuit Description

The 9400, when used as a frequency to voltage converter, generates an output voltage which is linearly proportional to the input frequency waveform.

Each zero crossing at the comparator's input causes a precise amount of charge (q = C_{REF} x V_{REF}) to be dispensed into the op amp's summing junction. This charge in turn flows through the feedback resistor generating voltage pulses at the output of the op amp. A capacitor (C_{INT}) across R_{INT} averages these pulses into a DC voltage which is linearly proportional to the input frequency.

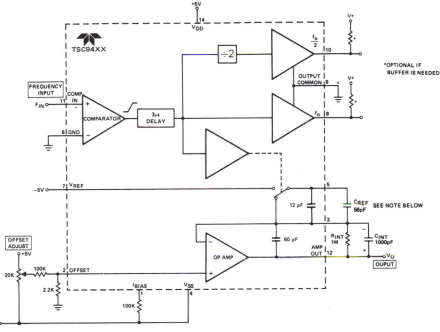

Figure 7. DC — 10KHz F/V Converter

F/V Design Information

Input/Output Relationships — The output voltage is related to the input frequency (F_{IN}) by the transfer equation:

$$V_{OUT} = \left[V_{REF} \, C_{REF} \, R_{INT} \right] F_{IN}$$

The response time to a change in F_{IN} is equal to $\left(R_{INT} \, C_{INT} \right)$. The amount of ripple on V_{OUT} is inversely proportional to C_{INT} and the Input Frequency.

C_{INT} can be increased to lower the ripple. 1µF to 100µF are perfectly acceptable values for low frequencies.

When 9400 is used in the single supply mode, V_{REF} is defined as the voltage difference between Pin 7 and Pin 2.

Input Voltage Levels — The input signal must cross through zero in order to trip the comparator. In order to overcome the hysteresis the amplitude must be greater than ±200mV.

If only a unipolar input signal (F_{IN}) is available, it is recommended that either an offset circuit using resistor be used or that the signal be coupled in via a capacitor.

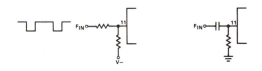

Note: C_{REF} should be increased for lower F_{IN} max. Adjust C_{REF} so that V_0 is approximately 2.5 to 3.0 volts for the maximum input frequency. When F_{IN} max is less than 1 kHz, the duty cycle should be greater than 20% to insure that C_{REF} is fully charged and discharged.

☆ TELEDYNE SEMICONDUCTOR

Courtesy of Teledyne Semiconductor, Mountain View, CA.

9400, 9401, 9402
F/V Design Information (Contd.)

For 100KHz maximum input R_{INT} should be decreased to 100KΩ.

Input Buffer — f_o and $f_o/2$ are not used in the F/V mode. However, these outputs may be useful for some applications, such as a buffer to feed additional circuitry. f_o will then follow the input frequency waveform; except that f_o will go high 3μs after F_{IN} goes high. $f_o/2$ will be square wave with a frequency of one half f_o.

If these outputs are not used, then Pins 8, 9 and 10 may be left floating or connected to ground.

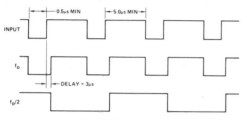

Figure 8. F/V Digital Outputs

The sawtooth ripple which is on the output of an F/V can be eliminated without affecting the F/V's response time by using the circuit in Figure 10. The circuit has a DC gain of +1. Any AC components such as a ripple are amplified both positively, via the lower path, and negatively, via the upper path. When both paths have the same gain, the AC ripple is cancelled. The amount of cancellation is directly proportional to gain matching. If the two paths are matched within 10%, then the ripple will be lowered by 1/10. For 1% matching, the ripple is lowered by 1/100. The 10K potentiometer is used to make the gain equal in both paths. This circuit is insensitive to both frequency changes and to signal wave shape.

Package Information

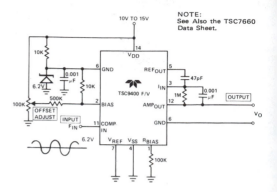

1. The input is now referenced to 6.2 V (Pin 6). The input signal must therefore be restricted to be greater than 4 volts (Pin 6 −2V) and less than 10 to 15V (V_{DD}).

If the signal is AC coupled then a resistor (100K to 10MΩ) must be placed between the input (Pin 11) and Pin 6.

2. The output will now be referenced to Pin 6 which is at 6.2 V(V_Z). For frequency meter applications a 1mA meter with a series scaling resistor can be placed across Pins 6 and 12.

Figure 9. F/V Single Supply

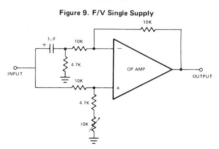

Figure 10. F/V Ripple Eliminator

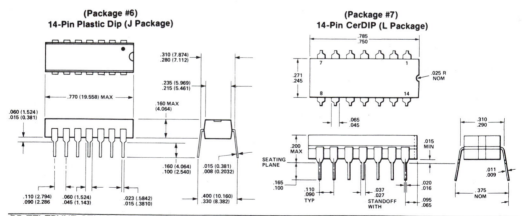

(Package #6)
14-Pin Plastic Dip (J Package)

(Package #7)
14-Pin CerDIP (L Package)

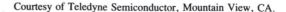

T TELEDYNE SEMICONDUCTOR

Courtesy of Teledyne Semiconductor, Mountain View, CA.

23 C&D SERIES

DESCRIPTION

C & D Series motors have a step angle of 1.8° and offer the user the most in selection and versatility. In addition to listed options, almost any mechanical **or** electrical variation can be built.

Differences between C and D motors are mechanical, **see specifications.**

ELECTRICAL SPECIFICATIONS

MODEL NUMBER*	RES./PHASE (ohms)	RATED CURRENT (amps)	RATED VOLTAGE (VDC)	INDUCTANCE (mh)
23D-6102 23C-6102	5	1.0	5	10.0
23D-6108 23C-6108	.33	3.9	1.3	.63
23D-6204 23C-6204	2.6	1.8	4.7	5.7
23D-6209 23C-6209	.37	4.7	1.7	.8
23D-6306 23C-6306	1.16	2.9	3.4	2.9
23D-6309 23C-6309	.48	4.6	2.2	1.2

*Add A for single shaft.
 C for double shaft extension.

MECHANICAL SPECIFICATIONS

STATIC TORQUE (oz. in.)	DETENT TORQUE (oz. in.)	ROTOR INERTIA (gm. cm.²)	WEIGHT (oz.)	LENGTH (in.)
53	5	115	20	2.0
53	5	115	20	2.0
100	8	235	32	3.25
100	8	235	32	3.25
150	10	320	44	4.0
150	10	320	44	4.0

DYNAMIC TORQUE CHARACTERISTICS — MEASURED WITH 28 VDC SOURCE — DUAL PHASE

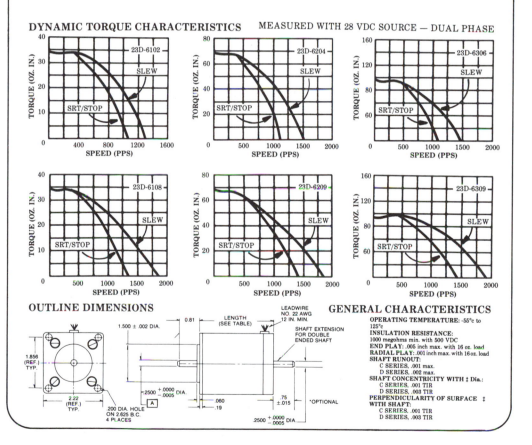

OUTLINE DIMENSIONS

GENERAL CHARACTERISTICS

OPERATING TEMPERATURE: -55°c to 125°c
INSULATION RESISTANCE: 1000 megohms min. with 500 VDC
END PLAY: .005 inch max. with 16 oz. load
RADIAL PLAY: .001 inch max. with 16 oz. load
SHAFT RUNOUT:
  C SERIES, .001 max.
  D SERIES, .002 max.
SHAFT CONCENTRICITY WITH ‡ Dia.:
  C SERIES, .001 TIR
  D SERIES, .003 TIR
PERPENDICULARITY OF SURFACE ‡ WITH SHAFT:
  C SERIES, .001 TIR
  D SERIES, .003 TIR

Courtesy of American Precision.

23E SERIES

DESCRIPTION

E Series motors are similar to **C/D Series** motors in performance. The primary difference is that a cast aluminum housing is utilized. This allows the motor to be lower in cost and also allows for a more even distribution of heat.

ELECTRICAL SPECIFICATIONS

MODEL NUMBER*	RES./PHASE (ohms)	RATED CURRENT (amps)	RATED VOLTAGE (VDC)	INDUCTANCE (mh)
23E-6102	5	1.0	5	10
23E-6108	.37	3.8	1.4	.63
23E-6137	25.5	.44	11.2	47

*Add A for single shaft.
C for double shaft extension.

MECHANICAL & PERFORMANCE SPECIFICATIONS

STEP ANGLE — 1.8°
STEP ACCURACY — ±5% NON CUMULATIVE[1]
DETENT TORQUE — 5 OZ-IN NOMINAL
HOLDING TORQUE — 53 OZ-IN[1]
ROTOR INERTIA — 115 GM-CM²
MAXIMUM THRUST LOAD — 25 LBS.
MAXIMUM OVERHANG LOAD — 15 LBS.
WEIGHT — 18 OZ.
MAX. TEMP RISE — 65°C[1]

[1]MEASURED WITH RATED CURRENT AND DUAL PHASE OPERATION

DYNAMIC TORQUE CHARACTERISTICS MEASURED WITH 24 VDC SOURCE — DUAL PHASE

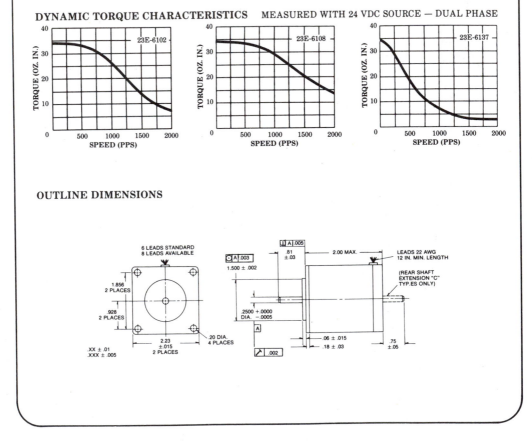

OUTLINE DIMENSIONS

Courtesy of American Precision.

DESCRIPTION

The 23E Series was developed to meet the requirements of O.E.M. users who need high quality components at a competitive price. This motor is available to large volume customers as a standard unit or modified to your specification.

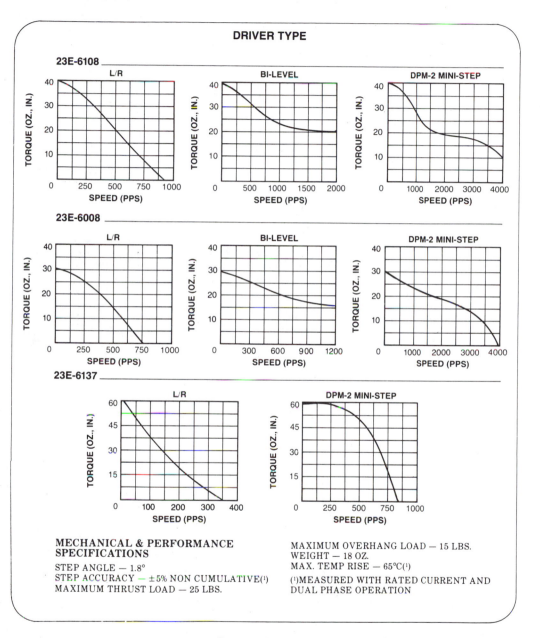

MECHANICAL & PERFORMANCE SPECIFICATIONS

STEP ANGLE — 1.8°
STEP ACCURACY — ±5% NON CUMULATIVE[1]
MAXIMUM THRUST LOAD — 25 LBS.

MAXIMUM OVERHANG LOAD — 15 LBS.
WEIGHT — 18 OZ.
MAX. TEMP RISE — 65°C[1]

[1]MEASURED WITH RATED CURRENT AND DUAL PHASE OPERATION

Courtesy of American Precision.

23H SERIES

DESCRIPTION

Utilizing a slightly longer stator/housing assembly and different **rotor** construction from the C/D Series, the **H Series** motors offer a higher torque to inertia ratio.

Specifically, because of a patented design on the -0X units, the rotor inertia is app. 20% of the -**5XX** motors.

ELECTRICAL SPECIFICATIONS

MODEL NUMBER*	RES./PHASE (ohms)	RATED CURRENT (amps)	RATED VOLTAGE (VDC)	INDUCTANCE (mh)
23H-02	80	.35	28	42
23H-05	5	1.2	6	5
23H-501	5	1.2	6	4.5
23H-502	20	.6	12	16
23H-503	80	.3	24	60
23H-700	3	2.0	6	5

*Add A for single shaft.
 C for double shaft extension.

MECHANICAL SPECIFICATIONS

STATIC TORQUE (oz. in.)	DETENT TORQUE (oz. in.)	ROTOR INERTIA (gm. cm.²)	WEIGHT (oz.)	LENGTH (in.)	CON-FIGURA-TION
30	1	30	17	2.0	R
30	1	30	17	2.0	R
60	2	120	20	2.25	S
60	2	120	20	2.25	S
60	2	120	20	2.25	S
90	4	200	30	3.25	S

DYNAMIC TORQUE CHARACTERISTICS

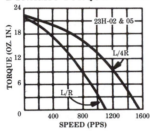

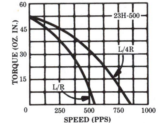

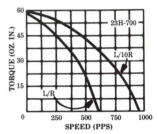

"S" CONFIGURATION

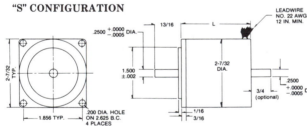

"R" CONFIGURATION

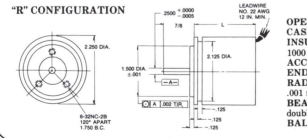

GENERAL CHARACTERISTICS

OPERATING TEMP.: -55°C to 125°C
CASE TEMP. AT 25°c AMBIENT: 50°c max.
INSULATION RESISTANCE AT 500 VDC: 1000 megohms min.
ACCURACY: ±5% max.
END PLAY WITH 16 OZ. LOAD: 005 max.
RADIAL PLAY WITH 16 OZ. LOAD: .001 max.
BEARINGS: Stainless steel, grease packed, double shielded ball bearings.
BALL BEARINGS

Courtesy of American Precision.

BULLETIN 14000
Revised July, 1985

PITMO® D-C SERVO MOTORS

Series 14000 - 2.125 in. O.D.
with Stall Torques from 160 to 286 oz.-in.

This family of permanent magnet field motors offers significantly higher performance than the Pittman® 13000 series through the use of an 11-slot armature lamination designed to use most advantageously the high air gap flux densities provided by radially oriented Ceramic 8 magnets. Series 14000 servo motors have been developed, produced and proved for long, maintenance-free operation. Premium quality materials coupled with the very latest manufacturing and assembly techniques provide excellent reliability. In addition, every motor is subjected to complete testing of all critical parameters under both load and no load conditions in the unique Pittman® computerized final testing station. A printout of test data is kept on file for any further reference.

Speed, voltage, current and torque characteristics can be varied over a wide range to meet specific needs. Please note that armature winding changes, and any relatively simple modifications that do not require extensive redesign or tooling alterations, may be specified for prototype quantities at only nominal costs.

PRIMARY DESIGN FEATURES OF THE SERIES 14000

PEAK TORQUE (STALL)
from 160 to 286 oz.-in.

NO LOAD SPEEDS
from about 3,000 to 3,700 rpm for standard motors at rated voltages

ARMATURES
11-slot design, skewed for reduction of reluctance torque. Laminations are silicon steel, with standard windings of film-insulated (class 200°C) magnet wire — impregnated with polyester resin and baked.

COMMUTATORS
diamond turned after armature assembly to ensure optimum concentricity and long brush life.

BRUSHES
copper-graphite standard.
Optional materials at additional costs include silver-graphite and other specified material compositions.

FIELD
Radially oriented strontium-ferrite magnets enclosed in heavy-gage steel return rings. End bells are zinc die castings.

BEARINGS
self-aligning sintered bronze, precisely sized to provide optimum journal clearance. Also equipped with felt wicks for reserve lubrication. Optional double shielded ball bearings available at additional cost.

Courtesy of Pittman, a division of Penn Engineering and Manufacturing Corp.

ITEM	MOTOR SIZE CONSTANTS	UNITS	SYMBOL	14203 VALUE	14204 VALUE	14205 VALUE	14206 VALUE
1	PEAK TORQUE (STALL)	OZ-IN	TPK	160	205	226	286
1 a	Peak torque (stall)	N•m	TPK	1.13	1.45	1.60	2.02
2	MOTOR CONSTANT	OZ-IN/$\sqrt{W}$	PKO	7.88	8.63	9.97	10.9
2 a	Motor constant	mN•m/$\sqrt{W}$	PKO	55.6	60.9	70.4	77.0
3	POWER FOR PEAK TORQUE	W	PWR	417	570	519	686
4	DAMPING (ZERO SOURCE IMPED.)	OZ-IN/(rad/s)	DPO	0.439	0.526	0.701	0.841
4 a	Damping (zero source imped.)	mN•m/(rad/s)	DPO	3.10	3.71	4.95	5.94
5	DAMPING (INFINITE SOURCE IMPED.)	OZ-IN/(rad/s)	DPI	8.5×10^{-4}	10.3×10^{-4}	11.2×10^{-4}	13.0×10^{-4}
5 a	Damping (infinite source imped.)	mN•m/(rad/s)	DPI	6.0×10^{-3}	7.3×10^{-3}	7.9×10^{-3}	9.2×10^{-3}
6	NO LOAD SPEED	REV/MIN	SNL	3420	3670	3040	3170
6 a	No load speed	rad/s	WNL	358	384	318	332
7	ELECTRICAL TIME CONSTANT	ms	TCE	1.64	1.58	1.63	1.62
8	MECHANICAL TIME CONSTANT	ms	TCM	6.8	7.0	6.3	6.2
9	FRICTION TORQUE	OZ-IN	TOF	2.0	2.0	2.0	2.0
9 a	Friction torque	mN•m	TOF	14.1	14.1	14.1	14.1
10	ARMATURE INERTIA	OZ-IN-s²	ERT	3.0×10^{-3}	3.7×10^{-3}	4.4×10^{-3}	5.2×10^{-3}
10 a	Armature inertia	kg•m²	ERT	21.2×10^{-6}	26.1×10^{-6}	31.1×10^{-6}	36.7×10^{-6}
11	MOTOR WEIGHT	OZ	WGT	31.2	35.2	39.5	45.4
11 a	Motor mass	kg	WGT	0.88	1.0	1.1	1.3
12	THEORETICAL ACCELERATION	rad/s²	CEL	53300	55400	51300	55000
13	THERMAL TIME CONSTANT	MIN	TCT	26.0	28.8	29.4	33.6
14	ULTIMATE TEMP. RISE/WATT	°C/W	TPR	8.1	7.7	7.3	6.8
15	MAXIMUM WINDING TEMPERATURE	°C	TMX	155	155	155	155

WINDING CONSTANTS (other windings available)

	MODEL 14203						
*	WINDING CONSTANTS	UNITS	SYMBOL	WDG #1	WDG #2	WDG #3	WDG #4
16	VOLTAGE	V	VLT	12.0	19.1	24.0	30.3
17	CURRENT (STALL)	A	AMP	34.8	21.8	17.4	13.7
18	TORQUE CONSTANT	OZ-IN/A	TPA	4.63	7.41	9.26	11.7
18 a	Torque constant	mN•m/A	TPA	32.7	52.3	65.4	82.8
19	TERMINAL RESISTANCE	OHMS	RTR	0.345	0.877	1.38	2.21
20	BACK EMF	V/(rad/s)	BEF	0.033	0.052	0.065	0.083
21	INDUCTANCE	mH	DUK	0.565	1.45	2.26	3.63
22	CURRENT (NO LOAD)	A	INL	0.390	0.244	0.195	0.154

	MODEL 14204						
*	WINDING CONSTANTS	UNITS	SYMBOL	WDG #1	WDG #2	WDG #3	WDG #4
16	VOLTAGE	V	VLT	12.0	19.1	24.0	30.3
17	CURRENT (STALL)	A	AMP	47.7	30.2	23.8	19.1
18	TORQUE CONSTANT	OZ-IN/A	TPA	4.34	6.86	8.67	10.8
18 a	Torque constant	mN•m/A	TPA	30.6	48.5	61.2	76.5
19	TERMINAL RESISTANCE	OHMS	RTR	0.251	0.633	1.01	1.59
20	BACK EMF	V/(rad/s)	BEF	0.031	0.048	0.061	0.076
21	INDUCTANCE	mH	DUK	0.400	1.00	1.60	2.50
22	CURRENT (NO LOAD)	A	INL	0.420	0.265	0.210	0.168

	MODEL 14205						
*	WINDING CONSTANTS	UNITS	SYMBOL	WDG #1	WDG #2	WDG #3	WDG #4
16	VOLTAGE	V	VLT	12.0	19.1	24.0	30.3
17	CURRENT (STALL)	A	AMP	43.4	27.4	21.6	17.3
18	TORQUE CONSTANT	OZ-IN/A	TPA	5.26	8.32	10.5	13.1
18 a	Torque constant	mN•m/A	TPA	37.1	58.7	74.1	92.7
19	TERMINAL RESISTANCE	OHMS	RTR	0.277	0.696	1.11	1.75
20	BACK EMF	V/(rad/s)	BEF	0.037	0.059	0.074	0.093
21	INDUCTANCE	mH	DUK	0.453	1.14	1.81	2.83
22	CURRENT (NO LOAD)	A	INL	0.363	0.229	0.181	0.145

* OTHER WINDINGS AVAILABLE

Courtesy of Pittman, a division of Penn Engineering and Manufacturing Corp.

WINDING CONSTANTS	UNITS	SYMBOL	WDG #1	WDG #2	WDG #3	WDG #4
		MODEL 14206				
16 VOLTAGE	V	VLT	12.0	19.1	24.0	30.3
17 CURRENT (STALL)	A	AMP	60.5	36.3	28.6	22.6
18 TORQUE CONSTANT	OZ-IN/A	TPA	4.76	7.93	10.0	12.7
18 a Torque constant	mN•m/A	TPA	33.4	55.7	70.6	89.2
19 TERMINAL RESISTANCE	OHMS	RTR	0.20	0.53	0.84	1.34
20 BACK EMF	V/(rad/s)	BEF	0.034	0.056	0.071	0.090
21 INDUCTANCE	mH	DUK	0.305	0.848	1.36	2.17
22 CURRENT (NO LOAD)	A	INL	0.464	0.279	0.220	0.174

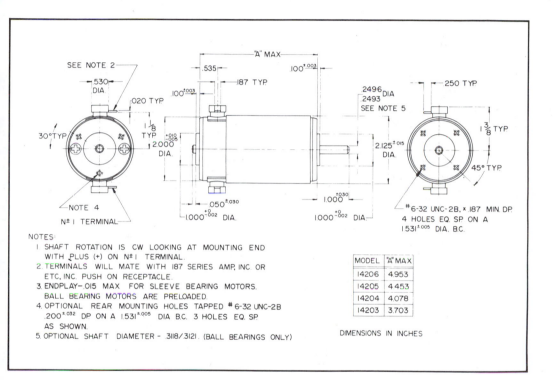

NOTES:
1. SHAFT ROTATION IS CW LOOKING AT MOUNTING END WITH PLUS (+) ON Nº1 TERMINAL.
2. TERMINALS WILL MATE WITH 187 SERIES AMP, INC. OR ETC, INC. PUSH ON RECEPTACLE.
3. ENDPLAY—.015 MAX FOR SLEEVE BEARING MOTORS. BALL BEARING MOTORS ARE PRELOADED.
4. OPTIONAL REAR MOUNTING HOLES TAPPED #6-32 UNC-2B .200±.032 DP ON A 1.531±.005 DIA B.C. 3 HOLES EQ. SP AS SHOWN.
5. OPTIONAL SHAFT DIAMETER - .3118/.3121. (BALL BEARINGS ONLY)

MODEL	"A" MAX
14206	4.953
14205	4.453
14204	4.078
14203	3.703

DIMENSIONS IN INCHES

TELEPHONE (215) 256-6601

PITTMAN
HARLEYSVILLE, PA 19438-0003 USA
A DIVISION OF PENN ENGINEERING & MANUFACTURING CORP.

TELEX 283668
TWX 510-661-8696

Courtesy of Pittman, a division of Penn Engineering and Manufacturing Corp.

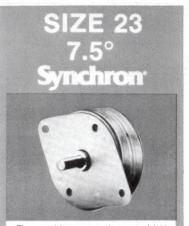

SIZE 23
7.5°
Synchron®

- • Sleeve Bearing
- • Roller Bearing
- • Ball Bearing

These precision power packages are of rigid construction allowing for various bearing configurations. Also available in other voltages - stepping angles - lead positioning - shaft and mounting configurations - B(135°C) & F(155°C) insulation.

Can be fitted with gear box if desired.

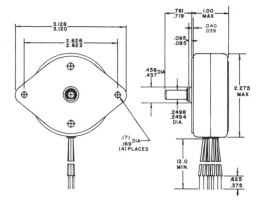

NOTES

1. Computed from holding torque & rotor inertia.
2. Computed from twice the rated current times the dc voltage.
3. Motor mounted on a 6"x6"x1/4" aluminum plate.
4. Motor mounted on a 8"x8"x1/4" aluminum plate.
5. Peak value of a sleeve bearing motor.
6. Peak to peak voltage.
7. Adjusted to 25°C.
8. Measured at 400Hz; small ac signal & 1/3 rated dc current applied.
9. Time for rotor to reach its initial crossover point (minimum reluctance) of its final step position.
∗ Measured at rated current; see graphs for further information.
∗∗ Contact IMC/Hansen for application notes or further information.

CALL 812/385-3415

BASIC MOTOR PARAMETERS

PARAMETER		SYMBOL	UNIT	S-114	S-388
HOLDING TORQUE	∗	THD	OZ-IN	14.2	19.8
TORQUE TO INERTIA RATIO	(1)	TOJ	RAD/SEC²	30400	42400
NO LOAD SPEED (PULL-IN)	∗	NNL	PULSES/SEC	328	450
INPUT POWER	(2)	PIN	WATTS	8.3	16.6
THERMAL RESISTANCE	(3)	RTS	°C/WATT	7.0	6.1
THERMAL RESISTANCE	(4)	RTE	°C/WATT	5.6	4.9
STEP ANGLE		STA	DEGREES	7.5	
STEP ANGLE ACCURACY	∗∗	SAA	DEGREES(%)	±0.5 (6.7)	
STEPS PER REVOLUTION		SPR		48	
ROTOR INERTIA		JRO	MOISS (GM-CM²)	467 (33)	
DETENT TORQUE	(5)	TDE	OZ-IN	2.0	
INSULATION RESISTANCE		INR	MEGOHMS	300x10³	
WEIGHT		WGT	OZ (GM)	8.4 (238)	
MAXIMUM UNIT TEMPERATURE		TMX	°C	105	

WINDING PARAMETERS

	DC VOLTAGE	MODEL NUMBER	UNIT BACK EMF V/KRPM (6)	RATED DC CURRENT PER PHASE (amps)	RESISTANCE PER PHASE (ohms) (7)	INDUCTANCE PER PHASE (mH) (8)	ELECTRICAL TIME CONSTANT (ms)	STEP TIME (ms) (9)
U N I P O L A R	5	S-342	16.7	.760	6.6	10.0	1.5	6.0
	12	S-114	40.0	.345	35.0	58.0	1.6	7.5
	24	S-382	80.0	.171	140.0	230.0	1.6	10.5
B I P O L A R	5	S-402	17.5	1.52	3.3	8.6	2.6	3.0
		S-411	26.5	.516	9.7	25.9	2.7	3.6
	12	S-388	43.0	.686	17.5	45.5	2.6	4.5
		S-412	66.5	.195	61.0	180.5	3.0	5.4
	24	S-355	85.0	.343	70.0	187.0	2.6	6.5
		S-413	72.0	.278	86.2	200.0	2.3	6.2

FOR MORE INFORMATION CONTACT IMC/HANSEN SALES DEPT.

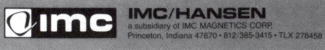

IMC/HANSEN
a subsidiary of IMC MAGNETICS CORP.
Princeton, Indiana 47670 • 812/385-3415 • TLX 278458

Courtesy of Hansen Manufacturing Company, Inc.

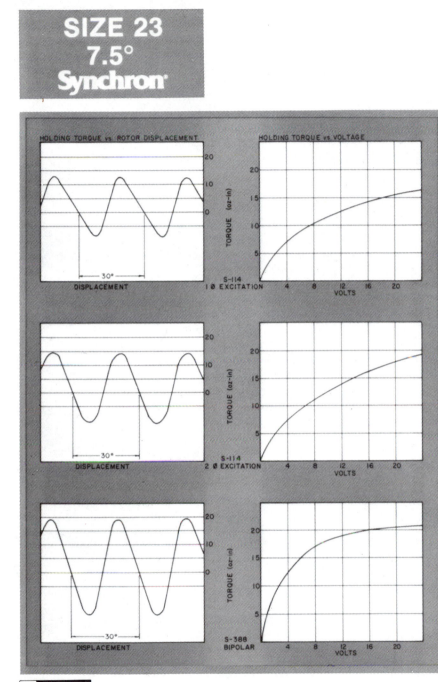

Courtesy of Hansen Manufacturing Company, Inc.

SIZE 23
7.5°
Synchron

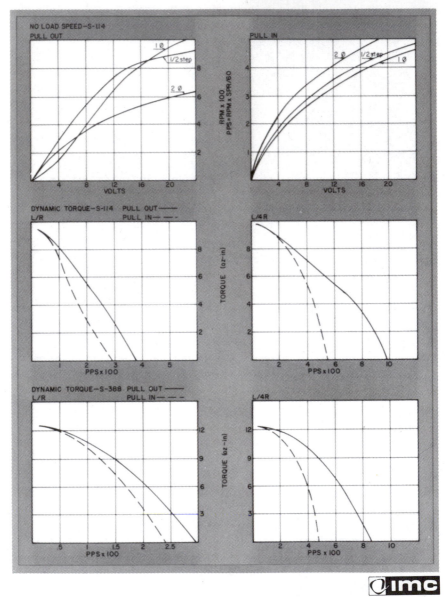

Courtesy of Hansen Manufacturing Company, Inc.

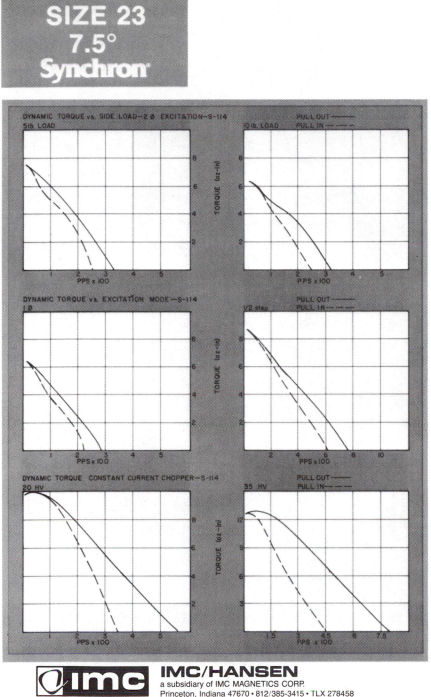

**SIZE 23
7.5°
Synchron®**

IMC/HANSEN
a subsidiary of IMC MAGNETICS CORP.
Princeton, Indiana 47670 • 812/385-3415 • TLX 278458

Courtesy of Hansen Manufacturing Company, Inc.

- **High Accuracy (± 1.3%** typical, ± 3% guaranteed)
- **Greater Torque Than Competitive Motors**
- **Withstands Over 5 Times Rated Current With No Demagnetization**
- **Low Cost (excellent cost/performance ratio)**
- **Rugged Construction**

MA61 Series SLO-SYN Stepping Motors combine unique design features and advanced manufacturing techniques to achieve higher performance than competitive stepping motors at no increase in price. Holding torque and pull-out torque ratings, for example, are high when compared with stepping motors of comparable physical size. Step accuracy is guaranteed to be ±3% maximum and is typically ±1.3%.

Motors of the MA61 Series are ruggedly constructed and incorporate ABEC-1 shaft bearings to provide long life with no deterioration in performance. Eight-lead construction is standard to allow the motors to be driven by a variety of stepping motor controls.

DIMENSIONS

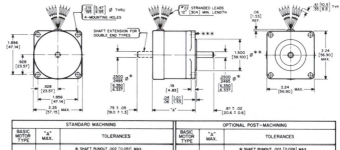

STANDARD MACHINING			OPTIONAL POST–MACHINING		
BASIC MOTOR TYPE	"A" MAX.	TOLERANCES	BASIC MOTOR TYPE	"A" MAX.	TOLERANCES
MA61FS	2.02 [51.30]	∗ SHAFT RUNOUT .002 [0.051] MAX. ∗∗ DIAMETER TOLERANCE ± .002 [±0.051] DIAMETER CONCENTRIC TO SHAFT DIAMETER WITHIN .003 [0.077] T.I.R. ∗∗∗ SURFACE SQUARE TO SHAFT DIAMETER WITHIN .003 [0.077] T.I.R. ONLY WITHIN A DIAM. OF 2.25 [57.2]. BEYOND THIS AREA THE SURFACE MAY BE DEPRESSED .003 [0.077] MAX.	MA61FC	2.02 [51.30]	∗ SHAFT RUNOUT .001 [0.026] MAX. ∗∗ DIAMETER TOLERANCE ± .001 [±0.026] DIAMETER CONCENTRIC TO SHAFT DIAMETER WITHIN .001 [0.026] T.I.R. ∗∗∗ SURFACE SQUARE TO SHAFT DIAMETER WITHIN .001 [0.026] T.I.R. ONLY WITHIN A DIAM. OF 2.25 [57.2]. BEYOND THIS AREA THE SURFACE MAY BE DEPRESSED .003 [0.077] MAX.

NOTES:---
1– Dimensions in brackets are millimeters.
2– Tolerance on decimals ---
.XXX = ± 0.005 [0.13] unless otherwise specified.
3– Dimensions shown apply before painting or plating.
4– This drawing shows only those features which are pertinent to the form, fit, and function of the motor.

 **SUPERIOR ELECTRIC**

Courtesy of the Superior Electric Company.

MA61 Series SLO-SYN® Stepping Motors

SPECIFICATIONS

PARAMETER	UNITS	MA61FS-80020 MA61FC-80020			MA61FS-80096 MA61FC-80096		
		MIN	TYP	MAX	MIN	TYP	MAX
Resistance per Coil	OHMS	4.5	5.0	5.5	0.207	0.230	0.253
Inductance per Coil	mh	8.6	10.7	12.8	0.29	0.36	0.43
Counter emf Constant	Vp/kRPM	—	35.0	—	—	6.7	—
Rated Current per Phase (2-ON)	AMPS	—	1.0	—	—	4.8	—
Rated Current per Phase (1-ON)	AMPS	—	1.4	—	—	6.8	—
Nominal Voltage per Phase (2-ON)	VOLTS	—	5.0	—	—	1.10	—
Nominal Voltage per Phase (1-ON)	VOLTS	—	7.0	—	—	1.56	—
Time for Single Step (2-ON, L/4R)	ms	—	2.35	—	—	1.90	—
Damping Time (90% decay)	ms	—	40	—	—	38	—
Holding Torque (2-ON)							
@ Rated Current	OZ. IN. (N-cm)	60 (42.5)	70 (50)	—	60 (42.5)	70 (50)	—
@ 2x Rated Current	OZ. IN. (N-cm)	90 (63.8)	106 (75)	—	90 (63.8)	106 (75)	—
Holding Torque (1-ON)							
@ Rated Current	OZ. IN. (N-cm)	50 (35)	59 (42)	—	50 (35)	59 (42)	—
Pull-Out Torque (2-ON, L/4R, 350 s/s)	OZ. IN. (N-cm)	42.5 (30)	50 (35)	—	42.5 (30)	50 (35)	—
Residual Torque (Detent Torque)	OZ. IN. (N-cm)	—	6.0 (4.3)	—	—	6.0 (4.3)	—
Rotor Polar Moment of Inertia	LB-IN.² (G-cm²)	.034 (100)	.038 (113)	.042 (125)	.034 (100)	.038 (113)	.042 (125)
Winding Temperature Limit	DEG. C	—	—	130	—	—	130
Frame Temperature Limit	DEG. C	—	—	100	—	—	100
Thermal Resistance							
Winding to Frame	DEG. C/Watt	—	1.25	—	—	1.25	—
Frame to Air	DEG. C/Watt	—	5.0	—	—	5.0	—
Step Angle	DEGREES	—	1.8	—	—	1.8	—
Step Angle Error (@ Rated Current)	± %	—	1.3	3.0	—	1.3	3.0
Step Position Error (@ Rated Current)	± %	—	1.3	3.0	—	1.3	3.0
Dielectric Withstand (winding to frame)	VRMS-1 sec	1000	—	—	1000	—	—
Dielectric Withstand (winding to winding)	VRMS-1 sec	500	—	—	500	—	—
Bearing Classification	ABEC-(XX)		1			1	
Radial Static Load Rating (w/o damage)	LBS (NEWTONS)	—	—	15 (67)	—	—	15 (67)
Axial Shaft Play @ 1.0 LB	INCH (mm)	—	—	0.005 (0.13)	—	—	0.005 (0.13)
Lead Wires (8)		#22 AWG, UL STYLE 3265, 150V, 125°C			#22 AWG, UL STYLE 3265, 150V, 125°C		
Weight (Mass)	LBS (GRAMS)	—	1.1 (500)	—	—	1.1 (500)	—

Options

MA61 motors can be supplied with special features to meet the needs of specific applications. The following features are typical of those which can be incorporated.

- **Special Windings**
- **Modified Shafts**
- **Special Lead Lengths**
- **Connectors**
- **Timing Pulleys or Gears Affixed to Shaft**
- **With Final Machining (MA61FC-XXXXX) Without Final Machining (MA61FS-XXXXX)**

THE
SUPERIOR ELECTRIC
COMPANY
HEADQUARTERS: 383 Middle Street
Bristol, Connecticut 06010
USA
TEL: (203) 582-9561   TELEX: 96-2446   TWX: 710-454-0682
Cable Address: SUPELEC   Printed in U.S.A.
SE-L68417

Courtesy of the Superior Electric Company.

ANSWERS TO ODD-NUMBERED PROBLEMS

appendix D_____

CHAPTER 1

No solutions.

CHAPTER 2

2–1. 500 Ω

2–3. 100

2–5. 6.6 V

2–9. **(a)** 4
   (b) 2^2 bit

2–11. **(a)** 7.5 V
   (b) 5 mV
   (c) 10.24 V
   (d) 10 bits

2–13. **(a)** 4000 rpm
   (b) 26.3 ft-lb
   (c) 525 ft-lb

CHAPTER 3

3–1. 100
3–3. **(a)** $K_t = 4$ oz-in./V,

$$K_i = 6 \text{ oz-in./A},$$
$$K_b = 4.5 \text{ V/1000 rpm},$$
$$B_m = 18 \text{ oz-in./1000 rpm},$$
$$K_m = 125 \text{ rpm/V},$$
$$\tau_m = 52.4 \text{ ms}$$

(b) $\dfrac{\Omega m(s)}{V_a(s)} = \dfrac{125}{(0.05)s + 1} \dfrac{\text{rpm}}{\text{V}}$

(c) 5.4 V

(d) 9.6 V

(e) $I_a(0) = 4$ A,
$I_{ass} = 10.5$ A

(f) $I_a(0) = 3.1$ A,
$I_{ass} = 1.33$ A,
$a(0) = 13.2 \text{ ft/s}^2$

(g) $x(0.5) = 8.9$ ft,
$v(0.5) = 4.8 \text{ ft/s}$,
$a(0.5) = 6.8 \text{ ft/s}^2$

3–7. (a) 960
(b) 2.4 s

3–9. 3 lb

CHAPTER 4

4–1.

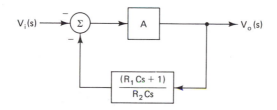

$$V_i(s) \longrightarrow \boxed{\dfrac{K}{s\tau + 1}} \longrightarrow V_o(s)$$

$K = R_2/R_1$ and $\tau = R_2 C$

The transfer function of the network in Figure E4–1 has a pole at the origin whereas the transfer function of the network in Figure P4–1 has a pole at $S = -\dfrac{1}{\tau}$. This difference is mainly due to the addition of R_2.

4–3. $\dfrac{V_o}{V_i}(s) = \dfrac{-A}{1 + \dfrac{A(R_1Cs + 1)}{R_2Cs}}$

$$V_i(s) \longrightarrow \Sigma \longrightarrow \boxed{A} \longrightarrow V_o(s)$$

$$\boxed{\dfrac{(R_1Cs + 1)}{R_2Cs}}$$

The transfer function obtained in Example 4–4 has a "zero" at the origin whereas the above transfer function has a "zero" at the origin and a finite "pole."

4-5. $\dfrac{V_o}{V_i}(s) = \dfrac{A}{s^2 + \dfrac{[R_1C_1 + R_2C_1 + R_1C_2(1+AK) - AR_1C_2]s}{R_1R_2C_1C_2(1+AK)} + \dfrac{1}{R_1R_2C_1C_2}}$

where $k = \left(\dfrac{R_3}{R_3 + R_4}\right)$

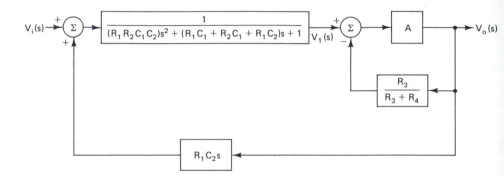

4-7. $\dfrac{C(s)}{R(s)} = \dfrac{G_1(s)G_2(s)G_3(s)}{1 + G_2(s)G_3(s)H_2(s) + G_1(s)G_2(s)H_1(s)}$

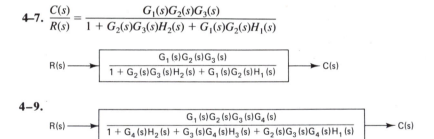

4-9.

$R(s) \longrightarrow \boxed{\dfrac{G_1(s)G_2(s)G_3(s)G_4(s)}{1 + G_4(s)H_2(s) + G_3(s)G_4(s)H_3(s) + G_2(s)G_3(s)G_4(s)H_1(s)}} \longrightarrow C(s)$

4-11. $C(s) = \dfrac{G_1(s)G_2(s)G_4(s)\ R(s)}{1 + G_1(s)G_2(s)G_4(s)H_1(s)} + \dfrac{G_3(s)G_4(s)\ D(s)}{1 + G_1(s)G_2(s)G_4(s)H(s)}$

4-13.

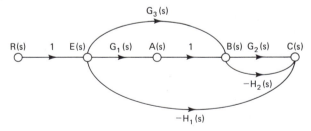

$\dfrac{C(s)}{R(s)} = \dfrac{G_1(s)G_2(s) + G_2(s)G_3(s)}{1 + G_2(s)H_2(s) + G_1(s)G_2(s)H_1(s) + G_2(s)G_3(s)H_1(s)}$

The above transfer function is the same as that in Problem 4–8. This is because there is only one transfer function for a given system regardless of the technique used to determine it.

CHAPTER 5

5–1. (a) Type 'O'

(b) 24.5

(c) 2

(d) $\tau_{\text{open loop}} = 100$ ms

$\tau_{\text{closed loop}} = 10$ ms

5–3. $E = 10.9$

$K_c = 5.6$

5–5. 30°

5–7. (a) $H = 1$

(b) $V_o = -0.32$ V; switch in position 1

$V_o = -1.6$ V, switch in position 2

(c) DR $= -3$

No

(d) $R_1 = 25$ kΩ

$R_2 = 16$ kΩ

(e) $R_2 = 99$ kΩ

(f) $R_2 = 49$ kΩ

5–9. (a) Gain 1 and gain 3

(b) 60° for gain 2

5–11. (a) All in LHS of the s-plane

(b) 2 on $j\omega$ axis,

1 in LHS

(c) All in LHS

(d) 2 in RHS,

3 in LHS

(e) 2 in LHS,

4 in RHS

5–13. (c) $K_2 < 4$

(d) 253.3 μS

CHAPTER 6

6–1. (a) 9.8

(b) 4.6

(c) $0.4\dot{c} + 51c(t) = 100r(t)$

6–3. $R = 40$ kΩ,

$C = .01$ μF

6–7. 1715 pF

6–9. $A = 4040$,

$B = 17.3$,

$D = 0.4$

6–11. (a) $x = 1.19$,

$y = 320.9$

(c) $\dfrac{C(s)}{R(s)} = \dfrac{3.5 \times 10^5}{s^2 + 642.1s + 3.5 \times 10^5}$

6–13. (a) $A = 2.5 \times 10^5$,
  $B = 600$,
  $D = 1$

CHAPTER 7

7–1.

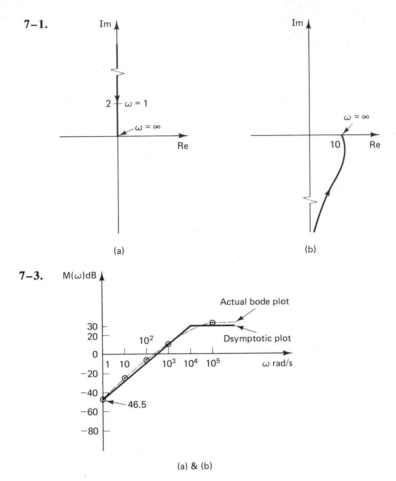

(a)   (b)

7–3.

(a) & (b)

(c) Crossover frequency is 212 Hz.

7–5. Crossover frequency is not applicable.

7–9. (a) GM is undefined,
  PM $= -270°$,
  system is stable;
  (b) GM is undefined,
  PM is $-275.7°$,
  system is stable.

7–11. (a) GM is undefined,
  PM is $-90°$,

system is stable. Although theoretically the system is stable, it is not a practical system. This is because in a practical system the number of poles is always larger than or equal to the number of zeros.

     (b) GM is 20dB,

        PM is $-84.15°$,

        system is unstable or conditionally stable. However, it can be made stable by reducing its gain until the GM is negative dB.

7–13. **(a)** GM is undefined,

        PM is undefined;

     (b) system is stable.

7–15. $K \cong 11,175$

7–17. $D = 5$ seconds

CHAPTER 8

8–1. $\omega_{ss} = 1724$ rpm (load),

     $\omega_{ss} = 1852$ rpm (no load),

     $\omega_d = 2000$ rpm

8–3. $I_a(0) = 33.3$ A,

     $I_{ass} = 3.79$ A,

     $I_a (0.02) = 11.6$ A

8–5. $\omega_m (1.02) = 573$ rpm,

     $\omega_m (1.1) = 136$ rpm

8–7. 56

8–9. 10.75 ms

8–11. **(a)** $t_2 = 69.3$ ms,

        $t_4 = 138.5$ ms

     (b) $t_1 = 38.4$ ms,

        $t_3 = 107.7$ ms

     (c) $55.2°$

     (d) 39.5 ms

     (e) 13.2 ms

8–13. $\dfrac{\Theta_o(s)}{\Theta_i(s)} = \dfrac{800.9}{s^2 + 15.3s + 800.9}$

     $(0.0645)\, \ddot{\theta}_o + \dot{\theta}_o + 52.4\, \theta_o(t) = 52.4\theta_i(t)$

8–15. 0.48 µF (overall)

     2.4 µF (local)

8–19. $POT = 44.4\%$

     $T_1 = 129.8$ ms

8–21. 6

8–23. 16.25

8–24. **(a)** $q_i (0) = 402.5$ Btu/s

        $q_o (0) = 2.5$ Btu/s

     (b) 3.61 s

     (c) 59.5 s

8–25. (a) Same as 8–24 (a)
  (b) Same as 8–24 (b)
  (c) 69.7 s

8–27. (a) 4 gal/s
  (b) 1.92 in.
  (c) 16.35 s
  (d) 26.4 s
  (e) 4.98 ft
  (f) 0.24 in.
  (g) 0.073 gal/s
  (h) 4.1 gal/s
  (i) 27.4 in./s
  (j) 0.18 in.

CHAPTER 9

9–1. 2996.7 mV or $\cong$ 3 V

9–3. $f_o = 87.52$ Hz

9–5. 30 inches

9–7.

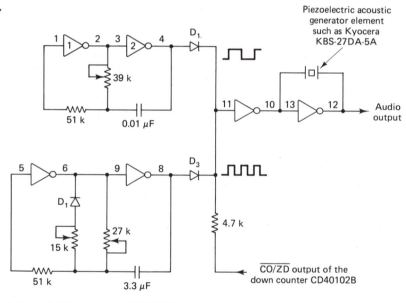

1 through 6 inverters are MC14069.
1 through 3 diodes are 1N914S.

CHAPTER 10

10–1. Refer to Figure 10–15 for the connection diagram.

10-3.

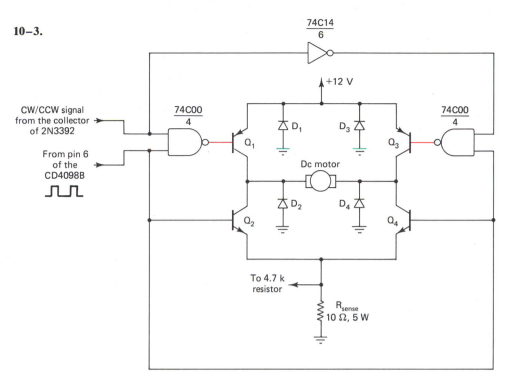

The remaining circuit is as shown in Figure 10–15.

10-5. Do the following modifications in the circuit of Figure 10–22:

1. Remove both 7407 noninverting open-collector buffers and the voltage divider network composed of 10 kΩ and 6.8 kΩ resistors.

2. Change the supply voltages of CD40102B and 2N3392 to 5 V and select an appropriate value for collect resistor of 2N3392 transistor, so that it can control the operation of the MOC3010.

10-7.

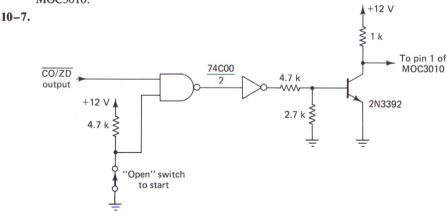

10-9. In Figure 10–29 the level of intensity is controlled by loading an appropriate number into the counter. However, to control the duration of the intensity one may use a time delay in software. A certain time delay can be set up in software using accumulators and an index register. When the time delay is up, a new number may be loaded into the counter and the above cycle can be repeated.

INDEX